20th SIGBioMed Workshop on Biomedical Language Processing (BioNLP 2021)

Online
11 June 2021

ISBN: 978-1-7138-3017-7

Printed from e-media with permission by:

Curran Associates, Inc.
57 Morehouse Lane
Red Hook, NY 12571

Association for Computational Linguistics
209 N. Eighth Street
Stroudsburg, Pennsylvania 18360

Phone: 1-570-476-8006
Fax: 1-570-476-0860

acl@aclweb.org

Additional copies of this publication are available from:

Curran Associates, Inc.
57 Morehouse Lane
Red Hook, NY 12571 USA
Phone: 845-758-0400
Fax: 845-758-2633
Email: curran@proceedings.com
Web: www.proceedings.com

NAACL-HLT 2021

Biomedical Language Processing (BioNLP)

Proceedings of the Twentieth Workshop

June 11, 2021

Stronger Biomedical NLP in the Face of COVID-19

Dina Demner-Fushman, Sophia Ananiadou, Kevin Bretonnel Cohen, Junichi Tsujii

This year marks the second virtual BioNLP workshop. BioNLP 2020 workshop was one of the community's first experiences in online conferences, BioNLP 2021 finds us as cohort of seasoned zoomers, webexers and users of other platforms that the conference organizers select in the hopes of finding an environment that will get us as close as possible to an in-person meeting. There is some light at the end of the tunnel: in many places the new SARS-CoV-2 infections are going down and the numbers of fully vaccinated people are going up, which allows us hoping for an in-person meeting in 2022. We believe that some of this success was enabled by our community: In 2020, BioNLP researchers contributed to development of efficient approaches to retrieval of pandemic-related information and developed approaches to clinical text processing that supported many tasks focused on containment of the pandemic and reduction of COVID-19 severity and complications.

Much of the language processing work related to COVID-19 was enabled by and built on the foundation established by the BioNLP community. This year, the community continued expanding BioNLP research that resulted in 43 submissions to the workshop and 16 additional submissions of the work describing innovative approaches to the MADIQA 2021 Shared Task described in the overview paper in this volume.

As always, we are deeply grateful to the authors of the submitted papers and to the reviewers (listed elsewhere in this volume) that produced three thorough and thoughtful reviews for each paper in a fairly short review period. The quality of submitted work continues growing and the Organizers are truly grateful to our amazing Program Committee that helped us determine which work is ready to be presented and which will benefit from additional experiments and analysis suggested by the reviewers. Based on the PC recommendations, we selected eight papers for oral presentations and 15 for poster presentations. These presentations include transformer-based approaches to such fundamental tasks as relation extraction and named entity recognition and normalization, creation of new datasets and exploration of knowledge-capturing abilities of deep learning models.

The keynote titled "Information Extraction from Texts Using Heterogeneous Information" will be presented by Dr. Makoto Miwa, an associate professor of Toyota Technological Institute (TTI). Dr. Miwa received his Ph.D. from the University of Tokyo in 2008. His research mainly focuses on information extraction from texts, deep learning, and representation learning. Specifically, the keynote will highlight the following:

With the development of deep learning, information extraction targeting sentences using only linguistic information has matured, and interest increases beyond the boundaries of sentences and languages. Labeled information is limited for such information extraction due to high annotation costs, and a variety of information must be used to complement them, such as language structure and external knowledge base information. In the talk, Dr Miwa will mainly introduce his recent efforts to extract information from texts using various heterogeneous information inside and outside the language and discuss the direction and prospects of information extraction in the future.

As always, we are looking forward to a productive workshop, and we hope that new collaborations and research will evolve, continuing contributions of our community to public health and well-being.

Organizing Committee

Dina Demner-Fushman, US National Library of Medicine
Kevin Bretonnel Cohen, University of Colorado School of Medicine, USA
Sophia Ananiadou, National Centre for Text Mining and University of Manchester, UK
Junichi Tsujii, National Institute of Advanced Industrial Science and Technology, Japan

Program Committee:

Sophia Ananiadou, National Centre for Text Mining and University of Manchester, UK
Emilia Apostolova, Language.ai, USA
Eiji Aramaki, University of Tokyo, Japan
Steven Bethard, University of Arizona, USA
Olivier Bodenreider, US National Library of Medicine
Leonardo Campillos Llanos, Universidad Autonoma de Madrid, Spain
Qingyu Chen, US National Library of Medicine
Fenia Christopoulou, National Centre for Text Mining and University of Manchester, UK
Kevin Bretonnel Cohen, University of Colorado School of Medicine, USA
Brian Connolly, Kroger Digital, USA
Jean-Benoit Delbrouck, Stanford University, USA
Dina Demner-Fushman, US National Library of Medicine
Bart Desmet, Clinical Center, National Institutes of Health, USA
Travis Goodwin, US National Library of Medicine
Natalia Grabar, CNRS, France
Cyril Grouin, LIMSI - CNRS, France
Tudor Groza, The Garvan Institute of Medical Research, Australia
Antonio Jimeno Yepes, IBM, Melbourne Area, Australia
William Kearns, UW Medicine, USA
Halil Kilicoglu, University of Illinois at Urbana-Champaign, USA
Ari Klein, University of Pennsylvania, USA
Andre Lamurias, University of Lisbon, Portugal
Alberto Lavelli, FBK-ICT, Italy
Robert Leaman, US National Library of Medicine
Ulf Leser, Humboldt-Universität zu Berlin, Germany
Timothy Miller, Children's Hospital Boston, USA
Claire Nedellec, INRA, France
Aurelie Neveol, LIMSI - CNRS, France
Mariana Neves, German Federal Institute for Risk Assessment, Germany
Denis Newman-Griffis, University of Pittsburgh, USA
Nhung Nguyen, The University of Manchester, UK
Karen O'Connor, University of Pennsylvania, USA
Yifan Peng, Weill Cornell Medical College, USA
Laura Plaza, UNED, Madrid, Spain
Francisco J. Ribadas-Pena, University of Vigo, Spain
Fabio Rinaldi, IDSIA (Dalle Molle Institute for Artificial Intelligence), Switzerland
Angus Roberts, King's College London, UK
Kirk Roberts, The University of Texas Health Science Center at Houston, USA
Roland Roller, DFKI GmbH, Berlin, Germany
Diana Sousa, University of Lisbon, Portugal
Karin Verspoor, The University of Melbourne, Australia
Davy Weissenbacher, University of Pennsylvania, USA
W John Wilbur, US National Library of Medicine

Shankai Yan, US National Library of Medicine
Chrysoula Zerva, National Centre for Text Mining and University of Manchester, UK
Ayah Zirikly, Johns Hopkins University, USA
Pierre Zweigenbaum, LIMSI - CNRS, France

Additional Reviewers:
Jaya Chaturvedi, King's College London, UK
Vani K, IDSIA (Dalle Molle Institute for Artificial Intelligence), Switzerland
Joseph Cornelius, IDSIA (Dalle Molle Institute for Artificial Intelligence), Switzerland
Shogo Ujiie, Nara Institute of Science and Technology, Japan

Shared Task MEDIQA 2021 Organizing Committee

Asma Ben Abacha, US National Library of Medicine
Chaitanya Shivade, Amazon
Yassine Mrabet, US National Library of Medicine
Yuhao Zhang, Stanford University, USA
Curtis Langlotz, Stanford University, USA
Dina Demner-Fushman, US National Library of Medicine

Shared Task MEDIQA 2021 Program Committee:
Asma Ben Abacha, US National Library of Medicine
Sony Bachina, National Institute of Technology Karnataka, India
Spandana Balumuri, National Institute of Technology Karnataka, India
Yi Cai, Chic Health, Shanghai, China
Duy-Cat Can, VNU University of Engineering and Technology, Hanoi, Vietnam
Songtai Dai, Baidu Inc., Beijing, China
Jean-Benoit Delbrouck, Stanford University, USA
Huong Dang, George Mason University, Virginia, USA
Deepak Gupta, US National Library of Medicine
Yifan He, Alibaba Group
Ravi Kondadadi, Optum
Jooyeon Lee, Christopher Newport University, Virginia, USA
Lung-Hao Lee, National Central University, Taiwan
Diwakar Mahajan, IBM Research, USA
Yassine Mrabet, US National Library of Medicine
Khalil Mrini, University of California, San Diego, La Jolla, CA, USA
Mourad Sarrouti, US National Library of Medicine
Chaitanya Shivade, Amazon
Mario Sänger, Humboldt-Universität zu Berlin, Germany
Quan Wang, Baidu Inc., Beijing, China
Leon Weber, Humboldt-Universität zu Berlin, Germany
Shweta Yadav, US National Library of Medicine
Yuhao Zhang, Stanford University, USA
Wei Zhu, East China Normal University, Shanghai, China

Table of Contents

Conference Program

Friday June 11, 2021

08:00–08:15 Opening remarks

08:15–09:15 Session 1: Information Extraction

08:15–08:30 *Improving BERT Model Using Contrastive Learning for Biomedical Relation Extraction*
Peng Su, Yifan Peng and K. Vijay-Shanker

08:30–08:45 *Triplet-Trained Vector Space and Sieve-Based Search Improve Biomedical Concept Normalization*
Dongfang Xu and Steven Bethard

08:45–09:00 *Scalable Few-Shot Learning of Robust Biomedical Name Representations*
Pieter Fivez, Simon Suster and Walter Daelemans

09:00–09:15 *SAFFRON: tranSfer leArning For Food-disease RelatiOn extractioN*
Gjorgjina Cenikj, Tome Eftimov and Barbara Koroušić Seljak

09:15–10:00 Session 2: Clinical NLP

09:15–09:30 *Are we there yet? Exploring clinical domain knowledge of BERT models*
Madhumita Sushil, Simon Suster and Walter Daelemans

09:30–09:45 *Towards BERT-based Automatic ICD Coding: Limitations and Opportunities*
Damian Pascual, Sandro Luck and Roger Wattenhofer

09:45–10:00 *emrKBQA: A Clinical Knowledge-Base Question Answering Dataset*
Preethi Raghavan, Jennifer J Liang, Diwakar Mahajan, Rachita Chandra and Peter Szolovits

10:00–10:30 *Coffee Break*

Friday June 11, 2021 (continued)

Session 3: MEDIQA 2021 Overview: Asma Ben Abacha

10:30–11:00 *Overview of the MEDIQA 2021 Shared Task on Summarization in the Medical Domain*
Asma Ben Abacha, Yassine Mrabet, Yuhao Zhang, Chaitanya Shivade, Curtis Langlotz and Dina Demner-Fushman

11:00–12:00 Session 4: MEDIQA 2021 Presentations

11:00–11:15 *WBI at MEDIQA 2021: Summarizing Consumer Health Questions with Generative Transformers*
Mario Sänger, Leon Weber and Ulf Leser

11:15–11:30 *paht_nlp @ MEDIQA 2021: Multi-grained Query Focused Multi-Answer Summarization*
Wei Zhu, Yilong He, Ling Chai, Yunxiao Fan, Yuan Ni, GUOTONG XIE and Xiaoling Wang

11:30–11:45 *BDKG at MEDIQA 2021: System Report for the Radiology Report Summarization Task*
Songtai Dai, Quan Wang, Yajuan Lyu and Yong Zhu

11:45–12:00 *damo_nlp at MEDIQA 2021: Knowledge-based Preprocessing and Coverage-oriented Reranking for Medical Question Summarization*
Yifan He, Mosha Chen and Songfang Huang

12:00–12:30 ***Coffee Break***

12:30–14:30 Session 5: Poster session 1

Stress Test Evaluation of Biomedical Word Embeddings
Vladimir Araujo, Andrés Carvallo, Carlos Aspillaga, Camilo Thorne and Denis Parra

BLAR: Biomedical Local Acronym Resolver
William Hogan, Yoshiki Vazquez Baeza, Yannis Katsis, Tyler Baldwin, Ho-Cheol Kim and Chun-Nan Hsu

Claim Detection in Biomedical Twitter Posts
Amelie Wührl and Roman Klinger

Friday June 11, 2021 (continued)

BioELECTRA:Pretrained Biomedical text Encoder using Discriminators
Kamal raj Kanakarajan, Bhuvana Kundumani and Malaikannan Sankarasubbu

Word centrality constrained representation for keyphrase extraction
Zelalem Gero and Joyce Ho

End-to-end Biomedical Entity Linking with Span-based Dictionary Matching
Shogo Ujiie, Hayate Iso, Shuntaro Yada, Shoko Wakamiya and Eiji ARAMAKI

Word-Level Alignment of Paper Documents with their Electronic Full-Text Counterparts
Mark-Christoph Müller, Sucheta Ghosh, Ulrike Wittig and Maja Rey

Improving Biomedical Pretrained Language Models with Knowledge
Zheng Yuan, Yijia Liu, Chuanqi Tan, Songfang Huang and Fei Huang

EntityBERT: Entity-centric Masking Strategy for Model Pretraining for the Clinical Domain
Chen Lin, Timothy Miller, Dmitriy Dligach, Steven Bethard and Guergana Savova

Contextual explanation rules for neural clinical classifiers
Madhumita Sushil, Simon Suster and Walter Daelemans

Exploring Word Segmentation and Medical Concept Recognition for Chinese Medical Texts
Yang Liu, Yuanhe Tian, Tsung-Hui Chang, Song Wu, Xiang Wan and Yan Song

BioM-Transformers: Building Large Biomedical Language Models with BERT, ALBERT and ELECTRA
Sultan Alrowili and Vijay Shanker

Semi-Supervised Language Models for Identification of Personal Health Experiential from Twitter Data: A Case for Medication Effects
Minghao Zhu and Keyuan Jiang

Context-aware query design combines knowledge and data for efficient reading and reasoning
Emilee Holtzapple, Brent Cochran and Natasa Miskov-Zivanov

Measuring the relative importance of full text sections for information retrieval from scientific literature.
Lana Yeganova, Won Gyu KIM, Donald Comeau, W John Wilbur and Zhiyong Lu

Friday June 11, 2021 (continued)

14:30–15:00 ***Coffee Break***

15:00–17:00 **Session 6: MEDIQA 2021 Poster Session**

UCSD-Adobe at MEDIQA 2021: Transfer Learning and Answer Sentence Selection for Medical Summarization
Khalil Mrini, Franck Dernoncourt, Seunghyun Yoon, Trung Bui, Walter Chang, Emilias Farcas and Ndapa Nakashole

ChicHealth @ MEDIQA 2021: Exploring the limits of pre-trained seq2seq models for medical summarization
Liwen Xu, Yan Zhang, Lei Hong, Yi Cai and Szui Sung

NCUEE-NLP at MEDIQA 2021: Health Question Summarization Using PEGASUS Transformers
Lung-Hao Lee, Po-Han Chen, Yu-Xiang Zeng, Po-Lei Lee and Kuo-Kai Shyu

SB_NITK at MEDIQA 2021: Leveraging Transfer Learning for Question Summarization in Medical Domain
Spandana Balumuri, Sony Bachina and Sowmya Kamath S

Optum at MEDIQA 2021: Abstractive Summarization of Radiology Reports using simple BART Finetuning
Ravi Kondadadi, Sahil Manchanda, Jason Ngo and Ronan McCormack

QIAI at MEDIQA 2021: Multimodal Radiology Report Summarization
Jean-Benoit Delbrouck, Cassie Zhang and Daniel Rubin

NLM at MEDIQA 2021: Transfer Learning-based Approaches for Consumer Question and Multi-Answer Summarization
Shweta Yadav, Mourad Sarrouti and Deepak Gupta

IBMResearch at MEDIQA 2021: Toward Improving Factual Correctness of Radiology Report Abstractive Summarization
Diwakar Mahajan, Ching-Huei Tsou and Jennifer J Liang

UETrice at MEDIQA 2021: A Prosper-thy-neighbour Extractive Multi-document Summarization Model
Duy-Cat Can, Quoc-An Nguyen, Quoc-Hung Duong, Minh-Quang Nguyen, Huy-Son Nguyen, Linh Nguyen Tran Ngoc, Quang-Thuy Ha and Mai-Vu Tran

MNLP at MEDIQA 2021: Fine-Tuning PEGASUS for Consumer Health Question Summarization
Jooyeon Lee, Huong Dang, Ozlem Uzuner and Sam Henry

UETfishes at MEDIQA 2021: Standing-on-the-Shoulders-of-Giants Model for Abstractive Multi-answer Summarization
Hoang-Quynh Le, Quoc-An Nguyen, Quoc-Hung Duong, Minh-Quang Nguyen, Huy-Son Nguyen, Tam Doan Thanh, Hai-Yen Thi Vuong and Trang M. Nguyen

Session 7: Invited Talk by Makoto Miwa

17:00–17:30 ***Makoto Miwa: Information Extraction from Texts Using Heterogeneous Information***

17:30–18:00 **Closing remarks**

Improving BERT Model Using Contrastive Learning for Biomedical Relation Extraction

Peng Su[†], **Yifan Peng**[‡,1], **K. Vijay-Shanker**[†,1]
[†] Department of Computer and Information Science, University of Delaware
[‡] Department of Population Health Sciences, Weill Cornell Medicine
{psu, vijay}@udel.edu, yip4002@med.cornell.edu

Abstract

Contrastive learning has been used to learn a high-quality representation of the image in computer vision. However, contrastive learning is not widely utilized in natural language processing due to the lack of a general method of data augmentation for text data. In this work, we explore the method of employing contrastive learning to improve the text representation from the BERT model for relation extraction. The key knob of our framework is a unique contrastive pre-training step tailored for the relation extraction tasks by seamlessly integrating linguistic knowledge into the data augmentation. Furthermore, we investigate how large-scale data constructed from the external knowledge bases can enhance the generality of contrastive pre-training of BERT. The experimental results on three relation extraction benchmark datasets demonstrate that our method can improve the BERT model representation and achieve state-of-the-art performance. In addition, we explore the interpretability of models by showing that BERT with contrastive pre-training relies more on rationales for prediction. Our code and data are publicly available at: https://github.com/udel-biotm-lab/BERT-CLRE.

1 Introduction

Contrastive learning is a family of methods to learn a discriminative model by comparing input pairs (Le-Khac et al., 2020). The comparison is performed between positive pairs of "similar" inputs and negative pairs of "dissimilar" inputs. The positive pairs can be generated in an automatic way by transforming the original data to variants without changing the key information (e.g., rotate an image). Contrastive learning can encode general properties (e.g. invariance) in the learned representation while it is relatively hard for other representation learning methods to achieve (Bengio et al., 2013; Le-Khac et al., 2020). Therefore, contrastive learning provides a powerful approach to learn representations in a self-supervised manner and has shown great promise and achieved the state of the art results in recent years (He et al., 2020; Chen et al., 2020).

Despite its advancement, contrastive learning has not been well studied in biomedical natural language processing (BioNLP), especially for relation extraction (RE) tasks. One obstacle lies in the discrete characteristics of text data. Compared to computer vision, it is more challenging to design a general and efficient data augmentation method to construct positive pairs. Instead, there have been significant advances in the development of pre-trained language models to facilitate downstream BioNLP tasks (Devlin et al., 2019; Radford et al., 2019; Peng et al., 2019). Therefore, leveraging contrastive learning in the large pre-trained language models to learn more general representation for RE tasks remains unexplored.

To bridge this gap, this paper presents an innovative method of contrastive pre-training to improve the language model representation for biomedical relation extraction. As the main difference from the existing contrastive learning framework, we augment the datasets for RE tasks by randomly changing the words that do not affect the relation expression. Here, we hypothesize that the shortest dependency path (SDP) between two entities (Bunescu and Mooney, 2005) captures the required knowledge for the relation expression. We hence keep words on SDP fixed during the data augmentation. In addition, we utilize external knowledge bases to construct more data to make the learned representation generalize better, which is a method that is frequently used in distant supervision (Mintz et al., 2009; Peng et al., 2016).

To verify the effectiveness of the proposed method, we use the transformer-based BERT model as a backbone (Devlin et al., 2019) and evaluate

[1]These authors contributed equally.

Proceedings of the BioNLP 2021 workshop, pages 1–10
June 11, 2021.

our method on three widely studied RE tasks in the biomedical domain: the chemical-protein interactions (ChemProt) (Krallinger et al., 2017), the drug-drug interactions (DDI) (Herrero-Zazo et al., 2013), and the protein-protein interactions (PPI) (Krallinger et al., 2008). The experimental results show that our method boosts the BERT model performance and achieves state-of-the-art results on all three tasks.

Interest has also grown in designing interpretable BioNLP models that are both plausible (accurate) and rely on a specific part of the input (faithful rationales) (DeYoung et al., 2020; Lei et al., 2016). Here rationale is defined as the supporting evidence in the inputs for the model to make correct predictions. In this direction, we propose a new metric, "prediction shift", to measure the sensitivity degree to which the small changes (out of the SDP) of the inputs will make model change its predictions. We show that the contrastively pre-trained model is more robust than the original model, suggesting that our model is more likely to make predictions based on the rationales of the inputs.

In sum, the contribution of this work is fourfold. (1) We propose a new method that utilizes contrastive learning to improve the BERT model on biomedical relation extraction tasks. (2) We utilize external knowledge to generate more data for learning more generalized text representation. (3) We achieve state-of-the-art performance on three benchmark datasets of relation extraction tasks. (4) We propose a new metric that aims to reveal the rationales that the model uses for predicting relations. The code and the new rationale test datasets are available at `https://github.com/udel-biotm-lab/BERT-CLRE`.

2 Related Work

The history of contrastive representation learning can be traced back to (Hadsell et al., 2006), in which the authors explore the method of representation learning that similar inputs are mapped to nearby points in the representation space. Recently, with the development of data augmentation techniques, deep neural network architectures, contrastive learning regains attention and achieves superior performance on visual representation learning (He et al., 2020; Chen et al., 2020). In (He et al., 2020), the Momentum Contrast (MoCo) framework is designed to learn representation using the mechanism of dictionary look-up: an encoded example (the query) should be similar to its matching key (augmented sample from the same data example) and dissimilar to others. In (Chen et al., 2020), the authors propose the SimCLR frame to learn the representations by maximizing the agreement between augmented views of the same data point.

The contrastive representation has all the properties that a good representation should have: 1) Distributed property; 2) Abstraction and invariant property; 3) Disentangled representation (Bengio et al., 2013; Le-Khac et al., 2020). The distributed property emphasizes the expressive aspect of the representation (different data points should have distinguishable representations). The capture of abstract concepts and the invariance to small and local changes are concerned in the abstraction and invariant property. From the disentangled representation's perspective, it should encode as much information as possible. In this work, we will show contrastive learning can improve the invariant aspect of the representation.

In the natural language processing (NLP) field, several works have utilized the contrastive learning technique. Fang et al. (2020) propose a pre-trained language representation model (CERT) using contrastive learning at the sentence level to benefit the language understanding tasks. Klein and Nabi (2020) employ contrastive self-supervised learning to solve the commonsense reasoning problem. Peng et al. (2020) propose a self-supervised pre-training framework for relation extraction to explore the encoded information for the textual context and entity type. Compared with the previous works, we employ different data augmentation techniques and utilize data from external knowledge bases in contrastive learning to improve the model for relation extraction tasks.

Relation extraction is usually seen as a classification problem when the entity mentions are given in the text. Many different methods have been proposed to solve the relation extraction problem (Culotta and Sorensen, 2004; Sierra et al., 2008; Sahu and Anand, 2018; Zhang et al., 2019; Su et al., 2019). However, the language model methods redefine this field with their superior performance (Dai and Le, 2015; Peters et al., 2018; Devlin et al., 2019; Radford et al., 2019; Su and Vijay-Shanker, 2020). Among all the language models, BERT (Devlin et al., 2019) –a language representation model based on bidirectional Transformer (Vaswani et al., 2017), attracts lots of attention in

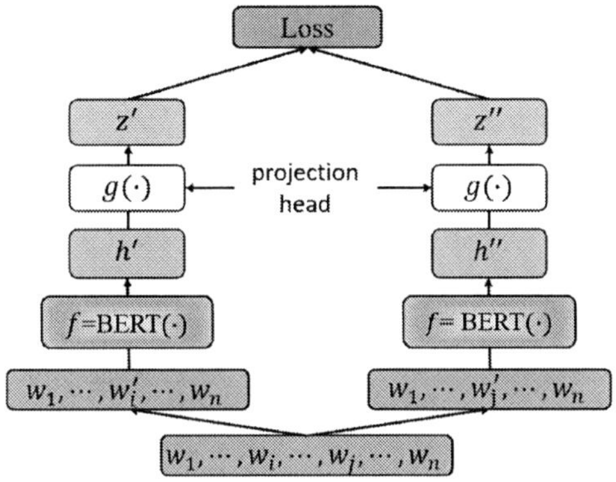

Figure 1: The framework of contrastive learning. For the data augmentation of relation extraction, we randomly replace some words that are not affecting the relation expression ($w_i \rightarrow w_i^{'}$ in the left sample, $w_j \rightarrow w_j^{'}$ in the right sample).

different fields. Several BERT models have been adapted for biomedical domain: BioBERT (Lee et al., 2020), SciBERT (Beltagy et al., 2019), BlueBERT (Peng et al., 2019) and PubMedBERT (Gu et al., 2021). BioBERT, SciBERT and BlueBERT are pre-trained based on the general-domain BERT using different pre-training data. In contrast, PubMedBERT (Gu et al., 2021) is pre-trained from scratch using PubMed abstracts.

In recent years, there is increasing interest in designing more interpretable NLP models that reveal the logic behind model predictions. In (DeYoung et al., 2020), multiple datasets of rationales (from human experts) are collected to facilitate the research on interpretable models in NLP. In (Lei et al., 2016), the authors propose an encoder-generator framework to automatically generate candidate rationales to justify the predictions of neural network models.

3 Methodology

3.1 The framework of contrastive learning

Our goal is to learn a text representation by maximizing agreement between inputs from positive pairs via a contrastive loss in the latent space and the learned representation can then be used for relation extraction. Figure 1 shows our framework of contrastive learning. Given a sentence $s = w_1, ...w_n$, we first produce two augmented views (a positive pair) $v' = w_1, ..., w_i', ...w_n$ and $v'' = w_1..., w_j', ...w_n$ $(i \neq j)$ from s by applying text augmentation technique (Section 3.1.1).

Our framework then uses one neural network to encode the two inputs, which consists of a neural network encoder f (Section 3.1.2) and a projection head g (Section 3.1.3). From the first augmented view v', we output a *representation* $h' \triangleq f(v')$ and a projection $z' \triangleq g(h')$. From the second augmented view v'', we output $h'' \triangleq f(v'')$ and another projection $z'' \triangleq g(h'')$.

The contrastive learning method learns the representation by comparing different samples in the training data (Section 3.1.4). The comparison is performed between both similar inputs and dissimilar inputs, and the similar inputs are positive pairs and the dissimilar inputs are negative pairs. During the training, the representations are learned by leading the positive pairs to have similar representations and making negative pairs have dissimilar representations. In applications, the positive pairs are usually from the augmented data of the same sample, and the negative pairs are generated by selecting augmented data from different samples.

At the end of training, we only keep the encoder f as in (Chen et al., 2020). For any text input x, $h = f(x)$ will be the representation of x from contrastive learning.

3.1.1 Data augmentation for relation extraction

The data augmentation module is a key component of contrastive learning, which needs to randomly generate two correlated views for the original data point. At the same time, the generated data should be different from each other to make them distinguishable (from the model's perspective), but should not be significantly different to change the structure and semantics of the original data. It is especially difficult to augment the text data of relation extraction. In this work, we only focus on binary relations. Given $< s, e_1, e_2, r >$, where e_1 and e_2 are two entity mentions in the sentence s with the relation type r, we keep e_1 and e_2 in the sentence and retain the relation expression between e_1 and e_2 in the augmented views.

Specifically, we propose a data augmentation method utilizing the shortest dependency path (SDP) between the two entities in the text. We hypothesize that the shortest dependency path captures the required information to assert the relationship of the two entities (Bunescu and Mooney, 2005). Therefore we fix the shortest dependency path, and randomly change the other tokens in the text to generate the augmented data. This idea is inspired by (Wei and Zou, 2019), which

Original	We further show that @PROTEIN$ directly interacts with @PROTEIN$ and Rpn4.
After SR	We further show that @PROTEIN$ **straight** interacts with @PROTEIN$ and Rpn4.
After RS	**Further we** show that @PROTEIN$ directly interacts with @PROTEIN$ and Rpn4.
After RD	We further show that @PROTEIN$ interacts with @PROTEIN$ and Rpn4.

Table 1: Examples after the three operations for data augmentation. The shortest dependency path between two proteins is "@PROTEIN$ interacts @PROTEIN$", which is marked with underline in the examples. The changed words are also marked with bold font.

employed easy data augmentation techniques to improve model performance on text classification tasks.

As the preliminary study, we experiment with three techniques to randomly replace the tokens to generate the augmented data and choose the best one for our contrastive learning method: 1) Synonym replacement (SR), 2) Random swap (RS), and 3) Random deletion (RD).

Table 1 gives some samples after applying the three operations on a sentence from the PPI task. For the synonym replacement, we randomly replace n words with their synonyms. To acquire the synonym of a word, we utilize the WordNet database (Miller, 1995) to extract a list of synonyms and randomly choose one from the list. For the random swap, we swap the positions of two words and repeat this operation n times. For the random deletion, we delete some words with the probability p. The probability p is set to 0.1 in our experiments and the parameter n for SR and RS is calculated by $p \times l$, where l is the length of the sentence.

To examine which operation performs better for relation extraction tasks, we train three BERT models using the three types of augmented data (combined with the original training data). Table 4 shows that the synonym replacement (SR) operation achieves the best performance on all three tasks and we will employ this operation in our data augmentation module in our contrastive learning experiments (We will further discuss it in Section 5.2).

3.1.2 The neural network encoder

In this work, we employ the BERT model (Devlin et al., 2019) as our encoder for the text data and the classification token ([CLS]) output in the last layer will be the representation of the input.

3.1.3 Projection head

As demonstrated in (Chen et al., 2020), adding a nonlinear projection head on the model output will improve the representation quality during training. Following the same idea, a multi-layer perceptron (MLP) will be applied to the model output h. Formally,

$$z = g(h) = W^2\phi(W^1h)$$

and ϕ is the ReLU activation function, W^1 and W^2 are the weights of the perceptron in the hidden layers.

3.1.4 Contrastive loss

Contrastive learning is designed to make similar representations be learned for the augmented samples (positive pairs) from the same data point. We follow the work of (Chen et al., 2020) to design the loss function (Algorithm 1). During contrastive learning, the contrastive loss is calculated based on the augmented batch derived from the original batch. Given N sentences in a batch, we first employ the data augmentation technique to acquire two views for each sentence in the batch. Therefore, we have $2N$ views from the batch. Given one positive pair (two views from the same sentence), we treat the other $2(N-1)$ within the batch as negative examples. Similar to (Chen et al., 2020), the loss for a positive pair is defined as:

$$l(z', z'') = -log\frac{exp(sim(z', z'')/\tau)}{\sum_{k=1}^{2N} \mathbb{1}_{[z_k \neq z']} exp(sim(z', z_k)/\tau)}$$

where $sim(\cdot, \cdot)$ is the cosine similarity function, $\mathbb{1}_{[z_k \neq z']}$ is the indicator function and τ is the temperature parameter. The final loss L is computed across all positive pairs, both (z', z'') and (z'', z'), in a batch.

For computation convenience, we arrange the $(2k-1)$-th example and the $2k$-th example in the batch are generated from the same sentence, a.k.a., $(2k-1, 2k)$ is a positive pair. Please see Algorithm 1 for calculating the contrastive loss in one batch. Then we can update the parameters of the BERT model and projection head g to minimize the loss L.

Algorithm 1: Contrastive loss in a batch

Input: encoder f (BERT), project head g, data augmentation module, data batch $\{s_k\}_{k=1}^{N}$;

for k=*1,...,N* **do**

  $v', v'' = data\_augment(s_k)$;

  $z_{2k-1} = g(f(v'))$;

  $z_{2k} = g(f(v''))$;

end

$L = \frac{1}{2N} \sum_{k=1}^{N} [l(z_{2k-1}, z_{2k}) + l(z_{2k}, z_{2k-1})]$

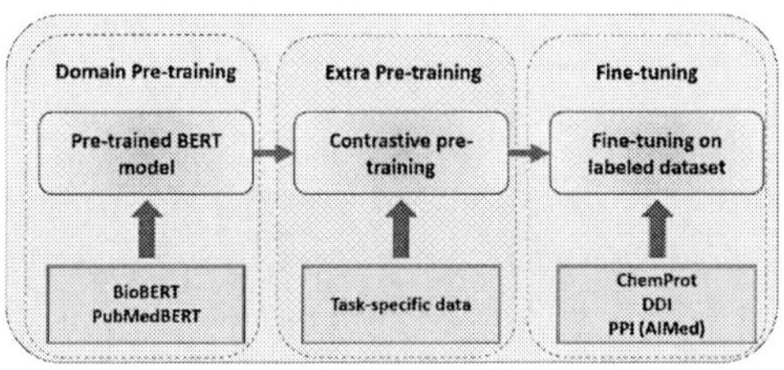

Figure 2: The pipeline of BERT model training with contrastive pre-training.

3.2 Training procedure

Figure 2 shows the training procedure of our framework. It consists of three stages. First, we pre-train the BERT model on a large amount of unlabeled data from a specific domain(e.g., biomedical domain). Second, we conduct contrastive pre-training on task-specific data as a continual pre-training step after the domain pre-training of BERT model. In this way, we retain the learned knowledge from general pre-training, and add the new features from contrastive learning. Finally, we fine-tune the model on the RE tasks to further gain task-specific knowledge through supervised training on the labeled datasets.

The domain pre-training stage follows that of the BERT using the masked language model and next sentence prediction technique (Devlin et al., 2019). In our experiments, we use two pre-trained versions for the biomedical domain: BioBERT (Lee et al., 2020) and PubMedBERT (Gu et al., 2021).

3.3 A knowledge-based method to enrich training dataset for contrastive learning

Contrastive pre-training requires a large-scale dataset to generalize the representation. Also, our data augmentation for contrastive learning needs SDP between two given entities, so we need to construct the augmented dataset with the entities

Task	Train	Dev	Test	EK
ChemProt	18,035	11,268	15,745	35,500
DDI	22,233	5,559	5,716	67,959
PPI*	5,251	-	583	97,853

Table 2: Statistics of datasets used for contrastive pre-training and fine-tuning. EK: datasets generated by external knowledge bases; *: since there is no standard split of training and test set for the PPI dataset (AIMed), we use 10-fold cross-validation and here we show number of the training and test in each fold.

mentioned in the text. For these purposes, we utilize external databases for the relations to acquire extra instances for contrastive learning.

Formally, assuming a curated database for relation r contains all the relevant entities and text, we consider every combination of the entity pairs in one sentence and use them as examples for this relation. For instance, there are three proteins in the sentence s: "Thus NIPP1 works as a molecular sensor for PP1 to recognize phosphorylated Sap155." We will generate three examples for PPI task from this sentence: <s,NIPP1,PP1,PPI>, <s,NIPP1,Sap155,PPI> and <s,PP1,Sap155,PPI>.

We use the IntAct database (Orchard et al., 2014) as the interacting protein pairs database for the PPI task. Similarly, DrugBank (Wishart et al., 2008) and BioGRID (Stark et al., 2006) are utilized for DDI and ChemProt, respectively. In the column "EK" of Table 2, we show the statistics of datasets for each task generated by external knowledge bases. We can see that the datasets from the external database are much larger than that of the human-labeled datasets.

4 Experiments

As discussed before, we will utilize the BERT model as the encoder for the inputs. In particular, we will employ two BERT models pre-trained for the biomedical domain in our experiments: BioBERT (Lee et al., 2020) and PubMedBERT (Gu et al., 2021).

4.1 Datasets and evaluation metrics

We will evaluate our method on three benchmark datasets. The statistics of these datasets is shown in Table 2. For ChemProt and DDI tasks, we employ the corpora in (Krallinger et al., 2017) and (Herrero-Zazo et al., 2013) respectively, and we use the same split of training, development and test sets with the

Model	ChemProt			DDI			PPI		
	P	R	F	P	R	F	P	R	F
BioBERT	74.3	**76.3**	75.3	79.9	78.1	79.0	79.0	**83.3**	81.0
BioBERT+CL	**77.0**	74.7	75.8	82.6	77.4	79.9	79.8	83.1	81.3
BioBERT+CLEK	76.6	76.0	**76.3**	**82.9**	**78.4**	**80.6**	**81.1**	83.2	**82.1**
PubMedBERT	78.8	75.9	77.3	82.6	81.9	82.3	**80.1**	84.3	82.1
PubMedBERT+CL	79.6	76.2	77.8	**83.3**	81.5	82.4	79.4	85.6	82.4
PubMedBERT+CLEK	**80.6**	**76.9**	**78.7**	**83.3**	**82.4**	**82.9**	79.9	**85.7**	**82.7**

Table 3: BERT model performance on ChemProt, DDI and PPI tasks. BioBERT/PubMedBERT: original BERT model; BioBERT/PubMedBERT+CL: BioBERT/PubMedBERT with contrastive pre-training on the training set of human-labeled dataset; BioBERT/PubMedBERT+CLEK: BioBERT/PubMedBERT with contrastive pre-training on the data from the external knowledge base.

PubMedBERT model (Gu et al., 2021) during the model evaluation. We utilize the AIMed corpus (Bunescu et al., 2005) for the PPI task, and we will employ 10-fold cross-validation on it since there is no standard split of training and test.

PPI is a binary classification problem, and we will use the standard precision (P), recall (R) and F1-score (F) to measure the model performance. However, the ChemProt and DDI tasks are multi-class classification problems. The ChemProt corpus is labeled with five positive classes and the negative class: CPR:3, CPR:4, CPR:5, CPR:6, CPR:9 and negative. Similar to the DDI corpus, there are four positive labels and one negative label: ADVICE, EFFECT, INT, MECHANISM and negative. The models for ChemProt and DDI will be evaluated utilizing micro precision, recall and F1 score on the non-negative classes.

4.2 Data pre-processing

One instance of relation extraction task contains two parts: the text and the entity mentions. In order to make the BERT model identify the positions of the entities, we replace the relevant entity names with predefined tags by following the standard pre-processing step for relation extraction (Devlin et al., 2019). Specifically, all the protein names are replaced with @PROTEIN$, drug names with @DRUG$, and chemical names with @CHEMICAL$. In Table 1, we show a pre-processed example of the PPI task.

4.3 Training setup

For the fine-tuning of the BioBERT models, we use the learning rate of 2e-5, batch size of 16, training epoch of 10, and max sequence length of 128. During the fine-tuning of PubMedBERT models, the learning rate of 2e-5, batch size of 8, training epoch of 10 and max sequence length of 256 are utilized.

In the contrastive pre-training step of the BERT models, we use the same learning rate with the fine-tuning, and the training epoch is selected from [2, 4, 6, 8, 10] based on the performance on the development set. If there is no development set (e.g., PPI task), we will use 6 as the default training epoch. Since contrastive learning benefits more from larger batch (Chen et al., 2020), we utilize the batch size of 256 and 128 for BioBERT and PubMedBERT respectively. In addition, the temperature parameter τ is set to 0.1 during the training.

5 Results and discussion

5.1 BERT model performance with contrastive pre-training

Table 3 demonstrates the experimental results using the BERT models with contrastive pre-training and external datasets. The first row is the BioBERT model performance without applying contrastive learning. The following two rows demonstrate the results after adding the contrastive pre-training step in BioBERT. The "BioBERT+CL" stands for the BioBERT model with contrastive pre-training on the training set of the human-labeled dataset, while "BioBERT+CLEK" is for the BioBERT model with contrastive pre-training on the data from the external knowledge base. Similarly, we give the PubMedBERT model performance of our method in the last three rows of Table 3.

We can see that the contrastive per-training improves the model performance in both cases. How-

Training data	ChemProt	DDI	PPI
Original	75.3	79.0	81.0
+RS	75.6	78.4	75.4
+RD	75.4	79.8	81.2
+SR	**76.0**	**80.1**	**81.9**

Table 4: BioBERT model performance (F1 score) using different types of augmented data. RS: random swap; RD: random deletion; SR: synonym replacement.

ever, contrastive pre-training on human-labeled dataset only improves the model with a small margin. We hypothesize that the limited improvement might be due to the poor generalization on small training set. Therefore, we include more data (EK data) in contrastive learning to enhance the model generalizability. The data generated from the external knowledge base are much more than the training data of the human-labeled dataset (column "EK" and "train" in Table 2). As shown in the third and sixth row in Table 3, contrastive learning with more external data can further boost the model performance. Compared with the BERT models without contrastive pre-training, we observe an averaged F1 score improvement (on the two BERT models) of 1.2%, 1.2%, and 0.85% on ChemProt, DDI, and PPI datasets, respectively.

Since PubMedBERT is the state-of-the-art (SOTA) model on these three tasks, we further improve its performance by adding contrastive pre-training. Thus, we achieve state-of-the-art performance on all three datasets.

5.2 Comparison of data augmentation techniques

Table 4 shows the BERT model performance after including three types of augmented data. We can see that the synonym replacement (SR) operation yields the best results on all three tasks. Therefore we use it as our default operation to generate augmented data in all our contrastive learning experiments. We also notice that the augmented data from the random swap (RS) operation hurt the model performance on the DDI and PPI tasks, which indicates that this operation might change the relation expression in the sentence. Thus it is necessary to verify the effectiveness of the operations before applying them on contrastive learning.

Input sentence	Prediction
(1) Instead, radiolabeled @CHEMICAL$ resulting from @PROTEIN$ hydrolysis were observed.	CPR:9
(2) **Or else**, radiolabeled @CHEMICAL$ resulting from @PROTEIN$ hydrolysis were observed.	False
(1) These results indicate that membrane @PROTEIN$ levels in N-38 neurons are dynamically autoregulated by @CHEMICAL$.	CPR:3
(2) These results indicate that membrane @PROTEIN$ levels in N-38 **nerve cell** are dynamically autoregulated by @CHEMICAL$.	False

Table 5: Examples of prediction shift. (1): Original sentence; (2): Augmented sentence.

Task	Model	Prediction Shift
ChemProt	BioBERT	246
	BioBERT+CLEK	191 (22% ↓)
	PubMedBERT	248
	PubMedBERT+CLEK	189 (24% ↓)
DDI	BioBERT	111
	BioBERT+CLEK	89 (20% ↓)
	PubMedBERT	90
	PubMedBERT+CLEK	75 (17% ↓)
PPI*	BioBERT	51
	BioBERT+CLEK	33 (35%↓)
	PubMedBERT	49
	PubMedBERT+CLEK	34 (31%↓)

Table 6: Count of prediction shift on the "augmented" test set. *: The sum of counts on the 10 folds.

5.3 Measurement of rationale faithfulness

As discussed previously, we hypothesize the words on the shortest dependency path (SDP) as the rationales in the input. Therefore, the model should make its predictions based on them. If the model predictions are all made based on a specific part of the input, we can define this specific part of the input to be the completely faithful rationales. In practice, the rationales are more faithful means they are more influential on the model predictions.

In this work, we define a new metric to measure the faithfulness of the rationales: "prediction shift". If the model predicts one test example (non-negative) with label L_t, but changes its prediction on its neighbor (the augmented data point) with another label $L_t^{'}$, we will say a "prediction shift" happens (In Table 5, we give two examples of prediction shift on PubMedBERT model). Fewer "prediction shift" indicates the information outside of SDP influences the prediction less, which means the rationales are more faithful.

To generate a similar set (with test set) for the measurement of "prediction shift", we apply the same synonym replacement (SR) technique on the original test data. Since we retain the words that are on the shortest dependency path between the two entities, the generated data should express the same relation with the original ones. The trained model should predict them with the same labels if the rationales of input are utilized during inference, and in that case, we say the rationales are faithful.

We compare the number of "prediction shift" on two types of BERT model: the original BERT and the BERT model with contrastive pre-training. Table 6 illustrates that the BERT models with contrastive pre-training dramatically reduce the number of "prediction shift". Those results indicate that the BERT models with contrastive pre-training rely more on the information of shortest dependency path for prediction, a.k.a., the rationales are more faithful. From another perspective, the results in Table 6 also demonstrate that the BERT models with contrastive pre-training are resilient to small changes of the inputs, which means the models are more robust.

6 Conclusion and Future Directions

In this work, we propose a contrastive pre-training method to improve the text representation of the BERT model. Our approach differs from previous studies in the choice of text data augmentation with linguistic knowledge and the use of the external knowledge bases to construct large-scale data to facilitate contrastive learning. The experimental results demonstrate that our method outperforms the original BERT model on three relation extraction benchmarks. Additionally, our method shows robustness to slightly changed inputs over the BERT models. In the future, we will investigate different settings of data augmentation and contrastive pre-training to exploit their capability on language models. We also hope that our work can inspire researchers to design better metrics and create high-quality datasets for the exploration of model interpretability.

Acknowledgment

Yifan Peng's research was supported by National Library of Medicine - National Institutes of Health (NIH) under award number R00LM013001. Peng Su's graduate studies were supported by a supplement of NIH grant U01GM125267.

References

Iz Beltagy, Kyle Lo, and Arman Cohan. 2019. SciBERT: a pretrained language model for scientific text. *arXiv:1903.10676 [cs]*.

Yoshua Bengio, Aaron Courville, and Pascal Vincent. 2013. Representation learning: a review and new perspectives. *IEEE Transactions on Pattern Analysis and Machine Intelligence*, pages 1798–1828.

Razvan Bunescu, Ruifang Ge, Rohit J. Kate, Edward M. Marcotte, Raymond J. Mooney, Arun K. Ramani, and Yuk Wah Wong. 2005. Comparative experiments on learning information extractors for proteins and their interactions. *Artificial intelligence in medicine*, 33(2):139–155.

Razvan C. Bunescu and Raymond J. Mooney. 2005. A shortest path dependency kernel for relation extraction. In *HLT-EMNLP*, pages 724–731, Stroudsburg, PA, USA.

Ting Chen, Simon Kornblith, Mohammad Norouzi, and Geoffrey Hinton. 2020. A simple framework for contrastive learning of visual representations. In *ICML*, pages 1597–1607.

Aron Culotta and Jeffrey Sorensen. 2004. Dependency tree kernels for relation extraction. In *ACL*, pages 1–7, Barcelona, Spain.

Andrew M Dai and Quoc V Le. 2015. Semi-supervised sequence learning. In *NIPS*, pages 3079–3087.

Jacob Devlin, Ming-Wei Chang, Kenton Lee, and Kristina Toutanova. 2019. BERT: pre-training of deep bidirectional transformers for language understanding. In *ACL*, pages 4171–4186.

Jay DeYoung, Sarthak Jain, Nazneen Fatema Rajani, Eric Lehman, Caiming Xiong, Richard Socher, and Byron C. Wallace. 2020. ERASER: a benchmark to evaluate rationalized NLP models. In *ACL*, pages 4443–4458.

Hongchao Fang, Sicheng Wang, Meng Zhou, Jiayuan Ding, and Pengtao Xie. 2020. CERT: Contrastive Self-supervised Learning for Language Understanding. *arXiv:2005.12766 [cs, stat]*.

Yu Gu, Robert Tinn, Hao Cheng, Michael Lucas, Naoto Usuyama, Xiaodong Liu, Tristan Naumann, Jianfeng Gao, and Hoifung Poon. 2021. Domain-specific language model pretraining for biomedical natural language processing. *arXiv:2007.15779*.

R. Hadsell, S. Chopra, and Y. LeCun. 2006. Dimensionality reduction by learning an invariant mapping. In *CVPR*, volume 2, pages 1735–1742.

K. He, H. Fan, Y. Wu, S. Xie, and R. Girshick. 2020. Momentum contrast for unsupervised visual representation learning. In *CVPR*, pages 9726–9735.

María Herrero-Zazo, Isabel Segura-Bedmar, Paloma Martínez, and Thierry Declerck. 2013. The DDI corpus: an annotated corpus with pharmacological substances and drug-drug interactions. *Journal of Biomedical Informatics*, 46(5):914–920.

Tassilo Klein and Moin Nabi. 2020. Contrastive self-supervised learning for commonsense reasoning. In *ACL*, pages 7517–7523.

Martin Krallinger, Florian Leitner, Carlos Rodriguez-Penagos, and Alfonso Valencia. 2008. Overview of the protein-protein interaction annotation extraction task of BioCreative II. *Genome Biology*, 9 Suppl 2:S4.

Martin Krallinger, Obdulia Rabal, Saber A. Akhondi, Martín Pérez Pérez, Jesús Santamaría, Gael Pérez Rodríguez, Georgios Tsatsaronis, Ander Intxaurrondo, José Antonio López1 Umesh Nandal, Erin Van Buel, Akileshwari Chandrasekhar, Marleen Rodenburg, Astrid Laegreid, Marius Doornenbal, Julen Oyarzabal, Analia Lourenço, and Alfonso Valencia. 2017. Overview of the BioCreative VI chemical-protein interaction track. In *Proceedings of the BioCreative workshop*, pages 141–146.

Phuc H. Le-Khac, Graham Healy, and Alan F. Smeaton. 2020. Contrastive representation learning: a framework and review. *IEEE Access*, 8:193907–193934.

Jinhyuk Lee, Wonjin Yoon, Sungdong Kim, Donghyeon Kim, Sunkyu Kim, Chan Ho So, and Jaewoo Kang. 2020. BioBERT: a pre-trained biomedical language representation model for biomedical text mining. *Bioinformatics (Oxford, England)*, 36(4):1234–1240.

Tao Lei, Regina Barzilay, and Tommi Jaakkola. 2016. Rationalizing neural predictions. In *EMNLP*, pages 107–117.

George A. Miller. 1995. WordNet: a lexical database for English. *Communications of the ACM*, 38(11):39–41.

Mike Mintz, Steven Bills, Rion Snow, and Dan Jurafsky. 2009. Distant supervision for relation extraction without labeled data. In *ACL*, pages 1003–1011.

Sandra Orchard, Mais Ammari, Bruno Aranda, Lionel Breuza, Leonardo Briganti, Fiona Broackes-Carter, Nancy H. Campbell, Gayatri Chavali, Carol Chen, Noemi del-Toro, Margaret Duesbury, Marine Dumousseau, Eugenia Galeota, Ursula Hinz, Marta Iannuccelli, Sruthi Jagannathan, Rafael Jimenez, Jyoti Khadake, Astrid Lagreid, Luana Licata, Ruth C. Lovering, Birgit Meldal, Anna N. Melidoni, Mila Milagros, Daniele Peluso, Livia Perfetto, Pablo Porras, Arathi Raghunath, Sylvie Ricard-Blum, Bernd Roechert, Andre Stutz, Michael Tognolli, Kim van Roey, Gianni Cesareni, and Henning Hermjakob. 2014. The MIntAct project–IntAct as a common curation platform for 11 molecular interaction databases. *Nucleic Acids Research*, 42(Database issue):D358–363.

Hao Peng, Tianyu Gao, Xu Han, Yankai Lin, Peng Li, Zhiyuan Liu, Maosong Sun, and Jie Zhou. 2020. Learning from context or names? An empirical study on neural relation extraction. In *EMNLP*, pages 3661–3672.

Yifan Peng, Chih-Hsuan Wei, and Zhiyong Lu. 2016. Improving chemical disease relation extraction with rich features and weakly labeled data. *Journal of Cheminformatics*, 8:53.

Yifan Peng, Shankai Yan, and Zhiyong Lu. 2019. Transfer learning in biomedical natural language processing: an evaluation of BERT and ELMo on ten benchmarking datasets. In *Proceedings of the Workshop on Biomedical Natural Language Processing (BioNLP)*, pages 58–65.

Matthew Peters, Mark Neumann, Mohit Iyyer, Matt Gardner, Christopher Clark, Kenton Lee, and Luke Zettlemoyer. 2018. Deep contextualized word representations. In *NAACL*, pages 2227–2237.

Alec Radford, Jeffrey Wu, Rewon Child, David Luan, Dario Amodei, and Ilya Sutskever. 2019. Language models are unsupervised multitask learners. *OpenAI Blog*, 1(8):9.

Sunil Kumar Sahu and Ashish Anand. 2018. Drug-drug interaction extraction from biomedical texts using long short-term memory network. *Journal of Biomedical Informatics*, 86:15–24.

Gerardo Sierra, Rodrigo Alarcón, César Aguilar, and Carme Bach. 2008. Definitional verbal patterns for semantic relation extraction. *Terminology. International Journal of Theoretical and Applied Issues in Specialized Communication*, 14(1):74–98.

Chris Stark, Bobby-Joe Breitkreutz, Teresa Reguly, Lorrie Boucher, Ashton Breitkreutz, and Mike Tyers. 2006. BioGRID: a general repository for interaction datasets. *Nucleic Acids Research*, 34(Database issue):D535–539.

Peng Su, Gang Li, Cathy Wu, and K. Vijay-Shanker. 2019. Using distant supervision to augment manually annotated data for relation extraction. *PloS One*, 14(7):e0216913.

Peng Su and K Vijay-Shanker. 2020. Investigation of bert model on biomedical relation extraction based on revised fine-tuning mechanism. In *2020 IEEE International Conference on Bioinformatics and Biomedicine (BIBM)*, pages 2522–2529. IEEE.

Ashish Vaswani, Noam Shazeer, Niki Parmar, Jakob Uszkoreit, Llion Jones, Aidan N Gomez, Łukasz Kaiser, and Illia Polosukhin. 2017. Attention is all you need. In *NIPS*, pages 5998–6008.

Jason Wei and Kai Zou. 2019. EDA: easy data augmentation techniques for boosting performance on text classification tasks. In *EMNLP-IJCNLP*, pages 6381–6387.

David S. Wishart, Craig Knox, An Chi Guo, Dean Cheng, Savita Shrivastava, Dan Tzur, Bijaya Gautam, and Murtaza Hassanali. 2008. DrugBank: a knowledgebase for drugs, drug actions and drug targets. *Nucleic Acids Research*, 36(Database issue):D901–906.

H. Zhang, R. Guan, F. Zhou, Y. Liang, Z. Zhan, L. Huang, and X. Feng. 2019. Deep residual convolutional neural network for protein-protein interaction extraction. *IEEE Access*, 7:89354–89365.

Triplet-Trained Vector Space and Sieve-Based Search Improve Biomedical Concept Normalization

Dongfang Xu and **Steven Bethard**
School of Information
University of Arizona
Tucson, AZ
{dongfangxu9,bethard}@email.arizona.edu

Abstract

Concept normalization, the task of linking textual mentions of concepts to concepts in an ontology, is critical for mining and analyzing biomedical texts. We propose a vector-space model for concept normalization, where mentions and concepts are encoded via transformer networks that are trained via a triplet objective with online hard triplet mining. The transformer networks refine existing pre-trained models, and the online triplet mining makes training efficient even with hundreds of thousands of concepts by sampling training triples within each mini-batch. We introduce a variety of strategies for searching with the trained vector-space model, including approaches that incorporate domain-specific synonyms at search time with no model retraining. Across five datasets, our models that are trained only once on their corresponding ontologies are within 3 points of state-of-the-art models that are retrained for each new domain. Our models can also be trained for each domain, achieving new state-of-the-art on multiple datasets.

1 Introduction

Concept normalization (aka. entity linking or entity normalization) is a fundamental task of information extraction which aims to map concept mentions in text to standard concepts in a knowledge base or ontology. This task is important for mining and analyzing unstructured text in the biomedical domain as the texts describing biomedical concepts have many morphological and orthographical variations, and utilize different word orderings or equivalent words. For instance, *heart attack*, *coronary attack*, *MI*, *myocardial infarction*, *cardiac infarction*, and *cardiovascular stroke* all refer to the same concept. Linking such terms with their corresponding concepts in an ontology or knowledge base is critical for data interoperability and the development of natural language processing (NLP) techniques.

Research on concept normalization has grown thanks to shared tasks such as disorder normalization in the 2013 ShARe/CLEF (Suominen et al., 2013), chemical and disease normalization in BioCreative V Chemical Disease Relation (CDR) Task (Wei et al., 2015), and medical concept normalization in 2019 n2c2 shared task (Henry et al., 2020), and to the availability of annotated data (Doğan et al., 2014; Luo et al., 2019). Existing approaches can be divided into three categories: rule-based approaches using string-matching or dictionary look-up (Leal et al., 2015; D'Souza and Ng, 2015; Lee et al., 2016), which rely heavily on hand-crafted rules and domain knowledge; supervised multi-class classifiers (Limsopatham and Collier, 2016; Lee et al., 2017; Tutubalina et al., 2018; Niu et al., 2019; Li et al., 2019), which cannot generalize to concept types not present in their training data; and two-step frameworks based on a non-trained candidate generator and a supervised candidate ranker (Leaman et al., 2013; Li et al., 2017; Liu and Xu, 2017; Nguyen et al., 2018; Murty et al., 2018; Mondal et al., 2019; Ji et al., 2020; Xu et al., 2020), which require complex pipelines and fail if the candidate generator does not find the gold truth concept.

We propose a vector space model for concept normalization, where mentions and concepts are encoded as vectors – via transformer networks trained via a triplet objective with online hard triplet mining – and mentions are matched to concepts by vector similarity. The online hard triplet mining strategy selects the hard positive/negative exemplars from within a mini-batch during training, which ensures consistently increasing difficulty of triplets as the network trains for fast convergence. There are two advantages of applying the vector space model for concept normalization: 1) it is computationally cheap compared with other supervised classification approaches as we only compute the representations for all concepts in ontology once

Proceedings of the BioNLP 2021 workshop, pages 11–22
June 11, 2021. ©2021 Association for Computational Linguistics

after training the network; 2) it allows concepts and synonyms to be added or deleted after the network is trained, a flexibility that is important for the biomedical domain where frequent updates to ontologies like the Unified Medical Language System (UMLS) Metathesaurus[1] are common. Unlike prior work, our simple and efficient model requires neither negative sampling before the training nor a candidate generator during inference.

Our work makes the following contributions:

- We propose a triplet network with online hard triplet mining for training a vector-space model for concept normalization, a simpler and more efficient approach than prior work.
- We propose and explore a variety of strategies for matching mentions to concepts using the vector-space model, with the most successful being a simple sieve-based approach that checks domain-specific synonyms before domain-independent ones.
- Our framework produces models trained on only the ontology – no domain-specific training – that can incorporate domain-specific concept synonyms at search time without retraining, and these models achieve within 3 points of state-of-the-art on five datasets.
- Our framework also allows models to be trained for each domain, achieving state-of-the-art performance on multiple datasets.

The code for our proposed framework is available at `https://github.com/dongfang91/Triplet-Search-ConNorm`.

2 Related work

Earlier work on concept normalization focuses on how to use morphological information to conduct lexical look-up and string matching (Kang et al., 2013; D'Souza and Ng, 2015; Leaman et al., 2015; Leal et al., 2015; Kate, 2016; Lee et al., 2016; Jonnagaddala et al., 2016). They rely heavily on handcrafted rules and domain knowledge, e.g., D'Souza and Ng (2015) define 10 types of rules at different priority levels to measure morphological similarity between mentions and candidate concepts in the ontologies. The lack of lexical overlap between concept mention and concept in domains like social media, makes rule-based approaches that rely on lexical matching less applicable.

Supervised approaches for concept normalization have improved with the availability of annotated data and deep learning techniques. When the number of concepts to be predicted is small, classification-based approaches (Limsopatham and Collier, 2016; Lee et al., 2017; Tutubalina et al., 2018; Niu et al., 2019; Li et al., 2019; Miftahutdinov and Tutubalina, 2019) are often adopted, with the size of the classifier's output space equal to the number of concepts. Approaches differ in neural architectures, such as character-level convolution neural networks (CNN) with multi-task learning (Niu et al., 2019) and pre-trained transformer networks (Li et al., 2019; Miftahutdinov and Tutubalina, 2019). However, classification approaches struggle when the annotated training data does not contain examples of all concepts – common when there are many concepts in the ontology – since the output space of the classifier will not include concepts absent from the training data.

To alleviate the problems of classification-based approaches, researchers apply learning to rank in concept normalization, a two-step framework including a non-trained candidate generator and a supervised candidate ranker that takes both mention and candidate concept as input. Previous candidate rankers have used point-wise learning to rank (Li et al., 2017), pair-wise learning to rank (Leaman et al., 2013; Liu and Xu, 2017; Nguyen et al., 2018; Mondal et al., 2019), and list-wise learning to rank (Murty et al., 2018; Ji et al., 2020; Xu et al., 2020). These learning to rank approaches also have drawbacks. Firstly, if the candidate generator fails to produce the gold truth concept, the candidate ranker will also fail. Secondly, the training of candidate ranker requires negative sampling beforehand, and it is unclear if these pre-selected negative samples are informative for the whole training process (Hermans et al., 2017; Sung et al., 2020).

Inspired by Schroff et al. (2015), we propose a triplet network with online hard triplet mining for concept normalization. Our framework sets up concept normalization as a one-step process, calculating similarity between vector representations of the mention and of all concepts in the ontology. Online hard triplet mining allows such a vector space model to generate triplets of (mention, true concept, false concept) within a mini-batch, leading to efficient training and fast convergence (Schroff et al., 2015). In contrast with previous vector space models where mention and candidate

[1] `https://www.nlm.nih.gov/research/umls/knowledge_sources/metathesaurus/index.html`

concepts are mapped to vectors via TF-IDF (Leaman et al., 2013), TreeLSTMs (Liu and Xu, 2017), CNNs (Nguyen et al., 2018; Mondal et al., 2019) or ELMO (Schumacher et al., 2020), we generate vector representations with BERT (Devlin et al., 2019), since it can encode both surface and semantic information (Ma et al., 2019).

There are a few similar works to our vector space model, CNN-triplet (Mondal et al., 2019), BIOSYN (Sung et al., 2020), RoBERTa-Node2Vec (Pattisapu et al., 2020), and TTI (Henry et al., 2020). CNN-triplet is a two-step approach, requiring a generator to generate candidates for training the triplet network, and requiring various embedding resources as input to CNN-based encoder. BIOSYN, RoBERTa-Node2Vec, and TTI are one-step approaches. BIOSYN requires an iterative candidate retrieval over the entire training data during each training step, requires both BERT-based and TF-IDF-based representations, and performs a variety of pre-processing such as acronym expansion. Both RoBERTa-Node2Vec and TTI use a BERT-based encoder to encode the mention texts into a vector space, but they differ in how to generate vector representations for medical concepts. Specifically, RoBERTa-Node2Vec uses a Node2Vec graph embedding approach to generate concept representations, and fixes such representations during training, while TTI randomly initializes vector representations for concepts, and keeps such representations learnable during training. Note that none of these works explore search strategies that allow domain-specific synonyms to be added without retraining the model, while we do.

3 Proposed methods

We define a concept mention m as a text string in a corpus D, and a concept c as a unique identifier in an ontology O. The goal of concept normalization is to find a mapping function f that maps each textual mention to its correct concept, i.e., $c = f(m)$. We define concept text t as a text string denoting the concept c, and $t \in T(c)$, where $T(c)$ is all the concept texts denoting concept c. Concept text may come from an ontology, $t \in O(c)$, where $O(c)$ is the synonyms of the concept c from the ontology O, or from an annotated corpus, $t \in D(c)$, where $D(c)$ is the mentions of the concept c in an annotated corpus D. $T(c)$ will allow the generation of tuples (t, c) such as (*MI,C0027051*) and (*Myocardial Infarction,C0027051*). Note that, for a

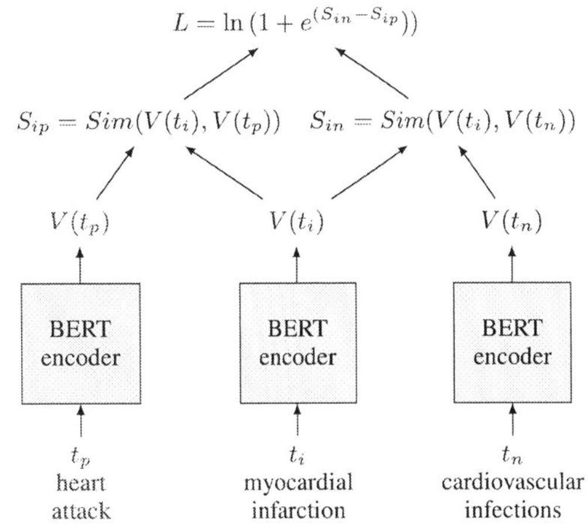

Figure 1: Example of loss calculation for a single instance of triplet-based training. The same BERT model is used for encoding t_i, t_p, and t_n.

concept c, it is common to have $|O(c)| > |D(c)|$, $O(c) \cap D(c) = \emptyset$, or even $D(c) = \emptyset$, i.e., it is common for there to be more concept synonyms in the ontology than the annotated corpus, it is common for the ontology and annotated corpus to provide different concept synonyms, and it is common that annotated corpus only covers a small subset of all concepts in an ontology.

We implement f as a vector space model:

$$f(m) = \underset{\substack{c \in O \\ t \in T(c)}}{\operatorname{argmax}} Sim(V(m), V(t)) \quad (1)$$

where $V(x)$ is a vector representation of text x and Sim is a similarity measure such as cosine similarity, inner product, or euclidean distance. We learn the vector representations $V(x)$ using a triplet network architecture (Hoffer and Ailon, 2015), which learns from triplets of (anchor text t_i, positive text t_p, negative text t_n) where t_i and t_p are texts for the same concept, and t_n is a text for a different concept. The triplet network attempts to learn V such that for all training triplets:

$$Sim(V(t_i), V(t_{ip})) > Sim(V(t_i), V(t_{in})) \quad (2)$$

The triplet network architecture has been adopted in learning representations for images (Schroff et al., 2015; Gordo et al., 2016) and text (Neculoiu et al., 2016; Reimers and Gurevych, 2019). It consists of three instances of the same sub-network (with shared parameters). When fed a (t_i, t_{ip}, t_{in}) triplet of texts, the sub-network outputs vector representations for each text, which are then fed into a triplet loss. We adopt PubMed-BERT (Gu et al.,

2020) as the sub-network, where the representation for the concept text is an average pooling of the representations for all sub-word tokens[2]. This architecture is shown in Figure 1. The inputs to our model are only the mentions or synonyms. We leave the resolution of ambiguous mentions, which will require exploration of contextual information, for future work.

3.1 Online hard triplet mining

An essential part of learning using triplet loss is how to generate triplets. As the number of synonyms gets larger, the number of possible triplets grows cubically, making training impractical. We follow the idea of online triplet mining (Schroff et al., 2015) which considers only triplets within a mini-batch. We first feed a mini-batch of b concept texts to the PubMed-BERT encoder to generate a d-dimensional representation for each concept text, resulting in a matrix $M \in \mathbb{R}^{b \times d}$. We then compute the pairwise similarity matrix:

$$S = Sim(M, M^T) \tag{3}$$

where each entry S_{ij} corresponds to the similarity score between the ith and j^{th} concept texts in the mini-batch. As the easy triplets would not contribute to the training and result in slower convergence (Schroff et al., 2015), for each concept text t_i, we only select a hard positive t_p and a hard negative t_n from the mini-batch such that:

$$p = \underset{j \in [1,b]: j \neq i \wedge C(j) = C(i)}{\text{argmin}} S_{ij} \tag{4}$$

$$n = \underset{k \in [1,b]: k \neq i \wedge C(k) \neq C(i)}{\text{argmax}} S_{ik} \tag{5}$$

where $C(x)$ is the ontology concept from which t_x was taken, i.e., if $t_x \in T(c)$ then $C(x) = c$.

We train the triplet network using batch hard soft margin loss (Hermans et al., 2017):

$$L(i) = \ln\left(1 + e^{(S_{in} - S_{ip})}\right) \tag{6}$$

where S, n, and p are as in eqs. (3) to (5), and the hinge function, $\max(\cdot, 0)$, in the traditional triplet loss is replaced by a softplus function, $\ln(1 + e^{(\cdot)})$.

3.2 Similarity search

Once our vector space model has been trained, we consider several options for how to find the most similar concept c to a text mention m. First, we must choose a search target: we can search over the concepts from the ontology, or the training data, or both. Second we must choose a representation type: we can compare m directly to each text (ontology synonym or training data mention) of each concept, or we can calculate a vector representation of each concept and then compare m directly to the concept vector. Table 1 summarizes these options.

[2]We also experimented with using the output of the *CLS*-token, and max-pooling of the output representations for the sub-word tokens as proposed by (Reimers and Gurevych, 2019), but neither resulted in better performance.

	Searching Over		Representation Type	
	Ontology	Training Data	Text	Concept
O-T	✓		✓	
O-C	✓			✓
D-T		✓	✓	
D-C		✓		✓
OD-T	✓	✓	✓	
OD-C	✓	✓		✓

Table 1: Names for similarity search modules.

We consider the following search targets:

Data We search over the concepts in the annotated data. These mentions will be more domain-specific (e.g., *PT* may refer to *patient* in clinical notes, but to *physical therapy* in scientific articles), but may be more predictive if the evaluation data is from the same domains. We search over the train subset of the data for dev evaluation, and train + dev subset for test evaluation.

Ontology We search over the concepts in the ontology. The synonyms will be more domain-independent, and the ontology will cover concepts never seen in the annotated training data.

Data and ontology We search over the concepts in both the training data and the ontology. For concepts in the annotated training data, their representations are averaged over mentions in the training data and synonyms in the ontology.

We consider the following representation types:

Text We represent each text (ontology synonym or training data mention) as a vector by running it through our triplet-fine-tuned PubMed-BERT encoder. Concept normalization then compares the mention vector to each text vector:

$$f(m) = \underset{\substack{c \in O \\ t \in T(c)}}{\text{argmax}}\, Sim(V(m), V(t)) \tag{7}$$

When a retrieved text t is present in more than one concept (e.g., *no appetite* appears in concepts *C0426579*, *C0003123*, *C1971624*), and thus we see the same Sim for multiple concepts, we pick a concept randomly to break ties.

First component	Second component
D-T	O-T
D-T	O-C
D-C	O-T
D-C	O-C
D-T	OD-T
D-T	OD-C
D-C	OD-T
D-C	OD-C

Table 2: Options for components in sieve-based search.

Concept We represent each concept as a vector by taking an average over the triplet-fine-tuned PubMed-BERT representations of that concept's texts (ontology synonyms and/or training data mentions). Concept normalization then compares the mention vector to each concept vector:

$$f(m) = \underset{c \in O}{\operatorname{argmax}}\, Sim\left(V(m), \underset{t \in T(c)}{\operatorname{mean}} V(t)\right) \quad (8)$$

The averages here mean that different concepts with some (but not all) overlapping synonyms (e.g., *C0426579*, *C0003123*, *C1971624* in UMLS all have the synonym *no appetite*) will end up with different vector representations.

3.2.1 Sieve-based search

Traditional sieve-based approaches for concept normalization (D'Souza and Ng, 2015; Jonnagaddala et al., 2016; Luo et al., 2019; Henry et al., 2020) achieved competitive performance by ordering a sequence of searches over dictionaries from most precise to least precise.

Inspired by this work, we consider a sieve-based similarity search that: 1) searches over the annotated training data, then 2) searches over the ontology (possibly combined with the annotated training data). Table 2 lists all possible combinations of first and second components in sieve-based search. For instance, in sieve-based search **D-T + O-C**, we first search over the annotated corpus using training-data-mention vectors (D-T), and then search over the ontology using concept vectors (O-C).

4 Experiments

4.1 Datasets

We conduct experiments on three scientific article datasets – NCBI (Doğan et al., 2014), BC5CDR-D and BC5CDR-C (Li et al., 2016) – and two clinical note datasets – MCN (Luo et al., 2019) and ShARe/CLEF (Suominen et al., 2013). The statistics of each dataset are described in table 3.

NCBI The NCBI disease corpus[3] contains 17,324 manually annotated disorder mentions from 792 PubMed abstracts. The disorder mentions are mapped to 750 MEDIC lexicon (Davis et al., 2012) concepts. We split the released training set into use 5,134 training mentions and 787 development mentions, and keep the 960 mentions from the original test set as evaluation. We use the 2012 version of MEDIC ontology which contains 11,915 concepts and 71,923 synonyms.

BC5CDR-D & BC5CDR-C These corpora were used in the BioCreative V chemical-induced disease (CID) relation extraction challenge[4]. BC5CDR-D and BC5CDR-C contain 12,850 disease mentions and 15,935 chemical mentions, respectively. The annotated disease mentions are mapped to 1075 unique concepts out of 11,915 concepts in the 2012 version of MEDIC ontology. The chemical mentions are mapped to 1164 unique concepts out of 171,203 concepts from the 2019 version of Comparative Toxicogenomics Database (CTD) chemical ontology. We use the configuration in the BioCreative V challenge to keep the same train/dev/test splits.

ShARe/CLEF The ShARe/CLEF corpus is from the ShARe/CLEF eHealth 2013 Challenge[5], where 11,167 disorder mentions in 298 clinical notes are annotated with their concepts mapping to the 12,6524 disorder concepts from the SNOMED-CT subset of the 2011AA version of UMLS. We take the 199 clinical notes consisting of 5,816 mentions as the train set and 5,351 mentions from the 99 clinical notes as test. Around 30.4% of the mentions in the corpus could not be mapped to any concepts in the ontology, and are assigned the *CUI-less* label.

MCN The MCN corpus from 2019 n2c2 Shared-Task track 3[6] consists of 13,609 concept mentions in 100 discharge summaries. The mentions are mapped to 3,792 unique concepts out of 434,056 possible concepts in the SNOMED-CT and RxNorm subset of UMLS version 2017AB.

[3]https://www.ncbi.nlm.nih.gov/CBBresearch/Dogan/DISEASE/

[4]https://biocreative.bioinformatics.udel.edu/tasks/biocreative-v/

[5]https://sites.google.com/site/shareclefehealth/data

[6]https://n2c2.dbmi.hms.harvard.edu/track3

Dataset	Scientific Articles			Clinical Notes	
	NCBI	BC5CDR-D	BC5CDR-C	ShARe/CLEF	MCN
Ontology	MEDIC	MEDIC	CTD-Chemical	SNOMED-CT	SNOMED-CT & RxNorm
# of Concepts (Ontology)	11,915	11,915	171,203	126,524	434,056
# of Synonyms (Ontology)	71,923	71,923	407,247	520,665	1,550,586
# of Documents (Datasets)	792	1,500	1,500	298	100
# of Concepts (Datasets)	750	1,075	1,164	1,313	3,792
# of Mentions (Datasets)	6,881	12,850	15,935	11,167	13,609

Table 3: Statistics of the five datasets in our experiments.

We take 40 clinical notes from the released data as training, consisting of 5,334 mentions, and the standard evaluation data with 6,925 mentions as our test set. Around 2.7% of mentions in MCN are assigned the *CUI-less* label.

4.2 Implementation details

Unless specifically noted otherwise, we use the same training procedure and hyper-parameter settings across all experiments and on all datasets. As the triplet mining requires at least one positive text in a batch for each anchor text, we randomly sample one positive text for each anchor text and group them into batches. Like previous work (Schroff et al., 2015; Hermans et al., 2017), we adopt euclidean distance to calculate similarity score during training, while at inference time, we compute cosine similarity as it is simpler to interpret. For the sieve-based search, if the cosine similarity score between the mention and the prediction of the first sieve is above 0.95, we use the prediction of first sieve, otherwise, we use the prediction of the second sieve.

When training the triplet network on the combination of the ontology and annotated corpus, we take all the synonyms from the ontology and repeat the concept texts in the annotated corpus such that $\frac{|D|}{|O|} = \frac{1}{3}$. In preliminary experiments we found that large ontologies overwhelmed small annotated corpora. We also experimented with three ratios $\frac{1}{3}$, $\frac{2}{3}$, and 1 between concept texts and synonyms of ontology on NCBI and BC5CDR-D datasets, and found that the ratio of $\frac{1}{3}$ achieves the best performance for Train:OD models. We then kept the same ratio setting for all datasets. We did not thoroughly explore other ratios and leave that to future work.

For all experiments, we use PubMed-BERT (Gu et al., 2020) as the starting point, which pre-trains a BERT-style model from scratch on PubMed abstracts and full texts. In our preliminary experiments, we also tried BioBERT (Lee et al., 2019) as the text encoder, but that resulted in worse performance across five datasets. We use the pytorch implementation of sentence-transformers[7] to train the Triplet Network for concept normalization. We use the following hyper-parameters during the training of the triplet network: sequence_length = 8, batch_size = 1500, epoch_size = 100, optimizer = Adam, learning_rate = 3e-5, warmup_steps = 0.

4.3 Evaluation metrics

The standard evaluation metric for concept normalization is accuracy, because the most similar concept in prediction is of primary interest. For composite mentions like *breast and ovarian cancer* that are mapped to more than one concept in NCBI, BC5CDR-D, and BC5CDR-C datasets, we adopt the evaluation strategy that composite entity is correct if every prediction for each separate mention is correct (Sung et al., 2020).

5 Model selection

We use the development data to choose whether to train the triplet network on just the ontology or also the training data, and to choose which among the similarity search strategies described in section 3.2. Table 4 shows the performance of all such systems across the five different corpora. The top half of the table focuses on settings where the triplet network only needs to be trained once, on the ontology, and the bottom half focuses on settings where the triplet network is retrained for each new dataset. For each half of the table, the last column gives the average of the ranks of each setting's performance across the five corpora. For example, when training the triplet network only on the ontology, the searching strategy D-C (search the training data using concept vectors) is almost always the worst performing,

[7] `https://github.com/UKPLab/sentence-transformers`

	Train	Search	NCBI	BC5CDR-D	BC5CDR-C	ShARe/CLEF	MCN	Avg. Rank
1	O	O-T	83.74	82.65	97.00	82.76	69.11	10.2
2	O	O-C	85.01	82.43	92.62	81.12	70.96	12
3	O	D-T	85.39	77.29	74.21	79.76	61.26	12.6
4	O	D-C	85.26	75.18	74.11	69.70	59.70	13.6
5	O	OD-T	89.58	88.87	97.75	88.12	72.67	4.8
6	O	OD-C	88.56	85.85	93.30	82.23	72.59	9.4
7	O	D-T + O-T	90.34	89.66	97.62	87.26	81.33	3.6
8	O	D-T + O-C	89.96	89.40	96.88	83.73	81.93	5
9	O	D-C + O-T	86.28	83.72	97.14	82.98	76.67	7.4
10	O	D-C + O-C	88.56	83.51	95.77	81.58	76.52	9.8
11	O	D-T + OD-T	91.36	**90.50**	97.64	**90.50**	81.85	2
12	O	D-T + OD-C	90.85	89.90	96.88	84.69	**82.15**	3.6
13	O	D-C + OD-T	**91.99**	89.47	**97.76**	86.83	79.19	3.2
14	O	D-C + OD-C	88.82	86.93	96.32	82.55	77.41	7.6
15	OD	O-T	89.58	87.82	96.71	86.62	72.37	9.8
16	OD	O-C	91.36	89.85	96.32	88.11	80.52	9.6
17	OD	D-T	86.40	79.01	74.23	79.87	63.33	13.2
18	OD	D-C	86.40	78.41	74.23	80.19	62.52	13.4
19	OD	OD-T	91.11	90.38	97.85	88.87	76.15	8.2
20	OD	OD-C	91.61	89.92	96.32	88.33	81.4	7.8
21	OD	D-T + O-T	91.25	91.10	97.81	90.15	84.37	4
22	OD	D-T + O-C	91.49	90.88	96.22	88.76	84.52	6.4
23	OD	D-C + O-T	92.25	90.71	97.87	89.61	83.78	4
24	OD	D-C + O-C	91.49	90.47	96.28	88.65	83.93	7.8
25	OD	D-T + OD-T	91.61	**91.22**	97.81	**90.21**	84.37	2.4
26	OD	D-T + OD-C	91.61	90.83	96.22	89.08	**84.67**	5.2
27	OD	D-C + OD-T	**92.25**	90.95	**97.91**	90.15	83.70	3.4
28	OD	D-C + OD-C	91.61	90.55	96.28	89.40	84.00	5.8

Table 4: Dev performances of the triplet network trained on ontology and ontology + data with different similarity search strategies. The last column *Avg. Rank* shows the average rank of each similarity search strategy across multiple datasets. Models with best average rank are highlighted in grey; models with best accuracy are bolded.

ranking 14th of 14 in four corpora and 12th of 14 in one corpus, for an average rank of 13.6.

Table 4 shows that the best models search over both the ontology and the training data. Models that only search over the training data (D-T and D-C) perform worst, with average ranks of 12.6 or higher regardless of what the triplet network is trained on, most likely because the training data covers only a fraction of the concepts in the test data. Models that only search over the ontology (O-T and O-C) are only slightly better, with average ranks between 9.6 and 12, though the models in the first two rows of the table at least have the advantage that they require no annotated training data (they train on and search over only the ontology). However, the performance of such models can be improved by adding domain-specific synonyms to the ontology, i.e., OD-T vs. O-T (rows 5 vs. 1), and OD-C vs. O-C (rows 6 vs. 2), or adding domain-specific synonyms and then searching in a sieve-based manner (rows 7-14).

Table 4 also shows that searching based on text (ontology synonyms or training data mentions) vectors typically outperforms searching based on concept (average of text) vectors. Each pair of rows in the table shows such a comparison, and only in rows 15-16 and 19-20 are the average ranks of the -C models higher than the -T models.

Table 4 also shows that models using mixed representation types (-T and -C) have worse ranks than the text-only models (-T). For instance, going from Train:O-Search:O-C to Train:O-Search:O-T improves the average rank from 12 to 10.2, going from Train:OD-Search:D-T+OD-C to Train:OD-Search:D-T+OD-T improves the average rank from 5.2 to 2.4, etc. There are a few exceptions to this on the MCN dataset. We analyzed the differences in the predictions of Train:OD-Search:D-T+OD-T (row 25) and Train:OD-Search:D-T+OD-C (row 26) on this dataset, and found that concept vectors sometimes helps to solve ambiguous mentions by averaging their concept texts. For instance, the OD-T model finds concepts *C0013144* and *C2830004* for mention *somnolent* as they have the overlapping synonym *somnolent*, while the OD-C model ranks *C2830004* higher as the other concept also has other synonyms such as *Drowsy*, *Sleepiness*.

Finally, table 4 shows that sieve-based models

Approach	NCBI	BC5CDR-D	BC5CDR-C	ShARe/CLEF	MCN
Sieve-based (D'Souza and Ng, 2015)	84.65	-	-	90.75	-
Sieve-based (Luo et al., 2019)	-	-	-	-	76.35
TaggerOne (Leaman and Lu, 2016)	88.80	88.9	94.1	-	-
CNN-based ranking (Li et al., 2017)	86.10	-	-	90.30	-
BERT-based ranking (Ji et al., 2020)	89.06	-	-	**91.10**	-
BERT-based ranking (Xu et al., 2020)	-	-	-	-	83.56
BIOSYN (Sung et al., 2020)	91.1	**93.2**	96.6	-	-
TTI (Henry et al., 2020)	-	-	-	-	**85.26**
PubMed-BERT + Search:O-T	76.56	76.60	91.78	73.64	59.97
PubMed-BERT + Search:D-T+OD-T	82.19	90.53	94.24	85.35	75.81
Train:O + Search:O-T	82.60	84.44	95.79	83.48	69.62
Train:O + Search:D-T+OD-T	89.48	92.30	96.67	89.19	82.19
Train:OD + Search:D-T+OD-T	88.96	92.92	96.81	90.41	83.23
Train:OD + Search:tuned	**91.15**	92.92	**96.91**	90.41	83.70

Table 5: Comparisons of our proposed approaches against the current state-of-the-art performances on *NCBI*, *BC5CDR-D*, *BC5CDR-C*, *ShARe/CLEF*, and *MCN* datasets. Approaches with best accuracy are bolded.

outperform their non-sieve-based counterparts. For example, D-T + O-T has better average ranks than O-T, D-T, or OD-T (rows 7 vs. 1, 3, and 5; and rows 21 vs. 15, 17, and 19).

From this analysis on the dev set, we select the following models to evaluate on the test set:

Train:O + Search:O-T This is the best approach that requires only the ontology; no annotated training data is used.

Train:O + Search:D-T+OD-T This is the best approach that only needs to be trained once (on the ontology), as the training data is only used to add extra concept text during search time. This is similar to a real-world scenario where a user manually adds some extra domain-specific synonyms for concepts they care about.

Train:OD + Search:D-T+OD-T This is the best approach that can be created from any combination of ontology and training data. The triplet network must be retrained for each new domain.

Train:OD + Search:tuned This is the bold models in the second half of table 4. It requires not only retraining the triplet network for each new domain, but also trying out all search strategies on the new domain and selecting the best one.

6 Results

Table 5 shows the results of our selected models on the test set, alongside the best models in the literature. Our Train:OD+Search:tuned model achieves new state-of-the-art on BC5CDR-C (p^8=0.0291), equivalent performance on NCBI (p=0.6753) and BC5CDR-D (p=0.1204), <1 point worse on ShARe (p=0.0375), and <2 points worse on MCN (p=0). Note that the performance of TTI is from an ensemble of multiple system runs. Yet this model is simpler than most prior work: it requires no two-step generate-and-rank framework (Li et al., 2017; Ji et al., 2020; Xu et al., 2020), no iterative candidate retrieval over the entire training data (Sung et al., 2020), no hand-crafted rules or features (D'Souza and Ng, 2015; Leaman and Lu, 2016; Luo et al., 2019), and no acronym expansion or TF-IDF transformations (D'Souza and Ng, 2015; Ji et al., 2020; Sung et al., 2020).

The PubMed-BERT rows in Table 5 demonstrate that the triplet training is a critical part of the success: if we use PubMed-BERT without triplet training, performance is 2 to 8 points worse than our best models, depending on the dataset. Yet, we can see that our proposed search strategies are also important, as on the BC5CDR datasets, PubMed-BERT can get within 3 points of the state-of-the-art using the D-T+OD-T search strategy (though it is much further away on the other datasets).

Perhaps most interestingly, our triplet network trained only on the ontology and no annotated training data, Train:O+Search:D-T+OD-T, achieves within 3 points of state-of-the-art on all datasets. We believe this represents a more realistic scenario: unlike prior work, our triplet network does not need to be retrained for each new dataset/domain if their concepts are from the same ontology. Instead, the model can be adapted to a new dataset/domain by simply pointing out any extra domain-specific synonyms for concepts, and the search can integrate these directly. Domain-specific synonyms do

[8] We used a one-sample bootstrap resampling test. The one sample is 10,000 runs of bootstrapping results of our system.

	PubMed-BERT + Search:OD-T			Train:O + Search:OD-T			Train:OD + Search:OD-T		
Rank	Text	Concept	Score	Text	Concept	Score	Text	Concept	Score
1	HNSCC	C535575	0.919	Hyperparathyroidism, Primary	D049950	0.767	Hyperparathyroidism, Primary	D049950	0.838
5	NPC2	C536119	0.903	Hyperparathyroidism 1	C564166	0.692	Primary Hyperparathyroidism	D049950	0.830
10	MPNST	D009442	0.900	HRPT1	C564166	0.611	HRPT1	C564166	0.672
15	HPNS	D006610	0.897	Hyperparathyroidism 2	C563273	0.595	Parathyroid Adenoma, Familial	C564166	0.644
20	PBC2	C567817	0.895	Hyperparathyroidism, Secondary	D006962	0.566	Hyperparathyroidisms, Secondary	D006962	0.608

Table 6: Top similar texts, their concepts, and similarity scores for mention *primary HPT (D049950)* predicted from models PubMed-BERT + Search:OD-T, Train:O + Search:OD-T and Train:OD + Search:OD-T.

seem to be necessary for all datasets; without them (i.e., Train:O+Search:O-T), performance is about 10 points below state-of-the-art.

As a small qualitative analysis of the models, Table 6 shows an example of similarity search results, where the systems have been asked to normalize the mention *primary HPT*. PubMed-BERT fails, producing unrelated acronyms, while both triplet network models find the concept and rank it with the highest similarity score.

7 Limitations and future research

Our ability to normalize polysemous concept mentions is limited by their context-independent representations. Although our PubMed-BERT encoder is a pre-trained contextual model, we feed in only the mention text, not any context, when producing a representation vector. This is not ideal for mentions with multiple meanings, e.g., *potassium* in clinical notes may refer to the substance (C0032821) or the measurement (C0202194), and only the context will reveal which one. A better strategy to generate the contextualized representation for the concept mention, e.g., Schumacher et al. (2020), may yield improvements for such mentions.

We currently train a separate triplet network for each ontology (one for MEDIC, one for CTD, one for SNOMED-CT, etc.) but in the future we would like to train on a comprehensive ontology like the UMLS Metathesaurus (Bodenreider, 2004), which includes nearly 200 different vocabularies (SNOMED-CT, MedDRA, RxNorm, etc.), and more than 3.5 million concepts. We expect such a general vector space model would be more broadly useful to the biomedical NLP community.

We explored one type of triplet training network, but in the future we would like to explore other variants, such as semi-hard triplet mining (Schroff et al., 2015) for generating samples, cosine similarity for measuring the similarity during training and inference, and multi-similarity loss (Wang et al., 2019) for calculating the loss.

8 Conclusions

We presented a vector-space framework for concept normalization, based on pre-trained transformers, a triplet objective with online hard triplet mining, and a new approach to vector similarity search. Across five datasets, our models that require only an ontology to train are competitive with state-of-the-art models that require domain-specific training.

Acknowledgements

Research reported in this publication was supported by the National Library of Medicine of the National Institutes of Health under Award Number R01LM012918. The computations were done in systems supported by the National Science Foundation under Grant No. 1228509. The content is solely the responsibility of the authors and does not necessarily represent the official views of the National Institutes of Health.

References

Olivier Bodenreider. 2004. The Unified Medical Language System (UMLS): integrating biomedical terminology. *Nucleic Acids Research*, 32(suppl_1):D267–D270.

Allan Peter Davis, Thomas C Wiegers, Michael C Rosenstein, and Carolyn J Mattingly. 2012. Medic: a practical disease vocabulary used at the comparative toxicogenomics database. *Database*, 2012.

Jacob Devlin, Ming-Wei Chang, Kenton Lee, and Kristina Toutanova. 2019. BERT: Pre-training of deep bidirectional transformers for language understanding. In *Proceedings of the 2019 Conference of the North American Chapter of the Association for Computational Linguistics: Human Language Technologies, Volume 1 (Long and Short Papers)*, pages 4171–4186, Minneapolis, Minnesota. Association for Computational Linguistics.

Rezarta Islamaj Doğan, Robert Leaman, and Zhiyong Lu. 2014. Ncbi disease corpus: a resource for disease name recognition and concept normalization. *Journal of biomedical informatics*, 47:1–10.

Jennifer D'Souza and Vincent Ng. 2015. Sieve-based entity linking for the biomedical domain. In *Proceedings of the 53rd Annual Meeting of the Association for Computational Linguistics and the 7th International Joint Conference on Natural Language Processing (Volume 2: Short Papers)*, pages 297–302, Beijing, China. Association for Computational Linguistics.

Albert Gordo, Jon Almazán, Jerome Revaud, and Diane Larlus. 2016. Deep image retrieval: Learning global representations for image search. In *European conference on computer vision*, pages 241–257. Springer.

Yu Gu, Robert Tinn, Hao Cheng, Michael Lucas, Naoto Usuyama, Xiaodong Liu, Tristan Naumann, Jianfeng Gao, and Hoifung Poon. 2020. Domain-specific language model pretraining for biomedical natural language processing. *arXiv preprint arXiv:2007.15779*.

Sam Henry, Yanshan Wang, Feichen Shen, and Ozlem Uzuner. 2020. The 2019 National Natural language processing (NLP) Clinical Challenges (n2c2)/Open Health NLP (OHNLP) shared task on clinical concept normalization for clinical records. *Journal of the American Medical Informatics Association*, 27(10):1529–1537.

Alexander Hermans, Lucas Beyer, and Bastian Leibe. 2017. In defense of the triplet loss for person re-identification. *arXiv preprint arXiv:1703.07737*.

Elad Hoffer and Nir Ailon. 2015. Deep metric learning using triplet network. In *International Workshop on Similarity-Based Pattern Recognition*, pages 84–92. Springer.

Zongcheng Ji, Qiang Wei, and Hua Xu. 2020. Bert-based ranking for biomedical entity normalization. *AMIA Summits on Translational Science Proceedings*, 2020:269.

Jitendra Jonnagaddala, Toni Rose Jue, Nai-Wen Chang, and Hong-Jie Dai. 2016. Improving the dictionary lookup approach for disease normalization using enhanced dictionary and query expansion. *Database*, 2016:baw112.

Ning Kang, Bharat Singh, Zubair Afzal, Erik M van Mulligen, and Jan A Kors. 2013. Using rule-based natural language processing to improve disease normalization in biomedical text. *Journal of the American Medical Informatics Association*, 20(5):876–881.

Rohit J. Kate. 2016. Normalizing clinical terms using learned edit distance patterns. *Journal of the American Medical Informatics Association*, 23(2):380–386.

André Leal, Bruno Martins, and Francisco Couto. 2015. ULisboa: Recognition and normalization of medical concepts. In *Proceedings of the 9th International Workshop on Semantic Evaluation (SemEval 2015)*, pages 406–411, Denver, Colorado. Association for Computational Linguistics.

Robert Leaman, Rezarta Islamaj Doğan, and Zhiyong Lu. 2013. DNorm: disease name normalization with pairwise learning to rank. *Bioinformatics*, 29(22):2909–2917.

Robert Leaman and Zhiyong Lu. 2016. Taggerone: joint named entity recognition and normalization with semi-markov models. *Bioinformatics*, 32(18):2839–2846.

Robert Leaman, Chih-Hsuan Wei, and Zhiyong Lu. 2015. tmchem: a high performance approach for chemical named entity recognition and normalization. *Journal of cheminformatics*, 7(S1):S3.

Hsin-Chun Lee, Yi-Yu Hsu, and Hung-Yu Kao. 2016. AuDis: an automatic CRF-enhanced disease normalization in biomedical text. *Database*, 2016. Baw091.

Jinhyuk Lee, Wonjin Yoon, Sungdong Kim, Donghyeon Kim, Sunkyu Kim, Chan Ho So, and Jaewoo Kang. 2019. BioBERT: a pre-trained biomedical language representation model for biomedical text mining. *Bioinformatics*. Btz682.

Kathy Lee, Sadid A. Hasan, Oladimeji Farri, Alok Choudhary, and Ankit Agrawal. 2017. Medical Concept Normalization for Online User-Generated Texts. In *2017 IEEE International Conference on Healthcare Informatics (ICHI)*, pages 462–469. IEEE.

Fei Li, Yonghao Jin, Weisong Liu, Bhanu Pratap Singh Rawat, Pengshan Cai, and Hong Yu. 2019. Fine-Tuning Bidirectional Encoder Representations From Transformers (BERT)–Based Models on Large-Scale Electronic Health Record Notes: An Empirical Study. *JMIR Med Inform*, 7(3):e14830.

Haodi Li, Qingcai Chen, Buzhou Tang, Xiaolong Wang, Hua Xu, Baohua Wang, and Dong Huang. 2017. Cnn-based ranking for biomedical entity normalization. *BMC bioinformatics*, 18(11):79–86.

Jiao Li, Yueping Sun, Robin J Johnson, Daniela Sciaky, Chih-Hsuan Wei, Robert Leaman, Allan Peter Davis, Carolyn J Mattingly, Thomas C Wiegers, and Zhiyong Lu. 2016. Biocreative v cdr task corpus: a resource for chemical disease relation extraction. *Database*, 2016.

Nut Limsopatham and Nigel Collier. 2016. Normalising medical concepts in social media texts by learning semantic representation. In *Proceedings of the*

54th Annual Meeting of the Association for Computational Linguistics (Volume 1: Long Papers)*, pages 1014–1023, Berlin, Germany. Association for Computational Linguistics.

Hongwei Liu and Yun Xu. 2017. A Deep Learning Way for Disease Name Representation and Normalization. In *Natural Language Processing and Chinese Computing*, pages 151–157. Springer International Publishing.

Yen-Fu Luo, Weiyi Sun, and Anna Rumshisky. 2019. MCN: A comprehensive corpus for medical concept normalization. *Journal of Biomedical Informatics*, pages 103–132.

Xiaofei Ma, Peng Xu, Zhiguo Wang, Ramesh Nallapati, and Bing Xiang. 2019. Universal text representation from bert: An empirical study. *arXiv preprint arXiv:1910.07973*.

Zulfat Miftahutdinov and Elena Tutubalina. 2019. Deep neural models for medical concept normalization in user-generated texts. In *Proceedings of the 57th Annual Meeting of the Association for Computational Linguistics: Student Research Workshop*, pages 393–399, Florence, Italy. Association for Computational Linguistics.

Ishani Mondal, Sukannya Purkayastha, Sudeshna Sarkar, Pawan Goyal, Jitesh Pillai, Amitava Bhattacharyya, and Mahanandeeshwar Gattu. 2019. Medical entity linking using triplet network. In *Proceedings of the 2nd Clinical Natural Language Processing Workshop*, pages 95–100, Minneapolis, Minnesota, USA. Association for Computational Linguistics.

Shikhar Murty, Patrick Verga, Luke Vilnis, Irena Radovanovic, and Andrew McCallum. 2018. Hierarchical losses and new resources for fine-grained entity typing and linking. In *Proceedings of the 56th Annual Meeting of the Association for Computational Linguistics (Volume 1: Long Papers)*, pages 97–109, Melbourne, Australia. Association for Computational Linguistics.

Paul Neculoiu, Maarten Versteegh, and Mihai Rotaru. 2016. Learning text similarity with Siamese recurrent networks. In *Proceedings of the 1st Workshop on Representation Learning for NLP*, pages 148–157, Berlin, Germany. Association for Computational Linguistics.

Thanh Ngan Nguyen, Minh Trang Nguyen, and Thanh Hai Dang. 2018. Disease Named Entity Normalization Using Pairwise Learning To Rank and Deep Learning. Technical report, VNU University of Engineering and Technology.

Jinghao Niu, Yehui Yang, Siheng Zhang, Zhengya Sun, and Wensheng Zhang. 2019. Multi-task Character-Level Attentional Networks for Medical Concept Normalization. *Neural Process Lett*, 49(3):1239–1256.

Nikhil Pattisapu, Sangameshwar Patil, Girish Palshikar, and Vasudeva Varma. 2020. Medical Concept Normalization by Encoding Target Knowledge. In *Proceedings of the Machine Learning for Health NeurIPS Workshop*, volume 116 of *Proceedings of Machine Learning Research*, pages 246–259. PMLR.

Nils Reimers and Iryna Gurevych. 2019. Sentence-BERT: Sentence embeddings using Siamese BERT-networks. In *Proceedings of the 2019 Conference on Empirical Methods in Natural Language Processing and the 9th International Joint Conference on Natural Language Processing (EMNLP-IJCNLP)*, pages 3982–3992, Hong Kong, China. Association for Computational Linguistics.

Florian Schroff, Dmitry Kalenichenko, and James Philbin. 2015. Facenet: A unified embedding for face recognition and clustering. In *Proceedings of the IEEE conference on computer vision and pattern recognition*, pages 815–823.

Elliot Schumacher, Andriy Mulyar, and Mark Dredze. 2020. Clinical concept linking with contextualized neural representations. In *Proceedings of the 58th Annual Meeting of the Association for Computational Linguistics*, pages 8585–8592, Online. Association for Computational Linguistics.

Mujeen Sung, Hwisang Jeon, Jinhyuk Lee, and Jaewoo Kang. 2020. Biomedical entity representations with synonym marginalization. In *Proceedings of the 58th Annual Meeting of the Association for Computational Linguistics*, pages 3641–3650, Online. Association for Computational Linguistics.

Hanna Suominen, Sanna Salanterä, Sumithra Velupillai, Wendy W Chapman, Guergana Savova, Noemie Elhadad, Sameer Pradhan, Brett R South, Danielle L Mowery, Gareth JF Jones, et al. 2013. Overview of the share/clef ehealth evaluation lab 2013. In *International Conference of the Cross-Language Evaluation Forum for European Languages*, pages 212–231. Springer.

Elena Tutubalina, Zulfat Miftahutdinov, Sergey Nikolenko, and Valentin Malykh. 2018. Medical concept normalization in social media posts with recurrent neural networks. *Journal of Biomedical Informatics*, 84:93–102.

Xun Wang, Xintong Han, Weilin Huang, Dengke Dong, and Matthew R. Scott. 2019. Multi-similarity loss with general pair weighting for deep metric learning. In *Proceedings of the IEEE/CVF Conference on Computer Vision and Pattern Recognition (CVPR)*.

Chih-Hsuan Wei, Yifan Peng, Robert Leaman, Allan Peter Davis, Carolyn J Mattingly, Jiao Li, Thomas C Wiegers, and Zhiyong Lu. 2015. Overview of the biocreative v chemical disease relation (cdr) task. In *Proceedings of the fifth BioCreative challenge evaluation workshop*, volume 14.

Dongfang Xu, Zeyu Zhang, and Steven Bethard. 2020. A generate-and-rank framework with semantic type regularization for biomedical concept normalization. In *Proceedings of the 58th Annual Meeting of the Association for Computational Linguistics*, pages 8452–8464, Online. Association for Computational Linguistics.

Scalable Few-Shot Learning of Robust Biomedical Name Representations

Pieter Fivez
CLiPS Research Centre
University of Antwerp
pieter.fivez@uantwerpen.be

Simon Šuster
Faculty of Engineering and Information Technology
University of Melbourne
simon.suster@unimelb.edu.au

Walter Daelemans
CLiPS Research Centre
University of Antwerp
walter.daelemans@uantwerpen.be

Abstract

Recent research on robust representations of biomedical names has focused on modeling large amounts of fine-grained conceptual distinctions using complex neural encoders. In this paper, we explore the opposite paradigm: training a simple encoder architecture using only small sets of names sampled from high-level biomedical concepts. Our encoder post-processes pretrained representations of biomedical names, and is effective for various types of input representations, both domain-specific or unsupervised. We validate our proposed few-shot learning approach on multiple biomedical relatedness benchmarks, and show that it allows for continual learning, where we accumulate information from various conceptual hierarchies to consistently improve encoder performance. Given these findings, we propose our approach as a low-cost alternative for exploring the impact of conceptual distinctions on robust biomedical name representations. Our code is open-source and available at www.github.com/clips/fewshot-biomedical-names.

1 Introduction

Recent research in biomedical NLP has focused on learning robust representations of biomedical names. To achieve robustness, an encoder should represent the semantic similarity and relatedness between different names (e.g. by their closeness in the embedding space), while its embeddings should also remain as transferable and generally applicable as self-supervised pretrained representations.

Prior research into robust representations has shown three distinct tendencies. Firstly, research typically focuses on encoders with complex neural architectures and a large amount of parameters. As compensation for this complexity, such models can be heavily regularized during training, e.g. by tying the output of a nested LSTM to a pooled embedding of its input representations (Phan et al., 2019), or by integrating a finetuned BERT model with sparse lexical representations (Sung et al., 2020).

Chapter V: Mental and behavioural disorders	
F34	**F63**
Persistent mood disorders	Habit and impulse disorders
F34.0	F63.0
Cyclothymia	*Pathological gambling*
F34.1	F63.1
Dysthymia	*Pyromania*

Table 1: Example of how reference names are grouped together within the ICD-10 hierarchy of disorders.

Secondly, encoders are typically trained on fine-grained concepts from biomedical ontologies such as the UMLS, i.e., concepts with no child nodes in the ontological directed graph. Small synonym sets of such fine-grained concepts are readily available as training data, and often serve as evaluation data for normalization tasks to which trained encoders can be applied.

Lastly, as a result of using fine-grained concepts, vast amounts of biomedical names are needed to model the large collection of fine-grained distinctions present in ontologies. For instance, Phan et al. (2019) train their encoder on 156K disorder names. These three tendencies share an underlying assumption: complex neural encoder architectures can learn biomedical semantics by generalizing in a bottom-up fashion from large amounts of fine-grained semantic distinctions, if provided with sufficient quantities of training data.

Proceedings of the BioNLP 2021 workshop, pages 23–29
June 11, 2021.

However, it is not self-evident that such an approach is the most effective way to achieve general-purpose biomedical name representations. For instance, it does not directly address what conceptual distinctions are actually *relevant* to improve representations for downstream NLP applications. Finding and exploiting relevant distinctions can be an empirical question, and as such require low-cost exploration of various conceptual hierarchies. Such a heuristic search is expensive in the current paradigm.

In this paper, we explore a scalable few-shot learning approach for robust biomedical name representations which is orthogonal to this paradigm. We investigate to what extent we can fit a simple encoder architecture using only a small selection of data, with a limited amount of concepts containing only a few samples each (i.e., few-shot learning). To this end, we don't use fine-grained concepts for training, but more general higher-level concepts which span a large range of fine-grained concepts. Table 1 gives an example of such a larger grouping of biomedical names.

This paper offers two main contributions. Firstly, our proposed approach offers an alternative for training biomedical name encoders with much lower computational cost, both for training and inference at test time. It is applicable to large-scale hierarchies containing at least ten thousands of names and is equally effective for different types of pretrained representations when tested on various biomedical relatedness benchmarks. Secondly, we show that this approach allows for low-cost continual learning from multiple concept hierarchies, and as such can help with the accumulation of relevant domain-specific information for downstream biomedical NLP tasks.

2 Approach

Our approach is similar to supervised post-processing techniques of word embeddings such as retrofitting and counterfitting (Faruqui et al., 2015; Mrkšić et al., 2016), but instead post-processes pretrained representations of biomedical names.

2.1 Encoder architecture

Our encoder architecture is a feedforward neural network with Rectified Linear Unit (ReLU) as non-linear activation function. This neural network transforms a pretrained representation of a biomedical name, after which this transformation is averaged with the pretrained representation:

$$f(n) = \frac{enc(u_n) + u_n}{2} \quad (1)$$

where $f(n)$ is the output representation for a biomedical name, u_n is its pretrained input representation, and enc is the feedforward neural network which transforms the input representation. The averaging step ensures that the encoder architecture learns to update the pretrained input representation rather than create an entirely new representation. This makes our model more robust against overfitting in few-shot learning settings.

	min	max	mean	stdev
ICD-10	247	40,519	3,414	8,693
SNOMED-CT	397	19,114	3,532	4,094
(+ ambiguous	1,108	23,915	4,990	5,134)

Table 2: Descriptive statistics about the number of names per concept for our training data.

2.2 Training objectives

Our training objectives are based on the state-of-the-art BNE model by Phan et al. (2019) and the DAN model by Fivez et al. (2021b), which generalizes the BNE model to any hierarchical level of biomedical concepts. Our framework requires a set of concepts C, where each concept $c \in C$ contains a set of concept names C_n. The set of biomedical names N contains the union of all those sets of concept names. We propose a simple multi-task training regime which applies two training objectives to each biomedical name $n \in N$. We use cosine distance as distance function d for both objectives.

Semantic similarity We enforce embedding similarity between names that are from the same concept by using a siamese triplet loss (Chechik et al., 2010). This loss forces the encoding of a biomedical name $f(n)$ to be closer to the encoding of a semantically similar name $f(n_{pos})$ than that of an encoded negative sample name $f(n_{neg})$, within a specified (possibly tuned) margin:

$$\begin{aligned} pos &= d(f(n), f(n_{pos})) \\ neg &= d(f(n), f(n_{neg})) \quad (2) \\ L_{sem} &= max(pos - neg + margin, 0) \end{aligned}$$

To select negative names during training we apply distance-weighted negative sampling (Wu et al.,

2017) over all training names, since this has been proven more effective than hard or random negative sampling.

Conceptually grounded regularization To prevent the model from overfitting on the semantic similarity objective, we regularize it by grounding the output representations to a stable and meaningful target. Simple approximations of prototypical concept representations can already be very effective as targets (Fivez et al., 2021a). Following the model by Fivez et al. (2021b), we use a grounding target which is applicable to any level of categorization, from fine-grained concept distinctions to higher-level groupings of names. This target is a compromise between the *contextual meaningfulness* and *conceptual meaningfulness* objectives of the BNE model. Rather than constraining a name encoding either to its pretrained name representation or to a pretrained representation of its concept, we minimize the distance to the average of both pretrained representations:

$$
\begin{aligned}
u_c &= \frac{1}{|C_n|} \sum_{n \in C_n} u_n \\
u_{ground} &= \frac{u_c + u_n}{2} \\
L_{ground} &= d(f(n), u_{ground})
\end{aligned}
\tag{3}
$$

where the concept representation u_c is approximated by averaging each pretrained embedding representation u_n from the set of names C_n belonging to the concept.

This constraint implies that the dimensionality of the encoder output should be the same as that of the input. However, if the input dimensionality is smaller than the desired output dimensionality, this could be solved using e.g. random projections, which work well for increasing the dimensionality of neural encoder inputs (Wieting and Kiela, 2019).

Multi-task loss Our multi-task loss sums the losses of the 2 training objectives:

$$L = \alpha L_{sem} + \beta L_{ground} \tag{4}$$

where α and β are possible weights for the individual losses. Since both losses directly reflect cosine distances, they are similarly scaled and don't require weighting to work properly. In our experiments, $\alpha = \beta = 1$ showed the most robust performance along all settings.

2.3 Training data

We extract sets of high-level concepts and their constituent names from 2 large-scale hierarchies of disorder concepts, ICD-10 and SNOMED-CT. Table 2 gives an overview of our data distributions.

ICD-10 We use the 2018 version of the ICD-10 coding system.[1] We select the 21 chapters as concept labels, and assign the reference name of each code in a chapter to its concept label. Table 1 gives an example of how such a grouping includes diverse semantic relations.

SNOMED-CT We use the 2018AB release of the UMLS ontology[2] to extract a directed ontological graph of SNOMED-CT concepts. We then select the first-degree child nodes of concept *C0012634*, which is the parent concept for all disorders. We then remove those children which are direct parents to other selected children, since they are redundant for our purpose.

This leaves us with 87 concepts, to which we assign the reference terms of all their child concepts in the ontological graph as biomedical names. To make this setup directly comparable to our ICD-10 setup, we select the 21 largest concepts. Finally, we leave out ambiguous names which belong to multiple concepts. Table 2 shows the impact on the data distribution.

3 Experiments and discussion

3.1 Pretrained representations

We experiment with 3 pretrained name representations. As a first baseline, we use 300-dimensional **fastText** (Bojanowski et al., 2017) word embeddings which we train on 76M sentences of preprocessed MEDLINE articles released by Hakala et al. (2016). We use average pooling (Shen et al., 2018) to extract a 300-dimensional name representation. As a second baseline, we average the 728-dimensional context-specific token activations of a name extracted from the publicly released **BioBERT** model (Lee et al., 2019).

As state-of-the-art reference, we extract 200-dimensional name representations using the publicly released pretrained **BNE** model with skipgram word embeddings, BNE + SG_W,[3] which was trained on approximately 16K synonym sets of disease

[1] https://www.cdc.gov/nchs/icd
[2] https://uts.nlm.nih.gov/home.html
[3] https://github.com/minhcp/BNE

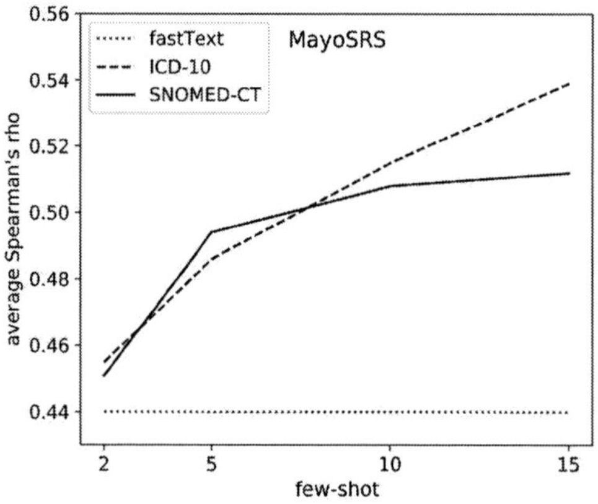

Figure 1: Few-shot performance for fastText encoders on MayoSRS, averaged over 5 random samples.

	EHR-RelB (rel)	MayoSRS (rel)	UMNSRS (rel)	(sim)
BioSyn	0.45	0.50	0.40	0.42
Fivez et al. (2021a)		**0.67**	**0.56**	0.56
fastText	0.39	0.44	0.47	0.48
BioBERT	0.34	0.23	0.18	0.26
BNE	0.47	0.63	0.54	<u>0.58</u>
SNOMED				
fastText	0.43	0.51	0.46	0.51
BioBERT	0.40	0.31	0.32	0.38
BNE	<u>0.53</u>	0.63	<u>0.55</u>	**0.60**
ICD-10				
fastText	0.43	0.55	0.52	0.56
BioBERT	0.35	0.34	0.32	0.38
BNE	0.51	<u>0.65</u>	**0.56**	**0.60**
S → I				
fastText	0.44	0.55	0.46	0.52
BioBERT	0.39	0.33	0.35	0.42
BNE	**0.54**	**0.67**	0.52	<u>0.58</u>
I → S				
fastText	0.45	0.54	0.46	0.51
BioBERT	0.39	0.33	0.37	0.42
BNE	**0.54**	**0.67**	0.53	<u>0.58</u>

Table 3: Spearman's rank correlation coefficient between human judgments and similarity scores of name embeddings, reported on semantic similarity (sim) and relatedness (rel) benchmarks. The highest score is denoted in bold; the second highest is underlined.

concepts in the UMLS, containing 156K disease names.

3.2 Training details

We randomly sample a small fixed amount of names from each concept in our training data as actual few-shot training names. We then randomly sample the same amount of names as validation data to calculate the multi-task loss as stopping criterion. This criterion is also used to finetune the size of the encoder network. Using only 1 hidden layer proved best in all settings, which leaves only the dimensionality of this layer to be tuned.

Our encoder network is implemented in PyTorch (Paszke et al., 2019). Adam optimization (Kingma and Ba, 2015) is performed on a batch size of 16, using a learning rate of 0.001 and a dropout rate of 0.5. Input strings are first tokenized using the Pattern tokenizer (Smedt and Daelemans, 2012) and then lowercased. We use a triplet margin of 0.1 for the siamese triplet loss L_{sem} defined in Equation 2.

3.3 Results

We evaluate our trained encoders on 3 biomedical benchmarks of semantic relatedness and similarity, which allow to compare similarity scores between name embeddings with human judgments of relatedness. MayoSRS (Pakhomov et al., 2011) contains multi-word name pairs of related but different fine-grained concepts. UMNSRS (Pakhomov et al., 2016) contains only single-word pairs, and makes a distinction between *relatedness* and *similarity*, which is a more narrow form of relatedness. Finally, EHR-RelB (Schulz et al., 2020) is much larger than the other benchmarks, and contains multi-word concept pairs which are chosen based on co-occurrence in electronic health records. This ensures that the evaluated concept pairs are actually relevant in function of downstream applications such as information retrieval.

We average all test results over 5 different random training samples. We use cosine similarity as similarity score for all baseline representations and trained encoders. Figure 1 shows the impact of the amount of few-shot training names on performance when using fastText representations. Our model already substantially improves over the baseline with only 5 names per concept (105 in total), and maintains consistent improvement up to 15 few-shot names. This confirms that our approach is well-suited to anticipate expected improvements from training on large-scale hierarchies.

Table 3 shows the results on all benchmarks for 15-shot learning. All encoders were tuned to 9,600 hidden dimensions. We include two state-of-the-art biomedical name encoders in our comparison. Firstly, BioSyn (Sung et al., 2020) sums the weighted inner products of fine-tuned BioBERT representations and sparse TF-IDF representations into one similarity score between two names. The pre-trained model[4] for which we report results was

[4] `https://github.com/dmis-lab/BioSyn`

Parent concept	C0042075	
Parent concept name	*disorder of the urinary system*	
Validation mention	**urinary hesitancy**	
	15-shot BNE	**BNE**
Top 10 ranking	nebulous urine	nebulous urine
	calculus of lower urinary tract (disorder)	calculus of lower urinary tract (disorder)
	urinary obstruction due to nodular prostate (disorder)	urinary obstruction due to nodular prostate (disorder)
	double kidney and/or pelvis	double kidney and/or pelvis
	covered exstrophy of bladder (disorder)	genital oedema
	nephropathy caused by aminoglycoside (disorder)	perineal laceration during delivery , nos
	renal vein thrombosis	abdominal hernia
	benign tumour of urethra	covered exstrophy of bladder (disorder)
	injury of male urethra	heart :[weak] or [failure nos] (disorder)
	postprocedural bulbous urethral stricture	hourglass contraction of uterus

Table 4: A comparison between the rankings of 315 SNOMED-CT training names for the validation mention *urinary hesitancy*. Non-matching names are underlined. While the pretrained BNE model makes various topical associations, our 15-shot model using the BNE representations as input has learned to cluster around the semantics of urinary tract disorders.

trained on the NCBI disease benchmark (Doğan et al., 2014) for biomedical entity normalization. Secondly, we include the results of the conceptually grounded Deep Averaging Network by Fivez et al. (2021a), which was trained on SNOMED-CT synonym sets mapped into larger ICD-10 categories.

The results show various trends. Firstly, almost all trained encoders improve over their input baselines for all benchmarks, regardless of the type of input representation. Secondly, the performance increase is consistent for both ICD-10 and SNOMED-CT, even as their conceptual hierarchies are substantially different. Lastly, we also look at continual learning from SNOMED-CT to ICD-10 (**S** $\rightarrow$ **I**) or vice versa (**I** $\rightarrow$ **S**), where we use the output of the first model as input representations to train the second model. This approach leads to systematic improvements for all representation types, including the state-of-the-art BNE representations. In other words, we provide tangible empirical evidence that few-shot robust representations can allow for continual specialization in biomedical semantics.

To better understand how our few-shot learning approach can have a visible impact on various relatedness benchmarks, Table 4 gives an example of nearest neighbor names from the training set of SNOMED-CT names for the validation mention *urinary hesitancy*. While the pretrained BNE model makes various topical associations, our 15-shot model using the BNE representations as input has learned to cluster around the semantics of urinary tract disorders. As this already generalizes to validation mentions, we can expect the model to transfer this information to downstream applications wherever urinary tract disorders are relevant. This applies to all 21 high-level topics which were simultaneously encoded for both the ICD-10 and SNOMED-CT ontologies.

4 Conclusion and future work

We have proposed a novel approach for scalable few-shot learning of robust biomedical name representations, which trains a simple encoder architecture using only small subsamples of names from higher-level concepts of large-scale hierarchies. Our model works for various pretrained input embeddings, including already specialized name representations, and can accumulate information over various hierarchies to systematically improve performance on biomedical relatedness benchmarks. Future work will investigate whether such improvements trickle down properly to downstream biomedical NLP tasks.

Acknowledgments

We would like to thank the anonymous reviewers for their feedback. This research was carried out in the framework of the Accumulate VLAIO SBO project, funded by the government agency Flanders Innovation & Entrepreneurship (VLAIO). This research also received funding from the Flemish Government (AI Research Program).

References

Piotr Bojanowski, Edouard Grave, Armand Joulin, and Tomas Mikolov. 2017. Enriching word vectors with subword information. *Transactions of the Association for Computational Linguistics*, 5:135–146.

Gal Chechik, Varun Sharma, Uri Shalit, and Samy Bengio. 2010. Large scale online learning of image similarity through ranking. *Journal of Machine Learning Research*, 11:1109–1135.

Rezarta Islamaj Doğan, Robert Leaman, and Zhiyong Lu. 2014. NCBI disease corpus: A resource for disease name recognition and concept normalization. *Journal of Biomedical Informatics*, 47:1–10.

Manaal Faruqui, Jesse Dodge, Sujay Kumar Jauhar, Chris Dyer, Eduard Hovy, and Noah A. Smith. 2015. Retrofitting word vectors to semantic lexicons. In *Proceedings of the 2015 Conference of the North American Chapter of the Association for Computational Linguistics: Human Language Technologies*, pages 1606–1615, Denver, Colorado. Association for Computational Linguistics.

Pieter Fivez, Simon Suster, and Walter Daelemans. 2021a. Conceptual grounding constraints for truly robust biomedical name representations. In *Proceedings of the 16th Conference of the European Chapter of the Association for Computational Linguistics: Main Volume*, pages 2440–2450, Online. Association for Computational Linguistics.

Pieter Fivez, Simon Suster, and Walter Daelemans. 2021b. Integrating higher-level semantics into robust biomedical name representations. In *Proceedings of the 12th International Workshop on Health Text Mining and Information Analysis*, pages 49–58, online. Association for Computational Linguistics.

Kai Hakala, Suwisa Kaewphan, Tapio Salakoski, and Filip Ginter. 2016. Syntactic analyses and named entity recognition for PubMed and PubMed Central — up-to-the-minute. *Proceedings of the 15th Workshop on Biomedical Natural Language Processing*, pages 102–107.

Diederik P. Kingma and Jimmy Ba. 2015. Adam: A method for stochastic optimization. In *3rd International Conference for Learning Representations (ICLR)*.

Jinhyuk Lee, Wonjin Yoon, Sungdong Kim, Donghyeon Kim, Sunkyu Kim, Chan Ho So, and Jaewoo Kang. 2019. BioBERT: a pre-trained biomedical language representation model for biomedical text mining. *Bioinformatics*, 36(4):1234–1240.

Nikola Mrkšić, Diarmuid Ó Séaghdha, Blaise Thomson, Milica Gašić, Lina M. Rojas-Barahona, Pei-Hao Su, David Vandyke, Tsung-Hsien Wen, and Steve Young. 2016. Counter-fitting word vectors to linguistic constraints. In *Proceedings of the 2016 Conference of the North American Chapter of the Association for Computational Linguistics: Human Language Technologies*, pages 142–148, San Diego, California. Association for Computational Linguistics.

Serguei V.S. Pakhomov, Greg Finley, Reed McEwan, Yan Wang, and Genevieve B. Melton. 2016. Corpus domain effects on distributional semantic modeling of medical terms. *Bioinformatics*, 32(23):3635–3644.

Serguei V.S. Pakhomov, Ted Pedersen, Bridget McInnes, Genevieve B. Melton, Alexander Ruggieri, and Christopher G. Chute. 2011. Towards a framework for developing semantic relatedness reference standards. *Journal of Biomedical Informatics*, 44:251–265.

Adam Paszke, Sam Gross, Francisco Massa, Adam Lerer, James Bradbury, Gregory Chanan, Trevor Killeen, Zeming Lin, Natalia Gimelshein, Luca Antiga, Alban Desmaison, Andreas Kopf, Edward Yang, Zachary DeVito, Martin Raison, Alykhan Tejani, Sasank Chilamkurthy, Benoit Steiner, Lu Fang, Junjie Bai, and Soumith Chintala. 2019. PyTorch: An imperative style, high-performance deep learning library. In H. Wallach, H. Larochelle, A. Beygelzimer, F. d'Alché-Buc, E. Fox, and R. Garnett, editors, *Advances in Neural Information Processing Systems 32*, pages 8024–8035. Curran Associates, Inc.

Minh C. Phan, Aixin Sun, and Yi Tay. 2019. Robust representation learning of biomedical names. In *Proceedings of the 57th Annual Meeting of the Association for Computational Linguistics*, pages 3275–3285, Florence, Italy. Association for Computational Linguistics.

Claudia Schulz, Josh Levy-Kramer, Camille Van Assel, Miklos Kepes, and Nils Hammerla. 2020. Biomedical concept relatedness – a large EHR-based benchmark. In *Proceedings of the 28th International Conference on Computational Linguistics*, pages 6565–6575, Barcelona, Spain (Online). International Committee on Computational Linguistics.

Dinghan Shen, Guoyin Wang, Wenlin Wang, Martin Renqiang Min, Qinliang Su, Yizhe Zhang, Chunyuan Li, Ricardo Henao, and Lawrence Carin. 2018. Baseline needs more love: On simple word-embedding based models and associated pooling mechanisms. In *Proceedings of the 56th Annual Meeting of the Association for Computational Linguistics (Long Papers)*, pages 440–450.

Tom De Smedt and Walter Daelemans. 2012. Pattern for Python. *Journal of Machine Learning Research*, 13:2031–2035.

Mujeen Sung, Hwisang Jeon, Jinhyuk Lee, and Jaewoo Kang. 2020. Biomedical entity representations with synonym marginalization. In *Proceedings of the 58th Annual Meeting of the Association for Computational Linguistics*, pages 3641–3650, Online. Association for Computational Linguistics.

John Wieting and Douwe Kiela. 2019. No training required: Exploring random encoders for sentence classification. In *International Conference on Learning Representations*.

Chao-Yuan Wu, R. Manmatha, Alexander J. Smola, and Philipp Krahenbuhl. 2017. Sampling matters in deep embedding learning. In *ICCV*.

SAFFRON: tranSfer leArning For Food-Disease RelatiOn extractioN

Gjorgjina Cenikj
Jožef Stefan International Postgraduate School
Ljubljana, Slovenia
Computer Systems Department
Jožef Stefan Institute
Ljubljana, Slovenia
gjorgjina.cenikj@ijs.si

Tome Eftimov
Computer Systems Department
Jožef Stefan Institute
Ljubljana, Slovenia
tome.eftimov@ijs.si

Barbara Koroušić Seljak
Computer Systems Department
Jožef Stefan Institute
Ljubljana, Slovenia
barbara.korousic@ijs.si

Abstract

The accelerating growth of big data in the biomedical domain, with an endless amount of electronic health records and more than 30 million citations and abstracts in PubMed, introduces the need for automatic structuring of textual biomedical data. In this paper, we develop a method for detecting relations between food and disease entities from raw text. Due to the lack of annotated data on food with respect to health, we explore the feasibility of transfer learning by training BERT-based models on existing datasets annotated for the presence of *cause* and *treat* relations among different types of biomedical entities, and using them to recognize the same relations between food and disease entities in a dataset created for the purposes of this study. The best models achieve macro averaged F1 scores of 0.847 and 0.900 for the *cause* and *treat* relations, respectively.

1 Introduction

The ongoing prevalence of malnutrition, the rising incidence of chronic diseases affected by diet, and the fact that even food that is generally considered to be healthy can be harmful to patients suffering from certain diseases or when ingested in combination with specific drugs, require a profound understanding of the role of nutrition in the complex environmental interactions that contribute to the development or treatment of different ailments. The effect of food on human health is the subject of numerous biomedical studies, however, the sheer volume and the predominantly unstructured form of newly published articles prevents medical professionals from keeping up with recent discoveries, and impedes the development of systems for knowledge-base construction, Decision Support, and Question-Answering (QA), which brings about the need for information extraction (IE) methods for structuring the newly published knowledge.

Knowledge graphs (KGs) are specialized data representation structures that store information as a collection of interlinked descriptions of entities. The development of Relation Extraction (RE) methods is necessary for automatic linking of the nodes in KGs and reducing the amount of work required by the experts in order to create and extend these resources.

A lot of research effort has been dedicated to extracting relations between different biomedical entities, however, the lack of annotated data impedes the development of food-disease RE methods, which are necessary for linking food entities to concepts from the biomedical domain, and understanding the impact of nutrition on human health.

Transfer learning (TL) (Weiss et al., 2016; Zhuang et al., 2019) is a potential solution for this problem, which involves improving a learner from one domain by transferring information from a related domain. The use of TL in this paper is two-fold. On the one hand, we use models that are pre-trained on large amounts of data, and fine-tune them for the RE task. On the other hand, we investigate the feasibility of re-purposing existing annotated IE resources in the biomedical domain as a potential strategy for making up for the deficit of such resources in the food domain.

We focus on the detection of *cause* and *treat* relations among food and disease entities, and represent the RE task as a binary classification problem, meaning that we train separate classifiers that detect the presence of each relation type. We perform fine-tuning of BERT (Devlin et al., 2018), BioBERT (Lee et al., 2019) and RoBERTa (Liu et al., 2019) models, which have achieved state of the art results in several Natural Language Processing (NLP) tasks.

To train the classifiers, we use the CrowdTruth (Dumitrache et al., 2017, 2015b,a) and Adverse Drug Events (ADE) (Gurulingappa et al., 2012) datasets, which contain sentences annotated

Proceedings of the BioNLP 2021 workshop, pages 30–40
June 11, 2021.

for the existence of relations between different types of biomedical entities. We then apply TL in order to use the classifiers trained on the source datasets to directly predict relations among food and disease entities. The reasoning behind the use of TL in this setting is that even though the sentences contain entities of different types, by masking the entity occurrences in the sentence, the models could use the context words around the entities and pick up on linguistic features such as keywords or sentence structure to detect the presence of a particular relation. Even though our goal is focused on detecting the relations between food and disease entities, we believe the method to be general enough to be applicable for entities of any type, as long as the relation is the same as the one the model was trained to recognize.

To evaluate the proposed models, we introduce a dataset of 608 sentences, which are extracted from abstracts of scientific articles from PubMed and are manually annotated for the presence of *cause* and *treat* relations between food and disease entities. To the best of our knowledge, this is the first English RE dataset in the food domain, and it is publicly available on GitHub [1], as an open-source resource that can be reused in future studies.

The rest of the paper is organized as follows. In the next section, we give an overview of the RE work in the domains of biomedicine, and food and nutrition. The data sources used for the experiments are described in Section 3. The text representation and classification models are presented in Section 4, while their evaluation is discussed in Section 5.

2 Related work

In the past decade, numerous methods have been developed for extracting biomedical relations, such as drug-drug (Dewi et al., 2017; Liu et al., 2016; Kim et al., 2015; Sahu and Anand, 2018), protein-protein (Koyabu et al., 2015; Fan et al., 2018; Zhou et al., 2018), drug-disease (Wang et al., 2017; Bchir and Karaa, 2013), chemical-gene (Lim and Kang, 2018) and chemical-protein (Lung et al., 2019) interactions.

In the domain of food and nutrition, the efforts directed at creating RE systems have been quite more limited in comparison. A gold standard for food RE has been generated for the German language (Wiegand et al., 2012b), and different methods such as distant supervision (DS), pattern-matching, and the use of co-occurrence measures have been investigated for the detection of food relations for customer advice (Wiegand et al., 2012a; Reiplinger et al., 2014). A Chinese food RE system (Miao et al., 2012) has also been developed, which treats RE as a sequence labeling task and adopts Conditional Random Fields (CRFs) models to extract relations between food and disease entities from Chinese biomedical data. However, resources in other languages are not easily re-purposed for the English language.

A related resource in the English language which contains extracted relations of food and disease entities is the NutriChem database (Jensen et al., 2014; Ni et al., 2017), which links plant-based foods with their small molecule components, drugs and disease phenotypes. A critical difference between NutriChem and the method we aim to develop in this work is the fact that NutriChem limits its scope to plant-based foods, while we do not pose a limitation on the type of foods or diseases between which the relations occur, and aim to extract relations from a broader range of food categories.

The benefits of TL have previously been investigated for the purposes of biomedical NER (Sun and Yang, 2019; Francis et al., 2019) and RE (Zhang et al., 2019; Peng et al., 2019; Hafiane et al., 2020). Recent work has been aimed at solving the challenges of imbalanced relation distribution, linguistic variation and partial transfer using relation-gated adversarial learning (Zhang et al., 2019), and capturing dependency tree information using TreeLSTM models (Legrand et al., 2018).

BERT has achieved state-of-the-art results on natural language processing (NLP) tasks, including RE between several types of biomedical entities, which is one of the tasks in the Biomedical Language Understanding Evaluation (BLUE) benchmark (Peng et al., 2019). A comparison of the performance of BERT models for detecting relations between proteins and chemicals, and genomic factors and drugs or drug responses (Hafiane et al., 2020), finds that depending on the target corpus, different variants of BERT may achieve the best results, and that fine-tuning the models is preferable over freezing the layers of the original model and only updating the weights of new layers added on top of the original ones. Guided by these findings, we perform fine-tuning of several BERT variants

[1] `https://github.com/gjorgjinac/food-disease-dataset`

for the RE task.

The Adverse Drug Events (ADE) corpus (Gurulingappa et al., 2012), which is one of the source datasets in our experiments, has been extensively used for training RE models, and more recently, for the exploitation of inter-task correlations for joint entity and relation extraction using different approaches, such as adversarial training (Bekoulis et al., 2018), Cross-Modal Attention Networks (Zhao et al., 2020) and BERT models (Eberts and Ulges, 2019). However, unlike the previous work done with this corpus, our goal is not to predict relations between the annotated entities, but to learn the context words used for expressing causal relations, so they can be recognized regardless of the entities between which they occur.

3 Data

TL usually involves the use of two types of datasets: source datasets and target datasets, where models are trained on the source datasets, and adapted to make predictions on the target datasets. We are specifically interested in extracting relations between food and disease entities, and we use the help of two existing source datasets, the CrowdTruth (Dumitrache et al., 2017) and the ADE dataset (Gurulingappa et al., 2012), in order to extract relations in the target FoodDisease dataset, which was created for the purposes of this study.

3.1 The CrowdTruth dataset

The CrowdTruth dataset (Dumitrache et al., 2017) for medical RE contains annotated data for *cause* and *treat* relations in sentences from abstracts of PubMed articles.

The dataset contains 4028 sentences annotated for the existence of a *cause* relation and 3983 sentences annotated for the existence of a *treat* relation. Every sample of the dataset contains the name of a relation, and a sentence containing two entities between which the relation may or may not occur. Each entity is further described by its UMLS name, its starting and ending position in the sentence, and the exact textual form in which it appears in the sentence. Apart from this, each sample is assigned several labels which indicate whether the relation is observed between the two terms.

The initial data (Wang and Fan, 2014) were collected using Distant supervision (DS) (Mintz et al., 2009), which is a inexpensive and straightforward way of labeling training data, but is also prone to producing noisy, low quality labels (Dumitrache et al., 2015b; Ji et al., 2017; Chen et al., 2021). Because of that, the annotations for the *cause* and *treat* relations collected using DS were further refined using the CrowdFlower platform where a multi-label annotation task was executed through crowdsourcing (Dumitrache et al., 2017, 2015b,a). Additionally, experts annotated sentences with binary labels, based on whether a specified seed relation discovered by DS is present between two given biomedical entities that occur in the sentence.

The *sentence relation* score given for each sample is computed as the cosine similarity between the vector containing the sum of the annotations of the non-expert workers, and the unit vector for the relation. Here, the unit vector refers to a one-hot vector where the value corresponding to the relation is equal to 1, and all other components are equal to 0. This score is in the range [0, 1]. The *crowd* score is calculated using the *sentence relation* score, by applying a threshold of 0.5 to separate positive from negative examples, and rescaling the obtained positive and negative samples in the ranges [0.5, 1], and [-1, -0.5], respectively.

The *expert* label is based on the experts' annotations and it takes values of either 1 or -1, indicating the presence or absence of the relation, respectively. However, due to the cost, limited time and availability of experts, the expert annotations were limited to 975 samples in the *cause* dataset and 621 samples in the *treat* dataset.

3.1.1 Target variable construction in the CrowdTruth dataset

The target variable is a binary indicator of the existence of the *cause* or *treat* relationship in the respective dataset. As the CrowdTruth dataset contains multiple indicators of these relations, we choose to rely on the labels assigned by experts, but since these are not available for all samples, we also use the *crowd* score, which has been shown to give reliable results in the original studies (Dumitrache et al., 2017, 2015b,a). To be more precise, if the sentence has been labeled by an expert, the target label is assigned a value of 1, if the score given by the expert is 1, or 0, if the score given by the expert is -1. If the sentence has not been labeled by an expert, the target label is assigned a value of 1, if the *crowd* score is positive, or 0, if the *crowd* score is negative.

3.2 The Adverse Drug Events (ADE) dataset

The ADE dataset (Gurulingappa et al., 2012) contains 6821 sentences expressing truthful relations between drugs and adverse effects they have been shown to cause, and 279 sentences with relations between drugs and dosages. Each sample consists of a sentence, the name of a drug, the name of a condition (if the relation expressed is *adverse effect*) or a dosage term (if the relation expressed is *dose*), and their starting and ending position in the sentence. The sentences were extracted from MEDLINE case reports, and were manually annotated by three annotators. There are 1319 unique drugs, 3341 unique conditions, and 130 unique dosage terms. In order to be consistent with the nomenclature in the other datasets, we refer to the *adverse effect* relation in the ADE dataset as a *cause* relation, and to the *condition* entities as *diseases*. The intuition behind using relations annotated as *adverse effect* to detect *cause* relations between food and disease entities is that one would use similar sentence structures to describe a disease occurring as a result of the ingestion of a particular drug or food.

3.3 The FoodDisease dataset

Since there was no data labeled for the existence of *cause* and *treat* relations between food and disease entities, for the purposes of this research we constructed a dataset containing 608 sentences from abstracts of PubMed articles. Fig. 1 depicts the steps taken in order to generate the dataset.

BuTTER (Cenikj et al., 2020) and SABER (Giorgi and Bader, 2019) were used for finding the food and disease entities in each abstract. Both are Named Entity Recognition (NER) methods based on Bidirectional Long Short-Term Memory and Conditional Random Fields. BuTTER extracts food entities from raw text, and is trained on the golden version of the FoodBase corpus (Popovski et al., 2019), which contains 1000 recipes annotated with food entities. In particular, we used the lexical lemmatized BuTTER model introduced in (Cenikj et al., 2020), which achieves a macro averaged F1 score of 0.946.

SABER is a biomedical NER tool, which provides several pre-trained NER models, from which we use the *DISO* model [2] to extract disease entities.

[2] https://baderlab.github.io/saber/resources

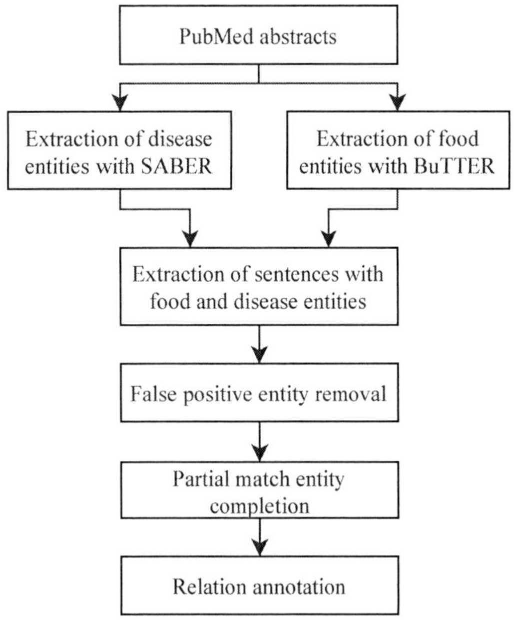

Figure 1: Steps taken to generate the FoodDisease dataset

The abstracts were filtered so that only sentences which contain at least one food and one disease entity were kept. The entities in each sentence were then manually checked and corrected in order to remove false positives and complete partial matches. Removing the false positive entities means that the tokens that were incorrectly extracted as food or disease entities by the BuTTER and SABER methods were discarded. Completing partial matches entails adding the missing words in entities which should contain multiple words, but some of them were not captured by the automatic annotators. Each sample contains a single food and a single disease entity, even if multiple such entities are mentioned in the sentence. Finally, each sentence was assigned binary labels to indicate if the *cause* and *treat* relations are present.

4 Methodology

In this section, we describe the proposed RE method, starting with the preprocessing applied to accomplish the generalization of the models trained on the source datasets to the target dataset. We then introduce the pre-trained transformer models used for text representation, and their fine-tuning for the RE task.

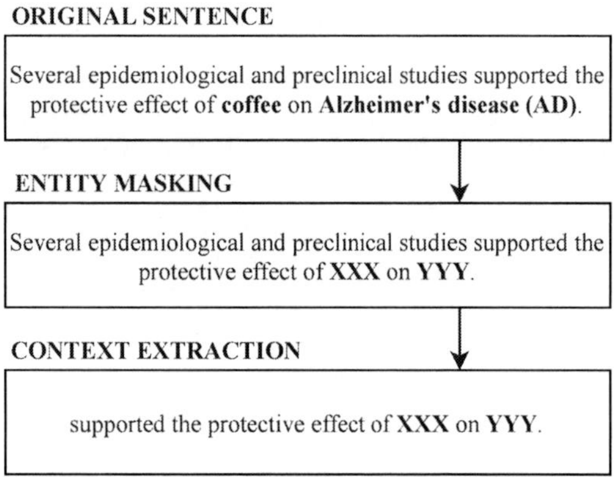

Figure 2: Application of the preprocessing steps on a sentence from the FoodDisease dataset

4.1 Data preprocessing

Since the datasets we are using are annotated with relations between different types of biomedical entities, and we would like the developed models to generalize, and be able to extract the same relations between different types of entities, we mask out the entity mentions in each sentence. The reasoning behind this is that the model would not learn to detect relations between the concrete entities, but instead, use the surrounding words to determine whether they express the particular relation.

Since there could be several relations present in one sentence, we use a context window of length 5, i.e. use the words whose positions in the sentence are in the range (i-5,j+5), where i is the word index of the first occurring entity in the sentence, and j is the word index of the second occurring entity in the sentence.

Fig. 2 shows an example of the preprocessing steps being applied on a sentence from the FoodDisease dataset. The bolded words in the original sentence are the food and disease entities, which get masked out in the *Entity Masking* step, where they are replaced by *XXX* and *YYY*, respectively. These masking tokens are chosen arbitrarily, since their only use is for the model to distinguish between the subject and object entity. In the *Context Extraction* step, the final preprocessed version of the sentence is generated by keeping only the words in between the entities, and the 5 words that precede the first entity, *coffee*. Had there been additional words after the second entity, *Alzheimer's disease (AD)*, the first 5 of them would also be included in the context.

4.2 Text representation

In order to represent the textual data in numerical format, we use 3 pre-trained transformed-based models: BERT, RoBERTa and BioBERT.

4.2.1 BERT

BERT (Bidirectional Encoder Representations from Transformers) (Devlin et al., 2018) is a bidirectional, contextual representation model that achieves state-of-the-art results in several natural language processing tasks. Following the principles of transductive TL, BERT is pre-trained on an unsupervised Mask Language Modeling (MLM) or Next Sentence Prediction (NSP) task, and then fine-tuned on another downstream task, such as NER, Natural Language Inference or Question Answering. The pre-trained BERT models can be finetuned without substantial modifications in their architecture. In the simplest case, only the output layer needs to be replaced, depending on the task that the model is intended to perform. We use the original BERT model, which is pre-trained on the BooksCorpus (Zhu et al., 2015) and English Wikipedia, and fine tune it for relation classification.

4.2.2 RoBERTa

RoBERTa (Robustly Optimized BERT Approach) (Liu et al., 2019) is a text representation model based on the original BERT architecture, with a number of improvements introduced in the pre-training phase, some of which include training on a larger amount of data, longer training, removal of the NSP task, and introduction of dynamic masking. Apart from the BooksCorpus and Wikipedia, which are used for the pretraining of BERT, RoBERTa is trained on data from 3 additional sources: the CommonCrawl News dataset (Nagel, 2016), the OpenWebText corpus (Gokaslan and Cohen, 2019) and Stories subset from the Common Crawl dataset (Trinh and Le, 2018).

4.2.3 BioBERT

BioBERT (Bidirectional Encoder Representations from Transformers for Biomedical Text Mining) (Lee et al., 2019) is a domain-specific version of the BERT model. Due to the fact that biomedical texts contain a considerable amount of domain-specific proper nouns and terms that do not appear in more general texts and would hence be unfamiliar to the original BERT, the data on which

BioBERT is trained is supplemented by PubMed abstracts and full-text articles from PubMed Central. As a result, BioBERT has been shown to outperform BERT in biomedical NER, RE, and QA (Lee et al., 2019).

4.3 Models

We perform end-to-end fine-tuning of the pre-trained BERT, RoBERTa and BioBERT models for the RE task. In order to adapt the original architecture to perform binary classification, the last layer of the models is replaced with a dropout and a linear layer which performs binary classification. During fine-tuning, the model parameters are initialized with the values from the pre-training step, and are fine-tuned using the labeled data from the source datasets. The input of a BERT model can unambiguously represent both a single sequence and a pair of text sequences (for example, a question and an answer) in one token sequence, by using a separator token [SEP] to mark the end of each sequence. We explore both types of inputs and construct two different models:

- Single Sequence Classifier (SSC) - The model takes a single sequence as an input and performs simple binary classification.

- Sequence Pair Classifier (SPC) - The model takes as input two sequences. The first sequence is the sequence that we want to classify (the one that is used on its own in the SSC), while the second sequence is a concatenation of 10 randomly sampled sequences which have positive labels for the relation we are aiming to detect. We refer to the first sequence as the *sequence of interest*, while we call the concatenation of 10 sequences a *ground truth* for the relation in question. The sentences used in the ground truth sequences are not used as sequences of interest.

  The intuition behind this approach is that we can reformulate the task *Does sequence X express relation Y?* as *Is sequence X similar to other sequences that contain relation Y?*. The task is still a binary classification, and the label remains the same as for the SSC.

  We construct 10 ground truth sequences for each relation, and pair each sequence of interest with each ground truth. The same generated ground truths are used at training and prediction time. For each sequence of interest in the test set, we generate 10 predictions (one for each ground truth) and assign the average of the predicted probabilities as the probability of the sequence of interest belonging to the positive class.

Table 1: Examples of inputs given to the SSC and SPC models when identifying the *treat* relation

Inputs given to the SSC model

Input	Label
supported the protective effect of XXX on YYY	1
XXX is known to cause YYY	0

Inputs given to the SPC model

Input	Label
Sequence of interest: supported the protective effect of XXX on YYY *Ground truth*: XXX has been used in the treatment of YYY; XXX is known to cure YYY; XXX is associated with a reduced incidence of YYY	1
Sequence of interest: XXX is known to cause YYY *Ground truth*: XXX has been used in the treatment of YYY; XXX is known to cure YYY; XXX is associated with a reduced incidence of YYY	0

Table 1 features examples of the inputs given to the SSC and SPC models that identify the *treat* relation. The first input sample expresses a *treat* relation, so the label is one, while the second input sample expresses a *cause* relation, so the label is zero. The inputs of the SSC model are the same as for the *sequences of interest* of the SPC model. For the sake of simplicity, for the SPC model in the examples, we demonstrate one ground truth, which is a concatenation of 3 sequences that represent a *treat* relation. In our experiments, we use 10 such *ground truths*, each being a concatenation of 10 sequences.

During the fine-tuning, the AdamW optimizer is used with a learning rate of $4 * 10^{-5}$. An early stopping strategy is applied to prevent overfitting. The models are trained for a maximum of 10 epochs, or until the improvement in validation loss of 2 consecutive epochs does not surpass $5 * 10^{-3}$.

The source codes for fine-tuning the SSC mod-

Table 2: Number of samples from the positive and negative class in each dataset

Dataset	CrowdTruth		ADE	FoodDisease	
Relation	Cause	Treat	Cause	Cause	Treat
Class					
Positive	1429	1406	6821	142	323
Negative	2555	2578	1685	466	285

els[3] and the SPC[4] models are publicly available on the Colab platform.

5 Evaluation

5.1 Evaluation on the source datasets

When applying TL, a model trained on a source dataset can experience some degradation in performance when evaluated on the target dataset. In order to get an idea about the upper bound of the performance expected on the target dataset, the models' performance is first evaluated on the same, source datasets they were trained on using 10-fold cross validation.

All 3 of the datasets are unbalanced, and the class distribution of each dataset is given in Table 2. For the ADE dataset, we only train classifiers for the detection of the *cause* relation, since that dataset does not contain annotations for the *treat* relations. We consider the sentences annotated with the *dose* relation in the ADE dataset to be negative samples for the *cause* relation. However, since there are only 279 such sentences, in order to avoid extreme class unbalance, we supplement the negative samples in the train portion of the ADE dataset by adding the samples that are annotated as positive for the *treat* relation in the CrowdTruth dataset. 10% of the training portion of each fold is removed and used for validation, preserving the ratio of the positive and negative samples.

Because of the unbalanced class distribution in all three datasets, we evaluate the models in terms of the macro averaged f1 scores, averaged from all folds, and these are depicted in Table 3. The models are both trained and evaluated on the datasets indicated in the table header. The SSC and SPC models combined with 3 different pretrained BERT models (BERT, RoBERTa and BioBERT) result in 6 models, which are evaluated on the 3 datasets. The first group of three rows of scores refers to the performance of the SSC model, while the second group refers to the SPC model. The underlined values refer to the highest f1 macro score in each column, and we can note that the BioBERT models give the best performance. The SSC models generally outperform the SPC models.

Table 3: Macro averaged F1 scores obtained from the evaluation on the source datasets when the proposed preprocessing is applied, averaged from 10 folds

Dataset	CrowdTruth		ADE	FoodDisease	
Relation	Cause	Treat	Cause	Cause	Treat
Model: SSC					
BERT	0.753	0.880	0.871	0.744	0.886
RoBERTa	0.740	0.879	0.866	0.711	0.884
BioBERT	0.750	<u>0.890</u>	<u>0.894</u>	<u>0.847</u>	0.871
Model: SPC					
BERT	0.745	0.873	0.822	0.478	0.835
RoBERTa	0.752	0.880	0.743	0.433	0.835
BioBERT	<u>0.771</u>	0.884	0.873	0.545	<u>0.900</u>

The performance of the SPC models which detect the *cause* relation in the FoodDisease dataset is notably lower than the rest of the models. Looking into the models' raw predictions, it is obvious that the models predict the negative class too often, which results in high recall for the negative class, but very low recall for the positive class. This can be attributed to the fact that from the 114 positive samples in the training portion of each fold, 100 are used for constructing the ground truth sequences used by the SPC models, leaving only 14 positive samples for training. Annotating more data, decreasing the number of ground truth sequences or the number of sentences in each ground truth sequence, and balancing the data are possible strategies which are expected to remedy this anomaly.

5.2 Transfer learning evaluation

In this subsection, we report the performance reached by the models trained on the CrowdTruth and ADE source datasets, when evaluated on the target FoodDisease dataset. In this case, the models are trained on balanced data, since the class distribution in the source datasets does not reflect the distribution in the target dataset, and are evaluated on the whole FoodDisease dataset.

[3] `https://colab.research.google.com/drive/1UOFuk6-_6c-za6P54SiIhdY9ZH6T5Xpo?usp=sharing`

[4] `https://colab.research.google.com/drive/1HA78g3YG90UuYT9SZPxw5brE6LqPKC2s?usp=sharing`

Table 4: Macro averaged F1 scores obtained from the evaluation on the target FoodDisease dataset, when the proposed preprocessing is applied

Dataset	CrowdTruth		ADE
Relation	Cause	Treat	Cause
Model: SSC			
BERT	0.727	0.841	0.750
RoBERTa	0.805	0.883	0.710
BioBERT	0.805	0.878	0.750
Model: SPC			
BERT	0.585	0.689	0.619
RoBERTa	0.701	0.838	0.648
BioBERT	0.636	0.872	0.639

Table 5: Macro averaged F1 scores obtained from the evaluation on the target FoodDisease dataset, when the entire sentence is being used as input

Dataset	CrowdTruth		ADE
Relation	Cause	Treat	Cause
Model: SSC			
BERT	0.595	0.828	0.568
RoBERTa	0.659	0.759	0.228
BioBERT	0.610	0.900	0.633
Model: SPC			
BERT	0.557	0.837	0.608
RoBERTa	0.594	0.844	0.587
BioBERT	0.657	0.881	0.625

Table 4 features the macro averaged F1 scores that the models achieve when the preprocessing introduced in subsection 4.1 is applied on the input.

When comparing the results in Table 3 and 4, we can observe that the SPC models and the models trained on the ADE dataset experience performance deterioration when they are evaluated on the target dataset, but the SSC models trained on the CrowdTruth dataset have a similar performance in both evaluations. This is expected to some extent, since the relations in the ADE dataset are originally annotated as *adverse effect*, which we loosely interpret as a *cause* relation, while the sentences in the CrowdTruth dataset are annotated for precisely *cause* and *treat* relations.

Additionally, we conduct experiments to evaluate the proposed preprocessing technique, which we compare to the scenario when no preprocessing is applied (neither the *Entity Masking* nor the *Context Extraction* step) and the entire sentences are given to the model. The macro averaged F1 scores obtained in such a setting are featured in Table 5. The best results are achieved by the RoBERTa and BioBERT models. Most of the models benefit from the preprocessing, which is especially noticable in the SSC models that identify the *cause* relation, where the proposed preprocessing leads to an improvement of the averaged macro f1 scores of at least 0.100. Looking into the metrics for the positive and negative class separately reveals that the lower performance of the models which do not use the proposed processing is due to their lower precision in identifying the positive class.

Interestingly, the SPC models that identify the *treat* relation seem to perform better without the preprocessing, even though only one the performance of the BERT model differs by a large margin, while the performances of the BioBERT and RoBERTa models differ by less than 0.010.

It is important to note that the evaluation on these models on the FoodDisease dataset may be somewhat flawed, since it may hide the possible disadvantage of using entire sentences as input, because all of the sentences in the FoodDisease dataset are unique. This would mean that if a sentence contains both relations, as for example *Nuts are known to reduce the risk of heart disease, but can also cause allergies*, the dataset would either contain the *(food, relation, disease)* triple *(nuts, treat, heart disease)* or the triple *(nuts, cause, allergies)*, but not both. The models that do not use the proposed preprocessing and get the entire sentence as input, would in this case produce an identical output for both triples, but when evaluated on the FoodDisease dataset, they would not be penalized for doing so.

Overall, the best models trained on the source datasets achieve a macro F1 scores of 0.805 and 0.900, for the detection of *cause* and *treat* relations, respectively, between food and disease entities in the target dataset. In comparison, the performance of the best models trained on the target FoodDisease dataset (the SSC-BioBERT and SPC-BioBERT in Table 3) is 0.847 and 0.900. This indicates that the application of TL using pretrained transformer models enables us to train models using a small amount of annotated data, but we can also obtain satisfactory results with no annotated data for the specific RE task, by repurposing annotations for the same relations between different entities.

6 Conclusion

In this paper, we propose Relation Extraction (RE) models for the detection of *cause* and *treat* relations between food and disease entities from raw text. To make up for the absence of annotated data for this task, we explore the feasibility of Transfer Learning (TL) by using the transformer models BERT, RoBERTa, and BioBERT, which are pretrained on large amounts of data, and fine-tuned for performing RE between various types of biomedical entities. The models are trained to recognize relations based on the context words used to express each relation, rather than the entities themselves, so they can successfully generalize to the task of recognizing the relations between food and disease entities, and likely, other types of entities, though this is not evaluated in the scope of this paper.

In order to evaluate the proposed approach, we introduce the FoodDisease dataset, which consists of 608 sentences annotated for the existence of the *cause* and *treat* relations between food and disease entities in sentences of PubMed abstracts. The dataset is released as an open-source resource, and is, to the best of our knowledge, the first annotated English RE dataset of such kind in the food domain.

The best models that are fine-tuned on this dataset achieve macro averaged F1 scores of 0.847 and 0.900 for the *cause* and *treat* relations, respectively. The best models which are fine-tuned using the data where the entities are not food-disease pairs, but other biomedical entities of various types, achieve macro averaged F1 score of 0.805 for the *cause* relation and 0.900 for the *treat* relation. This indicates that in the event where no experts are available to annotate data, the proposed method enables the repurposing of existing RE datasets for the training of models that can recognize the relation that the dataset is annotated for, between different types of entities.

The developed models will be used as part of an information extraction pipeline which will structure the findings of experts in biomedical scientific literature, with the aim of alleviating the process of linking knowledge graphs from the domain of biomedicine to the domain of food and nutrition.

Acknowledgements

This work was supported by the Ad Futura grant for postgraduate study; the Slovenian Research Agency (research core funding programmes P2-0098); the European Union's Horizon 2020 research and innovation programme under grant agreement 863059 (FNS-Cloud, Food Nutrition Security) and under grant agreement 101005259 (COMFOCUS).

References

Aida Bchir and Wahiba Ben Abdessalem Karaa. 2013. Extraction of drug-disease relations from medline abstracts. In *2013 World Congress on Computer and Information Technology (WCCIT)*, pages 1–3.

Giannis Bekoulis, Johannes Deleu, Thomas Demeester, and Chris Develder. 2018. Adversarial training for multi-context joint entity and relation extraction. In *Proceedings of the 2018 Conference on Empirical Methods in Natural Language Processing*, pages 2830–2836, Brussels, Belgium. Association for Computational Linguistics.

Gjorgjina Cenikj, Gorjan Popovski, Riste Stojanov, Barbara Koroušić Seljak, and Tome Eftimov. 2020. BuTTER: BidirecTional LSTM for Food Named-Entity Recognition. In *Proc. Big Food and Nutrition Data Management and Analysis at IEEE BigData 2020*, pages 3550–3556.

Tiantian Chen, Nianbin Wang, Hongbin Wang, and Haomin Zhan. 2021. Distant supervision for relation extraction with sentence selection and interaction representation. *Wireless Communications and Mobile Computing*, 2021:1–16.

Jacob Devlin, Ming-Wei Chang, Kenton Lee, and Kristina Toutanova. 2018. BERT: pre-training of deep bidirectional transformers for language understanding. *CoRR*, abs/1810.04805.

Ika Novita Dewi, Shoubin Dong, and Jinlong Hu. 2017. Drug-drug interaction relation extraction with deep convolutional neural networks. In *2017 IEEE International Conference on Bioinformatics and Biomedicine (BIBM)*, pages 1795–1802.

Anca Dumitrache, Lora Aroyo, and Chris Welty. 2015a. Achieving expert-level annotation quality with crowdtruth: The case of medical relation extraction. In *BDM2I@ISWC*.

Anca Dumitrache, Lora Aroyo, and Chris Welty. 2015b. Crowdtruth measures for language ambiguity: The case of medical relation extraction. In *LD4IE@ISWC*.

Anca Dumitrache, Lora Aroyo, and Chris Welty. 2017. Crowdsourcing ground truth for medical relation extraction. *CoRR*, abs/1701.02185.

Markus Eberts and Adrian Ulges. 2019. Span-based joint entity and relation extraction with transformer pre-training. *CoRR*, abs/1909.07755.

Ziling Fan, Luca Soldaini, Arman Cohan, and Nazli Goharian. 2018. Relation extraction for protein-protein interactions affected by mutations. In *Proceedings of the 2018 ACM International Conference on Bioinformatics, Computational Biology, and Health Informatics*, BCB '18, page 506–507, New York, NY, USA. Association for Computing Machinery.

Sumam Francis, Jordy Van Landeghem, and Marie-Francine Moens. 2019. Transfer learning for named entity recognition in financial and biomedical documents. *Information*, 10(8).

John Giorgi and Gary Bader. 2019. Towards reliable named entity recognition in the biomedical domain. *bioRxiv*.

Aaron Gokaslan and Vanya Cohen. 2019. Openwebtext corpus. `http://Skylion007.github.io/OpenWebTextCorpus`.

Harsha Gurulingappa, Abdul Mateen-Rajput, and Luca Toldo. 2012. Extraction of potential adverse drug events from medical case reports. *Journal of biomedical semantics*, 3(1):15–15.

Walid Hafiane, Joel Legrand, Yannick Toussaint, and Adrien Coulet. 2020. Experiments on transfer learning architectures for biomedical relation extraction. ArXiv:2011.12380.

Kasper Jensen, Gianni Panagiotou, and Irene Kouskoumvekaki. 2014. NutriChem: a systems chemical biology resource to explore the medicinal value of plant-based foods. *Nucleic Acids Research*, 43(D1):D940–D945.

Guoliang Ji, Kang Liu, Shizhu He, and Jun Zhao. 2017. Distant supervision for relation extraction with sentence-level attention and entity descriptions. In *Proceedings of the Thirty-First AAAI Conference on Artificial Intelligence (AAAI-17)*.

Sun Kim, Haibin Liu, Lana Yeganova, and W. John Wilbur. 2015. Extracting drug–drug interactions from literature using a rich feature-based linear kernel approach. *Journal of Biomedical Informatics*, 55:23–30.

Shun Koyabu, Thi Thanh Thuy Phan, and Takenao Ohkawa. 2015. Extraction of protein-protein interaction from scientific articles by predicting dominant keywords. *BioMed Research International*, 2015:928531.

Jinhyuk Lee, Wonjin Yoon, Sungdong Kim, Donghyeon Kim, Sunkyu Kim, Chan Ho So, and Jaewoo Kang. 2019. BioBERT: a pre-trained biomedical language representation model for biomedical text mining. *Bioinformatics*, 36(4):1234–1240.

Joël Legrand, Yannick Toussaint, Chedy Raïssi, and Adrien Coulet. 2018. Syntax-based transfer learning for the task of biomedical relation extraction. In *Proceedings of the Ninth International Workshop on Health Text Mining and Information Analysis*, pages 149–159, Brussels, Belgium. Association for Computational Linguistics.

Sangrak Lim and Jaewoo Kang. 2018. Chemical–gene relation extraction using recursive neural network. *Database*, 2018. Bay060.

Shengyu Liu, Buzhou Tang, Qingcai Chen, and Xiaolong Wang. 2016. Drug-drug interaction extraction via convolutional neural networks. *Computational and Mathematical Methods in Medicine*, 2016:6918381.

Yinhan Liu, Myle Ott, Naman Goyal, Jingfei Du, Mandar Joshi, Danqi Chen, Omer Levy, Mike Lewis, Luke Zettlemoyer, and Veselin Stoyanov. 2019. Roberta: A robustly optimized BERT pretraining approach. *CoRR*, abs/1907.11692.

Pei-Yau Lung, Zhe He, Tingting Zhao, Disa Yu, and Jinfeng Zhang. 2019. Extracting chemical–protein interactions from literature using sentence structure analysis and feature engineering. *Database*, 2019. Bay138.

Qingliang Miao, Shu Zhang, Bo Zhang, and Hao Yu. 2012. Extracting and visualizing semantic relationships from Chinese biomedical text. In *Proceedings of the 26th Pacific Asia Conference on Language, Information, and Computation*, pages 99–107, Bali, Indonesia. Faculty of Computer Science, Universitas Indonesia.

Mike Mintz, Steven Bills, Rion Snow, and Daniel Jurafsky. 2009. Distant supervision for relation extraction without labeled data. In *Proceedings of the Joint Conference of the 47th Annual Meeting of the ACL and the 4th International Joint Conference on Natural Language Processing of the AFNLP*, pages 1003–1011, Suntec, Singapore. Association for Computational Linguistics.

Sebastian Nagel. 2016. Cc news. `http://commoncrawl.org/2016/10/news-dataset-available/`. Accessed: 2021-03-10.

Yueqiong Ni, Kasper Jensen, Eirini Kouskoumvekaki, and Gianni Panagiotou. 2017. Nutrichem 2.0: exploring the effect of plant-based foods on human health and drug efficacy. *Database: The Journal of Biological Databases and Curation*, 2017(1).

Yifan Peng, Shankai Yan, and Zhiyong Lu. 2019. Transfer learning in biomedical natural language processing: An evaluation of BERT and elmo on ten benchmarking datasets. *CoRR*, abs/1906.05474.

Gorjan Popovski, Barbara Koroušić Seljak, and Tome Eftimov. 2019. FoodBase corpus: a new resource of annotated food entities. *Database*, 2019. Baz121.

Melanie Reiplinger, Michael Wiegand, and Dietrich Klakow. 2014. Relation extraction for the food domain without labeled training data – is distant supervision the best solution? In *Advances in Natural Language Processing*, pages 345–357, Cham. Springer International Publishing.

Sunil Kumar Sahu and Ashish Anand. 2018. Drug-drug interaction extraction from biomedical texts using long short-term memory network. *Journal of Biomedical Informatics*, 86:15–24.

Cong Sun and Zhihao Yang. 2019. Transfer learning in biomedical named entity recognition: An evaluation of BERT in the PharmaCoNER task. In *Proceedings of The 5th Workshop on BioNLP Open Shared Tasks*, pages 100–104, Hong Kong, China. Association for Computational Linguistics.

Trieu H. Trinh and Quoc V. Le. 2018. A simple method for commonsense reasoning. *CoRR*, abs/1806.02847.

Chang Wang and James Fan. 2014. Medical relation extraction with manifold models. In *Proceedings of the 52nd Annual Meeting of the Association for Computational Linguistics (Volume 1: Long Papers)*, pages 828–838, Baltimore, Maryland. Association for Computational Linguistics.

Pengwei Wang, Tianyong Hao, Jun Yan, and Lianwen Jin. 2017. Large-scale extraction of drug-disease pairs from the medical literature. *J. Assoc. Inf. Sci. Technol.*, 68(11):2649–2661.

Karl Weiss, Taghi Khoshgoftaar, and DingDing Wang. 2016. A survey of transfer learning. *Journal of Big Data*, 3(1):9.

Michael Wiegand, Benjamin Roth, and Dietrich Klakow. 2012a. Data-driven knowledge extraction for the food domain. In *Proceedings of KONVENS 2012*, pages 21–29. ÖGAI.

Michael Wiegand, Benjamin Roth, Eva Lasarcyk, Stephanie Köser, and Dietrich Klakow. 2012b. A gold standard for relation extraction in the food domain. In *Proceedings of the Eighth International Conference on Language Resources and Evaluation (LREC'12)*, pages 507–514, Istanbul, Turkey. European Language Resources Association (ELRA).

Ningyu Zhang, Shumin Deng, Zhanlin Sun, Jiaoyan Chen, Wei Zhang, and Huajun Chen. 2019. Transfer learning for relation extraction via relation-gated adversarial learning. *CoRR*, abs/1908.08507.

Shan Zhao, Minghao Hu, Zhiping Cai, and Fang Liu. 2020. Modeling dense cross-modal interactions for joint entity-relation extraction. In *Proceedings of the Twenty-Ninth International Joint Conference on Artificial Intelligence, IJCAI-20*, pages 4032–4038. International Joint Conferences on Artificial Intelligence Organization.

Huiwei Zhou, Zhuang Liu, Shixian Ning, Yunlong Yang, Chengkun Lang, Yingyu Lin, and Kun Ma. 2018. Leveraging prior knowledge for protein–protein interaction extraction with memory network. *Database*, 2018. Bay071.

Yukun Zhu, Ryan Kiros, Richard Zemel, Ruslan Salakhutdinov, Raquel Urtasun, Antonio Torralba, and Sanja Fidler. 2015. Aligning books and movies: Towards story-like visual explanations by watching movies and reading books. In *2015 IEEE International Conference on Computer Vision (ICCV)*, pages 19–27.

Fuzhen Zhuang, Zhiyuan Qi, Keyu Duan, Dongbo Xi, Yongchun Zhu, Hengshu Zhu, Hui Xiong, and Qing He. 2019. A comprehensive survey on transfer learning. *CoRR*, abs/1911.02685.

Are we there yet? Exploring clinical domain knowledge of BERT models

Madhumita Sushil[1] and **Simon Šuster**[2,] and **Walter Daelemans**[1]
[1] Computational Linguistics and Psycholinguistics Research Center (CLiPS), University of Antwerp, Belgium
firstname.lastname@uantwerpen.be

[2] Faculty of Engineering and Information Technology, University of Melbourne
simon.suster@unimelb.edu.au

Abstract

We explore whether state-of-the-art BERT models encode sufficient domain knowledge to correctly perform domain-specific inference. Although BERT implementations such as BioBERT are better at domain-based reasoning than those trained on general-domain corpora, there is still a wide margin compared to human performance on these tasks. To bridge this gap, we explore whether supplementing textual domain knowledge in the medical NLI task: a) by further language model pretraining on the medical domain corpora, b) by means of lexical match algorithms such as the BM25 algorithm, c) by supplementing lexical retrieval with dependency relations, or d) by using a trained retriever module, can push this performance closer to that of humans. We do not find any significant difference between knowledge supplemented classification as opposed to the baseline BERT models, however. This is contrary to the results for evidence retrieval on other tasks such as open domain question answering (QA). By examining the retrieval output, we show that the methods fail due to unreliable knowledge retrieval for complex domain-specific reasoning. We conclude that the task of unsupervised text retrieval to bridge the gap in existing information to facilitate inference is more complex than what the state-of-the-art methods can solve, and warrants extensive research in the future.

1 Introduction

Transformers-based neural architectures (Vaswani et al., 2017) currently hold the state-of-the-art performance on several NLP tasks and domains. In the biomedical domain itself, there exist several versions of transformers-based BERT models (Devlin et al., 2019) that have been shown to be successful. However, an analysis of the availability of medical knowledge to these models is incomplete. To facilitate better understanding, in our research, we analyze a sample of errors made by BioBERT (v1.1) model (Lee et al., 2019a) on a clinical language inference task (Romanov and Shivade, 2018). We find that the errors related to domain knowledge-based reasoning, such as the knowledge of treatments administered for certain diseases, are dominant (40%).

To address this limitation, we analyze a broad range of state-of-the-art methods to integrate medical knowledge in BERT models from textual medical corpora. These methods have previously been shown to excel at evidence retrieval in the generic domain. The goal of our study is to understand whether these methods can be successfully applied for knowledge integration in the more complex setup of finding *missing medical information* for supporting *sentence-pair* inference.

We explore both *implicit* and *explicit* knowledge integration, where *implicit* refers to indirect access to this knowledge by further language model pretraining on medical corpora, and *explicit* knowledge integration refers to the setup where a relevant sentence from external corpora is appended to the premise to support inference. For explicit knowledge integration, as the baseline method, we make use of the traditional best match 25 (BM25) algorithm (Robertson and Zaragoza, 2009) for finding the most relevant sentence in the medical corpora. As a modification of this method, we additionally incorporate syntactic knowledge in the retrieval step. We do so by restricting the retrieved sentence to the one that contains at least one dependency relation between premise and hypothesis medical entities. In the third setup, instead of using BM25 scores and dependency paths, we train an end-to-end model to first find the most relevant text block from Wikipedia for a given instance, and then append it to the instance for classification.

In both knowledge integration setups, we do not see any significant performance difference due to access to additional knowledge. On inspecting the sentences retrieved by the BM25 and dependency

Proceedings of the BioNLP 2021 workshop, pages 41–53
June 11, 2021.

relation-based methods, we find that these methods successfully shortlist sentences related to the topic, but it is difficult to then automatically rank the best candidate among the shortlisted options. This best candidate should fill the information gap between the sentence pairs to enable pairwise inference. We expect to overcome the ranking issue when we instead train an end-to-end model that learns to dynamically retrieve relevant supporting knowledge alongwith pairwise classification, as opposed to static heuristic-based retrieval. However, we find that although the blocks of text retrieved in the end-to-end setup provide medical context, they are often unrelated to the desired information and are insufficient for improving inference.

Although knowledge-integration methods are effective for evidence retrieval in open domain QA (Lee et al., 2019b), where the task is to retrieve a passage that mentions the correct entities, they are insufficient for the more complex task of augmenting missing information for pairwise domain knowledge-based reasoning in an unsupervised setup. Entity span-based supervision simplifies the problem statement in the first case, hence resulting in the documented success. However, the more realistic setup of retrieving the specific context that can fill the information gap between pairs of sentences without supervision is not yet solved.

2 Related work

Since the BERT models were found to be effective for a wide range of NLP tasks (Devlin et al., 2019), several efforts have been extended towards improving them by more efficient training strategies (Liu et al., 2019; Yang et al., 2019b; Sanh et al., 2019; Lan et al., 2019), training them for different domains (Beltagy et al., 2019; Lee et al., 2019a; Lee and Hsiang, 2019; Chalkidis et al., 2020; Gururangan et al., 2020) and languages (Devlin, 2018; de Vries et al., 2019; Le et al., 2020; Martin et al., 2020; Delobelle et al., 2020; Cañete et al., 2020). Within the clinical domain, different models include the BioBERT models pretrained on PubMed abstracts and PMC full-text articles (Lee et al., 2019a), SciBERT trained on scientific text (Beltagy et al., 2019), clinicalBERT models trained on patient notes from the MIMIC-III corpus (Johnson et al., 2016) (sometimes as a continuation of the BioBERT models) (Alsentzer et al., 2019), and BlueBERT models that also use Pubmed abstracts and MIMIC-III patient notes for training (Peng et al., 2019). These models hold promising performance for clinical language processing (Si et al., 2019; Lin et al., 2019) and have become a popular choice for several classification tasks that involve the medical data, spanning tasks such as literature search and question answering for assisting healthcare professionals (Jin et al., 2019; Wang et al., 2020; Möller et al., 2020), as well as patient outcome prediction such as diagnosis prediction (Franz et al., 2020; Rasmy et al., 2020). Despite being a popular choice, little is known about the medical knowledge of these models and their limitations when in-depth domain knowledge is required for correctly solving a task.

Much prior research has explored augmentation of background knowledge in neural models to make them more effective for downstream tasks. Most common approaches include adapting entity embeddings learned by models such as BERT by providing additional knowledge from different ontologies that define relations between entities. This can be done either by using templates to convert the relations to text before finetuning embeddings (Weissenborn et al., 2017; Lauscher et al., 2020; Chen et al., 2020), by combining relational information from knowledge graphs with text embeddings (Mihaylov and Frank, 2018; Chen et al., 2018; Zhang et al., 2019; Yang et al., 2019a; Liu et al., 2020), or by jointly learning knowledge graph and textual embeddings (Peters et al., 2019; Feng et al., 2020). These ontologies are either generic like WordNet (Miller, 1995), ConceptNet (Liu and Singh, 2004), and Wikidata (Vrandečić and Krötzsch, 2014), or more specific to a particular domain like the UMLS (Bodenreider, 2004). An advantage of using ontologies is that the semantics of entities gets encoded in the learned representations, thereby enhancing their effectiveness. However, they are expensive to construct and either are incomplete, or do not exist for specialized domains. Methods that make use of textual corpora for background knowledge integration are therefore more easily transferable to other domains. Talmor et al. (2020) have shown earlier that having explicit access to external information can often improve reasoning skills of the state-of-the-art models, which we investigate further.

Use of TF-IDF (Ullman, 2011) and BM25 scores has been frequently explored for evidence retrieval from Wikipedia for open domain QA (Chen et al., 2017; Wang et al., 2018; Glass et al., 2020). Another popular approach includes representation similarity-based evidence retrieval (Lee et al., 2018; Das et al., 2019). Recently, joint training of retriever for span identification and pretraining language models have also been analyzed by Hu et al. (2019); Lee et al. (2019b); Guu et al. (2020). Although the methods extensively explore QA, this line of work has not been explored much for language inference, especially in specialized domains.

Existing studies for augmenting medical knowledge for clinical language inference are limited to the use of UMLS knowledge graph embeddings (Sharma et al., 2019), interaction weighting between premise and hypothesis based on distance in the UMLS (Chopra et al., 2019), augmenting clinical concept definitions during representation learning (Lu et al., 2019) and adding domain knowledge by means of pretraining existing models further on different in-domain corpora and closely related tasks (Romanov and Shivade, 2018; Lee et al., 2019a; Alsentzer et al., 2019; Chopra et al., 2019). The closest work to ours is the contemporary work by He et al. (2020) that shows improvements when knowledge from Wikipedia is implicitly integrated by training BERT masked language models to predict disease names and their aspects (such as symptoms, treatments) given the corresponding context. In our work, we instead explore whether we can augment domain knowledge by dynamically fetching relevant context in an unsupervised manner to improve medical language inference.

3 Medical language inference

In medical language inference, given a pair of sentences, the goal is to describe a logical relation between them. We make use of the MedNLI dataset (Romanov and Shivade, 2018), where the premise is a sentence borrowed from patient notes in the MIMIC-III dataset (Johnson et al., 2016), and the hypothesis is written by medical experts such that the premise either entails or contradicts the hypothesis, or their relation cannot be established (neutral). Entailment refers to whether the meaning of the second sentence, also known as the 'hypothesis', is already contained in the first sentence called the 'premise'. We explore whether the BioBERT v(1.1) model encodes sufficient medical knowledge for this task. In the same manner as Peng et al. (2019), we model this task as a sentence pair classification task, where the final pooled BERT [CLS] representations of the premise and the hypothesis are processed through a dense neural layer to classify the correct class. We then perform manual analysis on a subset of 50 incorrectly classified instances in the development set to understand the type of errors made by the model. We eliminate ambiguity in the cause of errors by using an adversarial evaluation, where we modify an instance according to a potential cause of error, and monitor whether the output changes accordingly. In this manner, we obtain the distribution of errors presented in Table 2 and discussed in Section 5.1.

4 Medical knowledge augmentation

4.1 External medical corpora

Different versions of BERT that exist for biomedical tasks are either trained on journal abstracts and articles, or on patient notes. These articles and notes are written by and for an audience with an advanced level of domain knowledge. Fundamental domain-specific information, such as an understanding of domain terminology, commonly accepted clinical practices for specific medical conditions, human physiology and anatomy, etc. is often also required for clinical language understanding. We hypothesize that access to such fundamental domain knowledge during model training would complement training on more advanced information. To explore this, we create two corpora — one containing only the medical subset of Wikipedia (Wikimed), and one with contents of a popular medical textbook (Medbook). The Wikimed subset is parsed from the HTML sources of the medical Wikipedia dataset used in the Android app by the Kiwix team[1]. The medical subset of Wikipedia contains about 40 million tokens, and the medical textbook corpus contains nearly 3.6 million tokens.

4.2 Implicit knowledge integration

Starting from an existing BioBERT checkpoint that is already pretrained on a combination of Google books, Wikipedia, biomedical abstracts and journal articles (Lee et al., 2019a), we continue to train BERT language models on the Medbook and the Wikimed corpora. Our goal is to explore whether further training on corpora that contain fundamental domain knowledge can implicitly improve medical knowledge-based reasoning in the medical language inference task. Since Wikimed is the

[1]https://play.google.com/store/apps/details?id=org.kiwix.kiwixcustomwikimed&hl=en_US&gl=US

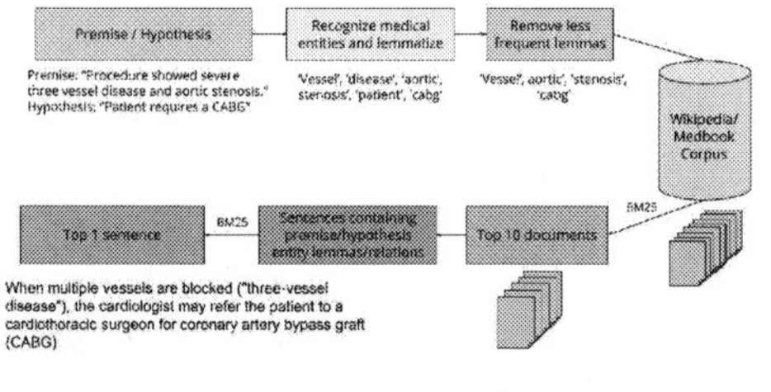

(a) Lexical knowledge retrieval using BM25 score between a query formulated from premise-hypothesis pair and sentences in the external corpora. These sentences are restricted to either those that mention a premise *and* a hypothesis medical entity term, or hold a dependency relation between them.

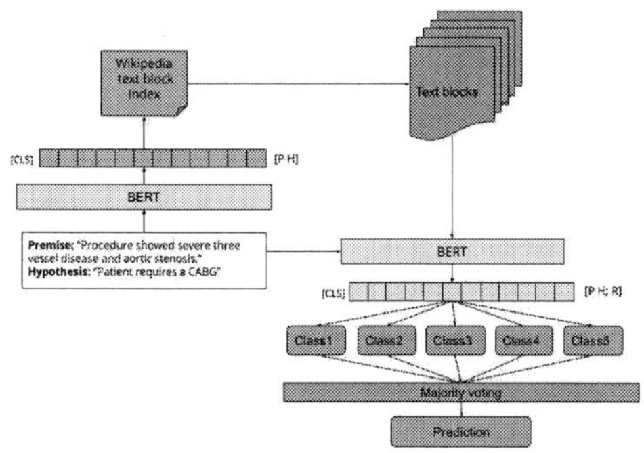

(b) Relevant knowledge retrieval in an end-to-end manner by training weights that compute similarity between sentences in an external corpus and a premise(P)-hypothesis(H) query during classification.

Figure 1: Explicit domain knowledge integration for the MedNLI task.

medical-only subset of Wikipedia, it was also included in the first phase of training of BERT models. We do not expect to see a significant difference in the classification performance here due to this reason. However, since the Medbook corpus is quite different from other corpora used earlier, we expect bigger differences in classification results.

4.3 Explicit knowledge integration

We explore methods to explicitly augment medical knowledge to the instances in the MedNLI dataset by retrieving and appending relevant text blocks from either the Wikimed corpus or the Medbook corpus before processing it through our BERT models for finetuning, as described next. We illustrate the methods pipeline in Figure 1.

4.3.1 Lexical retrieval

We first explore the use of TF-IDF based techniques for retrieving evidence from external textual corpora to support inference. Although these methods are fairly simple, they have been shown to be effective for several open domain QA tasks (Lee et al., 2019b). Our goal is to investigate whether these simple methods are also effective at more complex information retrieval in our setup.

To this end, we construct a query from premise and hypothesis by retaining only the lemmas that are a part of infrequent medical entities, and then use the best match 25 (BM25) algorithm (Robertson and Zaragoza, 2009) to find the most relevant sentences. As the first step, we recognize premise and hypothesis medical entities with the help of Scispacy (Neumann et al., 2019). We lemmatize these entities and retain only those lemmas that occur less than a thousand times in the external corpus[2]. These lemmas jointly form the query. We first rank the documents in the external corpora according to their BM25 scores to retain the top 10 documents. The query is then used again to find the best matching sentences from these documents.

Due to the manner in which the MedNLI data has been annotated, premise is longer and more varied than the hypothesis. Hence, premise entities often dominate the BM25 retrieval at the cost of hypothesis entities. To overcome this, we prune the retrieved sentences if they do not mention at least one premise and one hypothesis entity lemma.

The highest ranking sentence retrieved in this manner is then appended[3] to the premise before classification. If none of the sentences satisfy either the constraint or the threshold score, then the use of explicit knowledge is skipped.

4.3.2 Lexical and syntactic retrieval

In our previous setup, we add an entity-presence constraint to ensure that the retrieved sentence is about both the premise and the hypothesis. In order to ensure that the retrieved knowledge also establishes an explicit relation between the two, we modify the previous approach to rank sentences based on dependency paths between premise and hypothesis lemmas. In this setup, we find the top documents in the same manner as earlier. Once the top documents are found, we restrict to the set of sentences in these documents that have a dependency relation between a premise and a hypothesis lemma. Once we have established the set of sentences that hold this relation, we rank them either

[2]The threshold was decided based on preliminary results on the development set, where retaining less frequent lemmas provides more specific matches.

[3]Separated by a space.

using the minimum dependency path length, or using the BM25 score between the query and a sentence. The sentence with the highest score above the threshold is then appended to the instance in the same manner as described earlier.

4.3.3 Joint retrieval and classification model

By using lexical and syntactic approaches that we have discussed earlier, we ensure that the candidate and the retrieved sentences would be related to both the premise and the hypothesis. However, when we are confronted with a high number of relevant candidate sentences, shortlisting one sentences becomes challenging. Adding multiple sentences is also infeasible due to the limited input sequence length in BERT models. In order to overcome this challenge, in our third setup, we instead train an end-to-end model, where the weights of the retriever are updated along with classification. Hence, the retriever learns to select the sentence that provides information that can improve classification. This approach has been previously shown to be quite successful in open domain QA via span identification (Lee et al., 2019b) and in language model pretraining (Guu et al., 2020), since it provides access to a wider evidence space compared to the limited number of retrieved blocks when using lexical approaches. However, the use of such an end-to-end retriever has not been explored for augmenting knowledge from textual corpora to support reasoning in NLI tasks. Since we do not have data annotated specifically for retrieval of supporting evidence for NLI tasks, training the retriever becomes much more complex compared to span identification. However, given the success of the end-to-end approaches earlier, we are interested in investigating its feasibility for our setup and we build upon existing methods for this.

Retriever pretraining: We reuse the pretrained retrieval model shared by Lee et al. (2019b), trained in an inverse cloze task (ICT) setup on complete Wikipedia, for our experiments. In this setup, a sentence in Wikipedia is treated as the query, and the retriever is trained to retrieve its context[4] in the original text. This retrieval is performed by computing a weighted dot product between the pooled BERT [CLS] embeddings of the query and the text block. In 10% of the cases, the query is not removed from the context to ensure that the model learns to retrieve lexical as well as semantic matches. Although it is trained on entire Wikipedia instead of only a subset, we reuse it due to resource constraints for retraining the retriever. Since the medical portion of Wikipedia is only a subset of this data, we expect to still be able to retrieve the sentences relevant for the MedNLI task.

End-to-end-classification: In an end-to-end setup, the retriever module first returns the k[5] most similar blocks of text given a BERT-encoded premise and hypothesis pair, in the same manner as described earlier. We add these k retrieved blocks to the input along with the premise and the hypothesis to obtain k inputs corresponding to each instance. We then encode these inputs with BERT to obtain k different [CLS] representations. All of these k [CLS] representations are then individually used for classification by adding a dense layer on the top in the finetuning phase. In this manner, we obtain k different outputs for a given instance. We then aggregate these k outputs together by retaining the most frequent output among the k options. We also experimented with average pooling and selecting the most peaked softmax output distribution, but majority pooling provided more promising results on the development set.

Classification loss: We use the categorical cross entropy loss (Murphy, 2012). The gradients are backpropagated jointly to both the classifier and the weights used to compute the similarity between the query and the blocks of Wikipedia text.

Retriever loss: In the span identification setup developed by Lee et al. (2019b), mention of the correct entity in the text provides the retriever with an explicit feedback. This makes their training easier compared to our setup where we do not have this supervised signal. To make the training more feasible, we experiment with an additional retrieval loss. This loss quantifies the difference between the model performance with and without the retrieved text block, and uses this difference to improve the retriever. The objective of this loss is to reward the model when it is better if a retrieved text block is used as opposed to when only the premise and the hypothesis are used for inference. We define this loss in terms of pairwise retrieval loss, i.e.,

$$R = max(0, m - (L_{(P,H)} - L_{(P,H,R)})),$$

where R is the retrieval loss, $L_{(P,H)}$ is the categorical cross entropy loss without using the retrieved

[4] Blocks of at most 288 wordpiece tokens (Wu et al., 2016)

[5] We use $k = 5$ in our experiments

text block, and $L_{(P,H,R)}$ is the categorical cross entropy loss after adding the retrieved text block to the given instance, and m is the margin value that we treat as a hyperparameter. We use $m = 0.1$ based on the results on the development set. To explain this loss, we consider three different cases:

1. The model performs equivalently with and without the retrieved text block: In this case, the model ignores the retriever and optimizes for classification without it. This is undesirable, and we set the retriever loss to the margin value, which refers to the minimum desired difference between the two sets of losses.

2. The model is worse after adding the retrieved text block: This behavior is again undesirable since the goal of retrieval is to improve the model. Hence, along with the margin, we also add the difference between the two losses to compute the retrieval loss.

3. The model improves after adding the retrieved text block: If the model becomes better due to retrieval, it could either be better by chance (when the difference is lower than the minimum margin), or the difference could be substantial. In the first case, we quantify the retrieval loss as the margin value. The latter behavior is the desired behavior of the model, and we set the retrieval loss to be zero.

Here, the final loss function is computed as the sum of the classification loss and the retrieval loss.

5 Results and Discussion

5.1 Availability of domain knowledge

In the top section of Table 1, we present the results when we finetune BERT models for medical language inference. Here we can see that the BERT model which has been trained on in-domain Pubmed abstracts for the largest number of optimization steps is consistently the best on both development and test sets. As expected based on prior research, all other models trained on in-domain data are also significantly better than the BERT models that are not trained on in-domain data.

We investigate the errors made by the best model, BioBert (v1.1). As discussed in Section 3, in Table 2, we present the distribution of the first 50 errors made on the development set of the MedNLI dataset. Examples of these errors are illustrated in Table 3. Although we present the distribution of errors for one specific run here, we also analyzed this distribution across 3 different runs of the model. We found that the average pairwise Cohen's kappa agreement (McHugh, 2012) between the predictions on the development set across 3 different runs is 0.9, and the distribution of errors across these runs is comparable. In Table 2, we can see that 40% of the errors happen due to insufficient domain information. Some of these errors happen because of missing factual domain knowledge, some lack advance reasoning based on factual domain knowledge, and some are incorrect potentially because of model biases due to limited size of the training dataset, such as assumption that a certain treatment is always administered for a specific condition, although the treatment might be more diverse. This highlights the potential to improve the BioBERT model by providing access to additional fundamental domain information.

Other dominant category of errors are related to spurious correlations, numeric inference, negation, and temporal reasoning. These categories are important for understanding patient condition in medical notes, since test results are often expressed in a numeric manner, patient conditions are often hedged and negated, and patient information is usually longitudinal in nature. We limit the focus of this work to the more frequent error category of integrating domain information.

5.2 Domain knowledge integration

In Table 1, we see marginal improvements on the test set between the BioBERT (v1.1) models with and without additional domain knowledge — both when the integration is done implicitly via additional language model pretraining, and when relevant sentences are retrieved using lexical and syntactic methods. Knowledge integration from the Medbook corpus — both implicit and explicit, does not show any improvement in the results. Despite marginal improvements using the Wikimed corpus, a lack of consistent pattern across both development and test sets suggests a random effect rather than significant differences. When we train an end-to-end retrieval model instead of further language modeling or pre-selecting the most relevant sentence, we again see a marginal improvement on the test set. However, this improvement is again not visible on the development set. Furthermore, we see that the pairwise loss for more aggressive

Model	MedNLI (% Acc.)	
	Dev	Test
BERT-base-uncased	82.1	77.8
BERT-base-cased	79.9	78.8
BERT-base-cased + PMC + PubMed (BioBERT v1.0)	84.3	82.5
BERT-base-cased + Pubmed 1M (BioBERT v1.1)	84.8	82.9
SciBERT-base-uncased (SciBERT vocab)	81.5	82.2
He et al. (2020): BioBERT v1.1 + disease	NA	82.2
Sharma et al. (2019)	NA	79.0
BERT-base-cased + Pubmed 1M (BioBERT v1.1)	84.8	82.9
BioBERT v1.1 + Wikimed MLM	84.2	**83.3**
BioBERT v1.1 + Medbook MLM	83.2	80.1
BioBERT v1.1 + Wikimed (lexical)	84.3	**83.2**
BioBERT v1.1 + Medbook (lexical)	83.8	82.6
BioBERT v1.1 + Wikimed (lexical+syntactic)	83.9	**83.1**
BioBERT v1.1 + Medbook (lexical+syntactic)	83.8	82.5
BERT-base-uncased (Wikipedia+BooksCorpus)	82.1	77.8
BERT-base-uncased + trained Wiki retriever	79.4	78.5
BERT-base-uncased + trained Wiki retriever + retrieval loss	79.1	77.9

Table 1: Classification accuracy of BERT models and explicit and implicit domain knowledge integration methods on MedNLI development and test sets. MLM here refers to masked language modeling.

Error type	Count
Insufficient domain knowledge	20
Spurious correlations / dataset bias	6
Difficult instance	5
Incorrect numeric inference	4
Incorrect negation	3
Incorrect tense resolution	2
Incorrect temporal sequence inference	2
Lexical (P,H) overlap trick	2
Modifier ignored	2
Incorrect abbreviation understanding	2
Insufficient commonsense knowledge	1

Table 2: Analysis of the first 50 errors of the BioBERT (v1.1) model on the MedNLI development set.

retriever training along with the classification cross-entropy loss does not have any significant impact. Despite this additional signal, the classifier continues to learn the task by ignoring the retrieved context, thus indicating that the penalty for incorrect retrieval is still not aggressive enough.

Our joint models use the complete Wikipedia as the source of knowledge, and the improvement patterns here are consistent with using the Wikimed corpus both implicitly and explicitly, but contrary to using the Medbook corpus. This suggests that Wikipedia, both complete and the medical-only subset, functions as a better source of information for the MedNLI task as compared to the medical textbook that contains more fundamental domain information. We believe that the difference in results of the two corpora emerges from a difference in their sizes, since the medical subset of Wikipedia is 10 times in size compared to the textbook corpus. We could not scale the Medbook corpus to larger sizes due to copyright limitations.

When we analyze the retrieved text blocks for one example in the development set and compare it to the gold standard retrieval by humans (presented in Table 4), we see that none of the retrieval algorithms are capable of finding the desired missing information to improve semantic inference. Although the 'lexical + syntactic' retriever finds a sentence related to the topic as well as to the premise and the hypothesis, it doesn't bridge the knowledge gap for correct inference. Moreover, the end-to-end model with a trained retriever retrieves text block that is unrelated to the topic, although in the medical genre.

Hence, we find that none of the explored methods provide better access to medical information for domain knowledge-based reasoning, although the desired factual information is present in these external corpora. One reason why we do not see further improvements on the BioBERT (v1.1) model

Error type	Example
Insufficient domain knowledge	**P:** ... she was treated with **Benadryl** ... **H:** Patient has had an **allergic reaction** ~~Entailment~~ Neutral
Spurious correlations / dataset bias	**P:** She spoke with her **oncology team** ... **H:** The patient has **cancer**. ~~Neutral~~ Entailment
Incorrect numeric inference	**P:** ... an **ejection fraction of 69%** with normal wall motion. **H:** patient has **normal cardiac output** ~~Entailment~~ Contradiction
Incorrect negation resolution	**P:** ... **no** identified sepsis risk factors. **H:** ... has **multiple** risk factors for sepsis ~~Contradiction~~ Entailment
Incorrect tense resolution	**P:** ... he **had a CT of the chest** and CTA of his coronary arteries ... **H:** patient **will** go for **coronary angiography** ~~Neutral~~ Entailment
Incorrect temporal inference	**P:** ... biopsy ... showed signs of rejection ... **subsequently did well**. **H:** The patient **had transplant failure** ~~Contradiction~~ Entailment
Lexical (P, H) overlap trick	**P: Pt denies** any **recent** chills ... **H:** The **patient denies recent** illness ~~Neutral~~ Entailment
Modifier ignored	**P:** Left common femoral dorsalis **pedis** bypass graft. **H:** Patient has **CAD** ~~Neutral~~ Entailment
Incorrect abbreviation understanding	**P:** Her ... **PO** intake have been normal. **H:** She has been **NPO** since midnigh ~~Contradiction~~ Neutral
Insufficient commonsense knowledge	**P:** ... status post high speed motor **vehicle crash** ... **H:** Patient has recent **trauma** ~~Entailment~~ Neutral

Table 3: One example of each category of errors made by the BioBERT (v1.1) model on the MedNLI development set. ~~a~~ b refers to the fact that class a is the gold class, but the model predicts class b instead.

(that is a very strong baseline), despite the success of these methods in other tasks and domains, could be the complexity of the research question. Retrieval of relevant information for language inference demands a delicate balance between selecting a sentence that provides sufficient supporting information related to the given topic and instance to improve inference, and yet that is neither redundant nor superfluous. As we show in our results, in a limited computation setting as ours, current state-of-the-art methods are not capable of striking this balance in unsupervised setups and result in unreliable knowledge augmentation. He et al. (2020) also report similar results on the same task using the same BioBERT model. These results suggest that we either need more computation power to train these models for longer time to enable convergence, or we need to create large annotated corpora for retrieving missing facts to enable better performance of these algorithms with limited computation power. We need to direct our efforts towards investigating advanced evidence retrieval and knowledge integration setups such as this to improve knowledge-based reasoning of the current state-of-the-art models.

Method	Text
Example	**P**: Infusion stopped and she was treated with Benadryl 50 mg x 1, prednisone 40 mg x 1, ativan 1 mg. **H**: Patient has had an allergic reaction
Gold retrieval	Benadryl is a brand name for a number of different antihistamine medications used to stop allergies, including diphenhydramine, acrivastine and cetirizine.
Lexical retrieval	None
Lexical + syntactic retrieval	Prednisone is used for many different autoimmune diseases and inflammatory conditions, including asthma, COPD, CIDP, rheumatic disorders, allergic disorders, ulcerative colitis and Crohn's disease, granulomatosis with polyangiitis, adrenocortical insufficiency, hypercalcemia due to cancer, thyroiditis, laryngitis, severe tuberculosis, hives, lipid pneumonitis, pericarditis, multiple sclerosis, nephrotic syndrome, sarcoidosis, to relieve the effects of shingles, lupus, myasthenia gravis, poison oak exposure, Ménière's disease, autoimmune hepatitis, giant-cell arteritis, the Herxheimer reaction that is common during the treatment of syphilis, Duchenne muscular dystrophy, uveitis, and as part of a drug regimen to prevent rejection after organ transplant.
Trained Wiki retriever + retrieval loss	Gemeprost (16, 16-dimethyl-trans-delta2 PGE methyl ester) is an analogue of prostaglandin E. It is used as a treatment for obstetric bleeding. It is used with mifepristone to terminate pregnancy up to 24 weeks gestation. Vaginal bleeding, cramps, nausea, vomiting, loose stools or diarrhea, headache, muscle weakness; dizziness; flushing; chills; backache; dyspnoea; chest pain; palpitations and mild pyrexia. Rare: Uterine rupture, severe hypotension, coronary spasms with subsequent myocardial infarctions. **...**

Table 4: Text blocks retrieved by different methods from the (medical) Wikipedia corpus for one example in the development set that requires further domain knowledge for correct inference. Gold retrieval mentioned here is a manually retrieved sentence from Wikipedia, in presence of which the model corrects its output.

6 Conclusions and Future Work

On investigating the error categories of BioBERT (v1.1) models on the clinical language understanding task, we find that despite having a strong performance, the models still make several mistakes on examples that require medical domain knowledge. To this end, we explored multiple methods to improve access of these models to medical domain knowledge by implicit and explicit knowledge retrieval and augmentation. However, we see that these extensions do not show significant improvement on the test sets. We conclude that state-of-the-art solutions lead to unreliable knowledge augmentation for language inference, as is shown by a detailed analysis in our work. Future research should concentrate efforts towards developing methods to augment fundamental domain knowledge from textual corpora to solve the problem of advanced knowledge-based reasoning in these domains.

Acknowledgements

This research was carried out within the Accumulate strategic basic research project, funded by the government agency Flanders Innovation & Entrepreneurship (VLAIO) [grant number 150056]. It also received funding from the Flemish Government (AI Research Program). This research was conducted following an internship at Google Research in Zürich. The experience gained during the internship was instrumental in the research. We would like to thank everyone in the team, and particularly André Susano Pinto for several discussions related to BERT and Tensorflow, for exchange of ideas, and for feedback on the draft. We would like to additionally thank all the anonymous reviewers whose useful comments have ensured a better version of the paper.

References

Emily Alsentzer, John Murphy, William Boag, Wei-Hung Weng, Di Jindi, Tristan Naumann, and Matthew McDermott. 2019. Publicly available clinical BERT embeddings. In *Proceedings of the 2nd Clinical Natural Language Processing Workshop*, pages 72–78, Minneapolis, Minnesota, USA. Association for Computational Linguistics.

Iz Beltagy, Kyle Lo, and Arman Cohan. 2019. SciBERT: A pretrained language model for scientific text. In *Proceedings of the 2019 Conference on Empirical Methods in Natural Language Processing and the 9th International Joint Conference on Natural Language Processing (EMNLP-IJCNLP)*, pages 3615–3620, Hong Kong, China. Association for Computational Linguistics.

Olivier Bodenreider. 2004. The unified medical language system (umls): integrating biomedical terminology. *Nucleic acids research*, 32(suppl_1):D267–D270.

José Cañete, Gabriel Chaperon, Rodrigo Fuentes, and Jorge Pérez. 2020. Spanish pre-trained bert model and evaluation data. In *to appear in PML4DC at ICLR 2020*.

Ilias Chalkidis, Manos Fergadiotis, Prodromos Malakasiotis, Nikolaos Aletras, and Ion Androutsopoulos. 2020. Legal-bert: The muppets straight out of law school. *arXiv preprint arXiv:2010.02559*.

Danqi Chen, Adam Fisch, Jason Weston, and Antoine Bordes. 2017. Reading Wikipedia to answer open-domain questions. In *Proceedings of the 55th Annual Meeting of the Association for Computational Linguistics (Volume 1: Long Papers)*, pages 1870–1879, Vancouver, Canada. Association for Computational Linguistics.

Mingda Chen, Zewei Chu, Karl Stratos, and Kevin Gimpel. 2020. Mining knowledge for natural language inference from wikipedia categories. *arXiv preprint arXiv:2010.01239*.

Qian Chen, Xiaodan Zhu, Zhen-Hua Ling, Diana Inkpen, and Si Wei. 2018. Neural natural language inference models enhanced with external knowledge. In *Proceedings of the 56th Annual Meeting of the Association for Computational Linguistics (Volume 1: Long Papers)*, pages 2406–2417, Melbourne, Australia. Association for Computational Linguistics.

Sahil Chopra, Ankita Gupta, and Anupama Kaushik. 2019. MSIT_SRIB at MEDIQA 2019: Knowledge directed multi-task framework for natural language inference in clinical domain. In *Proceedings of the 18th BioNLP Workshop and Shared Task*, pages 488–492, Florence, Italy. Association for Computational Linguistics.

Rajarshi Das, Shehzaad Dhuliawala, Manzil Zaheer, and Andrew McCallum. 2019. Multi-step retriever-reader interaction for scalable open-domain question answering. In *International Conference on Learning Representations*.

Wietse de Vries, Andreas van Cranenburgh, Arianna Bisazza, Tommaso Caselli, Gertjan van Noord, and Malvina Nissim. 2019. Bertje: A dutch bert model. *arXiv preprint arXiv:1912.09582*.

Pieter Delobelle, Thomas Winters, and Bettina Berendt. 2020. Robbert: a dutch roberta-based language model. *arXiv preprint arXiv:2001.06286*.

Jacob Devlin. 2018. Multilingual bert readme document.

Jacob Devlin, Ming-Wei Chang, Kenton Lee, and Kristina Toutanova. 2019. BERT: Pre-training of deep bidirectional transformers for language understanding. In *Proceedings of the 2019 Conference of the North American Chapter of the Association for Computational Linguistics: Human Language Technologies, Volume 1 (Long and Short Papers)*, pages 4171–4186, Minneapolis, Minnesota. Association for Computational Linguistics.

Yanlin Feng, Xinyue Chen, Bill Yuchen Lin, Peifeng Wang, Jun Yan, and Xiang Ren. 2020. Scalable multi-hop relational reasoning for knowledge-aware question answering.

Leopold Franz, Yash Raj Shrestha, and Bibek Paudel. 2020. A deep learning pipeline for patient diagnosis prediction using electronic health records. *arXiv preprint arXiv:2006.16926*.

Michael Glass, Alfio Gliozzo, Rishav Chakravarti, Anthony Ferritto, Lin Pan, G P Shrivatsa Bhargav, Dinesh Garg, and Avi Sil. 2020. Span selection pre-training for question answering. In *Proceedings of the 58th Annual Meeting of the Association for Computational Linguistics*, pages 2773–2782, Online. Association for Computational Linguistics.

Suchin Gururangan, Ana Marasović, Swabha Swayamdipta, Kyle Lo, Iz Beltagy, Doug Downey, and Noah A. Smith. 2020. Don't stop pretraining: Adapt language models to domains and tasks. In *Proceedings of the 58th Annual Meeting of the Association for Computational Linguistics*, pages 8342–8360, Online. Association for Computational Linguistics.

Kelvin Guu, Kenton Lee, Zora Tung, Panupong Pasupat, and Ming-Wei Chang. 2020. Realm: Retrieval-augmented language model pre-training. *arXiv preprint arXiv:2002.08909*.

Yun He, Ziwei Zhu, Yin Zhang, Qin Chen, and James Caverlee. 2020. Infusing disease knowledge into bert for health question answering, medical inference and disease name recognition. *arXiv preprint arXiv:2010.03746*.

Minghao Hu, Yuxing Peng, Zhen Huang, and Dongsheng Li. 2019. Retrieve, read, rerank: Towards end-to-end multi-document reading comprehension.

In *Proceedings of the 57th Annual Meeting of the Association for Computational Linguistics*, pages 2285–2295, Florence, Italy. Association for Computational Linguistics.

Qiao Jin, Bhuwan Dhingra, Zhengping Liu, William Cohen, and Xinghua Lu. 2019. PubMedQA: A dataset for biomedical research question answering. In *Proceedings of the 2019 Conference on Empirical Methods in Natural Language Processing and the 9th International Joint Conference on Natural Language Processing (EMNLP-IJCNLP)*, pages 2567–2577, Hong Kong, China. Association for Computational Linguistics.

Alistair EW Johnson, Tom J Pollard, Lu Shen, H Lehman Li-Wei, Mengling Feng, Mohammad Ghassemi, Benjamin Moody, Peter Szolovits, Leo Anthony Celi, and Roger G Mark. 2016. Mimic-iii, a freely accessible critical care database. *Scientific data*, 3(1):1–9.

Zhenzhong Lan, Mingda Chen, Sebastian Goodman, Kevin Gimpel, Piyush Sharma, and Radu Soricut. 2019. Albert: A lite bert for self-supervised learning of language representations. *arXiv preprint arXiv:1909.11942*.

Anne Lauscher, Olga Majewska, Leonardo FR Ribeiro, Iryna Gurevych, Nikolai Rozanov, and Goran Glavaš. 2020. Common sense or world knowledge? investigating adapter-based knowledge injection into pretrained transformers. *arXiv preprint arXiv:2005.11787*.

Hang Le, Loïc Vial, Jibril Frej, Vincent Segonne, Maximin Coavoux, Benjamin Lecouteux, Alexandre Allauzen, Benoit Crabbé, Laurent Besacier, and Didier Schwab. 2020. FlauBERT: Unsupervised language model pre-training for French. In *Proceedings of the 12th Language Resources and Evaluation Conference*, pages 2479–2490, Marseille, France. European Language Resources Association.

Jieh-Sheng Lee and Jieh Hsiang. 2019. Patentbert: Patent classification with fine-tuning a pre-trained bert model. *arXiv preprint arXiv:1906.02124*.

Jinhyuk Lee, Wonjin Yoon, Sungdong Kim, Donghyeon Kim, Sunkyu Kim, Chan Ho So, and Jaewoo Kang. 2019a. BioBERT: a pre-trained biomedical language representation model for biomedical text mining. *Bioinformatics*, 36(4):1234–1240.

Jinhyuk Lee, Seongjun Yun, Hyunjae Kim, Miyoung Ko, and Jaewoo Kang. 2018. Ranking paragraphs for improving answer recall in open-domain question answering. In *Proceedings of the 2018 Conference on Empirical Methods in Natural Language Processing*, pages 565–569, Brussels, Belgium. Association for Computational Linguistics.

Kenton Lee, Ming-Wei Chang, and Kristina Toutanova. 2019b. Latent retrieval for weakly supervised open domain question answering. In *Proceedings of the 57th Annual Meeting of the Association for Computational Linguistics*, pages 6086–6096, Florence, Italy. Association for Computational Linguistics.

Chen Lin, Timothy Miller, Dmitriy Dligach, Steven Bethard, and Guergana Savova. 2019. A bert-based universal model for both within-and cross-sentence clinical temporal relation extraction. In *Proceedings of the 2nd Clinical Natural Language Processing Workshop*, pages 65–71.

Hugo Liu and Push Singh. 2004. Conceptnet—a practical commonsense reasoning tool-kit. *BT technology journal*, 22(4):211–226.

Weijie Liu, Peng Zhou, Zhe Zhao, Zhiruo Wang, Qi Ju, Haotang Deng, and Ping Wang. 2020. K-BERT: enabling language representation with knowledge graph. In *The Thirty-Fourth AAAI Conference on Artificial Intelligence, AAAI 2020, The Thirty-Second Innovative Applications of Artificial Intelligence Conference, IAAI 2020, The Tenth AAAI Symposium on Educational Advances in Artificial Intelligence, EAAI 2020, New York, NY, USA, February 7-12, 2020*, pages 2901–2908. AAAI Press.

Yinhan Liu, Myle Ott, Naman Goyal, Jingfei Du, Mandar Joshi, Danqi Chen, Omer Levy, Mike Lewis, Luke Zettlemoyer, and Veselin Stoyanov. 2019. Roberta: A robustly optimized bert pretraining approach. *arXiv preprint arXiv:1907.11692*.

M. Lu, Y. Fang, F. Yan, and M. Li. 2019. Incorporating domain knowledge into natural language inference on clinical texts. *IEEE Access*, 7:57623–57632.

Louis Martin, Benjamin Muller, Pedro Javier Ortiz Suárez, Yoann Dupont, Laurent Romary, Éric de la Clergerie, Djamé Seddah, and Benoît Sagot. 2020. CamemBERT: a tasty French language model. In *Proceedings of the 58th Annual Meeting of the Association for Computational Linguistics*, pages 7203–7219, Online. Association for Computational Linguistics.

Mary L McHugh. 2012. Interrater reliability: the kappa statistic. *Biochemia medica: Biochemia medica*, 22(3):276–282.

Todor Mihaylov and Anette Frank. 2018. Knowledgeable reader: Enhancing cloze-style reading comprehension with external commonsense knowledge. In *Proceedings of the 56th Annual Meeting of the Association for Computational Linguistics (Volume 1: Long Papers)*, pages 821–832, Melbourne, Australia. Association for Computational Linguistics.

George A Miller. 1995. Wordnet: a lexical database for english. *Communications of the ACM*, 38(11):39–41.

Timo Möller, Anthony Reina, Raghavan Jayakumar, and Lawrence Livermore. 2020. Covid-qa: A question & answering dataset for covid-19.

Kevin P Murphy. 2012. *Machine learning: a probabilistic perspective*. MIT press.

Mark Neumann, Daniel King, Iz Beltagy, and Waleed Ammar. 2019. ScispaCy: Fast and Robust Models for Biomedical Natural Language Processing. In *Proceedings of the 18th BioNLP Workshop and Shared Task*, pages 319–327, Florence, Italy. Association for Computational Linguistics.

Yifan Peng, Shankai Yan, and Zhiyong Lu. 2019. Transfer learning in biomedical natural language processing: An evaluation of bert and elmo on ten benchmarking datasets. In *Proceedings of the 2019 Workshop on Biomedical Natural Language Processing (BioNLP 2019)*, pages 58–65.

Matthew E. Peters, Mark Neumann, Robert Logan, Roy Schwartz, Vidur Joshi, Sameer Singh, and Noah A. Smith. 2019. Knowledge enhanced contextual word representations. In *Proceedings of the 2019 Conference on Empirical Methods in Natural Language Processing and the 9th International Joint Conference on Natural Language Processing (EMNLP-IJCNLP)*, pages 43–54, Hong Kong, China. Association for Computational Linguistics.

Laila Rasmy, Yang Xiang, Ziqian Xie, Cui Tao, and Degui Zhi. 2020. Med-bert: pre-trained contextualized embeddings on large-scale structured electronic health records for disease prediction. *arXiv preprint arXiv:2005.12833*.

Stephen Robertson and Hugo Zaragoza. 2009. *The probabilistic relevance framework: BM25 and beyond*. Now Publishers Inc.

Alexey Romanov and Chaitanya Shivade. 2018. Lessons from natural language inference in the clinical domain. In *Proceedings of the 2018 Conference on Empirical Methods in Natural Language Processing*, pages 1586–1596, Brussels, Belgium. Association for Computational Linguistics.

Victor Sanh, Lysandre Debut, Julien Chaumond, and Thomas Wolf. 2019. Distilbert, a distilled version of bert: smaller, faster, cheaper and lighter. *arXiv preprint arXiv:1910.01108*.

Soumya Sharma, Bishal Santra, Abhik Jana, Santosh Tokala, Niloy Ganguly, and Pawan Goyal. 2019. Incorporating domain knowledge into medical NLI using knowledge graphs. In *Proceedings of the 2019 Conference on Empirical Methods in Natural Language Processing and the 9th International Joint Conference on Natural Language Processing (EMNLP-IJCNLP)*, pages 6092–6097, Hong Kong, China. Association for Computational Linguistics.

Yuqi Si, Jingqi Wang, Hua Xu, and Kirk Roberts. 2019. Enhancing clinical concept extraction with contextual embeddings. *Journal of the American Medical Informatics Association*, 26(11):1297–1304.

Alon Talmor, Oyvind Tafjord, Peter Clark, Yoav Goldberg, and Jonathan Berant. 2020. Teaching pretrained models to systematically reason over implicit knowledge. In *Advances in Neural Information Processing Systems 34*.

Jeffrey Ullman. 2011. *Mining of massive datasets*. Cambridge University Press.

Ashish Vaswani, Noam Shazeer, Niki Parmar, Jakob Uszkoreit, Llion Jones, Aidan N Gomez, Łukasz Kaiser, and Illia Polosukhin. 2017. Attention is all you need. In *Advances in neural information processing systems*, pages 5998–6008.

Denny Vrandečić and Markus Krötzsch. 2014. Wikidata: a free collaborative knowledgebase. *Communications of the ACM*, 57(10):78–85.

Lucy Lu Wang, Kyle Lo, Yoganand Chandrasekhar, Russell Reas, Jiangjiang Yang, Darrin Eide, Kathryn Funk, Rodney Kinney, Ziyang Liu, William Merrill, et al. 2020. Cord-19: The covid-19 open research dataset. *ArXiv*.

Shuohang Wang, Mo Yu, Jing Jiang, Wei Zhang, Xiaoxiao Guo, Shiyu Chang, Zhiguo Wang, Tim Klinger, Gerald Tesauro, and Murray Campbell. 2018. Evidence aggregation for answer re-ranking in open-domain question answering. In *International Conference on Learning Representations*.

Dirk Weissenborn, Tomáš Kočiský, and Chris Dyer. 2017. Dynamic integration of background knowledge in neural nlu systems. *arXiv preprint arXiv:1706.02596*.

Yonghui Wu, Mike Schuster, Zhifeng Chen, Quoc V Le, Mohammad Norouzi, Wolfgang Macherey, Maxim Krikun, Yuan Cao, Qin Gao, Klaus Macherey, et al. 2016. Google's neural machine translation system: Bridging the gap between human and machine translation. *arXiv preprint arXiv:1609.08144*.

An Yang, Quan Wang, Jing Liu, Kai Liu, Yajuan Lyu, Hua Wu, Qiaoqiao She, and Sujian Li. 2019a. Enhancing pre-trained language representations with rich knowledge for machine reading comprehension. In *Proceedings of the 57th Annual Meeting of the Association for Computational Linguistics*, pages 2346–2357.

Zhilin Yang, Zihang Dai, Yiming Yang, Jaime Carbonell, Russ R Salakhutdinov, and Quoc V Le. 2019b. XLNet: Generalized autoregressive pretraining for language understanding. In H. Wallach, H. Larochelle, A. Beygelzimer, F. d'Alché-Buc, E. Fox, and R. Garnett, editors, *Advances in Neural Information Processing Systems 32*, pages 5753–5763. Curran Associates, Inc.

Zhengyan Zhang, Xu Han, Zhiyuan Liu, Xin Jiang, Maosong Sun, and Qun Liu. 2019. ERNIE: Enhanced language representation with informative entities. In *Proceedings of the 57th Annual Meet-*

ing of the Association for Computational Linguistics, pages 1441–1451, Florence, Italy. Association for Computational Linguistics.

Towards BERT-based Automatic ICD Coding: Limitations and Opportunities

Damián Pascual Sandro Luck Roger Wattenhofer
ETH Zurich, Switzerland
{dpascual,wattenhofer}@ethz.ch, sluck@student.ethz.ch

Abstract

Automatic ICD coding is the task of assigning codes from the International Classification of Diseases (ICD) to medical notes. These codes describe the state of the patient and have multiple applications, e.g., computer-assisted diagnosis or epidemiological studies. ICD coding is a challenging task due to the complexity and length of medical notes. Unlike the general trend in language processing, no transformer model has been reported to reach high performance on this task. Here, we investigate in detail ICD coding using PubMedBERT, a state-of-the-art transformer model for biomedical language understanding. We find that the difficulty of fine-tuning the model on long pieces of text is the main limitation for BERT-based models on ICD coding. We run extensive experiments and show that despite the gap with current state-of-the-art, pretrained transformers can reach competitive performance using relatively small portions of text. We point at better methods to aggregate information from long texts as the main need for improving BERT-based ICD coding.

1 Introduction

During patient stays in medical institutions, clinicians generate text notes that record the state of the patient as well as the diagnoses and the treatments administered. Typically, a code from the International Classification of Diseases (ICD) is assigned to these clinical notes, in order to provide standardized information about the patient condition. ICD codes are used for different purposes, such as billing, computer-assisted diagnosis or epidemiological studies (Choi et al., 2016; Denny et al., 2010; Avati et al., 2018). Assigning ICD codes to medical notes is usually done manually by clinicians. This is an error-prone and time-consuming procedure and therefore, automatic solutions have been studied for over two decades (Larkey and Croft, 1996; de Lima et al., 1998).

However, automatic ICD code assignment proves challenging for multiple reasons. First, there exists a very large number of ICD codes (17.000) and each clinical report may have associated more than one code. To deal with this large multi-label classification problem, it is common to reduce the number of codes to those that appear most frequently (Mullenbach et al., 2018). Second, medical text usually lacks structure, includes irrelevant passages, as well as abbreviations, misspellings, numbers and a very specific vocabulary. On top of that, medical notes are long, which makes it difficult for automatic coding models to draw relations between different sections of the reports.

Current state-of-the-art methods for automatic ICD coding from medical notes are based on deep learning (Wang et al., 2018b; Mullenbach et al., 2018; Vu et al., 2020). These methods use different configurations of convolutional (CNN) and recurrent (RNN) neural networks as well as attention modules(Bahdanau et al., 2014). This stands in contrast to most areas of natural language processing (NLP), where models based on the transformer architecture (Vaswani et al., 2017) dominate the state-of-the-art (Wang et al., 2019). One of the main strengths of transformer models is their ability to deal with long range dependencies. This is a desirable property in ICD coding, where an understanding of different parts of the document may be necessary to assign a code. The lack of transformer models for ICD coding is surprising, especially since there already exist BERT-based models (Devlin et al., 2019) (a type of bidirectional transformer) that are trained on medical text data (Lee et al., 2020; Alsentzer et al., 2019; Gu et al., 2020). These models have achieved state-of-the-art performance on other tasks such as named entity recognition or question answering on medical documents (Gu et al., 2020).

On the other hand, the complexity of transformers scales quadratically with the length of their in-

Proceedings of the BioNLP 2021 workshop, pages 54–63
June 11, 2021. ©2021 Association for Computational Linguistics

put, which restricts the maximum number of words that they can process at once. This limitation may be critical in ICD coding, since clinical notes usually exceed this maximum input length. In this work, we investigate in detail BERT-based ICD coding, and explore different strategies to overcome the constraint on the input length by using an encoder-decoder architecture. We use the MIMIC-III dataset (Johnson et al., 2016), a big and widely used dataset for the ICD coding task, in order that our results are directly comparable to other existing methods (Wang et al., 2018b; Mullenbach et al., 2018; Vu et al., 2020). By exposing the limitations and benefits of BERT-based models on this task our work sets a solid basis for further research on automatic ICD coding systems.

2 Related Work

Automatic ICD coding has been an active area of research for over two decades. Already Larkey and Croft (1996) and de Lima et al. (1998) proposed different strategies to extract features from medical documents in order to build classifiers for automatically assigning ICD codes to medical notes. More recently, Perotte et al. (2014) proposed a multi-level Support Vector Machine (SVM) model to predict ICD codes from the MIMIC-II dataset (Saeed et al., 2011), the precursor of the MIMIC-III dataset (Johnson et al., 2016) that we consider in this work. Similarly, Scheurwegs et al. (2017) presented a method to extract features from structured and unstructured text and evaluated it on the MIMIC-III dataset.

In the last years, the state-of-the-art of automatic ICD coding has been dominated by deep learning models. Shi et al. (2017) proposed an LSTM model that operates at the character-level combined with an attention mechanism (Bahdanau et al., 2014). Wang et al. (2018b) proposed an embedding model based on GloVE embeddings (Pennington et al., 2014) that maps text and labels to the same space, where predictions are made using the cosine similarity. Mullenbach et al. (2018) proposed a model that combined convolutions with a per-label attention mechanism. This model was further improved by Xie et al. (2019) and Li and Yu (2020). Vu et al. (2020), proposed a label-attention model that reached the current best performance for ICD coding on the MIMIC-III dataset. All of these works represent only a portion of the research carried out in this field (Karimi et al., 2017; Baumel et al., 2018; Song et al., 2020; Prakash et al., 2017; Cao et al., 2020).

Since the appearance of the Transformer model (Vaswani et al., 2017), transformer-based architectures (Brown et al., 2020; Lewis et al., 2020; Raffel et al., 2019) have become state-of-the-art in almost every area of Natural Language Processing (Wang et al., 2018a, 2019) thanks to their ability to handle long range dependencies. BERT (Devlin et al., 2019), a bidirectional transformer, is of particular importance since it is the basis of many other language understanding models. Nonetheless, given the specific characteristics of medical text, e.g., specialized vocabulary, models pretrained on generic language, like BERT, do not reach high performance on biomedical language understanding tasks. Therefore, specialized models, such as BioBERT (Lee et al., 2020) or ClinicalBERT (Alsentzer et al., 2019), pretrained on medical text have been proposed. In particular, the recent PubMedBERT model (Gu et al., 2020) is the state-of-the-art in the BLURB benchmark (Gu et al., 2020), a benchmark for biomedical language understanding which includes the following tasks: named entity recognition, question answering, document classification, relation extraction, sentence similarity and evidence-based medical information extraction. Despite its prominence in medical language understanding, automatic ICD coding escapes the set of tasks where BERT-based models excel. To the best of our knowledge, no BERT-based model has been proposed yet that reaches competitive performance on ICD coding on the MIMIC-III dataset. In this work, we investigate in detail BERT-based ICD coding and identify existing limitations and opportunities.

3 Background

In this section we present the BERT model used in our experiments as well as the evaluation metrics.

3.1 PubMedBERT

PubMedBERT (Gu et al., 2020) is a transformer model with the same architecture as BERT-base (Devlin et al., 2019), i.e., it has 12 transformer layers, 100 million parameters and it outputs vector representations of 768 elements. PubMedBERT is trained from scratch on PubMed text, on a dataset of 3.1 billion words (21 GB). Furthermore, Pub-

MedBERT has not been pretrained on the MIMIC datasets as ClinicalBERT (Alsentzer et al., 2019) or BlueBERT (Peng et al., 2019), and therefore, we can evaluate it on MIMIC-III without information leakage from the test set. We choose this model among the existing ones because it is currently the state-of-the-art in biomedical understanding tasks as measured by the BLURB benchmark[1]. We use the implementation from HuggingFace (Wolf et al., 2019).

3.2 Evaluation Metrics

Following previous work (Wang et al., 2018b; Mullenbach et al., 2018; Vu et al., 2020), we report the results of our experiments using macro- and micro-averaged AUC (Area Under the ROC Curve). In a multi-class classification problem, the macro-average computes the metric (AUC in our case) for each class independently and then averages it across classes. This gives the same weight to all classes regardless of possible imbalances in the data. Micro-averaging, on the other hand, computes the average score over all samples, giving the same weight to each sample rather than to each class.

4 Dataset

In this work, we use the widely-used MIMIC-III dataset (Johnson et al., 2016). This dataset contains medical information in various forms, however, as in previous studies (Wang et al., 2018b; Mullenbach et al., 2018; Vu et al., 2020), we consider exclusively the discharge summaries for ICD coding. Discharge summaries are medical notes created by doctors at the end of a stay in a medical facility and contain all the information about the stay. In the MIMIC-III dataset, the length of the discharge summaries after tokenization ranges from 78 to $18,429$ tokens with a mean length of $2,740$ tokens and a median of $2,500$. Each of these discharge summaries has associated to it one or more ICD codes from the ICD-9 taxonomy, with an average of 13.15 ICD codes per summary. Therefore, ICD coding is a multi-label classification task.

The MIMIC-III dataset consists of $52,722$ discharge summaries with a total of $8,921$ unique ICD codes. However, most of the codes are very infrequent, and therefore, existing work (Wang et al., 2018b; Mullenbach et al., 2018; Vu et al., 2020) narrows down the task to finding only the 50 most frequent ICD codes. We follow this strategy and use the reduced dataset, sometimes referred to as MIMIC-III-50. This dataset consists of a training set of $8,067$ samples, a validation set of $1,574$ samples and a test set of $1,730$ samples. This data split is aligned with previous work, and thus, our results are directly comparable to those in the existing literature.

4.1 Pre-processing

We pre-process the discharge summaries from the MIMIC-III dataset following the method proposed by Mullenbach et al. (2018), which is also used by other recent work (Vu et al., 2020). This way, we convert all the text to lower case and we remove all numbers. However, we do not remove infrequent words as in (Mullenbach et al., 2018) since BERT uses WordPiece for tokenizing and hence, it does not suffer from out-of-vocabulary terms.

5 Model

Discharge summaries are longer than the maximum length accepted by PubMedBERT such that it fits in the memory of a modern GPU and thus, we need to split the summaries into pieces of text. In order to process more than one piece of text per summary we adopt an encoder-decoder structure, where the encoder and the decoder are trained separately. This way, the encoder is the BERT model that maps the different pieces of text to vector representations. These vector representations are then combined and decoded into ICD codes by the decoder, which can be any kind of model.

5.1 Encoder

We use PubMedBERT as the encoder of our model, as described in Section 3. We run our experiments on TITAN RTX GPUs with 24 GB of memory, where we can fit PubMedBERT with a maximum sequence length of 512 tokens.[2] We devise five different strategies to split the text of the discharge summaries:

- *Front*: First 512 tokens of the summary.
- *Back*: Last 512 tokens of the summary.
- *Mixed*: First 256 and the last 256 tokens of the summary.

[1] `https://microsoft.github.io/BLURB/leaderboard.html`

[2] Note that even if we could fit sequences of 1024 or 2048 tokens, they would still be shorter than the mean and median sequence length of the summaries.

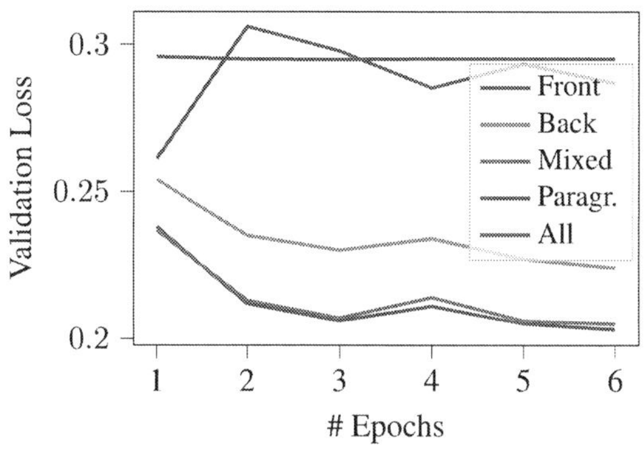

Figure 1: Validation losses for PubMED-BERT trained on different parts of the text.

- *All*: Split the whole discharge summary into consecutive chunks of 512 tokens; since summaries are of different length, each summary is split in a different number of chunks with the last chunk being possibly shorter.
- *Paragraph*: Given that the discharge summaries consist of named paragraphs, we select the 200 most frequent paragraphs, i.e., those that are present most often in the discharge summaries, each with a maximum length of 512 tokens.

PubMedBERT has been pretrained on the masked language modeling task, and therefore, it can produce generic representations of the input text. To fine-tune this model for the ICD coding task without exceeding the memory constraints we can feed only one chunk of text at a time. This way, we fine-tune five different instances of the PubMedBERT model, one per splitting strategy, using a batch size of 1 (to ensure the model fits in memory) and a learning rate of $5e^{-4}$. In each case, the model receives as input a piece of text of a maximum length of 512 tokens and it is trained to predict the ICD codes of the corresponding discharge summary. Note that while the text of *front*, *back* and *mixed* corresponds always to the same part of the discharge summary, when fine-tuning the model on the *paragraph* and *all* splits, each training example consists of only one paragraph or chunk, respectively. Therefore, there is no alignment across training examples (each training example comes from a different section of a discharge summary), which introduces noise to the training.

Figure 1 depicts the validation losses after 6 epochs of training for each of the trained models. For *front*, *back* and *mixed*, we see that the validation loss decreases quickly during the first three epochs and then, it slowly stabilizes. However, for *paragraph* and *all*, the validation loss stays constant, which indicates that the model is failing to learn; in other words, the lack of alignment between training samples makes the task of ICD coding too challenging for the model to learn meaningful representations of the input text.

5.2 Decoder

If we consider only one part of the text at a time, PubMedBERT can directly make a prediction on the ICD codes for the corresponding summary, as done during fine-tuning. However, in order to use the information from different pieces of text, we need a decoder capable of combining the information from several encodings. This way, the decoder receives as input one or several encoded representations (from the same discharge summary) generated by PubMedBERT during the encoding stage and outputs a vector of probabilities for the 50 ICD codes. For the decoder architecture, we consider a linear layer, multi-layer perceptrons (MLPs) and transformers.

In all cases, the decoders are trained with binary cross entropy loss with logits. We use a batch size of 32, a learning rate of $1e^{-4}$ with linear decay for 30 epochs and weight decay with $\lambda = 1e^{-3}$. We train for a maximum of 100 epochs with early stopping on the validation set.

Linear layer Our simplest decoder consists of a linear layer that takes as input a concatenation of the encoding vectors (of size 768 each); when only one chunk is considered, the input is just one encoding vector. The output of this linear layer is the probability vector for the ICD codes.

Multi Layer Perceptron We consider two variants of MLP-architectures, flat and parallel. In the flat architecture, the input is the concatenation of the encodings, as for the linear layer. This vector is passed through two non-linear layers, which produce intermediate representation of size 768 and 512 respectively, and then to a final linear layer that outputs the probabilities of the 50 ICD codes. In the parallel architecture, each of the input encodings is processed by a different dense layer, each of which produces an output of size $768/n$, where n is the number of input encodings. These intermediate representations are concatenated and passed through two additional non-linear layers, with the same sizes as in the flat architecture.

Each of the non-linear layers includes layer normalization (Ba et al., 2016), PReLU activation (He et al., 2015), and dropout (Srivastava et al., 2014) with $p = 0.1$.

Transformer This decoder takes as input the encodings and treats each of them as a token of dimensionality 768. These tokens are passed through a transformer layer with 8 attention heads. The output of this transformer layer is of the same size as the input, i.e., a set of tokens of 768 elements. The tokens are then concatenated and passed through an MLP of the same structure as the *flat* MLP described above.

6 Results

We pose six research questions regarding the different strategies to encode and decode discharge summaries using a BERT-based encoder. In our experiments, we fix the random seed so that all the results are comparable.

How much does fine-tuning the encoder help decoding?

Here, we consider only the PubMedBERT models fine-tuned on *front*, *back* and *mixed* data, since they were the only ones to learn during fine-tuning, as shown in Section 5.1. To investigate the impact of this fine-tuning step on decoding performance, we use a simple linear layer which receives as input the concatenation of the encodings of the *front*, *back* and *mixed* chunks. Each of these pieces of text is encoded by the PubMedBERT model trained on that piece of text, i.e., we use three different encoders. We study the difference in performance for three different training points of the encoders: not fine-tuned, fine-tuned for three epochs and fine-tuned for six epochs. The results are detailed in Table 1.

Epochs	Macro AUC	Micro AUC
None	55.76	69.55
3	81.47	86.00
6	**83.00**	**86.98**

Table 1: Performance for different number of training epochs when combining the *front*, *back* and *mixed* chunks with a linear decoder.

These results show that fine-tuning the encoder significantly improves the decoding performance and that the best performance is obtained after six epochs. In fact, the difference between fine-tuning for six epochs and not fine-tuning is as large as 27.24 points for the Macro AUC score and 17.43 points for the Micro AUC score. We observed the same pattern in all of our experiments, and therefore, in the following we will only present results with the encoder fine-tuned for six epochs, unless stated otherwise.

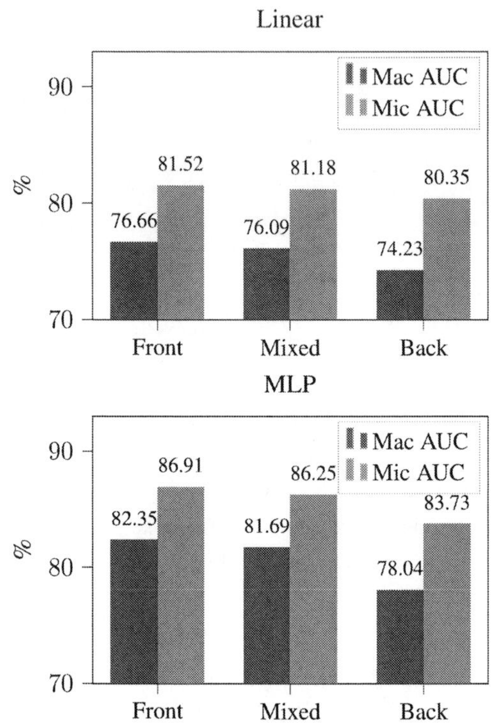

Figure 2: Performance of a linear layer (top) and a non-linear MLP (bottom) on the *front*, *back* and *mixed* encodings.

Which of the three pieces of text, *front*, *back* or *mixed*, contains the most relevant information for ICD coding?

We experiment with a linear and a flat MLP decoder and apply these models to the encodings of each of the three chunks of text separately, i.e., *front*, *back* and *mixed*. We report the results in Figure 2.

We see that *front*, i.e., the first 512 tokens of the discharge summary yields the best performance, both when the decoder is a linear layer and an MLP. Although slightly inferior, the *mixed* chunk produces competitive scores while when using an MLP the AUC scores are more than 3 points lower for *back* than for *front*. Furthermore, using as decoder an MLP improves the performance significantly over using a linear layer; with the *front* non-linear model performing comparably to the combination of the three chunks with a linear decoder, as reported in the previous section, Table 1.

This naturally raises the question of whether the combination of the chunks yields an improvement. To study this, we use the same non-linear MLP architecture as in Figure 2 (bottom) on 1) the concatenation of the encodings of *front* and *back* and 2) the concatenation of the three encodings, *front*, *back* and *mixed*. We report the results in Table 2.

Model	Mac. AUC	Mic. AUC
Front-Back	83.70	88.11
Front-Back-Mixed	**84.42**	**88.58**

Table 2: Performance of combining the *front*, *back* and *mixed* chunks using a two-layer flat MLP decoder.

These results show that combining *front* and *back* improves performance in comparison to using only *front*. As it may be expected, adding the mixed paragraph, which contains redundant information, produces only a small improvement. Overall, the combination of the three chunks produces an improvement of 2.07 points for Macro AUC and 1.67 points for Micro AUC over using only *front*. Given the larger input, these models have more parameters than the ones using only one of the chunks, which could partly explain the improvement, especially when adding redundant information, i.e., the *mixed* chunk. This result leads us to investigate the influence of the decoder architecture.

How does the architecture of the decoder impact performance?

Here, we consider flat MLP, parallel MLP and transformer decoders on the combination of *front*, *back* and *mixed*. For each of these architectures, we evaluate three different sizes: Base, Large and X-Large, where the difference between these sizes is only the number of layers and the size of the internal representations. This way, our experiments aim at discerning whether the structure of the decoder, the number of parameters, or both, influence the performance of the ICD coding model. Table 3 details the results of these experiments.

None of the models considered obtains a performance significantly higher than the others, with the largest difference across Macro and Micro AUC scores being of only 0.28 and 0.57 points, respectively. This result is surprising since, given the complexity of the task, it could be expected that larger

Model	AUC Mac.	AUC Mic.
Flat ($1.5M$)	84.42	88.58
Flat L ($3M$)	84.30	88.45
Flat XL ($7M$)	84.30	88.47
Parallel ($1M$)	84.45	88.65
Parallel L ($2M$)	84.23	88.48
Parallel XL ($3M$)	84.51	88.49
Transformer ($6.5M$)	84.30	88.49
Transformer L ($14M$)	84.27	88.45
Transformer XL ($18M$)	84.29	88.08

Table 3: Performance of different decoder architectures for the combination of *front*, *back* and *mixed*, the number of parameters of each model is specified in parenthesis.

and more sophisticated decoders would perform better. Notwithstanding, the saturation in performance suggests that all the information available in the input of the decoder is successfully extracted by every model, regardless of its complexity. This in turn indicates that the performance of the whole encoder-decoder model is limited by the reduced amount of text that is given as input (only the beginning and the end of the discharge summaries). Therefore, we next consider providing larger portions of text from the discharge summaries as input.

Is dividing the discharge summaries by paragraphs a good splitting strategy?

By splitting the discharge summaries into paragraphs we take into account information from a larger body of text than by using the front and the back. The main disadvantage of this approach is that the encoder fails to converge during fine-tuning. Here, we test the hypothesis of whether the decoder can compensate the lack of fine-tuning of the encoder and, by leveraging the larger amount of information available, reach competitive performance. We encode the 200 most frequent paragraphs using the PubMedBERT model fine-tuned on *paragraph* data, although due to lack of convergence during fine-tuning, we observed very similar results when using the not fine-tuned version.

Since not all the discharge summaries contain the same paragraphs, there is a misalignment between samples. For this reason, here we consider only the transformer decoder; the self-attention modules of the transformer should be able to cope with the misalignment better than the other architectures. We consider the transformer decoders (Base, Large and X-Large) from the previous section. Now, the

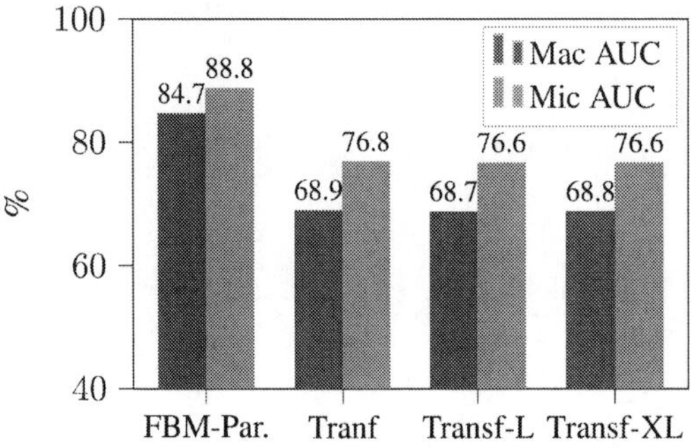

Figure 3: Comparison of front-back-mixed parallel (FBM-Par.) and three sizes of transformer decoders (Transf) on *paragraph* data.

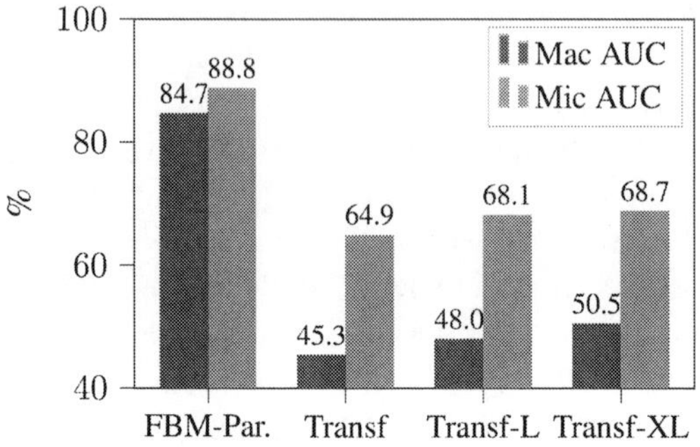

Figure 4: Comparison of front-back-mixed parallel (FBM-Par.) and three sizes of transformer decoders (Transf) on *all* data.

transformer decoder receives 200 encoded representations, one per paragraph. Given this large number of input representations or tokens, we aggregate the output of the transformers by taking the mean over the representations produced for all the paragraphs[3].

In Figure 3, we compare these *paragraph* decoders to the Parallel MLP model on the *front*, *back* and *mixed* chunks from the previous section.

We see that dividing the discharge summaries into paragraphs greatly under-performs in comparison to using the beginning and end of the summaries encoded by fine-tuned PubMedBERT models. This result partly rejects the hypothesis that the decoder can benefit from a larger unstructured input. Next, we continue investigating this hypothesis by feeding the decoder with the complete discharge summaries following the *all* strategy.

How does splitting the complete summaries in consecutive chunks perform?

We split the whole text of each discharge summary into consecutive chunks of 512 tokens (the last chunk of each summary may be smaller). We encode these chunks using the PubMedBERT model fine-tuned on *all* data; as before, we observed very similar results with the not fine-tuned model. The encodings are then fed into the decoder. Again, the varying size of the discharge summaries produces misalignment across examples. Therefore, we consider only the transformer decoders (Base, Large and X-Large). We report the results of this experiment in Figure 4.

The largest transformer model (XL) performs the best of the three models on *all* data. Nevertheless, its 50.5% Macro and 68.7% Micro AUC scores are much lower than the results obtained by the *front-back-mixed*. In fact, splitting the text into chunks of the same size performs the worst among all the methods that we have investigated. These results confirm that the decoder cannot compensate the lack of convergence during the fine-tuning of the encoder and points at the encoder as the main responsible for the model's performance.

How do our results compare to the state-of-the-art?

Finally, in Table 4 we compare one of our best performing BERT-ICD models, the *front-back-mixed* Parallel model, with the existing state-of-the art models for ICD coding on the MIMIC-III dataset. In particular, we compare against the condensed memory networks (C-MemNN) by Prakash et al. (2017), LEAM by Wang et al. (2018b), CAML and DR-CAML by Mullenbach et al. (2018), MSATT-KG by Xie et al. (2019) and the Label Attention model by Vu et al. (2020). We report the performance of each model as provided in the corresponding original work.

Model	AUC Mac.	AUC Mic.
C-MemNN	83.3	-
LEAM	88.1	91.2
CAML	87.5	90.9
DR-CAML	88.0	90.2
MSATT-KG	91.4	93.6
Label Attention	**92.1**	**94.6**
BERT-ICD	84.45	88.65

Table 4: Comparison of different state-of-the-art models for ICD coding.

[3]We experimented with other aggregation techniques like max pooling and large MLPs obtaining very similar results.

We see that although our BERT-based ICD coding model is competitive with some of the state-of-the-art models, there is still a substantial gap between the best performing model from Vu et al. (2020), and our BERT-ICD model.

7 Discussion

Automatic ICD coding from discharge summaries using transformer models has proven to be challenging. Discharge summaries are very long documents and thus, they need to be divided into chunks in order to be entirely processed by BERT-like models.This way, a decoder is necessary to combine the representations of each chunk, which are independently generated by the BERT encoder. We have shown that for these representations to be meaningful the encoder needs to be fine-tuned on the ICD coding task. It is straight-forward to fine-tune a BERT encoder such as PubMedBERT using specific parts of the summary, e.g., the beginning or the end. However, in our experiments, fine-tuning PubMedBERT on excerpts extracted from different parts of the text, i.e., *paragraph* and *all*, prevented convergence due to the lack of alignment between samples, i.e., due to each training sample containing information from a different section of a discharge summary. Furthermore, our results show that the decoder, regardless of its architecture, cannot compensate for lack of convergence during the fine-tuning of the encoder.

On the other hand, our best BERT-ICD model reaches competitive performance using only $1,024$ tokens (*front* and *back*), which represents a significantly smaller portion of text than state-of-the-art models, based on CNNs and RNNs. Unlike BERT, CNN and RNN models can extract information from texts of any length without needing to split them, which allows for end-to-end training over long pieces of text. Mullenbach et al. (2018) found that the performance of their convolutional attention model benefits from longer input texts until a length of between $2,500$ and $6,500$ words, and Vu et al. (2020) use up to $4,000$ words as input. Our model combines encodings from the beginning and the end of the discharge summary, and reaches better performance in that case than when it processes either of those portions of text alone. This supports the statement that including more text improves ICD coding. All of these results suggest that the difficulty of fine-tuning a BERT encoder on long pieces of text is the main bottleneck for performance and the reason for the existing gap with state-of-the-art models for ICD coding.

One of the main advantages of transformer models over CNNs and RNNs is that they can handle long-range dependencies. Hence, if longer text could be fed at once into a BERT encoder it would be possible to find relationships and patterns over longer spans of text. It is therefore likely that advances either in terms of hardware, i.e., larger GPU memories allowing for longer pieces of text to be processed at once; or in compressing BERT-like models, e.g., distillation, will progressively close the gap with the state-of-the-art, following the same trend of other areas of NLP. On top of that, we consider that the two most promising directions for future research on BERT-based ICD coding are: 1) devising strategies to fine-tune the encoder over longer spans of text, e.g., building an ensemble of models where each of them is trained on one section of the text; 2) improving the methods to aggregate encodings from different parts of the document.

Finally, to deploy automatic ICD coding systems in the real world, it is important that their decisions can be explained. Explaining transformer models is currently a field of active research, and although there exist important concerns about the interpretability of attention distributions in transformers (Brunner et al., 2019; Pruthi et al., 2020), methods based on gradient attribution (Pascual et al., 2020) or on attention flow (Abnar and Zuidema, 2020) can provide insights on their decision-making. A BERT-based ICD coding system could directly benefit from this field of research and eventually provide explanations together with its ICD code predictions.

8 Conclusion

Contrary to what is common in most NLP tasks, the transformer architecture is not the state-of-the-art in assigning ICD codes to discharge summaries. In this work, we have presented a thorough study of the performance of BERT-based models on this task and we have identified the length of the discharge summaries as the main obstacle holding back their performance. Our work sets a solid foundation for further research on ICD coding and suggests that overcoming the exposed limitations of BERT-based models is likely to lead to a new state-of-the-art. Furthermore, we believe that the interpretability of ICD coding models is an interesting avenue for future work, which can benefit from a large body of existing research.

References

Samira Abnar and Willem Zuidema. 2020. Quantifying attention flow in transformers. In *Proceedings of the 58th Annual Meeting of the Association for Computational Linguistics*, pages 4190–4197.

Emily Alsentzer, John R Murphy, Willie Boag, Wei-Hung Weng, Di Jin, Tristan Naumann, and Matthew McDermott. 2019. Publicly available clinical bert embeddings. *arXiv preprint arXiv:1904.03323*.

Anand Avati, Kenneth Jung, Stephanie Harman, Lance Downing, Andrew Ng, and Nigam H Shah. 2018. Improving palliative care with deep learning. *BMC medical informatics and decision making*, 18(4):55–64.

Jimmy Lei Ba, Jamie Ryan Kiros, and Geoffrey E Hinton. 2016. Layer normalization. *arXiv preprint arXiv:1607.06450*.

Dzmitry Bahdanau, Kyunghyun Cho, and Yoshua Bengio. 2014. Neural machine translation by jointly learning to align and translate. *arXiv preprint arXiv:1409.0473*.

Tal Baumel, Jumana Nassour-Kassis, Raphael Cohen, Michael Elhadad, and Noémie Elhadad. 2018. Multi-label classification of patient notes: Case study on icd code assignment. In *AAAI Workshops*.

Tom B Brown, Benjamin Mann, Nick Ryder, Melanie Subbiah, Jared Kaplan, Prafulla Dhariwal, Arvind Neelakantan, Pranav Shyam, Girish Sastry, Amanda Askell, et al. 2020. Language models are few-shot learners. *arXiv preprint arXiv:2005.14165*.

Gino Brunner, Yang Liu, Damian Pascual, Oliver Richter, Massimiliano Ciaramita, and Roger Wattenhofer. 2019. On identifiability in transformers. In *International Conference on Learning Representations*.

Pengfei Cao, Yubo Chen, Kang Liu, Jun Zhao, Shengping Liu, and Weifeng Chong. 2020. Hypercore: Hyperbolic and co-graph representation for automatic icd coding. In *Proceedings of the 58th Annual Meeting of the Association for Computational Linguistics*, pages 3105–3114.

Edward Choi, Mohammad Taha Bahadori, Andy Schuetz, Walter F Stewart, and Jimeng Sun. 2016. Doctor ai: Predicting clinical events via recurrent neural networks. In *Machine learning for healthcare conference*, pages 301–318. PMLR.

Luciano RS de Lima, Alberto HF Laender, and Berthier A Ribeiro-Neto. 1998. A hierarchical approach to the automatic categorization of medical documents. In *Proceedings of the seventh international conference on Information and knowledge management*, pages 132–139.

Joshua C Denny, Marylyn D Ritchie, Melissa A Basford, Jill M Pulley, Lisa Bastarache, Kristin Brown-Gentry, Deede Wang, Dan R Masys, Dan M Roden, and Dana C Crawford. 2010. Phewas: demonstrating the feasibility of a phenome-wide scan to discover gene–disease associations. *Bioinformatics*, 26(9):1205–1210.

Jacob Devlin, Ming-Wei Chang, Kenton Lee, and Kristina Toutanova. 2019. Bert: Pre-training of deep bidirectional transformers for language understanding. In *Proceedings of the 2019 Conference of the North American Chapter of the Association for Computational Linguistics: Human Language Technologies, Volume 1 (Long and Short Papers)*, pages 4171–4186.

Yu Gu, Robert Tinn, Hao Cheng, Michael Lucas, Naoto Usuyama, Xiaodong Liu, Tristan Naumann, Jianfeng Gao, and Hoifung Poon. 2020. Domain-specific language model pretraining for biomedical natural language processing. *arXiv preprint arXiv:2007.15779*.

Kaiming He, Xiangyu Zhang, Shaoqing Ren, and Jian Sun. 2015. Delving deep into rectifiers: Surpassing human-level performance on imagenet classification. In *Proceedings of the IEEE international conference on computer vision*, pages 1026–1034.

Alistair EW Johnson, Tom J Pollard, Lu Shen, H Lehman Li-Wei, Mengling Feng, Mohammad Ghassemi, Benjamin Moody, Peter Szolovits, Leo Anthony Celi, and Roger G Mark. 2016. Mimic-iii, a freely accessible critical care database. *Scientific data*, 3(1):1–9.

Sarvnaz Karimi, Xiang Dai, Hamed Hassanzadeh, and Anthony Nguyen. 2017. Automatic diagnosis coding of radiology reports: a comparison of deep learning and conventional classification methods. In *BioNLP 2017*, pages 328–332.

Leah S Larkey and W Bruce Croft. 1996. Combining classifiers in text categorization. In *Proceedings of the 19th annual international ACM SIGIR conference on Research and development in information retrieval*, pages 289–297.

Jinhyuk Lee, Wonjin Yoon, Sungdong Kim, Donghyeon Kim, Sunkyu Kim, Chan Ho So, and Jaewoo Kang. 2020. Biobert: a pre-trained biomedical language representation model for biomedical text mining. *Bioinformatics*, 36(4):1234–1240.

Mike Lewis, Yinhan Liu, Naman Goyal, Marjan Ghazvininejad, Abdelrahman Mohamed, Omer Levy, Veselin Stoyanov, and Luke Zettlemoyer. 2020. Bart: Denoising sequence-to-sequence pre-training for natural language generation, translation, and comprehension. In *Proceedings of the 58th Annual Meeting of the Association for Computational Linguistics*, pages 7871–7880.

Fei Li and Hong Yu. 2020. Icd coding from clinical text using multi-filter residual convolutional neural network. In *Proceedings of the AAAI Conference on Artificial Intelligence*.

James Mullenbach, Sarah Wiegreffe, Jon Duke, Jimeng Sun, and Jacob Eisenstein. 2018. Explainable prediction of medical codes from clinical text. In *NAACL-HLT*.

Damian Pascual, Gino Brunner, and Roger Wattenhofer. 2020. Telling bert's full story: from local attention to global aggregation. *arXiv preprint arXiv:2004.05916*.

Yifan Peng, Shankai Yan, and Zhiyong Lu. 2019. Transfer learning in biomedical natural language processing: An evaluation of bert and elmo on ten benchmarking datasets. In *Proceedings of the 18th BioNLP Workshop and Shared Task*, pages 58–65.

Jeffrey Pennington, Richard Socher, and Christopher D Manning. 2014. Glove: Global vectors for word representation. In *Proceedings of the 2014 conference on empirical methods in natural language processing (EMNLP)*, pages 1532–1543.

Adler Perotte, Rimma Pivovarov, Karthik Natarajan, Nicole Weiskopf, Frank Wood, and Noémie Elhadad. 2014. Diagnosis code assignment: models and evaluation metrics. *Journal of the American Medical Informatics Association*, 21(2):231–237.

Aaditya Prakash, Siyuan Zhao, Sadid Hasan, Vivek Datla, Kathy Lee, Ashequl Qadir, Joey Liu, and Oladimeji Farri. 2017. Condensed memory networks for clinical diagnostic inferencing. In *Proceedings of the AAAI Conference on Artificial Intelligence*.

Danish Pruthi, Mansi Gupta, Bhuwan Dhingra, Graham Neubig, and Zachary C Lipton. 2020. Learning to deceive with attention-based explanations. In *Proceedings of the 58th Annual Meeting of the Association for Computational Linguistics*, pages 4782–4793.

Colin Raffel, Noam Shazeer, Adam Roberts, Katherine Lee, Sharan Narang, Michael Matena, Yanqi Zhou, Wei Li, and Peter J Liu. 2019. Exploring the limits of transfer learning with a unified text-to-text transformer. *arXiv preprint arXiv:1910.10683*.

Mohammed Saeed, Mauricio Villarroel, Andrew T Reisner, Gari Clifford, Li-Wei Lehman, George Moody, Thomas Heldt, Tin H Kyaw, Benjamin Moody, and Roger G Mark. 2011. Multiparameter intelligent monitoring in intensive care ii (mimic-ii): a public-access intensive care unit database. *Critical care medicine*, 39(5):952.

Elyne Scheurwegs, Boris Cule, Kim Luyckx, Léon Luyten, and Walter Daelemans. 2017. Selecting relevant features from the electronic health record for clinical code prediction. *Journal of biomedical informatics*, 74:92–103.

Haoran Shi, Pengtao Xie, Zhiting Hu, Ming Zhang, and Eric P Xing. 2017. Towards automated icd coding using deep learning. *arXiv preprint arXiv:1711.04075*.

Congzheng Song, Shanghang Zhang, Najmeh Sadoughi, Pengtao Xie, and Eric P Xing. 2020. Generalized zero-shot text classification for icd coding. In *IJCAI*.

Nitish Srivastava, Geoffrey Hinton, Alex Krizhevsky, Ilya Sutskever, and Ruslan Salakhutdinov. 2014. Dropout: a simple way to prevent neural networks from overfitting. *The journal of machine learning research*, 15(1):1929–1958.

Ashish Vaswani, Noam Shazeer, Niki Parmar, Jakob Uszkoreit, Llion Jones, Aidan N Gomez, Łukasz Kaiser, and Illia Polosukhin. 2017. Attention is all you need. In *Proceedings of the 31st International Conference on Neural Information Processing Systems*, pages 6000–6010.

Thanh Vu, Dat Quoc Nguyen, and Anthony Nguyen. 2020. A label attention model for icd coding from clinical text. *arXiv preprint arXiv:2007.06351*.

Alex Wang, Yada Pruksachatkun, Nikita Nangia, Amanpreet Singh, Julian Michael, Felix Hill, Omer Levy, and Samuel R Bowman. 2019. Superglue: A stickier benchmark for general-purpose language understanding systems. *Advances in Neural Information Processing Systems*, 32.

Alex Wang, Amanpreet Singh, Julian Michael, Felix Hill, Omer Levy, and Samuel R Bowman. 2018a. Glue: A multi-task benchmark and analysis platform for natural language understanding. *arXiv preprint arXiv:1804.07461*.

Guoyin Wang, Chunyuan Li, Wenlin Wang, Yizhe Zhang, Dinghan Shen, Xinyuan Zhang, Ricardo Henao, and Lawrence Carin. 2018b. Joint embedding of words and labels for text classification. In *Proceedings of the 56th Annual Meeting of the Association for Computational Linguistics (Volume 1: Long Papers)*, pages 2321–2331.

Thomas Wolf, Lysandre Debut, Victor Sanh, Julien Chaumond, Clement Delangue, Anthony Moi, Pierric Cistac, Tim Rault, Rémi Louf, Morgan Funtowicz, et al. 2019. Huggingface's transformers: State-of-the-art natural language processing. *arXiv preprint arXiv:1910.03771*.

Xiancheng Xie, Yun Xiong, Philip S Yu, and Yangyong Zhu. 2019. Ehr coding with multi-scale feature attention and structured knowledge graph propagation. In *Proceedings of the 28th ACM International Conference on Information and Knowledge Management*, pages 649–658.

emrKBQA: A Clinical Knowledge-Base Question Answering Dataset

Preethi Raghavan[1,3,*], Diwakar Mahajan[1,3,#], Jennifer Liang[1,3,§], Rachita Chandra[1,3,†], Peter Szolovits[2,3,‡]

[1]IBM Research, [2]MIT CSAIL, [3]MIT-IBM Watson AI Lab

{*praghav, #dmahaja, §jjliang, †rachitac}@us.ibm.com, ‡psz@mit.edu

Abstract

We present emrKBQA, a dataset for answering physician questions from a structured patient record. It consists of questions, logical forms and answers. The questions and logical forms are generated based on real-world physician questions and are slot-filled and answered from patients in the MIMIC-III KB (Johnson et al., 2016) through a semi-automated process. This community-shared release consists of over 940000 question, logical form and answer triplets with 389 types of questions and ≈7.5 paraphrases per question type. We perform experiments to validate the quality of the dataset and set benchmarks for question to logical form learning that helps answer questions on this dataset.

1 Introduction

The last decade has seen widespread adoption of electronic health records (EHRs) across hospitals and clinics in the US (Jha et al., 2006; Evans, 2016). Physicians often seek answers to questions from a patient's EHR to support clinical decision-making (Demner-Fushman et al., 2009). It is not too hard to imagine a future where a physician interacts with an EHR system and asks it complex questions and expects precise answers, with adequate context, from a patient's record (Pampari et al., 2018). Central to such a world is a medical question answering system that processes natural language questions asked by physicians and finds answers to the questions in structured and unstructured sources in the patient's record.

However, the longitudinal, domain specific nature of patient records along with privacy concerns makes it difficult to develop large-scale annotated datasets for training machine learning models. This motivated Pampari et al. (2018) to develop the first community-shared patient QA dataset, emrQA, using a semi-automated process and create a large-scale dataset with over 1M question-answer and question-logical form pairs. They templated and slot-filled physician questions and logical forms on clinical notes and extracted corresponding answers from annotations on clinical notes for tasks like entity extraction and relation learning in the i2b2 challenges (Uzuner et al., 2011).

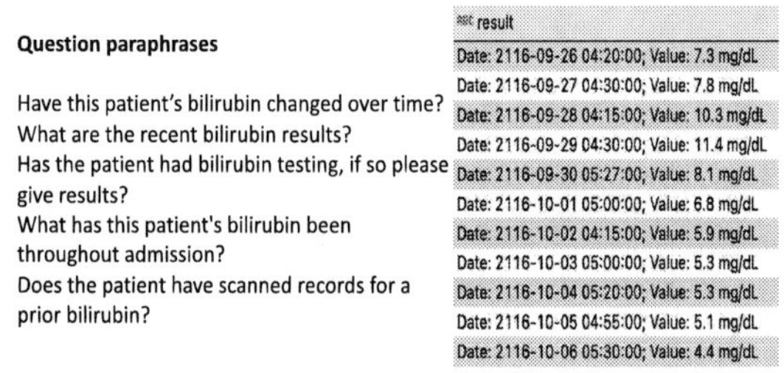

Figure 1: Questions (and paraphrases) with answers from MIMIC-III

However, emrQA is restricted to answers within or across clinical notes. Clinical notes are known to capture relations between entities (treatments for problems, side-effects of a drug), signs or symptoms (palpitations), temporal and causal events. On the other hand, structured data in the EHR is considered more reliable for labs results, prescriptions, vitals and other measurements (Hanauer et al., 2015). Hence, a complete EHR QA system should consider data across both these sources in answering a question.

Thus, we propose emrKBQA, a dataset for answering natural language questions from the structured portion of EHR data by mapping questions to logical forms. We demonstrate an instance of using this dataset for question answering using the MIMIC-III KB (a set of question paraphrases and answers from MIMIC shown in Figure 1). The resultant dataset consists of 940,713 question answer pairs from 389 question types (unique instances of questions, i.e., templates) and 52 question/logical

Proceedings of the BioNLP 2021 workshop, pages 64–73
June 11, 2021. ©2021 Association for Computational Linguistics

forms groups (where questions within the group are paraphrases) from 100 patients. We benchmark semantic parsing and answering results on this dataset by learning to map natural language questions to logical forms and retrieving the answer from a KB of patient records. The main contributions of this work are as follows: (1) We develop and release emrKBQA, the first large-scale community-shared dataset for patient-specific QA on structured patient records[1]. (2) emrKBQA will help train models for semantic parsing and answering questions from the structured EHR. This will help us progress towards answering on the EHR as a whole (in conjunction with emrQA). (3) We benchmark state of the art semantic parsing models on the dataset for QA on structured patient records.

2 Related Work

The question answering (QA) problem is usually defined over unstructured texts or structured knowledge bases (KB QA). In case of KB QA, questions are usually mapped to logical forms (or a query language using SQL, SPARQL, etc.) (Zettlemoyer and Collins, 2005; Berant and Liang, 2014) that are then used to retrieve the answer. In the medical domain, there is limited prior work on answering patient-specific questions over structured clinical data.

Roberts and Demner-Fushman (2016, 2015) introduce target logical form definitions and present a rule based method for converting natural language questions over structured data in the EHR into logical forms. They work with a dataset of 446 questions collected during clinician ICU visits and propose an approach using question decomposition, concept recognition and normalization, and rule based semantic parsing. However, the questions and logical forms were not publicly released. In contrast, we present a large-scale community-shared dataset of over 900k generated questions from 52 unique question templates, logical forms and answers.

More recently, Wang et al. (2020) create a new large-scale Question-SQL pair dataset (MIMIC-SQL) on the MIMIC-III dataset, again using the generation process as in Pampari et al. (2018). They propose a deep learning based TRanslate-Edit Model for Question-to-SQL generation that adapts the widely used sequence-to-sequence model to directly generate the SQL query for a given question, and also performs edits using an attentive-copying mechanism. The questions in the dataset are always asked over a patient-cohort such as "how many patients had the diagnosis icd9 code 53190?". However, the questions in emrQA are specific to a patient. This makes a big difference as the corpus for answering is smaller (limited to the patient's record, which may include several admissions), the answers may be viewed in conjunction with answers from the unstructured record, the type of questions asked varies, and redundancy and variability in answers to the same question may affect model performance.

Park et al. (2020) construct an EHR QA dataset from MIMIC-III where the question-answer pairs are represented in SQL (table-based) and SPARQL (graph-based). Here again, the questions are defined over patient cohorts; e.g., "What number of married patients suffered from other convulsions?", making it inherently different from the emrKBQA task. They construct a knowledge graph by relating tables in the database and explore both table-based and graph-based QA (using SPARQL). emrKBQA maps questions to logical forms based on a schema of entities and relations. The tables and columns in the KB are mapped to the entities and attributes in the schema. Logical forms capturing the information need expressed in the question are then instantiated from this schema. Thus, emrKBQA instantiates logical forms from a relational schema (representing entities and relations typically found in the EHR) and facilitates a query language/ resource independent way of representing questions and answering them beyond just individual tables in the KB.

KB-based QA datasets (question semantic parsing) use annotated question and logical form pairs for supervision where the logical forms (that can be then easily be mapped to any query language) are used to retrieve answers from a database (Bordes et al., 2014; Zettlemoyer and Collins, 2005; Berant and Liang, 2014). emrKBQA provides a dataset that can be used to train models to retrieve answers to natural language questions (by mapping them to logical forms) from the structured part of the EHR. The logical forms are instantiated from a schema that captures domain entities, attributes and relations proposed in emrQA (Pampari et al., 2018). We demonstrate the value of the dataset by answering natural language questions posed by physicians

[1] `https://github.com/emrQA/emrKBQA` scripts to generate emrKBQA from MIMIC data.

as follows. We first train state of the art sequence models for semantic parsing to map questions to (query-language agnostic) logical forms. We then map the learned logical forms to the desired query language (SQL) using a deterministic process.

3 Dataset Creation

emrKBQA is generated using a process similar to emrQA. We begin with the same initial question, logical form and template pool as emrQA. However, the question template groups, corresponding logical forms and what constitutes an answer have all been updated by a medical expert to better reflect answering needs.

Questions. emrKBQA contains natural language questions posed by physicians at the Veteran's Administration (VA), Mayo Clinic and Cleveland Clinic on patient records (Raghavan et al., 2018). These questions have been transformed into templates by replacing entities with entity-type placeholders (same as emrQA). The dataset consists of 389 such question templates. The placeholders are then slot-filled with appropriate entities from a KB. For instance, "Is the patient on *lisinopril*?" is transformed to: "Is the patient on |*medication*|?" The |*medication*| placeholder is then slot-filled with different medication names from a KB. While the slot-filling is done indiscriminately in emrQA, we constrain the slot-filling by constraining the entity types, wherever possible, with the help of a medical expert. E.g., we filter *Prescriptions* (table) with drug_type (table column) *base* (column value) in slot-filling medication questions. We also filter out certain icd_codes from the diagnoses_icd table in questions with conditions. We process the date field (yyyy-mm-dd, hh:mm:ss)

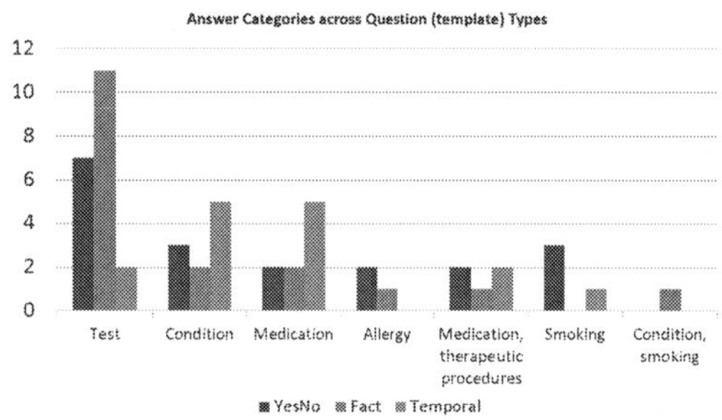

Figure 2: Distribution of answer categories against question template types. Some questions have multiple categories like medication and therapeutic procedure or condition and smoking .

to also insert instances of just month and day, or date without time when slot-filling (along with using the original format). Doing so ensures that the questions are more likely to be naturally asked.

As in this example, the questions are patient-specific and the expected answer is in the structured part of the patient record. Each question template is also assigned one or more question types, which is a new field (not in emrQA) to further categorize question templates in emrKBQA. Question type can take one or more of the following values:

- YesNo = yes/no questions, e.g., "Is |test| value abnormal", "Is the patient on |medication|"
- Temp = temporal or when questions, e.g., "date last |test|"
- Fact = factual or what questions, e.g., "Range of |test|"

A side-effect of the generation process (slot-filling) is that all YesNo questions have a Yes answer. We counter this by also generating questions where the answer will be No. We do this by slot-filling |problem|, |test|, |medication|, |treatment| based on the question and using top 50 most frequently occurring entities in appropriate tables (based on the entity type). Some of these questions are now bound to have No as the answer when applied to our patient set.

The types of questions are a consequence of the questions provided by the physicians who were polled for the initial question set. This was independent of any underlying data and simply based on what they would want to know about their own patients. While several other questions may be answerable on any underlying KB (like MIMIC), we wanted the question set to reflect what an actual physician may want to know from a patient record.

Logical Forms. Logical forms are a structured representation that capture the information need expressed in the question through entities, relations and attributes and are generated as a by-product of the emrQA generation process. They provide a human-comprehensible symbolic representation, linking questions to answers, and help build interpretable models critical to the medical domain (Davis et al., 1977; Vellido et al., 2012). They are formally defined by Pampari et al. (2018) in emrQA. They encapsulate how we are answering a question (since that can be subjective). They are instantiated from a schema representing entities and relations found in the EHR. We use the same

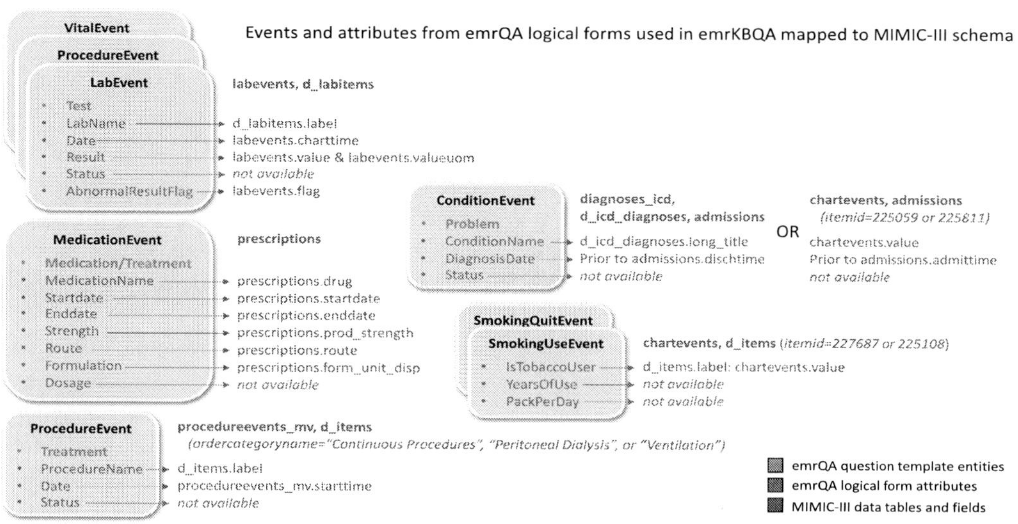

Figure 3: Mapping between emrKBQA schema entities, attributes and tables (yellow boxes) and columns in MIMIC (shown in blue). See MIMIC schema for a description of MIMIC table and column names(Johnson et al., 2016)

schema as Pampari et al. (2018) and map the tables and columns in MIMIC to the schema entities and attributes (see Figure 3).

The schema entities (yellow boxes in Figure 3) represent entities of interest in patient records. In emrQA these are derived from the annotated entities in i2b2 (since emrQA was slot-filled from i2b2 annotations). We use the same entities for emrKBQA as our question set is a subset of emrQA. The structured MIMIC KB does not contain any semantic relations (relates, conducted/reveals, improves, worsens, causes, given/not given (Pampari et al., 2018)). Thus, Figure 3 does not show any of the relations defined in the emrQA schema. An example of the mapping between a schema entity and MIMIC table is as follows. The MedicationEvent (entity that corresponds to Medication and Treatment in our logical form templates) from the schema maps to the Prescriptions table in MIMIC. The entity attributes (shown in red) correspond to the columns in the Prescription table (shown in blue) as illustrated in the figure.

In our example, the logical form for question template "Is the patient on |*medication*|?" would be annotated as "MedicationEvent |medication|", where |medication| would be slot-filled with medication names from the KB. The logical form helps identify appropriate tables, entities and values required from the KB.

Structured data typically factually records lab values, vitals, conditions on admission, and medications but rarely records relations between these entities. In case of emrKBQA, none of the questions that involve resolving relations to answer a question in emrQA are answerable from structured data in MIMIC. However, answering questions about schema entities and attributes requires querying and combining information from multiple related tables in MIMIC.

While logical forms are an outcome of the process used to generate emrQA, they are not essential to answering questions over unstructured data like clinical notes. The more traditional use of logical forms is in answering natural language questions from a structured KB. It is easier to convert a question to logical form than to SQL (which is longer and more complex for most questions, often including multiple nested queries and joins). They provide a query-language agnostic intermediate representation that captures information need expressed in the question using a representation that is perhaps more annotator friendly. Moreover, since logical forms are defined over a schema that captures domain-specific entities and relations, they are independent of the underlying database type or query language.

Question Paraphrase Groups. Question paraphrases are different ways of asking the same thing. The emrKBQA dataset is paraphrase rich with an average of 7.5 paraphrases per question. In emrKBQA, questions that map to the same logical form and share the same question type are considered paraphrases. The dataset has 52 question template groups where each group maps to the same logical form. This is because the answer to a question may vary based on question type even if they map to the same logical form. E.g., Consider the questions in Table 1; the first set of questions are paraphrases since their question type is Fact and they map to the same logical form. So the expected answer is the lab values and date. However, in case of the last question, where the question type is YesNo, the expected answer is a Yes or a No along with the lab values and date. The paraphrases were a natural outcome of the question collection process, where the physicians who were polled phrased the same information need in different ways. Paraphrases may be syntactic variations (word re-ordering) or substitution based (word/ phrase substitution) or a combination of the two.

Paraphrases	Ques Type
Previous |test| levels?	Fact
What is |test| value?	Fact
What is the patient's |test| levels?	Fact
How is his |test| trending?	Fact
Show me a trend of his |test|?	Fact
Has |test| been measured before	YesNo

Table 1: Example question paraphrases that map to the same logical form LabEvent (|test|) [date=x, result=x, sortBy(date)] OR VitalEvent (|test|) [date=x, result=x, sortBy(date)], the first set that also share question type are considered paraphrases.

Answers. Answers in emrKBQA are cell values from a table(s) in the KB. Broadly the answer categories in emrKBQA are Test, Medication, Allergy, Therapeutic Procedures, Conditions and Smoking. Figure 2 shows the distribution of questions across different answer categories. Most questions asking about Test are factual or YesNo whereas Condition and Medication have more questions that are Temporal in nature.

As in emrQA, the answers to questions are derived in a semi-automated manner. Each question is mapped to a logical form that captures the entities and relations that are required to adequately answer the question. This mapping is done by a medical expert. The expert uses an ontology that captures entities, entity attributes and relations in the patient record to define the logical form for a question (we use the same schema as emrQA). The slot-filled logical forms such as, "MedicationEvent|lisinopril|", are mapped to an underlying query language using a deterministic procedure (like SQL) that help retrieve the answer from the KB. The answer to this question would be evidence in the structured data that records the patient taking lisinopril along with some contextual details about the medication. "Yes/No, Start date, End date".

Dataset Generation Process. We use the question/logical form templates from emrQA and filter out templates that cannot be mapped to MIMIC structured data. We then map entity placeholders in the templates to MIMIC columns and populate the placeholders with MIMIC data corresponding to the placeholder entity type. The mapping between entity placeholders and the MIMIC tables and columns[2] is shown in Figure 4. Finally, we extract answers from MIMIC. In the example below, the entity |test| is populated by joining the labevents table with d_labitems (dictionary mapping lab itemids to labels) and retrieving the label field (Hemoglobin), which is used to slot fill the question template and the logical form template. The result for this question is a concatenation of value and valueuom (unit of measurement) from the labevents table; these are sorted by the charttime field. Example questions, logical forms, question type and answer categories are shown in Table 2.

4 Dataset Creation Results

emrKBQA consists of 940,713 question answer pairs over 100 patients, generated from 389 question templates and 52 question type-specific logical form templates[3]. emrKBQA contains an average of 7.5 paraphrases per question type-specific logical form template (ranging from 1 to 55), where a paraphrase is defined as question templates sharing the same question type that map to the same logical form template. Of the generated question answer pairs, 90.9% are test results, 7.8% relate to medications, 1.2% to conditions, and the remaining to other topics (e.g., allergies, tobacco use). The limited size of the medication data can be attributed to the use of emrQA questions as the starting point. emrQA questions are based on an outpatient setting where medication data is available while emrKBQA is from an ICU setting where prescription data is available. Thus several questions about adherence, dosage and frequency of medication were not part of emrKBQA. Only 1% (3,429 rows) of the generated dataset were condition related results since fields such as diagnosis time and relationships

[2]https://mit-lcp.github.io/mimic-schema-spy/

[3]the process can be applied to any number of patients

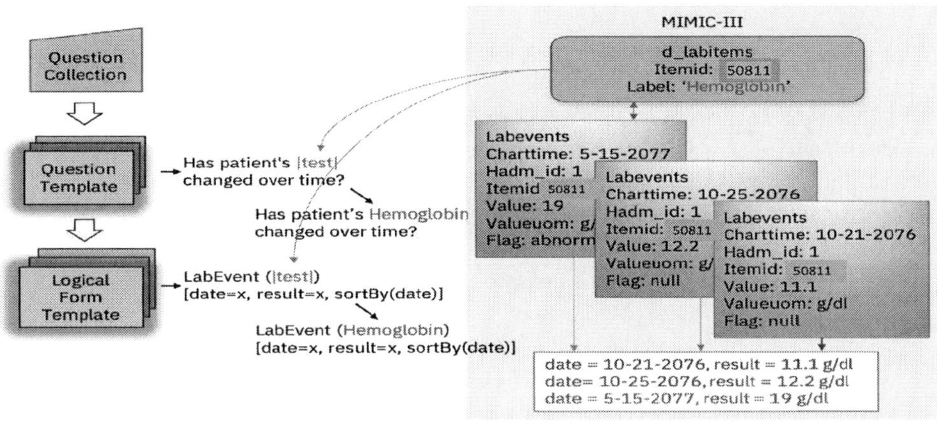

Figure 4: emrKBQA generation process

between treatments and conditions or between medications and conditions are unavailable in MIMIC.

5 Task Definition and Models

Each instance in emrKBQA consists of the follwing elements - question, question paraphrase group, question type, logical form, answer - defined in Section 3. Our goal is to build a model that when presented with a test question on the KB, provides an answer. We achieve this by first modeling the question to logical form learning problem as a semantic parsing task. Here, given an input natural language question, we predict its logical form. Next, we map the predicted logical form to a SQL query in a deterministic manner to retrieve the answer from the KB. The answer is the set of cell values from the underlying KB that answer the question. We detail these two steps in the following sections.

5.1 Semantic Parsing

The task setup for semantic parsing is as follows: given a question in emrKBQA, predict the logical form for that question. As emrKBQA contains several question paraphrases that map to the same logical form, the learning task can be set up in two ways, (1) naive splitting scheme, where input instances are split at random between train and test data, and (2) paraphrase-level splitting scheme, where a question paraphrase seen during train time is not observed in the test set. Thus, the model is tested on whether it can infer the meaning of this question only from its paraphrased forms seen during training. While the paraphrase-level split is more challenging than the naive one, the setting is more realistic. Since the test instances are paraphrases of some training instance, the model is expected to generalize to unseen test instance.

In a previous work, Min et al. (2020) have shown state-of-the-art performance on model generalization for sequence to sequence tasks. They handle unseen sentential paraphrases at test time by incorporating paraphrase detection and generation as auxiliary tasks. In case of paraphrase generation (ParaGen), they sample a question paraphrase during training and learn to generate it along with the main task of logical form prediction. In the paraphrase detection model (ParaDetect), they sample a paraphrase and learn to identify if the sample and the input question are paraphrases by looking at their embeddings in the auxiliiary task. We use the best performing model reported in Min et al. (2020) and perform the following experiments across both splitting schemes: (1) Naive splitting scheme with a baseline model - seq2seq model with copy mechanism (Gu et al., 2016), (2) Paraphrase splitting scheme with a baseline model - seq2seq model with copy mechanism, and (3) Paraphrase splitting scheme with the best-performing ParaGen+ParaDetect model.

5.2 Predicted Logical form to Answer

Finally, the predicted logical form is now mapped to a SQL query to retrieve an answer from the KB. Each question template maps to a logical form template and for each logical form template, we have a corresponding SQL query template. While this mapping is deterministic, the errors in the predicted logical forms require us to use approximate matching functions to map the predicted logical form (template) to the correct logical form template. We

Question	Logical Form	QType	ACat
What were the results of abnormal |test| in |date|?	LabEvent(|test|) [abnormalResultFlag=Y, date=|date|, result=x] OR [{LabEvent(|test|) [date=|date|, abnormalResultFlag=Y]	F	Test
What is the patients |problem| history?	ConditionEvent(|problem|) [diagnosisdate=x] OR SymptomEvent(|problem|) [onsetdate=x]	F	Cond
How long has patient been on |medication|?	MedicationEvent(|medication|) [startdate=x, enddate=x]	T	Med
Has the patient ever been diagnosed or treated for |problem|?	ConditionEvent(|problem|) [diagnosisdate=x] OR [{MedicationEvent(x) OR ProcedureEvent(x)} given ConditionEvent(|problem|)]	YN	Cond

Table 2: Example questions and logical forms across question types Fact(F), Temporal(T), YesNo (YN) and answer categories Test, Condition, Medication

achieve this by matching the by using string similarity measures like edit distance. We then extract the slot filled entity from the predicted logical form and slot fill the SQL query. This query is then run to derive the answer. This answering accuracy is captured in the denotation accuracy metric.

5.3 Experimental Settings

We split emrKBQA dataset according to our two splitting schemes, naive and paraphrase-level, and create two sets of train (70%), dev (10%) and test (20%) datasets. We evaluate the performance of our semantic parsing step using Exact Match (EM) (Min et al., 2020), and our logical form to answer step using Denotation Accuracy (Lin et al., 2019) metrics. EM only considers model outputs that are identical to the labeled ones as correct, while denotation accuracy considers logical forms that return the label answer from the database as correct. We utilize Min et al. (2020)'s public implementation[4] for executing the experiments. We used the default hyperparameters.

5.4 Results

Table 3 presents results of the experiments[5]. The baseline seq2seq with copy model gives high performance in the naive splitting scheme, however the performance drops when we evaluate the model with the paraphrase-level splits. In our experiments, the ParaGen+ParaDetect model provides similar performance to the baseline seq2seq with copy model. This may be attributed to a lack of hyperparameter tuning on out emrKBQA dataset.

[4]https://github.com/jointparalearning/AdvancingSeq2Seq

[5]Results will vary with different initialization seeds

Splitting Scheme	Model	EM	Denotation Accuracy
Naive	Seq2seq with copy	0.95	0.96
Paraphrase	Seq2seq with copy	0.83	0.84
Paraphrase	ParaGen + ParaDetect	0.82	0.82

Table 3: Semantic parsing results on paraphrase splits.

For error analysis, we randomly sampled 100 error instances from our best performing seq2seq with copy model predictions. We present the major error categories with examples in Table 4. Almost half of the errors were attributed to questions with multiple entities. In the first example, the two entities "white blood cells" and date "2139-04-01 06:23:00" are merged to "white 06:23:00" in the predicted logical form, leading to an error. Another big chunk of errors can be attributed to incorrect recognition of the entity types present in the question, e.g., whether the entity is of type lab or procedure, or condition or symptom (example 2). To resolve this error, pretraining the model with a named entity recognition objective might be useful. A next set of errors are due to identification of incorrect span of entities (example 3). This error can be attributed to the fact that the the model has not seen the question form in train data (due to paraphrase-level splits). For the remaining error categories, 7% are caused due to attribute errors like min, max, and finally 4% of the errors are

Question Form	Predicted LF	GT Logical Form	Error category	Perc
what were the results of the abnormal **white blood cells** in **2139-04-01 06:23:00**	labevent (white blood cells) [abnormalresultflag=y, date=2139-04-01 06:23:00, result=x] or procedureevent(white blood cells) [abnormalresultflag=y, date=2139-04-01 06:23:00,result=x] or vitalevent(white blood cells) [date=**white 06:23:00** (result=x)>vital.refhigh] or......	labevent (white blood cells) [abnormalresultflag=y, date=2139-04-01 06:23:00, result=x] or procedureevent(white blood cells) [abnormalresultflag=y, date=2139-04-01 06:23:00,result=x] or vitalevent(white blood cells) [date=**2139-04-01 06:23:00**, (result=x)>vital.refhigh] or	multiple entities	47%
has the patient had a previous **intracerebral hemorrhage**	labevent (intracerebral hemorrhage) [date=x] or procedureevent (intracerebral hemorrhage) [date=x]	conditionevent (intracerebral hemorrhage) [diagnosisdate=x] or symptomevent (intracerebral hemorrhage) [onsetdate=x]	confusion between the entity type	28%
has this patient ever had a documented **chest x-ray** at another va	labevent (documented chest) [date=x] or procedureevent (documented chest) [date=x] or vitalevent (documented chest) [date=x]	labevent (chest x-ray) [date=x] or procedureevent (chest x-ray) [date=x] or vitalevent (chest x-ray) [date=x]	wrong entity span (paraphrase split)	12%
date of **acute bronchitis**	conditionevent (acute bronchitis) [min(diagnosisdate=x)] or symptomevent (acute bronchitis)	conditionevent (acute bronchitis) [diagnosisdate=x] or symptomevent (acute bronchitis) [onsetdate=x]	attribute error	7%
has the patient had a previous **unspecified viral hepatitis c without hepatic coma**	conditionevent (unspecified hepatitis c without hepatic coma) [diagnosisdate=x] or symptomevent (unspecified viral hepatitis c without hepatic coma) [onsetdate=x]]	conditionevent (unspecified hepatitis c without hepatic coma) [diagnosisdate=x] or symptomevent (unspecified viral hepatitis c without hepatic coma) [onsetdate=x]	semantic errors (extra brackets)	4%

Table 4: Error analysis of randomly chosen 100 error instances in the semantic parsing model.

caused due to a long tail of semantic errors like extra brackets, etc.

6 Discussion

Advantages of emrKBQA. emrKBQA is the first large-scale community shared patient-specific QA dataset for answering physician questions from structured patient records. It follows a semi-automated process similar to emrQA (which releases QA pairs on clinical notes), where logical forms are the only expert-provided input. These logical forms lend credibility to the dataset as they capture entities, attributes, and relations required to answer a question and enable slot filling and answer generation. Some highlights of emrKBQA are **(1) Question Quality.** Unlike emrQA, emrKBQA slot-fills entities with discretion by filtering out certain entities based on their attributes (like certain diagnoses based on ICD codes, medications based on drug type). This results in more realistic realization of question instances. **(2) Question Diversity.** The dataset is rich in paraphrases (paraphrase groups have been updated from emrQA) **(3) Dataset Difficulty.** We provide paraphrase-level splits that helps train models that can generalize to unseen paraphrases of the train questions at test time. This is useful in practical settings. As described in the error analysis, in learning to map questions to logical forms, the challenges include recognizing the correct entity spans and types from the question, learning to predict long logical forms, and generating multiple attributes and constraints in the logical form. **(4) Logical forms generated from the same schema as emrQA**, allowing the schema to be a unifying factor across structured and unstructured QA. This allows for future updates in a uniform manner.

Limitations of emrKBQA. (1) Since we wanted the question set to comprise of actual questions asked by physicians, the question set is limited to the initial pool collected from the polled physicians. (2) The dataset is generated in a semi-automated manner that leads to some slot-filled questions that are unlikely to be asked in a real setting. (3) Redundancy of "question form" due to slot filling. Several instances of the same template with different slot-filled entities.

In future versions of the dataset, some of the planned updates include the following: increasing the range of question types, the granularity of questions asked, infuse the need for domain knowledge in understanding a question (using word/ phrase synonyms in slot-filling), better classification of temporal questions based on TimeML, (Pustejovsky et al., 2003), generating more question paraphrases using automated methods (Soni and Roberts, 2019; Min et al., 2020; Neuraz et al., 2018; Dong et al., 2017). While this version of the dataset is generated on randomly sampled 100 patients, we could apply the dataset generation process to any number of patients in MIMIC. It may be interesting to include patient's chosen as per some criteria and contrast answers to similar questions across the chosen cohort.

Differences between emrQA and emrKBQA. emrKBQA is best suited for answering factoid questions such as test results as seen from the results discussed; 87% of emrKBQA (vs 11% of emrQA) comprises test results since test value columns are rarely null. Also, emrKBQA is not limited by annotated clinical notes, which may be a problem if there are very few sources to obtain them. The benefit of emrQA is that it includes questions and answers about medications for problems, response to treatments, temporal constraints and etiology, all of which are unavailable in emrKBQA.

The benefit of a structured dataset such as MIMIC is that explicit values are captured well in tables. Unstructured data may have the answer implicitly stated and may have to be inferred. It also might be incomplete in terms of certain types of crucial information like dates. The limitation of structured data is that it may not capture all types of information. Typically, structured data is unlikely to store symptoms, relations between conditions and symptoms or relations between conditions and treatments. These relations are more likely to be captured by unstructured data.

Question Answering on the entire EHR. emrKBQA is a step in the direction of being able to answer a question anywhere in the EHR, since it utilizes the same schema as emrQA that is used to instantiate logical forms that capture information needs expressed in natural language questions. The answer could now be derived from the structured KB, clinical notes or from both sources in a complementary manner.

7 Conclusion

We create a new large-scale dataset, emrKBQA, for answering patient-specific physician questions from structured patient records. This community-shared release is created in a semi-automated manner and consists of over 900k question-logical form-answer triples, 389 question types (templates), with $\approx$7.5 paraphrases per question type. We benchmark the dataset and quantify its usefulness in answering questions by training models for semantic parsing of questions to logical forms.

Acknowledgement

This work is supported by MIT-IBM Watson AI Lab, Cambridge, MA USA. We thank the anonymous reviewers for their feedback.

References

Jonathan Berant and Percy Liang. 2014. Semantic parsing via paraphrasing. In *Proceedings of the 52nd Annual Meeting of the Association for Computational Linguistics (Volume 1: Long Papers)*, pages 1415–1425, Baltimore, Maryland. Association for Computational Linguistics.

Antoine Bordes, Jason Weston, and Nicolas Usunier. 2014. Open question answering with weakly supervised embedding models. In *Proceedings of the 2014th European Conference on Machine Learning and Knowledge Discovery in Databases - Volume Part I*, ECMLPKDD'14, pages 165–180, Berlin, Heidelberg. Springer-Verlag.

Randall Davis, Bruce Buchanan, and Edward Shortliffe. 1977. Production rules as a representation for a knowledge-based consultation program. *Artificial intelligence*, 8(1):15–45.

Dina Demner-Fushman, Wendy Webber Chapman, and Clement J. McDonald. 2009. What can natural language processing do for clinical decision support? *Journal of Biomedical Informatics*, 42(5):760–772.

Li Dong, Jonathan Mallinson, Siva Reddy, and Mirella Lapata. 2017. Learning to paraphrase for question answering. In *Proceedings of the 2017 Conference on Empirical Methods in Natural Language Processing*, pages 875–886, Copenhagen, Denmark. Association for Computational Linguistics.

R Scott Evans. 2016. Electronic health records: then, now, and in the future. *Yearbook of medical informatics*, Suppl 1:S48.

Jiatao Gu, Zhengdong Lu, Hang Li, and Victor O.K. Li. 2016. Incorporating copying mechanism in sequence-to-sequence learning. In *Proceedings of the 54th Annual Meeting of the Association for Computational Linguistics (Volume 1: Long Papers)*, pages 1631–1640, Berlin, Germany. Association for Computational Linguistics.

David A Hanauer, Qiaozhu Mei, James Law, Ritu Khanna, and Kai Zheng. 2015. Supporting information retrieval from electronic health records: A report of university of michigan's nine-year experience in developing and using the electronic medical record search engine (emerse). *Journal of biomedical informatics*, 55:290–300.

Ashish K Jha, Timothy G Ferris, Karen Donelan, Catherine DesRoches, Alexandra Shields, Sara Rosenbaum, and David Blumenthal. 2006. How common are electronic health records in the united states? a summary of the evidence: About one-fourth of us physician practices are now using an ehr, according to the results of high-quality surveys. *Health Affairs*, 25(Suppl1):W496–W507.

Alistair EW Johnson, Tom J Pollard, Lu Shen, H Lehman Li-Wei, Mengling Feng, Mohammad Ghassemi, Benjamin Moody, Peter Szolovits, Leo Anthony Celi, and Roger G Mark. 2016. Mimic-iii, a freely accessible critical care database. *Scientific data*, 3(1):1–9.

Kevin Lin, Ben Bogin, Mark Neumann, Jonathan Berant, and Matt Gardner. 2019. Grammar-based neural text-to-sql generation. *arXiv preprint arXiv:1905.13326*.

So Yeon Min, Preethi Raghavan, and Peter Szolovits. 2020. Advancing seq2seq with joint paraphrase learning. In *Proceedings of the 3rd Clinical Natural Language Processing Workshop*, pages 269–279.

Antoine Neuraz, Leonardo Campillos Llanos, Anita Burgun, and Sophie Rosset. 2018. Natural language understanding for task oriented dialog in the biomedical domain in a low resources context. *arXiv preprint arXiv:1811.09417*.

Anusri Pampari, Preethi Raghavan, Jennifer Liang, and Jian Peng. 2018. emrQA: A large corpus for question answering on electronic medical records. In *Proceedings of the 2018 Conference on Empirical Methods in Natural Language Processing*, pages 2357–2368, Brussels, Belgium. Association for Computational Linguistics.

Junwoo Park, Youngwoo Cho, Haneol Lee, Jaegul Choo, and Edward Choi. 2020. Knowledge graph-based question answering with electronic health records. *arXiv preprint arXiv:2010.09394*.

James Pustejovsky, José M Castano, Robert Ingria, Roser Sauri, Robert J Gaizauskas, Andrea Setzer, Graham Katz, and Dragomir R Radev. 2003. Timeml: Robust specification of event and temporal expressions in text. *New directions in question answering*, 3:28–34.

Preethi Raghavan, Siddharth Patwardhan, Jennifer J Liang, and Murthy V Devarakonda. 2018. Annotating electronic medical records for question answering. *arXiv preprint arXiv:1805.06816*.

Kirk Roberts and Dina Demner-Fushman. 2015. Toward a natural language interface for ehr questions. *AMIA Summits on Translational Science Proceedings*, 2015:157.

Kirk Roberts and Dina Demner-Fushman. 2016. Annotating logical forms for ehr questions. In *LREC... International Conference on Language Resources & Evaluation:[proceedings]. International Conference on Language Resources and Evaluation*, volume 2016, page 3772. NIH Public Access.

Sarvesh Soni and Kirk Roberts. 2019. A paraphrase generation system for ehr question answering. In *Proceedings of the 18th BioNLP Workshop and Shared Task*, pages 20–29.

Özlem Uzuner, Brett R South, Shuying Shen, and Scott L DuVall. 2011. 2010 i2b2/va challenge on concepts, assertions, and relations in clinical text. *Journal of the American Medical Informatics Association*, 18(5):552–556.

Alfredo Vellido, José David Martín-Guerrero, and Paulo JG Lisboa. 2012. Making machine learning models interpretable. In *ESANN*, volume 12, pages 163–172. Citeseer.

Ping Wang, Tian Shi, and Chandan K Reddy. 2020. Text-to-sql generation for question answering on electronic medical records. In *Proceedings of The Web Conference 2020*, pages 350–361.

Luke S. Zettlemoyer and Michael Collins. 2005. Learning to map sentences to logical form: Structured classification with probabilistic categorial grammars. In *Proceedings of the Twenty-First Conference on Uncertainty in Artificial Intelligence*, UAI'05, pages 658–666, Arlington, Virginia, United States. AUAI Press.

Overview of the MEDIQA 2021 Shared Task on Summarization in the Medical Domain

Asma Ben Abacha
NLM/NIH
benabachaa@nih.gov

Yassine Mrabet
NLM/NIH
mrabety@mail.nih.gov

Yuhao Zhang
Stanford University
zyh@stanford.edu

Chaitanya Shivade
Amazon
shivadc@amazon.com

Curtis Langlotz
Stanford University
langlotz@stanford.edu

Dina Demner-Fushman
NLM/NIH
ddemner@mail.nih.gov

Abstract

The MEDIQA 2021 shared tasks at the BioNLP 2021 workshop addressed three tasks on summarization for medical text: (i) a question summarization task aimed at exploring new approaches to understanding complex real-world consumer health queries, (ii) a multi-answer summarization task that targeted aggregation of multiple relevant answers to a biomedical question into one concise and relevant answer, and (iii) a radiology report summarization task addressing the development of clinically relevant impressions from radiology report findings. Thirty-five teams participated in these shared tasks with sixteen working notes submitted (fifteen accepted) describing a wide variety of models developed and tested on the shared and external datasets. In this paper, we describe the tasks, the datasets, the models and techniques developed by various teams, the results of the evaluation, and a study of correlations among various summarization evaluation measures. We hope that these shared tasks will bring new research and insights in biomedical text summarization and evaluation.

1 Introduction

Text summarization aims to create natural language summaries that represent the most important information in a given text. Extractive summarization approaches tackle the task by selecting content from the original text without any modification (Nallapati et al., 2017; Xiao and Carenini, 2019; Zhong et al., 2020), while abstractive approaches extend the summaries' vocabulary to out-of-text words (Rush et al., 2015; Gehrmann et al., 2018; Chen and Bansal, 2018).

Several past challenges and shared tasks have focused on summarization. The Document Understanding Conference[1] (DUC) organized seven challenges from 2000 to 2007 and the Text Analysis Conference[2] (TAC) ran four shared tasks (2008-2011) on news summarization. The last TAC 2014 summarization task tackled biomedical article summarization with referring sentences from external citations. Recent efforts in summarization have focused on neural methods (See et al., 2017; Gehrmann et al., 2018) using benchmark datasets compiled from news articles, such as the CNN-DailyMail dataset (CNN-DM) (Hermann et al., 2015). However, despite its importance, fewer efforts have tackled text summarization in the biomedical domain for both consumer and clinical text and its applications in Question Answering (QA) (Afantenos et al., 2005; Mishra et al., 2014; Afzal et al., 2020).

While the 2019 BioNLP-MEDIQA[3] edition focused on question entailment and textual inference and their applications in medical Question Answering (Ben Abacha et al., 2019), MEDIQA 2021[4] addresses the gap in medical text summarization by promoting research on summarization for consumer health QA and clinical text. Three shared tasks are proposed for the summarization of (i) consumer health questions, (ii) multiple answers extracted from reliable medical sources to create one answer for each question, and (iii) textual clinical findings in radiology reports to generate radiology impression statements.

For the first two tasks, we created new test sets for the official evaluation using consumer health questions received by the U.S. National Library of Medicine (NLM) and answers retrieved from reliable sources using the Consumer Health Question Answering system CHiQA[5]. For the third task, we created a new test set by combining public radiology reports in the Indiana Univer-

[1]www-nlpir.nist.gov/projects/duc

[2]tac.nist.gov/tracks

[3]sites.google.com/view/mediqa2019

[4]sites.google.com/view/mediqa2021

[5]chiqa.nlm.nih.gov

Proceedings of the BioNLP 2021 workshop, pages 74–85
June 11, 2021. ©2021 Association for Computational Linguistics

sity dataset (Demner-Fushman et al., 2016) and newly released chest x-ray reports from the Stanford Health Care.

Through these tasks, we focus on studying:

- The best approaches according to the summarization task objective and the language/vocabulary (consumers' questions, patient-oriented medical text, and professional clinical reports);
- The impact of medical data scarcity on the development and performance of summarization methods in comparison with open-domain summarization;
- The effects of different summary evaluation measures including lexical metrics such as ROUGE (Lin, 2004), embedding-based metrics such as BERTScore (Zhang et al., 2019), and hybrid ensemble-oriented metrics such as HOLMS (Mrabet and Demner-Fushman, 2020).

2 MEDIQA 2021 Task Descriptions

2.1 Consumer Health Question Summarization (QS)

Consumer health questions tend to contain peripheral information that hinders automatic Question Answering (QA). Empirical studies based on manual expert summarization of these questions showed a substantial improvement of 58% in QA performance (Ben Abacha and Demner-Fushman, 2019a). Effective automatic summarization methods for consumer health questions could therefore play a key role in enhancing medical question answering. The goal of this task is to promote the development of new summarization approaches that address specifically the challenges of long and potentially complex consumer health questions. Relevant approaches should be able to generate a condensed question expressing the minimum information required to find correct answers to the original question (Ben Abacha and Demner-Fushman, 2019b).

2.2 Multi-Answer Summarization (MAS)

Different answers can bring complementary perspectives that are likely to benefit the users of QA systems. The goal of this task is to promote the development of multi-answer summarization approaches that could solve simultaneously the aggregation and summarization problems posed by multiple relevant answers to a medical question (Savery et al., 2020).

2.3 Radiology Report Summarization (RRS)

The task of radiology report summarization aims to promote the development of clinical summarization models that are able to generate the concise impression section (i.e., summary) of a radiology report conditioned on the free-text findings and background sections (Zhang et al., 2018). The resulting systems have significant potential to improve the efficiency of clinical communications and accelerate the radiology workflow. While state-of-the-art techniques in language generation have enabled the generation of fluent summaries, these models occasionally generate spurious facts limiting the clinical validity of the generated summaries (Zhang et al., 2020b). It is therefore important to develop systems that are able to summarize the radiology findings in a consistent manner.

3 Data Description

3.1 QS Datasets

The MeQSum dataset of consumer health questions and their summaries (Ben Abacha and Demner-Fushman, 2019b) was suggested as a training dataset. It consists of 1,000 consumer health questions and their associated summaries. Participants were encouraged to use available external resources including, but not limited to, medical QA datasets and question focus and type recognition datasets. For instance, the Consumer Health Questions dataset (Kilicoglu et al., 2018) contains annotations of medical entities, focus, and type of the MeQSum questions and additional NLM questions[6].

The new QS validation and test sets[7] cover a wide range of topics and question types such as *Treatment*, *Information*, *Side effects*, *Cause*, *Effect*, *Person-Organization*, *Diet-Lifestyle*, *Complications*, *Contraindications*, *Diagnosis*, *Usage*, *Interaction*, *Ingredients*, *Prognosis*, *Susceptibility*, *Transmission*, and *Toxicity*. They consist of manually de-identified consumer health questions received by the U.S. National Library of Medicine and gold summaries created by medical experts. The validation set includes 50 NLM questions and

[6] https://bionlp.nlm.nih.gov/CHIQAcollections/CHQA-Corpus-1.0.zip

[7] https://github.com/abachaa/MEDIQA2021/tree/main/Task1

Example 1 (QID: 139)
NLM Question: *did anyone have this and does it require surgery? my mri says forminal stenosis from bone spurs c4,5,6. my nerve test shows severe nerve compression c7,8. i'm in so much pain, mostly my arm and shoulder and leg. waiting to see the pain specialist to see what's next. would love to know what you guys think is required.* **Question Summary:** *How can I get rid of pain caused by foraminal stenosis and nerve compression?*
Example 2 (QID: 111)
NLM Question: *covid-19 how long to quarantine after being positive how long are you contagious if i tested positive for covid-19. how long before i can safely return to work after a positive covid 19 test* **Question Summary:** *How long will I remain contagious after testing positive for COVID-19?*

Table 1: Test set examples for the QS task.

their summaries with additional annotations of the question focus and type. The test set contains 80 consumer health questions. Table 1 presents two examples from the QS test set.

3.2 MAS Datasets

The MEDIQA-AnS dataset (Savery et al., 2020) was suggested as a training set for the MAS task. Participants were allowed to use available external resources (e.g. existing medical QA datasets) as well as data creation, selection, and augmentation methods. To create the MAS validation and test sets[8], we used 130 consumer health questions received by NLM. In order to retrieve more accurate answers, we created question summaries that we used to query the medical QA system CHiQA that searches for answers from only trustworthy medical information sources (Ben Abacha and Demner-Fushman, 2019c; Demner-Fushman et al., 2020).

The answer summaries were manually created by medical experts. We provided both extractive and abstractive gold summaries, and encouraged the use of all types of summarization approaches (extractive, abstractive, and hybrid). The MAS validation set contains 192 answers to 50 medical questions. The test set contains 303 answers to 80 medical questions. Each question has at least two answers, one extractive multi-answer summary, and one abstractive multi-answer summary. Table 2 presents an example from the test set.

Original NLM question: *I have dementia like symptoms and wanted to know where is the best source to be tested for diagnosis? I have been prescribed Anticholinergic medicine since 2008...since I have been diagnosed with, Celiac disease and Obstructive Sleep Apnea. I think I have Frontal Temporal lobe atrophy. I'm going to try to get tested...any references on which process is easiest would be much appreciated. I can't take my Nasalcrom allergy spay any more nor, valium or prozac, benadryl and glutamate additives in meats because it sends me straight into cognitive emergency state and irrational thinking*
Question summary used in answer retrieval: *What tests are used to diagnose dementia?*
CHiQA's Answer #1: *Dementia is not a specific disease. It is a descriptive term for a collection of symptoms that can be caused by a number of disorders that affect the brain. People with dementia have significantly impaired intellectual functioning that interferes with normal activities and relationships. They also lose their ability to solve problems and maintain emotional control, and they may experience personality changes and behavioral problems, such as agitation, delusions, and hallucinations (...).*
CHiQA's Answer #2: *To diagnose dementia, doctors first assess whether a person has an underlying treatable condition such as depression, abnormal thyroid function, normal pressure hydrocephalus, or vitamin B12 deficiency. Early diagnosis is important, as some causes for symptoms can be treated. In many cases, the specific type of dementia a person has may not be confirmed until after the person has died and the brain is examined.A medical assessment for dementia generally includes:- Patient history (...) - Physical exam (...) - Neurological tests (...).*
Reference Extractive Summary: *Dementia is not a specific disease. It is a descriptive term for a collection of symptoms that can be caused by a number of disorders that affect the brain. Doctors diagnose dementia only if two or more brain functions - such as memory and language skills – are significantly impaired without loss of consciousness. To diagnose dementia, doctors first assess whether a person has an underlying treatable condition such as depression, abnormal thyroid function, normal pressure hydrocephalus, or vitamin B12 deficiency. Early diagnosis is important, as some causes for symptoms can be treated. In many cases, the specific type of dementia a person has may not be confirmed until after the person has died and the brain is examined.A medical assessment for dementia generally includes:- Patient history (...) - Physical exam (...) - Neurological tests (...).*
Reference Abstractive Summary: *Dementia could be caused by many different diseases of the brain. it is diagnosed if at least two brain functions are effected, for example, if people experience memory loss and changes in behavior and personality. Diagnostic tests for dementia include family history, physical examination, and neurological tests to asses balance, sensory functions, reflexes, vision, eye movements, and cognitive functions. In many cases, the type of dementia is confirmed after the person dies.*

Table 2: Test set example for the MAS task (QID:105).

[8] https://github.com/abachaa/MEDIQA2021/tree/main/Task2

3.3 RRS Datasets

We focus on the summarization of chest radiography reports for the RRS task, since chest radiography represents the most common study type in radiology, and public resources for chest studies are easily accessible. For training, we sampled a collection of 91,544 reports from the MIMIC-CXR chest X-ray report dataset[9] based on simple criteria such as the acceptable length of each section. For validation, we combined another 2,000 reports from the MIMIC-CXR dataset and 2,000 reports from the Indiana University chest X-ray dataset[10](Demner-Fushman et al., 2016). We sampled the reports such that there is no overlapping patients in the validation and training sets.

For the official test set, we used a combination of 300 reports from the Indiana dataset and 300 newly released chest X-ray reports drawn from the Stanford Health Care system. We intentionally designed the test set to be partially from a hospital system different from the training set (out-of-domain) to test the generalizability of the participating systems.

4 Evaluation

4.1 Evaluation Measures

Several new metrics for evaluating text generation systems were studied in recent years (Mao et al., 2020; Bhandari et al., 2020a,b; Zhang et al., 2019; Sellam et al., 2020), with a focus on evaluating text generation based on deep and contextualized representations. To understand these metrics in the context of summarization, Fabbri et al. (2020) have compared 34 traditional and recent model-based metrics on a manually annotated subset from the CNN-DM dataset. Although the study relied only on one correlation factor (Kendall's Tau) and one dataset, it highlighted the (continued) general relevance of ROUGE variants (Lin, 2004) and the challenge of designing or determining the best measure to use. Specifically, the study found that a different measure obtained the best score in each of the four considered evaluation dimensions: *coherence*, *consistency*, *fluency*, and *relevance*, with substantial discrepancies in rankings.

In parallel, HOLMS was recently proposed as an ensemble measure combining both contextualized similarity and a lexical ROUGE component through a multi-dimensional Gaussian function (Mrabet and Demner-Fushman, 2020). HOLMS was evaluated on multiple DUC and TAC datasets, and three correlation factors (Pearson's, Spearman's, and Kendall's), and was shown to benefit from the complementary strengths of lexical and language model-based similarity measurements for evaluating summarization systems.

In this shared task, we chose ROUGE-2 as our official ranking metric following its superiority observed by Owczarzak et al. (2012) on multiple TAC summarization datasets, and by Bhandari et al. (2020c) on the CNN-DM dataset.

We chose two additional metrics for the three tasks: (1) BERTScore for its wider adoption as a language model-based text generation metric, and (2) HOLMS for its hybrid and ensemble-oriented approach. For the RRS task we also considered an additional evaluation metric based on the hamming similarity on the labels produced by the CheXbert labeler (Smit et al., 2020) when applied to both the system and reference summaries, similar to the approach by Zhang et al. (2020b).

4.2 Baseline Systems

Our baseline system for the QS task relied on a distilled PEGASUS model (Zhang et al., 2020a) trained on the CNN-DM dataset and fine-tuned on a combination of biomedical answer-to-question data and question summarization data from MeQSum, LiveQA-Med data (Ben Abacha et al., 2017), a collection of clinical questions (Ely et al., 2000), and Quora question pairs dataset (Iyer et al., 2017). For the Quora and clinical questions datasets, we extracted only the question pairs with a minimum token reduction ratio of 33%.

Our extractive baseline for the MAS task relied on sentence clustering and selection. We used our fine-tuned question summarization model to generate a short question from each sentence, and then clustered the sentences using a word-based cosine distance between the generated questions and a distance threshold set to 0.7. Intersecting clusters were merged. For each cluster, we selected the sentence that was the best cumulative TF-IDF answer to all other sentences as a representative.

For the RRS task, we prepared three baselines: a base pointer-generator model without modeling the background section of a radiology report, a full pointer-generator model with background model-

[9] https://physionet.org/content/mimic-cxr/2.0.0/

[10] openi.nlm.nih.gov/faq#collection

ing (Zhang et al., 2018), and a zero-shot T5-base summarization model (Raffel et al., 2020).

5 Official Results

We published three AIcrowd projects (one for each task) to release the datasets and manage team registration, submission, and leaderboard ranking[11].

5.1 Participating Teams

In total, 35 teams participated in the MEDIQA shared tasks and submitted 310 individual runs (with a limit of ten runs per team per task). Table 3 presents the participating teams with accepted working notes papers. The results of all 35 teams are available on AIcrowd and on the MEDIQA 2021 website.

5.2 Summarization Approaches & Results

A vast majority of the approaches submitted to the QS and RRS tasks were abstractive and relied on fine-tuning of pre-trained generative language models and encoders-decoders architectures. For the MAS task, most submitted approaches were extractive and used a wide spectrum of sentence selection techniques.

Question Summarization. Table 4 presents the official results of the teams with accepted working notes papers from the 22 teams that participated in the QS task.

All approaches submitted to the question summarization task were abstractive methods relying on the fine-tuning of pretrained transformer models (Vaswani et al., 2017). A wide variety of fine tuning, knowledge-based, and ensemble methods was investigated by the participating teams to achieve higher performance (Mrini et al., 2021; Xu et al., 2021; Zhu et al., 2021; Sänger et al., 2021; Lee et al., 2021b; Balumuri et al., 2021; Yadav et al., 2021; He et al., 2021; Lee et al., 2021a). A first interesting insight from the overview is that building ensemble models with deep neural networks such as discriminators is not a trivial task, and achieves results that stay on par with the best single model (Sänger et al., 2021). In contrast, heuristic, downstream ensembles of the models outputs led to substantial improvements when compared to its components/single models (He et al., 2021). The best performing approach relied on such an ensemble by ranking the outputs of PEGASUS, T5, and BART models according to hand-picked features based on the contents of the input question and lengths of the outputs. Spell checking was also a performance boost factor in the question summarization task with some teams using a knowledge base to correct misspelling errors in the original long questions (He et al., 2021), and others relying on third party tools such as CSpell (Yadav et al., 2021; Lu et al., 2019). The datasets used for transfer learning or fine-tuning also played a major role in the achieved performance as demonstrated, for instance, by the combination of datasets from HealthCareMagic, question entailment recognition and question summarization in (Mrini et al., 2021). Moving forward, we think that the overview of the question summarization task revealed two key challenges that need to be addressed to enhance the relevance and performance of existing systems:

1. a relevant learning-based ensemble method that could rely either on the textual outputs or the logits of single models.

2. a more systemic way to select the most relevant datasets for both pretraining and fine tuning.

Multi-Answer Summarization. Both extractive and abstractive approaches were used by the 17 teams that submitted runs to MAS task (Zhu et al., 2021; Can et al., 2021; Xu et al., 2021; Mrini et al., 2021; Yadav et al., 2021; Le et al., 2021; Lee et al., 2021a). Table 5 and Table 6 present official results of the teams with extractive and abstractive systems when evaluated, respectively, on extractive gold summaries and abstractive gold summaries.

The best MAS run (Zhu et al., 2021) relied on an ensemble method and a recent multi-document summarization approach (Xu and Lapata, 2020) using a Roberta model to rank locally the candidate sentences and a Markov chain to evaluate them globally. A similar approach was also used by the ChicHealth team (Xu et al., 2021) without a downstream ensemble method. Participating teams used transfer learning (e.g. (Mrini et al., 2021)) as well as answer sentence selection methods. Sentence selection was used in building extractive summaries (e.g. (Can et al., 2021)) and as an intermediate step in abstractive summarization to provide more concise inputs to generative models (e.g. (Le et al., 2021)). Different models, such

[11] `www.aicrowd.com/challenges/mediqa-2021`

Team	Institution	QS	MAS	RRS
BDKG (Dai et al., 2021)	Baidu, Inc			✓
ChicHealth (Xu et al., 2021)	Chic Health		✓	✓
damo_nlp (He et al., 2021)	Alibaba Group	✓		✓
IBMResearch (Mahajan et al., 2021)	IBM Research			✓
MNLP (Lee et al., 2021a)	George Mason University	✓	✓	
NCUEE-NLP (Lee et al., 2021b)	National Central University	✓		
NLM (Yadav et al., 2021)	U.S. National Library of Medicine	✓	✓	
optumize (Kondadadi et al., 2021)	Optum			✓
paht_nlp (Zhu et al., 2021)	ECNU & Pingan Health Tech	✓	✓	✓
QIAI (Delbrouck et al., 2021)	Stanford University	✓		✓
SB_NITK (Balumuri et al., 2021)	National Institute of Technology Karnataka	✓		
UCSD-Adobe (Mrini et al., 2021)	UC San Diego & Adobe Research	✓	✓	
UETfishes (Le et al., 2021)	VNU University of Engineering and Technology		✓	
UETrice (Can et al., 2021)	VNU University of Engineering and Technology		✓	
WBI (Sänger et al., 2021)	Humboldt University of Berlin	✓		

Table 3: Participating teams with accepted working notes papers at BioNLP-MEDIQA 2021

Rank	Team	ROUGE-2	ROUGE-1	ROUGE-L	HOLMS	BERTScore
1	damo_nlp	**0.1608**	**0.3514**	0.3131	0.5677	0.6898
2	WBI	0.1599	0.3340	**0.3149**	0.5767	0.6996
3	NCUEE-NLP	0.1597	0.3352	0.3090	**0.5787**	0.6960
4	NLM	0.1514	0.3556	0.3110	0.5649	0.6892
5	UCSD-Adobe	0.1414	0.3463	0.3065	0.5586	0.6942
6	ChicHealth	0.1398	0.3403	0.2962	0.5551	0.6810
7	SB_NITK	0.1393	0.3331	0.3077	0.5663	**0.7025**
–	*QS Baseline*	0.1373	0.3203	0.2962	0.5672	0.6277
8	MNLP	0.1114	0.2840	0.2587	0.5455	0.6732
9	paht_nlp	0.0935	0.2486	0.2331	0.5428	0.6591
10	QIAI	0.0385	0.1514	0.1356	0.4898	0.5101

Table 4: Official results of the MEDIQA-QS task.

Rank	Team	ROUGE-2	ROUGE-1	ROUGE-L	HOLMS	BERTScore
1	paht_nlp	**0.5076**	0.5848	0.4354	0.7047	**0.8038**
2	UETrice	0.5040	**0.6110**	**0.4412**	0.7383	0.7958
3	ChicHealth	0.4893	0.5776	0.4261	0.7033	0.7916
4	UCSD-Adobe	0.4720	0.6073	0.4289	**0.7612**	0.7753
5	NLM	0.4677	0.5470	0.3276	0.6575	0.7645

Table 5: Official results of the MEDIQA-MAS task (1): **Extractive Approaches**.

Team	Rank	ROUGE-2	ROUGE-1	ROUGE-L	HOLMS	BERTScore
paht_nlp	**1**	**0.5076**	0.5848	**0.4354**	0.7047	**0.8038**
	(1)	**0.1621**	0.3215	0.1910	0.4220	**0.6528**
UETfishes	**2**	0.4698	0.5720	0.4001	0.6970	0.7821
	(3)	0.1495	0.3124	0.1885	0.4213	0.6466
UCSD-Adobe	**3**	0.4595	**0.5921**	0.4170	**0.7502**	0.7689
	(2)	0.1604	**0.3843**	**0.2117**	0.4937	0.6326
MNLP	4	0.2594	0.4220	0.2954	0.6568	0.6479
	(4)	0.1167	0.3490	0.2047	**0.5269**	0.5763

Table 6: Official results of the MEDIQA-MAS task (2): **Abstractive Approaches.** Ranks in bold and in parenthesis correspond to evaluation on extractive gold summaries and on abstractive gold summaries, respectively.

Rank	Team	R-2	R-1	R-L	HOLMS	BERTScore	CheXbert
1	BDKG	**0.4362**	**0.5572**	**0.5365**	**0.7402**	**0.7184**	**0.6927**
2	IBMResearch	0.4082	0.5328	0.5134	0.7185	0.7115	0.6774
3	optumize	0.3918	0.5185	0.4957	0.7087	0.6975	0.6773
4	QIAI	0.3778	0.4954	0.4793	0.7132	0.5328	0.5565
5	ChicHealth	0.3236	0.4606	0.4410	0.6822	0.6768	0.6261
6	damo_nlp	0.2763	0.4329	0.4115	0.6604	0.6576	0.6343
–	*baseline (PG-full)*	0.2734	0.4182	0.4041	0.6647	0.6194	0.6014
–	*baseline (PG-base)*	0.2639	0.4026	0.3885	0.6553	0.6103	0.5537
7	paht_nlp	0.1987	0.3400	0.3053	0.5915	0.5985	0.6705
–	*baseline (T5)*	0.0945	0.2108	0.1831	0.4432	0.4921	0.5245

Table 7: Official results of the MEDIQA-RRS task on the full test set.

Rank	Team	ROUGE-2		CheXbert	
		Stanford	Indiana	Stanford	Indiana
1	BDKG	**0.2768**	**0.5955**	0.6547	**0.7052**
2	ChicHealth	0.2690	0.3781	0.6291	0.5873
3	damo_nlp	0.2687	0.2839	0.6645	0.5517
4	optumize	0.2654	0.5182	0.6474	0.6592
5	QIAI	0.2516	0.5039	0.5508	0.4970
6	paht_nlp	0.2491	0.1483	**0.6834**	0.6148
–	*baseline (PG-full)*	0.2414	0.3054	0.6216	0.5466
–	*baseline (PG-base)*	0.2408	0.2870	0.5892	0.4754
7	IBMResearch	0.2283	0.5880	0.6472	0.6937
–	*baseline (T5)*	0.1280	0.0610	0.5067	0.5609

Table 8: Official results of the MEDIQA-RRS task on the Stanford and Indiana test splits.

as BART and T5, and datasets (e.g. MEDIQA-AnS, MSMARCO, MEDIQA-2019) have been used for single and multiple answer summarization (Yadav et al., 2021; Mrini et al., 2021; Zhu et al., 2021; Can et al., 2021).

Radiology Report Summarization. 14 teams participated in the RRS task. Table 7 presents the official results of the teams (with accepted papers) on the full test set, and Table 8 presents the results on the Stanford and Indiana subsets of the test set.

Similar to the previous tasks, participating teams for the RRS task have extensively used pretrained transformer models: out of the 7 teams that submitted papers describing their systems, 6 reported the use of pretrained language models such as BART or PEGASUS in their submissions (Xu et al., 2021; Zhu et al., 2021; Kondadadi et al., 2021; Dai et al., 2021; Mahajan et al., 2021; He et al., 2021). Among them, Xu et al. (2021); Zhu et al. (2021); Dai et al. (2021) reported that best results were achieved with pretrained PEGASUS models, while Kondadadi et al. (2021) reported better results from BART. Xu et al. (2021) and Zhu et al. (2021) reported that using PEGASUS models pretrained on the PubMed corpus yielded worse results than using the general PEGASUS models, potentially due to the domain difference of the RRS task with the PubMed text.

In addition to the use of pretrained models, the highest-ranked systems from Dai et al. (2021) made effective use of a dedicated domain adaptation module, an ensemble module, and text normalization heuristics. Zhu et al. (2021) reported that freezing the embedding layer in the pretrained models helps the model generalize at test time. Kondadadi et al. (2021) reported that adding the background section as input improves performance at validation time, but not test time, suggesting that the model performance is sensitive to the different text styles of the background sections from different splits. Mahajan et al. (2021) focused their study on the factual consistency of generated summaries, and proposed a specialized fact-aware re-ranking approach based on the predicted disease values from the findings section with a transformer model. As a result, their submissions achieved competitive rankings under the CheXbert metric. Lastly, Delbrouck et al. (2021) studied the use of image features for the RSS task: they retrieved and linked images for each study to the report at training and validation time, and combined a visual encoder with a text encoder for the summarization task. They found that at validation time the multi-modal setting is beneficial to the summarization of MIMIC reports, but not to the Indiana reports, potentially due to the distribution shift in the images.

6 Correlations among the Evaluation Measures

In this section, we discuss correlations between the different evaluation metrics that we used in the challenge. Table 9 shows Pearson correlations between the F1 scores of the three lexical measures (ROUGE-1, ROUGE-2, and ROUGE-L) and the two language model-based and ensemble-based measures (i.e., HOLMS and BERTScore).

Over all three tasks the HOLMS metric had a better Pearson correlation with ROUGE, ranging from 0.734 to 0.755, while also maintaining a high correlation of 0.736 with BERTScore. This observation supports the findings from the experiments in (Mrabet and Demner-Fushman, 2020), which suggested that lexical measures such as ROUGE and language model-based measures bring different and complementary perspectives to summary-evaluation.

Table 10 shows Pearson correlations for the RRS task. HOLMS is substantially closer than CheXbert and BERTScore in its correlation with ROUGE for the RRS task, while maintaining high correlation of respectively 0.645 and 0.702 with CheXbert and BERTScore.

In contrast, BERTScore is substantially closer than HOLMS in its correlation with the ROUGE metrics for both the MAS task (cf. table 11) and the QS task (see Table 12). Two factors that could explain these correlations are (i) the predominance of extractive runs in the MAS task and (ii) the sequential n-gram-based modeling in HOLMS that takes into account the order of the n-grams, while BERTScore relies on a cosine distance between two given sets of token embeddings.

Both language model-based measures had positive correlations with ROUGE for the QS task, but the level of correlation was substantially lower when compared to the MAS and RRS tasks, going from a Pearson coefficient range between 0.663 and 0.958 to a range between 0.193 and 0.372. As all submitted QS runs were described as abstractive or hybrid approaches, this discrepancy might be due to a stronger disagreement on summary assessment due to semantically-close but lexically distant summaries. It is also likely that the lexical distance between paraphrases was more pronounced due to the lengths of the question summaries, which are shorter than the summaries in the MAS task.

7 Conclusion

We presented an overview of the MEDIQA 2021 shared tasks on summarization in the medical domain. We presented the results for the three tasks on Question Summarization, Multi-Answer Summarization and Radiology Reports Summarization, and discussed the impact of summarization approaches and automatic evaluation methods. We find that pre-trained transformer models, fine-tuning on the carefully selected domain-specific text and ensemble methods worked well for all three summarization tasks. The results encourage future research to include in-depth exploration of ensemble methods, systematic approaches to selection of datasets for pre-training and fine-tuning, as well as a thorough assessment of the quality and relevance of different evaluation measures for summarization. We hope that the MEDIQA 2021 shared tasks will encourage further research efforts in medical text summarization and evaluation.

Acknowledgments

This research was partially supported by the Intramural Research Program of the National Library of Medicine, National Institutes of Health. We thank Anna Ripple (NLM/NIH) for her help with the manual annotation and Soumya Gayen (NLM/NIH) for his help with the summarization interface.

References

Stergos D. Afantenos, Vangelis Karkaletsis, and Panagiotis Stamatopoulos. 2005. Summarization from medical documents: A survey. *Artif Intell Med*, 33(2):157–77.

Muhammad Afzal, Fakhare Alam, Khalid Mahmood Malik, and Ghaus M Malik. 2020. Clinical context–aware biomedical text summarization using deep

Measure	ROUGE-1	ROUGE-2	ROUGE-L	HOLMS	BERTScore
ROUGE-1	1.000				
ROUGE-2	0.966	1.000			
ROUGE-L	0.813	0.762	1.000		
HOLMS	**0.734**	**0.722**	**0.755**	1.000	
BERTScore	0.546	0.519	0.409	**0.736**	1.000

Table 9: Pearson correlations between metrics aggregated over all three tasks. For ROUGE and BERTScore we use their F1 scores. Best correlations with the ROUGE metrics are highlighted in bold.

Measure	ROUGE-1	ROUGE-2	ROUGE-L	CheXbert	HOLMS	BERTScore
ROUGE-1	1.000					
ROUGE-2	0.970	1.000				
ROUGE-L	0.998	0.975	1.000			
CheXbert	0.777	0.667	0.749	1.000		
HOLMS	**0.951**	**0.938**	**0.958**	0.645	1.000	
BERTScore	0.752	0.663	0.743	**0.719**	**0.702**	1.000

Table 10: Pearson correlations between metrics for the RRS task. For ROUGE and BERTScore we used the F1 scores. Best correlations with the lexical measures are highlighted in bold.

Measure	ROUGE-1	ROUGE-2	ROUGE-L	HOLMS	BERTScore
ROUGE-1	1.000				
ROUGE-2	0.960	1.000			
ROUGE-L	0.951	0.946	1.000		
HOLMS	0.812	0.823	0.873	1.000	
BERTSCore	**0.913**	**0.924**	**0.889**	**0.784**	1.000

Table 11: Pearson correlations between metrics for the MAS task. Extractive runs were evaluated on extractive gold summaries. Abstractive runs were evaluated on both extractive and abstractive gold summaries. All evaluation scores were concatenated to compute correlations. For ROUGE and BERTScore we used the F1 scores. Best correlations with the lexical measures are highlighted in bold.

Measure	ROUGE-1	ROUGE-2	ROUGE-L	HOLMS	BERTScore
ROUGE-1	1.000				
ROUGE-2	0.951	1.000			
ROUGE-L	0.944	0.981	1.000		
HOLMS	0.193	0.204	0.259	1.000	
BERTSCore	**0.292**	**0.332**	**0.372**	0.972	1.000

Table 12: Pearson correlations between metrics for the QS task. For ROUGE and BERTScore we used the F1 scores. Best correlations with the lexical measures are highlighted in bold.

neural network: Model development and validation. *J Med Internet Res*, 22(10):e19810.

Spandana Balumuri, Sony Bachina, and Sowmya Kamath S. 2021. Sb_nitk at mediqa 2021: Leveraging transfer learning for question summarization in medical domain. In *Proceedings of the 20th SIGBioMed Workshop on Biomedical Language Processing, NAACL-BioNLP 2021*. Association for Computational Linguistics.

Asma Ben Abacha, Eugene Agichtein, Yuval Pinter, and Dina Demner-Fushman. 2017. Overview of the medical question answering task at trec 2017 liveqa. In *TREC 2017*.

Asma Ben Abacha and Dina Demner-Fushman. 2019a. On the role of question summarization and information source restriction in consumer health question answering. In *Proceedings of the AMIA 2019 Informatics Summit, San Francisco, CA, USA, 2019*.

Asma Ben Abacha and Dina Demner-Fushman. 2019b. On the summarization of consumer health questions. In *Proceedings of the 57th Conference of the Association for Computational Linguistics, ACL 2019, Florence, Italy, July 28- August 2, 2019, Volume 1: Long Papers*, pages 2228–2234. Association for Computational Linguistics.

Asma Ben Abacha and Dina Demner-Fushman. 2019c. A question-entailment approach to question answering. *BMC Bioinform.*, 20(1):511:1–511:23.

Asma Ben Abacha, Chaitanya Shivade, and Dina Demner-Fushman. 2019. Overview of the MEDIQA 2019 shared task on textual inference, question entailment and question answering. In *Proceedings of the 18th BioNLP Workshop and Shared Task, BioNLP@ACL 2019, Florence, Italy, August 1, 2019*, pages 370–379. Association for Computational Linguistics.

Manik Bhandari, Pranav Narayan Gour, Atabak Ashfaq, and Pengfei Liu. 2020a. Metrics also disagree in the low scoring range: Revisiting summarization evaluation metrics. In *Proceedings of the 28th International Conference on Computational Linguistics, COLING 2020, Barcelona, Spain (Online), December 8-13, 2020*, pages 5702–5711. International Committee on Computational Linguistics.

Manik Bhandari, Pranav Narayan Gour, Atabak Ashfaq, Pengfei Liu, and Graham Neubig. 2020b. Reevaluating evaluation in text summarization. In *Proceedings of the 2020 Conference on Empirical Methods in Natural Language Processing, EMNLP 2020, Online, November 16-20, 2020*, pages 9347–9359. Association for Computational Linguistics.

Manik Bhandari, Pranav Narayan Gour, Atabak Ashfaq, Pengfei Liu, and Graham Neubig. 2020c. Reevaluating evaluation in text summarization. In *Proceedings of the 2020 Conference on Empirical Methods in Natural Language Processing (EMNLP)*.

Duy-Cat Can, Quoc-An Nguyen, Quoc-Hung Duong, Minh-Quang Nguyen, Huy-Son Nguyen, Cam-Van Thi Nguyen, Quang-Thuy Ha, and Mai-Vu Tran. 2021. Uetrice at mediqa 2021: A prosper-thy-neighbour extractive multi-document summarization model. In *Proceedings of the 20th SIGBioMed Workshop on Biomedical Language Processing, NAACL-BioNLP 2021*. Association for Computational Linguistics.

Yen-Chun Chen and Mohit Bansal. 2018. Fast abstractive summarization with reinforce-selected sentence rewriting. In *Proceedings of the 56th Annual Meeting of the Association for Computational Linguistics, ACL 2018, Melbourne, Australia, July 15-20, 2018, Volume 1: Long Papers*, pages 675–686. Association for Computational Linguistics.

Songtai Dai, Quan Wang, Yajuan Lyu, and Yong Zhu. 2021. Bdkg at mediqa 2021: System report for the radiology report summarization task. In *Proceedings of the 20th SIGBioMed Workshop on Biomedical Language Processing, NAACL-BioNLP 2021*. Association for Computational Linguistics.

Jean-Benoit Delbrouck, Cassie Zhang, and Daniel Rubin. 2021. Qiai at mediqa 2021: Multimodal radiology report summarization. In *Proceedings of the 20th SIGBioMed Workshop on Biomedical Language Processing, NAACL-BioNLP 2021*. Association for Computational Linguistics.

Dina Demner-Fushman, Marc D Kohli, Marc B Rosenman, Sonya E Shooshan, Laritza Rodriguez, Sameer Antani, George R Thoma, and Clement J McDonald. 2016. Preparing a collection of radiology examinations for distribution and retrieval. *Journal of the American Medical Informatics Association*, 23(2):304–310.

Dina Demner-Fushman, Yassine Mrabet, and Asma Ben Abacha. 2020. Consumer health information and question answering: helping consumers find answers to their health-related information needs. *J. Am. Medical Informatics Assoc.*, 27(2):194–201.

John W. Ely, Jerome A. Osheroff, Paul N. Gorman, Mark H. Ebell, M. Lee Chambliss, Eric A. Pifer, and P. Zoe Stavri. 2000. A taxonomy of generic clinical questions: classification study. *British Medical Journal*, 321:429–432.

Alexander R Fabbri, Wojciech Kryściński, Bryan McCann, Caiming Xiong, Richard Socher, and Dragomir Radev. 2020. Summeval: Reevaluating summarization evaluation. *arXiv preprint arXiv:2007.12626*.

Sebastian Gehrmann, Yuntian Deng, and Alexander M. Rush. 2018. Bottom-up abstractive summarization. In *Proceedings of the 2018 Conference on Empirical Methods in Natural Language Processing, Brussels, Belgium, October 31 - November 4, 2018*, pages 4098–4109.

Yifan He, Mosha Chen, and Songfang Huang. 2021. damo_nlp at mediqa 2021: Knowledge-base preprocessing and coverage-oriented reranking for medical question summarization. In *Proceedings of the 20th SIGBioMed Workshop on Biomedical Language Processing, NAACL-BioNLP 2021*. Association for Computational Linguistics.

Karl Moritz Hermann, Tomás Kociský, Edward Grefenstette, Lasse Espeholt, Will Kay, Mustafa Suleyman, and Phil Blunsom. 2015. Teaching machines to read and comprehend. In *NIPS*, pages 1693–1701.

Shankar Iyer, Nikhil Dandekar, and Kornél Csernai. 2017. First quora dataset release: Question pairs.

Halil Kilicoglu, Asma Ben Abacha, Yassine Mrabet, Sonya E. Shooshan, Laritza Rodriguez, Kate Masterton, and Dina Demner-Fushman. 2018. Semantic annotation of consumer health questions. *BMC Bioinform.*, 19(1):34:1–34:28.

Ravi Kondadadi, Sahil Manchanda, Jason Ngo, and Ronan McCormack. 2021. Optum at mediqa 2021: Abstractive summarization of radiology reports using simple bart finetuning. In *Proceedings of the 20th SIGBioMed Workshop on Biomedical Language Processing, NAACL-BioNLP 2021*. Association for Computational Linguistics.

Hoang-Quynh Le, Quoc-An Nguyen, Quoc-Hung Duong, Minh-Quang Nguyen, Huy-Son Nguyen, Tam Doan Thanh, Hai-Yen Thi Vuong, and Trang M. Nguyen. 2021. Uetfishes at mediqa 2021: Standing-on-the-shoulders-of-giants model for abstractive multi-answer summarization. In *Proceedings of the 20th SIGBioMed Workshop on Biomedical Language Processing, NAACL-BioNLP 2021*. Association for Computational Linguistics.

Jooyeon Lee, Huong Dang, Ozlem Uzuner, and Sam Henry. 2021a. Mnlp at mediqa 2021: Fine-tuning pegasus for consumer health question summarization. In *Proceedings of the 20th SIGBioMed Workshop on Biomedical Language Processing, NAACL-BioNLP 2021*. Association for Computational Linguistics.

Lung-Hao Lee, Po-Han Chen, Yu-Xiang Zeng, Po-Lei Lee, and Kuo-Kai Shyu. 2021b. Ncuee-nlp at mediqa 2021: Health question summarization using pegasus transformers. In *Proceedings of the 20th SIGBioMed Workshop on Biomedical Language Processing, NAACL-BioNLP 2021*. Association for Computational Linguistics.

Chin-Yew Lin. 2004. ROUGE: A package for automatic evaluation of summaries. In *Text Summarization Branches Out*, pages 74–81, Barcelona, Spain. Association for Computational Linguistics.

Chris J Lu, Alan R Aronson, Sonya E Shooshan, and Dina Demner-Fushman. 2019. Spell checker for consumer language (cspell). *Journal of the American Medical Informatics Association*, 26(3):211–218.

Diwakar Mahajan, Ching-Huei Tsou, and Jennifer J Liang. 2021. Ibmresearch at mediqa 2021: Toward improving factual correctness of radiology report abstractive summarization. In *Proceedings of the 20th SIGBioMed Workshop on Biomedical Language Processing, NAACL-BioNLP 2021*. Association for Computational Linguistics.

Yuning Mao, Liyuan Liu, Qi Zhu, Xiang Ren, and Jiawei Han. 2020. Facet-aware evaluation for extractive summarization. In *Proceedings of the 58th Annual Meeting of the Association for Computational Linguistics, ACL 2020, Online, July 5-10, 2020*, pages 4941–4957. Association for Computational Linguistics.

Rashmi Mishra, Jiantao Bian, Marcelo Fiszman, Charlene R. Weir, Siddhartha Jonnalagadda, Javed Mostafa, and Guilherme Del Fiol. 2014. Text summarization in the biomedical domain: A systematic review of recent research. *Journal of Biomedical Informatics*, 52:457–467. Special Section: Methods in Clinical Research Informatics.

Yassine Mrabet and Dina Demner-Fushman. 2020. HOLMS: alternative summary evaluation with large language models. In *Proceedings of the 28th International Conference on Computational Linguistics, COLING 2020, Barcelona, Spain (Online), December 8-13, 2020*, pages 5679–5688. International Committee on Computational Linguistics.

Khalil Mrini, Franck Dernoncourt, Seunghyun Yoon, Trung Bui, Walter Chang, Emilias Farcas, and Ndapa Nakashole. 2021. Ucsd-adobe at mediqa 2021: Transfer learning and answer sentence selection for medical summarization. In *Proceedings of the 20th SIGBioMed Workshop on Biomedical Language Processing, NAACL-BioNLP 2021*. Association for Computational Linguistics.

Ramesh Nallapati, Feifei Zhai, and Bowen Zhou. 2017. Summarunner: A recurrent neural network based sequence model for extractive summarization of documents. In *Proceedings of the Thirty-First AAAI Conference on Artificial Intelligence, February 4-9, 2017, San Francisco, CA, USA.*, pages 3075–3081.

Karolina Owczarzak, John M. Conroy, Hoa Trang Dang, and Ani Nenkova. 2012. An assessment of the accuracy of automatic evaluation in summarization. In *Proceedings of Workshop on Evaluation Metrics and System Comparison for Automatic Summarization@NACCL-HLT 2012, Montrèal, Canada, June 2012, 2012*, pages 1–9. Association for Computational Linguistics.

Colin Raffel, Noam Shazeer, Adam Roberts, Katherine Lee, Sharan Narang, Michael Matena, Yanqi Zhou, Wei Li, and Peter J Liu. 2020. Exploring the limits of transfer learning with a unified text-to-text transformer. *Journal of Machine Learning Research*, 21:1–67.

Alexander M. Rush, Sumit Chopra, and Jason Weston. 2015. A neural attention model for abstractive sentence summarization. In *Proceedings of the 2015 Conference on Empirical Methods in Natural Language Processing, EMNLP 2015, Lisbon, Portugal, September 17-21, 2015*, pages 379–389.

Max E. Savery, Asma Ben Abachaand Soumya Gayen, and Dina Demner-Fushman. 2020. Question-driven summarization of answers to consumer health questions. *Scientific Data, Nature*, 7.

Abigail See, Peter J. Liu, and Christopher D. Manning. 2017. Get to the point: Summarization with pointer-generator networks. In *Proceedings of the 55th Annual Meeting of the Association for Computational Linguistics, ACL 2017, Vancouver, Canada, July 30 - August 4, Volume 1: Long Papers*, pages 1073–1083.

Thibault Sellam, Dipanjan Das, and Ankur P Parikh. 2020. Bleurt: Learning robust metrics for text generation. *arXiv preprint arXiv:2004.04696*.

Akshay Smit, Saahil Jain, Pranav Rajpurkar, Anuj Pareek, Andrew Ng, and Matthew Lungren. 2020. Combining automatic labelers and expert annotations for accurate radiology report labeling using BERT. In *Proceedings of the 2020 Conference on Empirical Methods in Natural Language Processing (EMNLP)*.

Mario Sänger, Leon Weber, and Ulf Leser. 2021. Wbi at mediqa 2021: Summarizing consumer health questions with generative transformers. In *Proceedings of the 20th SIGBioMed Workshop on Biomedical Language Processing, NAACL-BioNLP 2021*. Association for Computational Linguistics.

Ashish Vaswani, Noam Shazeer, Niki Parmar, Jakob Uszkoreit, Llion Jones, Aidan N Gomez, Lukasz Kaiser, and Illia Polosukhin. 2017. Attention is all you need. *arXiv preprint arXiv:1706.03762*.

Wen Xiao and Giuseppe Carenini. 2019. Extractive summarization of long documents by combining global and local context. In *Proceedings of the 2019 Conference on Empirical Methods in Natural Language Processing and the 9th International Joint Conference on Natural Language Processing, EMNLP-IJCNLP 2019, Hong Kong, China, November 3-7, 2019*, pages 3009–3019. Association for Computational Linguistics.

Liwen Xu, Yan Zhang, Lei Hong, Yi Cai, and Szui Sung. 2021. Chichealth @ mediqa 2021: Exploring the limits of pre-trained seq2seq models for medical summarization. In *Proceedings of the 20th SIGBioMed Workshop on Biomedical Language Processing, NAACL-BioNLP 2021*. Association for Computational Linguistics.

Yumo Xu and Mirella Lapata. 2020. Coarse-to-fine query focused multi-document summarization. In *Proceedings of the 2020 Conference on Empirical Methods in Natural Language Processing (EMNLP)*, pages 3632–3645.

Shweta Yadav, Mourad Sarrouti, and Deepak Gupta. 2021. Nlm at mediqa 2021: Transfer learning-based approaches for consumer question and multi-answer summarization. In *Proceedings of the 20th SIGBioMed Workshop on Biomedical Language Processing, NAACL-BioNLP 2021*. Association for Computational Linguistics.

Jingqing Zhang, Yao Zhao, Mohammad Saleh, and Peter Liu. 2020a. PEGASUS: Pre-training with extracted gap-sentences for abstractive summarization. In *International Conference on Machine Learning*.

Tianyi Zhang, Varsha Kishore, Felix Wu, Kilian Q Weinberger, and Yoav Artzi. 2019. Bertscore: Evaluating text generation with bert. *arXiv preprint arXiv:1904.09675*.

Yuhao Zhang, Daisy Yi Ding, Tianpei Qian, Christopher D. Manning, and Curtis P. Langlotz. 2018. Learning to summarize radiology findings. In *EMNLP 2018 Workshop on Health Text Mining and Information Analysis*.

Yuhao Zhang, Derek Merck, Emily Bao Tsai, Christopher D. Manning, and Curtis Langlotz. 2020b. Optimizing the factual correctness of a summary: A study of summarizing radiology reports. In *Proceedings of the 58th Annual Meeting of the Association for Computational Linguistics*.

Ming Zhong, Pengfei Liu, Yiran Chen, Danqing Wang, Xipeng Qiu, and Xuanjing Huang. 2020. Extractive summarization as text matching. In *Proceedings of the 58th Annual Meeting of the Association for Computational Linguistics*.

Wei Zhu, Yilong He, Ling Chai, Yunxiao Fan, Yuan Ni, GUOTONG XIE, and Xiaoling Wang. 2021. paht_nlp at mediqa 2021: Multi-grained query focused multi-answer summarization. In *Proceedings of the 20th SIGBioMed Workshop on Biomedical Language Processing, NAACL-BioNLP 2021*. Association for Computational Linguistics.

WBI at MEDIQA 2021: Summarizing Consumer Health Questions with Generative Transformers

Mario Sänger[*,♣] **, Leon Weber**[*,♣,†]**, Ulf Leser**[♣]
[♣]Computer Science Department, Humboldt-Universität zu Berlin
[†]Max Delbruck Center for Molecular Medicine
{saengema,weberple,leser}@informatik.hu-berlin.de

Abstract

This paper describes our contribution for the MEDIQA-2021 Task 1 question summarization competition. We model the task as conditional generation problem. Our concrete pipeline performs a finetuning of the large pretrained generative transformers PEGASUS (Zhang et al., 2020a) and BART (Lewis et al., 2020). We used the resulting models as strong baselines and experimented with (i) integrating structured knowledge via entity embeddings, (ii) ensembling multiple generative models with the generator-discriminator framework and (iii) disentangling summarization and interrogative prediction to achieve further improvements. Our best performing model, a fine-tuned vanilla PEGASUS, reached the second place in the competition with an ROUGE-2-F1 score of 15.99. We observed that all of our additional measures hurt performance (up to 5.2 pp) on the official test set. In course of a post-hoc experimental analysis which uses a larger validation set results indicate slight performance improvements through the proposed extensions. However, further analysis is need to provide stronger evidence.

1 Introduction

The internet provides a wealth of information on health topics through specialised websites, forums, blogs and social networks. Increasingly, consumers are using these information sources to answer their medical and health-related questions. In the course of this development, also the consumers' expectations regarding search engine functionalities have become much more demanding. Instead of reading through a list of relevant articles returned by a classical search engine, short and precise passages are now expected to answer questions. This transformation also has an impact on the technologies used to fulfill the user's information needs. In particular, approaches for automatic questions answering as well as automatic summarization and simplification of (long) articles has received a lot of attention by researchers in recent years (Allahyari et al., 2017; Kwiatkowski et al., 2019; Narayan et al., 2018b; See et al., 2017; Weber et al., 2019). This trend is also addressed by Task 1 of the MEDIQA 2021 shared task (Ben Abacha et al., 2021) through investigating consumer health-questions asked on the (experimental) medical question answering system CHiQA[1]. As we participated only in this task, we refer to it as Shared Task (ST) in the following.

The goal of Task 1 was to foster the development of new summarization approaches, specifically designed for the challenges of long and potentially complex consumer health questions. One major challenge of CHiQA is the extraction of the user's main concern from the question text. The given questions are often lengthy and contain a lot of peripheral information, which makes automatic processing and answering (much more) difficult. Recent studies highlight that expert-based summarizations of such questions can lead to significant enhancements of the overall QA process (Ben Abacha and Demner-Fushman, 2019). Effective automatic summarization methods could therefore play a key role for improving medical question answering.

We contribute to this task by first building a baseline using the general conditional generation framework and then investigating three modifications to summarize the consumer health questions. Our baseline relies on finetuning the large pretrained generative transformers PEGASUS (Zhang et al., 2020a) and BART (Lewis et al., 2020). We explore three different strategies to improve the performance of these baseline models, i.e. (i) integrating structured knowledge via entity embeddings, (ii) ensembling multiple generative models with the generator-discriminator framework and (iii) dis-

[*] These authors contributed equally. Author order was determined by coin flip.

[1]https://chiqa.nlm.nih.gov/

Proceedings of the BioNLP 2021 workshop, pages 86–95
June 11, 2021. ©2021 Association for Computational Linguistics

entangling summarization and question word prediction. Our best performing model, a fine-tuned vanilla PEGASUS, reached the second place in the competition. We observed that all measures hurt performance (up to 5.2 pp) on the evaluation set. However, a post-hoc experimental analysis (see Section 3), using a larger validation set, indicates slight improvements through the model extensions.

The remainder of the paper is organized as follows: the next section introduces our baseline and the three extension strategies in detail. Section 3 highlights and discusses the experiments and results we obtained in our own evaluation as well as in the official assessment. The paper concludes which a summary of the main findings.

2 Methods

2.1 Data & Baselines

The shared task provides only an official validation and test set as data. For training data, we follow the tasks' organizers suggestion to use the MeQSum corpus which consists of 1,000 consumer health questions and their summaries.

We model the summarization task as conditional generation, in which a model is prompted with the original question and then generates the summary in an autoregressive fashion. We base our implementation[2] on the huggingface transformers library (Wolf et al., 2020) and experiment with the included pretrained generative transformers *bart-base*[3], *bart-large*[4], *pegasus-large*[5] and *pegasus-xsum*[6]. *pegasus-xsum* is a version of PEGASUS that was already finetuned for summarization on the Xsum dataset (Narayan et al., 2018a). For all models, we use a learning rate of $3e-5$ and train for 10 epochs. We use beam search for decoding and tune the search parameters on the validation set. We independently evaluated $\{1, 10\}$ as the number of beams and the $\{0.7, 0.8, 0.9, 1.0\}$ for the length penalty and found 10 and 0.8 to be optimal.

[2]Our code is publicly available under `https://github.com/leonweber/bionlp21_summarize`

[3]`https://huggingface.co/facebook/bart-base`

[4]`https://huggingface.co/facebook/bart-large`

[5]`https://huggingface.co/google/pegasus-large`

[6]`https://huggingface.co/google/pegasus-xsum`

2.2 Integration of structured knowledge via entity embeddings

In initial analyses, we noticed that most question summaries revolve around a few central entities such as specific diseases or medications which are almost always mentioned in the source text. Furthermore, all of the generative transformers that we used were trained on texts from the general domain, in which such entities presumably are rare. We conjectured that it could be beneficial to explicitly provide entity information to the model. We approach this by first applying a domain-specific NER model to the source text and then enriching the input embeddings of the transformer with the found entities. Formally, we extend the computation of the i'th input embedding in the transformer to:

$$e_i = w_i + p_i + s_i + n_i, \quad (1)$$

where w_i, p_i, s_i are the standard subword, position and sequence type embeddings which are initialized with the weights of the pretrained transformer. n_i is a randomly initialized embedding, which represents the type of the named entity to which the token i belongs (including *None*) and has the same dimensionality as the other transformer embeddings. Note, that s_i is set to zero for transformers which do not use sequence type embeddings such as BART.

We experiment with two different NER models: (i) *HunFlair* (Weber et al., 2021), a state-of-the-art BioNER tagger and (ii) a custom *Flair* (Akbik et al., 2019) model trained on the CHQA corpus (Kilicoglu et al., 2018) consisting of manual annotations for the central entities of consumer health questions. Specifically, we use the *Disease* and *Chemical* models of *HunFlair* and the PC-harmonization of the CHQA corpus.

2.3 Ensembling multiple generative transformers

In preliminary experiments, we found that ensembling generative transformers by simply averaging the logits of different models hurt performance. Thus, we investigate a different strategy for ensembling generative models. We first use each model m of the ensemble to generate n summaries $\{s_{m1}, \ldots, s_{mn}\}$ conditioned on the original question q and then use a discriminative model to select the question-summary pair with the highest probability. The n different summaries are generated by simply taking the final generations of the top-n

scoring beams. We implement the discriminator as a BERT (Devlin et al., 2019) model that receives both the original question q and a question summary s produced by one of the ensembled models and predicts the ROUGE-L-F1 score between both $\mathit{ROUGE\text{-}L\text{-}F1}(s, q)$ using a tanh output layer. The model is trained via an L2-loss. More formally,

$$\mathbf{h} = \mathit{BERT}_{[\mathrm{CLS}]}(s, q) \tag{2}$$

$$o = 0.5 \cdot \mathit{tanh}(\mathbf{W} \cdot \mathbf{h} + \mathbf{b}) \tag{3}$$

$$\mathcal{L} = \|\mathit{ROUGE\text{-}L\text{-}F1}(s, q) - o\|_2, \tag{4}$$

where $\mathit{BERT}_{[\mathrm{CLS}]}$ is the BERT-embedding of the special [CLS] token, $\mathbf{W}$ and $\mathbf{b}$ are trainable parameters and $\mathcal{L}$ is the loss value.

For training the discriminator, we require generated summaries that are close to the generated summaries on the test data. We cannot simply use the training data of the generators to create the training data for the discriminator, because we expect the distributions of the generated summaries for seen and unseen data to be significantly different. Thus, we split MeQSum training data in a 75% / 25% fashion and use the first chunk for training the generators and the combination of both to train the discriminators. The full training process is illustrated in Figure 1a.

2.4 Disentangling summarization and interrogative prediction

We observed that the consumer questions cover different categories of health-related issues in the ST data, e.g. possible side-effects of certain drugs, suitable treatments for specific diseases or food-related questions. We conjectured that providing the putative category of the question to the summarization model could guide the generator towards a better summary. Moreover, we recognized that the different categories are aligned to some extent with the interrogative of the target questions summaries. Based on these two observations, we designed a third modification by creating a separate model to predict the putative interrogative, which acts as a surrogate for the different question categories.

To this end, we implement a BERT-based classification model which gets the original user question as input and predicts the interrogative of the target question summary. We combine the classification model with the output of our baseline method using a three-step approach: (i) we generate m question summaries using a generative transformer, (ii) we predict the interrogative given the original user question based on the trained classification model and (iii) selected the highest ranked candidate questions which starts with the predicted interrogative as target summary. The process is illustrated in Figure 1b. To train the classification models we use the data from the MeQSum corpus but just take the first word of the summaries as goldstandard interrogative. Because in this model there is no dependency between generative and classification models (as opposed to our generator-discriminator framework), the classification model can be trained on the complete training data.

3 Results

3.1 Evaluation setting

We evaluate our models in two different settings.

Setting 1 For our ten submissions to the shared task, we typically use some combination of MeQSum and the validation data for training. For model selection and evaluation of our modifications, we use the official validation set of the shared task. Finally, we report scores of our models on the shared tasks' hidden test set.

Setting 2 While preparing our runs, we noticed that the variance of the results on the validation and test set is rather high, which probably has to do with the small amount of validation and test data (50 and 100 questions respectively). To evaluate the performance impact of our modifications in a more stable manner, we devised a second evaluation setting after the ST submissions were closed. For this, we combine the MeQSum data and the shared task validation data in a single dataset and then split it into a train and validation set, reserving 200 questions for validation, which leaves 850 questions for training. We ensure that for each split the ratio of original MeQSum and validation data is equal. For each result, we compute three different runs with different random seeds and report the average and standard deviation.

Table 1 highlights the used splits of the two different data settings and provides basic statistics for them. The results for both settings differ significantly and thus, we report results for both settings in the following sections. In the official evaluation of the shared task, the approaches were ranked according to the achieved ROUGE-2-F1 score.

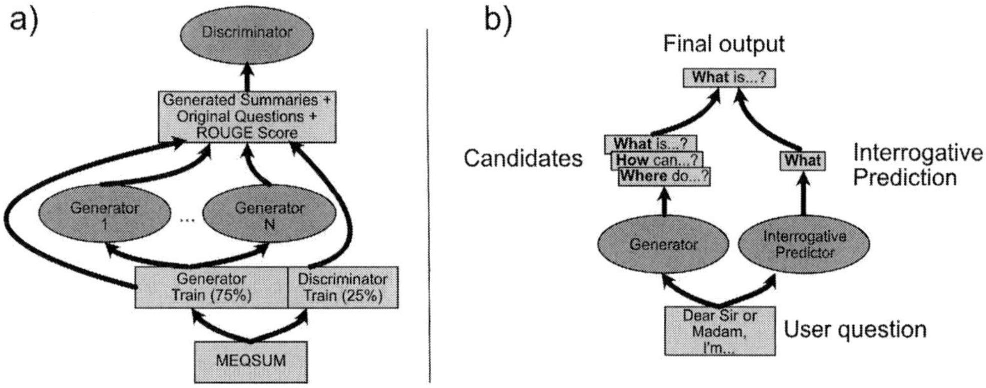

Figure 1: **(a)** Training an ensemble of multiple generators together with a discriminator. Resources are depicted as yellow rectangles and trained models as green ellipses. **(b)** Predicting summaries with the interrogative predictor. Resources are drawn as yellow rectangles and models as green ellipses.

Setting	Split	Questions	Tokens / Question			Tokens / Summary		
			Mean	Min	Max	Mean	Min	Max
Setting 1	Training (MeQSum)	1000	60.78	5	378	10.04	3	26
	Validation	50	64.16	9	234	9.34	4	19
Setting 2	Training	850	59.60	8	348	9.70	3	26
	Validation	200	66.64	5	378	10.18	3	26

Table 1: Overview about the data sets and splits used for training and evaluation in Setting 1 and 2. For Setting 2, we use all instances from the official training data (MeQSum) and validation data and randomly assign them to the two splits. We ensure that for each split the ratio of original training and validation data is equal.

3.2 Final evaluation results

Our best performing model achieved a ROUGE-2-F1 score of 15.99% on the hidden test set, leading to a second place in the competition. However, all top-5 models achieve results that are very close, and ranks change when different metrics are used. The top five of the official leaderboard is reproduced in Table 2. This best performing model is one of our baselines based on *pegasus-large* fine-tuned on the combination of MeQSum and the ST validation set. The results of our ten runs on the official hidden test set together with a description of each run can be found in Table 5.

3.3 Baseline results

In preliminary experiments on the ST validation set, we found that *pegasus-large* works better than *bart-large* when the model is fine-tuned on MeQSum and evaluated on the ST validation set (ROUGE-L-F1 of 33.32 vs. 32.82). Based on this result, we opted to select *pegasus-large* as baseline model for our submissions (refer to Section 3.7 for a discussion of challenges in model selection). In the official evaluation (i.e. Setting 1) the vanilla *pegasus-large* model achieves the best performance of all our submitted runs with an ROUGE-2-F1 score of 15.99 (see Run 1 in Table 5). In a post-hoc analysis, we noticed that in the consumer questions spelling errors for crucial pieces of information such as diseases are common and that the models tend to copy those spelling errors into the summary of the question. Thus, our approach probably could have benefited from incorporating a spell-checking tool that corrects the spelling errors in the health questions.

Setting 2 uses the same basic models, but relies on a different training setup. Table 3 shows the performance scores. The best performance is achieved by *bart-large* with ROUGE-1-, ROUGE-2 and ROUGE-L-F1 scores of 52.91, 34.06 and 49.88. This represents an improvement of 0.55pp concerning ROUGE-2-F1 to the next best model (*bart-base*). In this setting, the BART-based models achieve better results than the PEGASUS models.

3.4 Entity embedding results

We evaluate the addition of entity embeddings to a generative transformer using bart-base. For detecting entities, we experiment with the two different NER models HunFlair and a custom Flair model trained on the *PC*-harmonization (Passonneau and Carpenter, 2014) of the CHQA corpus. The results for Setting 2 can be found in Table 3. Adding entity embeddings to the input representation improves results consistently, leading to a gain of 0.3pp and 1.01pp in ROUGE-2-F1 over our bart-base baseline. However, we did not observe any gains in our preliminary experiments on the ST validation set and thus did not evaluate the models with entity embeddings in Setting 1. The submission of new runs was not possible at the time of writing.

3.5 Ensemble results

All results for the generator-discriminator ensembles in Setting 1 (on the hidden test set) can be found in Table 5, where each row with Type 'GD' corresponds to one configuration of a generator-discriminator ensemble. Considering ROUGE-2-F1, the best generator-discriminator result (run 7) still performs 1.4 pp worse than our best baseline model. This run used only one generator based on pegasus-large to produce ten candidates per question and a bert-large discriminator to select the most promising summary. The only setting in which a generator-discriminator model outperforms our strongest baseline on the hidden test set is run 8 which gains 0.2 pp under the BERTScore metric (Zhang et al., 2020b), making it the overall top ranking run of the ST under this metric. This run uses a single pegasus-large generator proposing ten candidate summaries per question and an ensemble of three different bert-large discriminators.

In Setting 2, we observed considerable gains by using an ensemble of bart-base, bart-large, pegasus-large and pegasus-xsum, while using a single bert-base as the discriminator, using only the most probable output sequence per model as candidate. Compared to pegasus-large, this configuration leads to an improvement of 2.16pp in ROUGE-1-F1, 1.46pp in ROUGE-2-F1 and 2.27pp in ROUGE-L-F1.

We also investigated the performance ceiling for our ensembling approach by evaluating the ensemble under a perfect discriminator, which always selects the summary yielding the highest Rouge-L-F1 score. Under this setting, our ensemble achieved a Rouge-2-F1 score of 44.87 which is an improvement of 10.9 pp. This shows the promise of our ensembling approach and suggests that a worthwhile path to obtain better results would be to improve the discriminator.

Rank	Team name	ROUGE-1-F1	ROUGE-2-F1	ROUGE-L-F1	HOLMS	BERTScore-F1
1	damo_nlp (summc)	35.14	16.08	31.31	56.77	68.98
2	**WBI**	**33.40**	**15.99**	**31.49**	**57.67**	**69.96**
3	NCUEE-NLP	33.52	15.97	30.90	57.87	69.60
4	yamr	32.80	15.25	30.38	57.86	68.77
5	Saama	33.33	15.18	29.50	57.72	69.38

Table 2: Top five of the official results for subtask one (ranked by ROUGE-2-F1). All scores are given in percent. In total 23 teams participated in this subtask. Our contribution is displayed in bold. These numbers correspond to our evaluation Setting 2.

Model type	Gen. model(s)	Add-on	ROUGE-1-F1	ROUGE-2-F1	ROUGE-L-F1
Baseline	bart-large	-	52.91 (± 0.91)	34.06 (± 1.01)	49.88 (± 0.66)
	bart-base	-	52.17 (± 0.14)	33.49 (± 0.84)	49.36 (± 0.32)
	pegasus-large	-	51.06 (± 0.78)	32.51 (± 0.72)	48.28 (± 0.68)
	pegasus-xsum	-	51.47 (± 0.28)	32.65 (± 0.58)	48.90 (± 0.30)
Entity embeddings	bart-base	HunFlair	52.16 (± 0.45)	33.79 (± 0.46)	49.24 (± 0.27)
	bart-base	CHQA flair model	53.17 (± 1.58)	34.5 (± 1.30)	50.22 (± 1.43)
Generator-discriminator	bart-base bart-large pegasus-large pegasus-xsum	bert-base	53.22 (± 1.81)	33.97 (± 1.40)	50.55 (± 1.75)
Interrogative prediction	pegasus-large	bert-base	52.11 (± 0.36)	33.71 (± 0.85)	49.21 (± 0.66)
	pegasus-large	bio-bert	52.22 (± 0.60)	33.42 (± 0.70)	49.26 (± 0.53)
	pegasus-large	biomed-roberta	52.66 (± 0.67)	33.71 (± 0.81)	49.58 (± 0.85)
	pegasus-large	bio-bert biomed-roberta	52.28 (± 0.58)	33.47 (± 0.69)	49.40 (± 0.67)

Table 3: Overview of Setting 2 evaluation results. For each experiment, we list the used generative transformer(s) and (if applicable) utilized complementary models (Add-on). For entity embeddings add-on models are named entity recognition models. In case of the generator-discriminator framework it's the discriminator model and regarding interrogative prediction it defines the applied classification model(s). For each experiment, we compute three different runs with different random seeds and report the average and standard deviation.

3.6 Interrogative-predictor results

For evaluating our interrogative prediction approach we experimented with different transformer-based models, pre-trained on either general domain or biomedical data, for classification: BERT[7], BioBERT (Lee et al., 2020)[8], BioMed-RoBERTa (Gururangan et al., 2020)[9] and multiple of these models arranged in an ensemble. All models are learned on the training portion (for each evaluation setting). For all models we use *pegasus-large* as generative model and produce 10 candidate summaries per user question.

As shown in Table 3 we observe clear performance improvements of this approach compared to the baseline when evaluated in Setting 2. Here, the best results are achieved with the BioMed-RoBERTa model. In this configuration, the model achieves a ROUGE-2-F1 score of 33.71 which represents an increase of 1.20 pp compared to the vanilla *pegasus-large* result. Again, the results achieved in the official evaluation (Setting 1) show a different picture. In this setting, the usage of an ensemble of three interrogative classification models lowers the performance by 2.6 pp (see Run 3 in Table 5).

We also investigated the accuracy of the interrogative prediction models. Table 4 highlights the achieved accuracy and macro $F1$-scores of the three models. All models predict the correct interrogative for only half of the consumer questions. An analysis of the predictions showed that all models are biased towards the majority classes, i.e. interrogatives with a high support in the training data.

Like in the generative ensemble setting, we further checked the potential performance gains of the interrogative prediction using a perfect classifier. For this, we took the gold standard interrogative and use the first generated summary candidate which starts with this interrogative as prediction. If no generated summary starts with the gold interrogative we use the highest ranked candidate. Using this selection scheme we reached an ROUGE-2-F1 score of 39.72 in Setting 2 which represents an increase by 7.21 pp over the baseline *pegasus-large* model. Again, this accentuates the suitability of the proposed approach.

[7]https://huggingface.co/bert-base-cased

[8]https://huggingface.co/dmis-lab/biobert-v1.1

[9]https://huggingface.co/allenai/biomed_roberta_base

Model	Accuracy	$F1$
bert-base	0.530	0.103
bio-bert	0.525	0.095
biomed-roberta	**0.555**	**0.228**

Table 4: Overview of the performance of the three interrogative classification models. For each model we report accuracy and macro $F1$ score. Bold figures highlight the highest value per column.

3.7 Discussion of result differences between Setting 1 and Setting 2

Tables 2 and 3 reveal enormous performance differences between Setting 1 (the official evaluation results) and Setting 2 (our post-hoc experimental analysis). In Setting 1, none of our proposed extensions leads to consistent quantitative improvements of the results and the best performance is achieved by an vanilla generative transformer. In contrast in Setting 2, we see (at least) slight benefits from all three strategies.

Explaining these results and differences is difficult for several reasons. Concerning Setting 2, the high variance of the results (see Table 3) prevents a clear conclusion. Results of the methods vary with different random initializations and are also quite sensitive to hyperparameter settings. Often the differences of the methods lie within the range of the standard deviation making it unclear whether the findings would hold up in further analysis or other contexts.

Regarding Setting 1, the small size of the evaluation data (only 100 instances) puts any conclusions about the quality of the proposed methods into question. In Setting 2, we tried to mitigate the problem of small test data by increasing the number of test instances, however the results remain unstable. Furthermore, weaknesses of the ROUGE metric, e.g. handling of synonyms, abbreviations or enumerations, must be taken into account in the result interpretation (Schluter, 2017; Kané et al., 2019). The automatic evaluation of generated summaries remains a research field in itself (Zhang et al., 2020b). In summary, we neither believe that the results from Setting 1 provide strong evidence of the extension's inappropriateness, nor that the results from Setting 2 allow a convincing statement about their positive effects. To this end, further investigation is necessary in order to draw definitive conclusions about our proposed modifications.

Run	Type	Description	ROUGE-2	HOLMS	BERTScore-F1
1	B	pegasus-large finetuned on MeQSum and validation data	**16.0**	**57.7**	70.0
2	B	pegasus-large first finetuned on MeQSum and then on validation data	12.4	55.5	69.3
3	IP	pegasus-large finetuned on MeQSum and validation data with ensemble of interrogative predictors consisting of two biobert and one biomed-roberta model	13.4	56.4	69.0
4	GD	Generator ensemble of bart-base, bart-large, pegasus-large and pegasus-xsum with one candidate summary per model and bert-base as discriminator	11.8	55.5	68.4
5	B	pegasus-xsum finetuned on MeQSum and validation data	12.4	55.5	68.7
6	GD	Same configuration as in run 4 but with an ensemble of discriminators consisting of bert-base, roberta-base and biobert	11.4	55.4	68.2
7	GD	pegasus-large trained on MeQSum with ten candidate summaries and a bert-large discriminator trained on MeQSum to select the best one	14.6	57.3	69.8
8	GD	Same configuration as in run 7 but with an ensemble of three different bert-large discriminators trained on MeQSum	14.2	57.0	**70.2**
9	GD	Same configuration as in run 7 but the bert-large discriminator is trained on MeQSum and validation data	12.0	55.4	68.9
10	GD	Same configuration as in run 8 but the the discriminators are trained on MeQSum and validation data	12.0	55.4	69.5

Table 5: Official results for our submitted runs for subtask one. In total we submitted 10 runs. The runs can be categorized according to their type into baseline models (B), models using interrogative prediction (IP) or the generator-discriminator framework (GD). The highest value per metric is highlighted in bold. This corresponds to our evaluation Setting 1.

4 Conclusion

In this work we investigate the large-scale pre-trained generative transformers PEGASUS and BART for the task of health-related consumer question summarization in the context of the MEDIQA 2021 shared task (Task 1). We propose and evaluate three different strategies, i.e. integrating structured knowledge via entity embeddings, utilizing a generator-discriminator framework and applying interrogative prediction, to extend these strong baseline models. Our best performing model, a fine-tuned pegasus-large transformer, reaches an ROUGE-2-F1 score of 15.99 and is ranked second place in the competition. Experimental results for our proposed extensions show a mixed picture and further analysis is needed to assess the quality of these extensions.

References

Alan Akbik, Tanja Bergmann, Duncan Blythe, Kashif Rasul, Stefan Schweter, and Roland Vollgraf. 2019. FLAIR: An easy-to-use framework for state-of-the-art NLP. In *Proceedings of the 2019 Conference of the North American Chapter of the Association for Computational Linguistics (Demonstrations)*, pages 54–59, Minneapolis, Minnesota. Association for Computational Linguistics.

Mehdi Allahyari, Seyedamin Pouriyeh, Mehdi Assefi, Saeid Safaei, Elizabeth D Trippe, Juan B Gutierrez, and Krys Kochut. 2017. Text summarization techniques: a brief survey. *arXiv preprint arXiv:1707.02268*.

Asma Ben Abacha and Dina Demner-Fushman. 2019. On the role of question summarization and information source restriction in consumer health question answering. *AMIA Summits on Translational Science Proceedings*, 2019:117.

Asma Ben Abacha, Yassine Mrabet, Yuhao Zhang, Chaitanya Shivade, Curtis Langlotz, and Dina Demner-Fushman. 2021. Overview of the mediqa 2021 shared task on summarization in the medical domain. In *Proceedings of the 20th SIG-BioMed Workshop on Biomedical Language Processing, NAACL-BioNLP 2021*. Association for Computational Linguistics.

Jacob Devlin, Ming-Wei Chang, Kenton Lee, and Kristina Toutanova. 2019. BERT: Pre-training of deep bidirectional transformers for language understanding. In *Proceedings of the 2019 Conference of the North American Chapter of the Association for Computational Linguistics: Human Language Technologies, Volume 1 (Long and Short Papers)*, pages 4171–4186, Minneapolis, Minnesota. Association for Computational Linguistics.

Suchin Gururangan, Ana Marasović, Swabha Swayamdipta, Kyle Lo, Iz Beltagy, Doug Downey, and Noah A Smith. 2020. Don't stop pretraining: Adapt language models to domains and tasks. In *Proceedings of the 58th Annual Meeting of the Association for Computational Linguistics*, pages 8342–8360.

Hassan Kané, Yusuf Kocyigit, Pelkins Ajanoh, Ali Abdalla, and Mohamed Coulibali. 2019. Towards neural similarity evaluator. In *Workshop on Document Intelligence at NeurIPS 2019*.

Halil Kilicoglu, Asma Ben Abacha, Yassine Mrabet, Sonya E. Shooshan, Laritza Rodriguez, Kate Masterton, and Dina Demner-Fushman. 2018. Semantic annotation of consumer health questions. *BMC Bioinform.*, 19(1):34:1–34:28.

Tom Kwiatkowski, Jennimaria Palomaki, Olivia Redfield, Michael Collins, Ankur Parikh, Chris Alberti, Danielle Epstein, Illia Polosukhin, Jacob Devlin, Kenton Lee, et al. 2019. Natural questions: a benchmark for question answering research. *Transactions of the Association for Computational Linguistics*, 7:453–466.

Jinhyuk Lee, Wonjin Yoon, Sungdong Kim, Donghyeon Kim, Sunkyu Kim, Chan Ho So, and Jaewoo Kang. 2020. Biobert: a pre-trained biomedical language representation model for biomedical text mining. *Bioinformatics*, 36(4):1234–1240.

Mike Lewis, Yinhan Liu, Naman Goyal, Marjan Ghazvininejad, Abdelrahman Mohamed, Omer Levy, Veselin Stoyanov, and Luke Zettlemoyer. 2020. BART: denoising sequence-to-sequence pre-training for natural language generation, translation, and comprehension. In *Proceedings of the 58th Annual Meeting of the Association for Computational Linguistics, ACL 2020, Online, July 5-10, 2020*, pages 7871–7880. Association for Computational Linguistics.

Shashi Narayan, Shay B. Cohen, and Mirella Lapata. 2018a. Don't give me the details, just the summary! topic-aware convolutional neural networks for extreme summarization. In *Proceedings of the 2018 Conference on Empirical Methods in Natural Language Processing, Brussels, Belgium, October 31 - November 4, 2018*, pages 1797–1807. Association for Computational Linguistics.

Shashi Narayan, Shay B Cohen, and Mirella Lapata. 2018b. Don't give me the details, just the summary! topic-aware convolutional neural networks for extreme summarization. In *Proceedings of the 2018 Conference on Empirical Methods in Natural Language Processing*, pages 1797–1807.

Rebecca J. Passonneau and Bob Carpenter. 2014. The benefits of a model of annotation. *Trans. Assoc. Comput. Linguistics*, 2:311–326.

Natalie Schluter. 2017. The limits of automatic summarisation according to rouge. In *Proceedings of the 15th Conference of the European Chapter of the Association for Computational Linguistics: Volume 2, Short Papers*, pages 41–45.

Abigail See, Peter J Liu, and Christopher D Manning. 2017. Get to the point: Summarization with pointer-generator networks. In *Proceedings of the 55th Annual Meeting of the Association for Computational Linguistics (Volume 1: Long Papers)*, pages 1073–1083.

Leon Weber, Pasquale Minervini, Jannes Münchmeyer, Ulf Leser, and Tim Rocktäschel. 2019. Nlprolog: Reasoning with weak unification for question answering in natural language. In *Proceedings of the 57th Annual Meeting of the Association for Computational Linguistics*, pages 6151–6161.

Leon Weber, Mario Sänger, Jannes Münchmeyer, Maryam Habibi, Ulf Leser, and Alan Akbik. 2021. HunFlair: an easy-to-use tool for state-of-the-art biomedical named entity recognition. *Bioinformatics*. Btab042.

Thomas Wolf, Lysandre Debut, Victor Sanh, Julien Chaumond, Clement Delangue, Anthony Moi, Pierric Cistac, Tim Rault, Rémi Louf, Morgan Funtowicz, Joe Davison, Sam Shleifer, Patrick von Platen, Clara Ma, Yacine Jernite, Julien Plu, Canwen Xu, Teven Le Scao, Sylvain Gugger, Mariama Drame, Quentin Lhoest, and Alexander M. Rush. 2020. Transformers: State-of-the-art natural language processing. In *Proceedings of the 2020 Conference on Empirical Methods in Natural Language Processing: System Demonstrations*, pages 38–45, Online. Association for Computational Linguistics.

Jingqing Zhang, Yao Zhao, Mohammad Saleh, and Peter J. Liu. 2020a. PEGASUS: pre-training with extracted gap-sentences for abstractive summarization. In *Proceedings of the 37th International Conference on Machine Learning, ICML 2020, 13-18 July 2020, Virtual Event*, volume 119 of *Proceedings of Machine Learning Research*, pages 11328–11339. PMLR.

Tianyi Zhang, Varsha Kishore, Felix Wu, Kilian Q. Weinberger, and Yoav Artzi. 2020b. Bertscore: Evaluating text generation with BERT. In *8th International Conference on Learning Representations, ICLR 2020, Addis Ababa, Ethiopia, April 26-30, 2020*. OpenReview.net.

paht_nlp @ MEDIQA 2021: Multi-grained Query Focused Multi-Answer Summarization

Wei Zhu[1] [*], **Yilong He**[2], **Ling Chai**[2], **Yunxiao Fan**[2],
Yuan Ni[2], **Guotong Xie**[2], **Xiaoling Wang**[1]
[1] East China Normal University, Shanghai, China
[2] Pingan Health Tech, Shanghai/Beijing, China

Abstract

In this article, we describe our systems for the MEDIQA 2021 Shared Tasks. First, we will describe our method for the second task, Multi-Answer Summarization (MAS). For extractive summarization, two series of methods are applied. The first one follows Xu and Lapata (2020). First a RoBERTa model is first applied to give a local ranking of the candidate sentences. Then a Markov Chain model is applied to evaluate the sentences globally. The second method applies cross-sentence contextualization to improve the local ranking and discard the global ranking step. Our methods achieve **the 1st Place** in the MAS task. For the question summarization (QS) and radiology report summarization (RRS) tasks, we explore how end-to-end pre-trained seq2seq model perform. A series of tricks for improving the fine-tuning performances are validated.

1 Introduction

Automatic summarization is an essential task in the medical domain. It is time consuming for users to read a lot of medical documents when they use a search engine like Google, Medline, etc, about some topic and obtain a list of documents which are potential answers. First, the contents might be too specialized for layman to understand. Second, one document may not answer the query completely, and the users might have to summarize the conclusions across multiple documents, which may lead to waste of time or misunderstanding. In order to improve the users' experiences when using medical applications, automatic summarization techniques are required.

The MEDIQA 2021 shared tasks are held to investigate the current state of the art summarization models, especially how they perform in the medical domains. Three tasks are held. The first one is Question Summarization (QS), which summarizes long and potentially complex consumer health questions into simple ones, which are proven to be beneficial for automatic question answering. Empirical QA studies based on manual expert summarization of these questions showed a substantial improvement of 58% in performance (Abacha and Demner-Fushman, 2019). The second task is Multi-Answer Summarization (MAS) (Savery et al., 2020). Different answers can bring complementary perspectives that are likely to benefit the users of QA systems. The goal of this task is to develop a system that can aggregate and summarize the answers scattered in multiple documents. The third task is Radiology Reports Summarization (RRS) (Zhang et al., 2018, 2020b), which is to generate radiology impression statements by summarizing textual findings written by radiologists. which have several applications. First, it can speed up the technicians' workflow. Second, a system can extract the information in the reports and summarize into sentences that a layman can understand.

In the MAS task, we improve upon (Xu and Lapata, 2020) via three methods. First, during the coarse ranking of a sentence in one of the given documents, we also add the surrounding sentences as input and use two special tokens marking the positions of the sentence. This modification improves the coarse ranking with a large margin. Second, during fine-grained re-ranking, instead of incorporating a inverse sentence frequency (IFS) score based similarity matrix between sentences in the Markov chain model, we find that directly using semantic similarity scores to form the similarity matrix performs better. Third, due to the low resource settings of this task, we find that applying a RoBERTa (Liu et al., 2019) model which is already fine-tuned on the MS-MACRO task (Campos et al., 2016) can be beneficial.

For the other two tasks, we mainly explore how the pre-trained seq2seq model like BART (Lewis et al., 2020), PEGASUS (Zhang et al., 2020a), etc, can perform in these tasks. Two take-aways can

[*] Contact: 52205901018@stu.ecnu.edu.cn.

Proceedings of the BioNLP 2021 workshop, pages 96–102
June 11, 2021. ©2021 Association for Computational Linguistics

be made. First, for tasks with small dataset size, freezing a part of the transformer blocks can be beneficial. Second, for the RRS task, we find that controlling the maximum output sequence length can improve the ROUGE score on the test set.

Our team PAHT_NLP participate in all the three tasks, and won the 1st place in the MAS task. Experiments will show that our modifications are beneficial for both stage of the MAS task. We also report extensive experiments for task 1 and task 3.

2 Multi-grained Multi-Answer Summarization

2.1 problem formulation

Let Q denote a query, and $D = \{d_1, d_2, ..., d_M\}$ a set of documents returned by the search engine or a question answering system (e.g., the ChiQA system ((Demner-Fushman et al., 2020))). It is often assumed (e.g., in our MAS task) that Q consists of a short question (e.g., Will influenza be the next pandemic?).

We implement the multi-grained MDS following Xu and Lapata (2020). We first decompose documents into segments, i.e., sentences. Then, a trained RoBERTa model quantifies the semantic similarities between a selected sentence and the query, which give importance estimations of the sentences based the sentence itself or their local contexts (Local Estimator). Third, to give a global estimations of the importance of the segments to the summary, we apply a Markov Chain (Erkan and Radev, 2004) based estimator (Global Estimator).

2.2 Local Estimator

We leverage fine-tuned pretrained language models as our evidence estimator, and use the trained estimators to rank the answer candidates.

Let Q denote a query sequence and $\{S_1, S_2, ..., S_N\}$ the set of candidate answers. Our training objective is to find the correct answers within this set. We leverage RoBERTa as our sequence encoder. We concatenate query Q and candidate sentence S into a sequence $< s >$, Q, $< /s >$, $< /s >$, S, $< /s >$ as the input to the RoBERTa encoder (we pad each sequence in a mini-batch of L tokens). The starting $< s >$ token's vector representations t serves as input to a single layer feed forward layer to obtain the distribution over positive and negative classes:

$$p_k = \frac{1}{Z} \exp{(t^T W_{:,k})}, \quad (1)$$

where $k = 0, 1$, 1 denoting that a sentence contains the answer and 0 otherwise. Z is the normalizing factor, and matrix $W = [W_{:,0}; W_{:,1}] \in \mathbf{R}^{d \times 2}$ is a learn-able parameter. We use a cross entropy loss as the training objective:

$$L = -\sum_{i=1}^{N} (y \log p_1^i + (1-y) \log p_0^i). \quad (2)$$

After finetuning, the probability of the positive class is regarded as the local evidence score and we will use it to rank all the sentences for each query.

2.3 Global Estimator

Although our local estimator measures the semantic relevance between the query and the candidate segments, these estimation is done locally. To obtain a global estimation of the scores for each segment, we apply a Global Estimator following (Xu and Lapata, 2020). The centrality estimator essentially is an extension of the well-known LexRank algorithm (Erkan and Radev, 2004).

For each document cluster, i.e., the collections of documents for each query in our tasks, LexRank builds a graph $G = (V; E)$ with nodes V corresponding to sentences and undirected edges E whose weights are computed based on a certain similarity metric. The original LexRank algorithm uses TF-IDF (Term Frequency Inverse Document Frequency). (Xu and Lapata, 2020) proposes to use TF-ISF (Term Frequency Inverse Sentence Frequency), which is similar to TF-IDF but operates at the sentence level.

Following ((Xu and Lapata, 2020)), we integrate our evidence estimator into the similarity matrix E, that is,

$$\tilde{E} = w * [\tilde{q}; ...; \tilde{q}] + (1-w) * E, \quad (3)$$

where $w \in (0, 1)$ controls the extent to which the evidence estimator can influence the final summarization, and $\tilde{q}$ is obtained by normalizing the evidence scores,

$$\tilde{q} = \frac{q}{\sum_v^{|V|} q_v}. \quad (4)$$

Note the similarity matrix E can be seen as the transition probabilities. If the similarity score $E_{i,j}$ between sentence i and j is higher, it is more likely that sentence i and j are both selected in the finally summary or are discarded at the same time. We can see selecting the sentences into summaries as a Markov chain process, and we will leverage the final stationary distribution $\tilde{q}^*$ of this Markov chain as the final scores of each segment. $\tilde{q}^*$ is obtained by solving this equation:

$$\tilde{q}^* = \tilde{q}^* \tilde{E} \tag{5}$$

Note that with our evidence estimator and centrality estimator, $\tilde{q}^*$ can simultaneously expresses the importance of a sentence in the document and its semantic relation to the query. Thus, to formulate the final summary, we rank the sentences based on $\tilde{q}^*$ and select the top k^{sum} ones.

3 Contextualized evidence estimation

The previous section describe a two-step method for extractive MDS. However, it does not fully exploit the advantages of pretrained sentence encoders, since it only compares the query to single sentences which suffers from losing the contexts. In this section, we provide a simple method to conduct extractive MDS in one step, and promote the performances.

Let Q denote a query sequence and $\{S_1, S_2, ..., S_N\}$ the set of candidate answers. And we put each sentence S_i back into its contexts by concatenating the sentences surrounding it. Denote the S_i with its contexts as $C_i = [N_i^L; S_i; N_i^R]$. For implementation, we limit the sequence length of N_i by L_{max}, which is 512 for RoBERTa. For formulating the input of RoBERTa, we concatenate C_i following its sequential order, so that its contexts is not corrupted. Thus the sequence input should be like $< s >, Q, < /s >, < s >, N_i^L, < /s >, < s >, S_i, < /s >, < s >, N_i^R, < /s >$.

The above operation adds the contextual information of S_i, but the position of S_i is not emphasized, and the model might focus on N_i^R or N_i^L instead of S_i. Thus, we add a pair of special tokens before and after S_i to address the position of the sentence we are concerning. Thus, the input sequence becomes $< s >, Q, < /s >, < s >, N_i^L, < /s >, < s >, < t1 >, S_i, < t2 >, < /s >, < s >, N_i^R, < /s >$.

The RoBERTa will encode the above sequence and outputs the semantic relevance score, which we will use as the final semantic score of the sentence regarding summarization.

4 End-to-end abstractive summarization

Pre-trained models. In this section, we experiment on applying pretrained Seq2Seq models to obtain abstractive summarizations, after finetuning their on our datasets. We mainly investigate two types of models, BART ((Lewis et al., 2020)) and PEGASUS ((Zhang et al., 2020a)).

In terms of architecture, BART adopts a standard transformer seq2seq architecture ((Vaswani et al., 2017)) with some small changes. It uses GeLU (xxx,) rather than ReLU (xxx,) as activation function and initiates paramaters with normal distribution. For pre-training tasks, BART allows arbitrary noising transformations of input texts and learns a model to rebuild original text. BART achieves the state-of-the-art (SOTA) results on a wide range of tasks, including summarization and machine translation.

PEGASUS uses pre-training objectives tailored for abstractive text summarization. During pre-training, the text inputs are documents with several important missing sentences and the output is the predicted missing sentence sequences. PEGASUS can perform quite well on summarization tasks with low resources, e.g., when the training sets only contains only hundreds of samples.

Finetuning techniques. For finetuning the pre-trained seq2seq models, we experiment a few methods/techniques which can improve the downstream task performances:

- Freezing parameters. For tasks like QS and MAS, the training dataset is quite small and the large pre-trained models can be easily overfitting. We alleviate the overfitting problem by freezing the lower layers of the models.
- We use the advarsarial training method, i.e., Projected Gradient Descent (PGD, (Madry et al., 2018)) for more robust fine-tuning.
- Back translation from English to Chinese, and Chinese to English is applied for data augmentation.

5 Experiments on MAS

In task 2, We used two methods to deal with the problem of low resource data. The first method is to add muti-ext-summary and single-ext-summary as targets to the training data. Since some sentences in the summary are not exactly the same as the sentences in the article, the Jaccard similarity is used to align the sentences in article to the sentences in the extractive summary. Because the final target is multi-text-summary, in order to increase its

model	ROUGE-1	ROUGE-2	ROUGE-L	ROUGE-Lsum
		dev set		
roberta-large	56.95	48.11	41.36	56.29
+marco	57.08	48.15	42.10	56.33
+marco+reverse	57.62	49.47	41.90	56.99
+marco+lexrank	57.06	48.31	42.04	56.07
+marco+context	57.57	48.62	42.06	56.75
electra-large+marco	58.53	49.46	42.35	57.84
ensemble-model	59.29	51.09	43.80	58.88
		test set		
ensemble-model	58.5	50.8	43.5	-

Table 1: Comparison of different models on dev set in Task 2. Marco means using ms-marco data pretrain model, reverse means inverting Q and S on the input refer to (Su et al., 2020) , lexrank means using lexrank to get the global score of the sentence described in section 2.3, context means adding Contextual information described in section 3

weight, we repeatedly sampled sentences in multi-ext-summary and added it to the training set. The second method, public dataset ms-marco is used to pre-train the RoBERTa model.

Finally, the top 20 sentences based on the model score are selected and we restore their relative positions by recording the position of each sentence in the article in advance as the target. The result is shown in Table 1. As roberta-large as a baseline model, both resampling and pretraining by ms-marco have slightly improved the result of the model because of the increasing of training set. Although the lexRank method described in section 2.3 has made a improvement, the weight of model score must be a large value compared to the TF-ISF, for example 0.99 in our model. For contextualized evidence estimation described in section 3, we selected the two sentences before and after as the context and this method greatly improves the model. Referring to (Su et al., 2020), we tried to concat the question and the sentence like <s>, S, </s>,</s>,Q,</s>, this method has achieved competitive results in validation set, but the result in test set has slightly decreased. In addition, we also tried the ELECTRA (Clark et al., 2020) model and achieved a competitive results in validation set compared to RoBERTa. Ensemble model uses all models mentioned above, and weighted sum all scores of model for one sentence based on the results normalized ROUGE-2 score in validation set. The ensemble model achieves the best results on the validation set.

Our model is optimized with Adam on one Tesla V100 GPU using the following parameters: learning rate = 1e-5 batch size = 16, maximum length = 128. The learning rate is warmed up over the first 1 epoch. Early stopping strategy for 5 epoch is used to select the optimal model

In the end, we submitted the results of ensemble model and achieved the first place, as shown in Table 1

6 Experiments on QS

At first, we compare the end-to-end abstractive methods on an 8:2 split at the train set, shown in Table 2. The result shows that the PEGASUS-large model with 3-freezed-layer encoder and 3-freezed-layer decoder gains the highest score. Training on the whole training set and evaluating on the official validation set, the model performs shown in Table 3, without the question type nor question focus given. We try to do data augmentation, like translating the train data to Chinese and German and then translating back to English, but have failed to improve the result. When concatenating the two kinds of information with the original message, we find that the result has been improved (Table 3).

Over CHQA datasets, we train a span prediction model based on the pointer networks and a question type classification model to predict the question focus and question type, respectively. The span prediction model obtains the performance of 83% exact match F1, and the question type classification model achieves 78% F1. Based on those two models, we process train, valid and test set to the same pattern as the input: "SUBJECT:{question_focus};{question_type} MESSAGE:{message}". Table 4 indicates the results with different parameters.

By checking the generated sentences, we find

model	ROUGE-1	ROUGE-2	ROUGE-L	ROUGE-Lsum
BART-base	52.33	34.93	49.91	49.90
BART-large	54.25	36.28	51.56	51.51
PEGASUS-large	51.30	34.28	49.33	49.37
PEGASUS-large(freeze=3)	56.97	38.74	54.03	54.07

Table 2: Comparison of different end-to-end models on 80% train set in Task 1

valid set	ROUGE-1	ROUGE-2	ROUGE-L	ROUGE-Lsum
NO type&focus(baseline)	36.17	16.39	35.23	35.32
data augmentation	34.50	13.73	34.03	33.85
WITH type&focus	38.58	12.47	38.42	38.42

Table 3: Results of PEGASUS-large model on valid set in Task 1

the questions are highly like to be predicted as two sentence patterns: "what the treatments for ..." and "where can I find information on ...". We find these patterns appear more than 300 of 1000 train data, so we do the re-sampling for train data according to the frequency of the first four word of target questions. We train model on this re-sampled train set and get the result on valid set (Table 5). Although the score on valid set has decreased but the final score in the test set has increased. We conclude that the improvement are due to the higher diversity of the sentence patterns.

7 Experiments on RRS

Table 6 reports the main results on 80% training set with the most popular end-to-end models for summarization task currently. When using a 8:2 split at official training set, we find that PEGASUS-large model outperforms all other models with a 2% difference of ROUGE-1. We also test PEGASUS-pubmed but find suprising low performances, indicating that pubmed corpus does not fit to our tasks.

Table 7 analyses how different freezing strategies influence model performances. We consider freezing two different kinds of layers in structure: embedding layers and encoder layers. So, there are four combinations of strategies. As for BART-base model, we can see that models with frozen encoder layers fall far behind models freezing none of encoder layers, indicating that encoder layers are more important than embedding layers. It is interesting that freezing embeding layers sometimes helps BART models perform better while other models worse. As a result, We than use stratgies of freezing embedding layers or freezing no layers to our subsequent trainging settings.

According to the results of table1, we choose PEGASUS as our best model. PEGASUS models stand out from other popular models due to their specially designed pretrain tasks. We test how different optimizers influence performances. Table 8 also reveals that using adafactor will raise the ROUGE-2 metric by 2%. From the data we have, private information of patients will be replaced by token "___", which absolutely will not appear in the vocabulary of PEGASUS. Considering the fact that summaries also contain this special token, we test whether adding this to vocabulary will help models perform better. The results show that this operation decreases the performance a little bit, possibly because of not having a good initial value for the added token in embedding space.

By analysing data carefully, we find that almost half of the summaries start with pattern like "No acute ..." or "No evidence of ...". A simple idea is that we can separate the data according to the pattern into two kinds, one with pattern of starting from "No", one with other patterns, and train models separately. When predicting, we also need a classifier to classify samples and send samples into according models. We label samples of which summaries start with "No ..." as label 1, and label other samples as label 0. We than train PEGASUS-large models to generate summaries and BERT-base model to classify. The results are shown on Table 9.

Considering our classifier does make mistakes when predicting, we set a threshold of 0.75. Only when the classifier give samples probabilities higher than this, will we use the separately trained models. Otherwise, we will use the wholly trained model to predict.

model	ROUGE-1	ROUGE-2	ROUGE-L	ROUGE-Lsum
PEGASUS-large(freeze=3)	42.83	23.50	41.47	41.33
PEGASUS-large(freeze=0)	42.97	23.93	41.73	41.57

Table 4: Results of PEGASUS-large model on valid set with question type and focus in Task 1

model	ROUGE-1	ROUGE-2	ROUGE-L	ROUGE-Lsum
PEGASUS-large(freeze=0)	38.30	19.68	36.68	36.94

Table 5: Results of PEGASUS-large model fine-tuned on re-sampled data in Task 1

model	ROUGE-1	ROUGE-2	ROUGE-L	ROUGE-Lsum
BERT-abs	49.79	35.51	46.68	46.72
BART-base	61.90	49.39	58.86	60.29
BART-large(freeze)	60.10	47.38	57.01	58.55
PEGASUS-large	**63.61**	**51.86**	**60.51**	**62.28**
PEGASUS-pubmed	30.61	19.28	26.91	29.12
T5-small	57.08	45.13	54.65	55.47
T5-base	61.77	49.30	58.72	60.34
T5-large	61.85	50.81	59.19	60.56

Table 6: a comparison of different end-to-end models on 80% training set in Task 3.

model	freeze encoder	freeze embedding	ROUGE-1	ROUGE-2	ROUGE-L	ROUGE-Lsum
BART-base	yes	yes	48.68	33.78	45.88	47.37
BART-base	no	yes	**61.90**	**49.39**	**58.86**	**60.29**
BART-base	yes	no	57.48	45.57	54.75	56.10
BART-base	no	no	61.30	49.31	58.45	60.01
PEGASUS-large	no	yes	53.68	42.58	51.57	52.45
PEGASUS-large	no	no	**63.61**	**51.86**	**60.51**	**62.28**
PEGASUS-pubmed	no	yes	26.83	15.83	23.79	24.41
PEGASUS-pubmed	no	no	**30.61**	**19.28**	**26.91**	**29.12**

Table 7: a comparison of same models using different freezing strategies

model	optimizer	add vocab	ROUGE -1	ROUGE -2	ROUGE -L	ROUGE -Lsum
PEGASUS-large	adam	no	62.29	49.15	59.30	60.62
PEGASUS-large	adafactor	no	**63.07**	**51.18**	**60.06**	**61.42**
PEGASUS-large	adafactor	yes	62.99	51.10	59.97	61.34

Table 8: a comparison of PEGASUS using different optimizer and adding special token in Task 3.

pipeline part	model	acc	ROUGE-1	ROUGE-2
classification	BERT-base	88.2		
label 0	PEGASUS-large		54.02	37.34
label 1	PEGASUS-large		76.81	69.73
ensemble			61.97	50.02

Table 9: pipeline results on task3

8 Conclusion

In this work, we elaborate on the methods we employed for the three tasks in the MEDIQA 2021 shared tasks. For the extractive summarization of MAS task, we build upon Xu and Lapata (2020), and achieve improvements by adding contexts and sentence position markers. For generating abstractive summaries, we leverage the pre-trained seq2seq models. To improve the fine-tuning performances on the downstream tasks, we implement a few techniques, like freezing part of the models, adversarial training and back-translation. Our team achieves the 1st place for the MAS task.

References

Asma Ben Abacha and Dina Demner-Fushman. 2019. On the role of question summarization and information source restriction in consumer health question answering. *AMIA Joint Summits on Translational Science proceedings. AMIA Joint Summits on Translational Science*, 2019:117–126.

Daniel Fernando Campos, T. Nguyen, M. Rosenberg, Xia Song, Jianfeng Gao, Saurabh Tiwary, Rangan Majumder, L. Deng, and Bhaskar Mitra. 2016. Ms marco: A human generated machine reading comprehension dataset. *ArXiv*, abs/1611.09268.

Kevin Clark, Minh-Thang Luong, Quoc V. Le, and Christopher D. Manning. 2020. Electra: Pre-training text encoders as discriminators rather than generators. *ArXiv*, abs/2003.10555.

Dina Demner-Fushman, Yassine Mrabet, and Asma Ben Abacha. 2020. Consumer health information and question answering: helping consumers find answers to their health-related information needs. *Journal of the American Medical Informatics Association : JAMIA*.

Günes Erkan and Dragomir R. Radev. 2004. Lexrank: Graph-based lexical centrality as salience in text summarization. *J. Artif. Intell. Res.*, 22:457–479.

M. Lewis, Yinhan Liu, Naman Goyal, Marjan Ghazvininejad, A. Mohamed, Omer Levy, Ves Stoyanov, and Luke Zettlemoyer. 2020. Bart: Denoising sequence-to-sequence pre-training for natural language generation, translation, and comprehension. *ArXiv*, abs/1910.13461.

Y. Liu, Myle Ott, Naman Goyal, Jingfei Du, Mandar Joshi, Danqi Chen, Omer Levy, M. Lewis, Luke Zettlemoyer, and Veselin Stoyanov. 2019. Roberta: A robustly optimized bert pretraining approach. *ArXiv*, abs/1907.11692.

A. Madry, Aleksandar Makelov, L. Schmidt, D. Tsipras, and Adrian Vladu. 2018. Towards deep learning models resistant to adversarial attacks. *ArXiv*, abs/1706.06083.

Max E. Savery, Asma Ben Abacha, Soumya Gayen, and Dina Demner-Fushman. 2020. Question-driven summarization of answers to consumer health questions. *Scientific Data*, 7.

Dan Su, Yan Xu, Tiezheng Yu, Farhad Bin Siddique, Elham Barezi, and Pascale Fung. 2020. CAiRE-COVID: A Question Answering and Query-focused Multi-Document Summarization System for COVID-19 Scholarly Information Management. *ArXiv*.

Ashish Vaswani, Noam Shazeer, Niki Parmar, Jakob Uszkoreit, Llion Jones, Aidan N. Gomez, L. Kaiser, and Illia Polosukhin. 2017. Attention is all you need. *ArXiv*, abs/1706.03762.

Yumo Xu and Mirella Lapata. 2020. Coarse-to-fine query focused multi-document summarization. In *EMNLP*.

Jingqing Zhang, Y. Zhao, Mohammad Saleh, and Peter J. Liu. 2020a. Pegasus: Pre-training with extracted gap-sentences for abstractive summarization. In *ICML*.

Yuhao Zhang, D. Ding, Tianpei Qian, Christopher D. Manning, and C. Langlotz. 2018. Learning to summarize radiology findings. In *Louhi@EMNLP*.

Yuhao Zhang, Derek Merck, E. Tsai, Christopher D. Manning, and C. Langlotz. 2020b. Optimizing the factual correctness of a summary: A study of summarizing radiology reports. In *ACL*.

BDKG at MEDIQA 2021: System Report for the Radiology Report Summarization Task

Songtai Dai, Quan Wang, Yajuan Lyu, Yong Zhu
Baidu Inc., Beijing, China
{daisongtai,wangquan05,lvyajuan,zhuyong}@baidu.com

Abstract

This paper presents our winning system at the Radiology Report Summarization track of the MEDIQA 2021 shared task. Radiology report summarization automatically summarizes radiology findings into free-text impressions. This year's task emphasizes the generalization and transfer ability of participating systems. Our system is built upon a pre-trained Transformer encoder-decoder architecture, i.e., PEGASUS, deployed with an additional domain adaptation module to particularly handle the transfer and generalization issue. Heuristics like ensemble and text normalization are also used. Our system is conceptually simple yet highly effective, achieving a ROUGE-2 score of 0.436 on test set and ranked the 1st place among all participating systems.

1 Introduction

Radiology reports are documents that record and interpret radiological examinations. A typical radiology report usually consists of three sections: (1) a *background* section that describes general information about the patient and exam, (2) a *findings* section that presents details of the examination, and (3) an *impression* section that summarizes the findings against the background (Kahn Jr et al., 2009). Figure 1 provides an example of such a radiology report. In a standard radiology reporting process, a radiologist first dictates detailed findings into the report, and then summarizes the findings into a concise impression based also on general background of the patient (Zhang et al., 2018). The impression section, which provides the most valuable information to make clinical decisions, is the most crucial part of a radiology report for both doctors and patients. However, manually summarizing radiology findings into impressions are time-consuming and error-prone (Gershanik et al., 2011), which necessitates the need to automatically generate radiology impressions.

Background: Examination: chest (portable AP) indication: history: ___m with acute coronary syndrome technique: upright AP view of the chest comparison: chest radiograph ___ **Findings**: Patient is status post median sternotomy and CABG. Heart size remains mildly enlarged. The aorta is tortuous. Mild pulmonary edema is new in the interval. Small bilateral pleural effusions are present. Patchy bibasilar airspace opacities likely reflect areas of atelectasis ...
Impression: Mild pulmonary edema and trace bilateral pleural effusions.

Figure 1: A radiology report sampled from MEDIQA 2021 training set, where the impression is a summarization of the findings taking the background into account.

The MEDIQA 2021 shared task (Abacha et al., 2021) at the NAACL-BioNLP workshop sets up a *Radiology Report Summarization* subtask, the aim of which is to build advanced systems to automatically summarize radiology findings (along with the background) into concise impressions. A key feature of this task is that radiology reports used for training and evaluation are collected from different sources, e.g., training instances are sampled from the MIMIC-CXR database (Johnson et al., 2019) and some evaluation instances come from the Indiana chest X-ray collection (Demner-Fushman et al., 2016). This inevitably results in significant discrepancies between training and evaluation, posing new challenges to the generalization and transfer ability of participating systems.

Zhang et al. (2018) presented the first sequence-to-sequence attempt at automatic summarization of radiology findings into natural language impressions. After that, several extensions and improvements have been proposed, e.g., to take into account the factual correctness (Zhang et al., 2019) or the ontologies (MacAvaney et al., 2019; Gharebagh

Proceedings of the BioNLP 2021 workshop, pages 103–111
June 11, 2021. ©2021 Association for Computational Linguistics

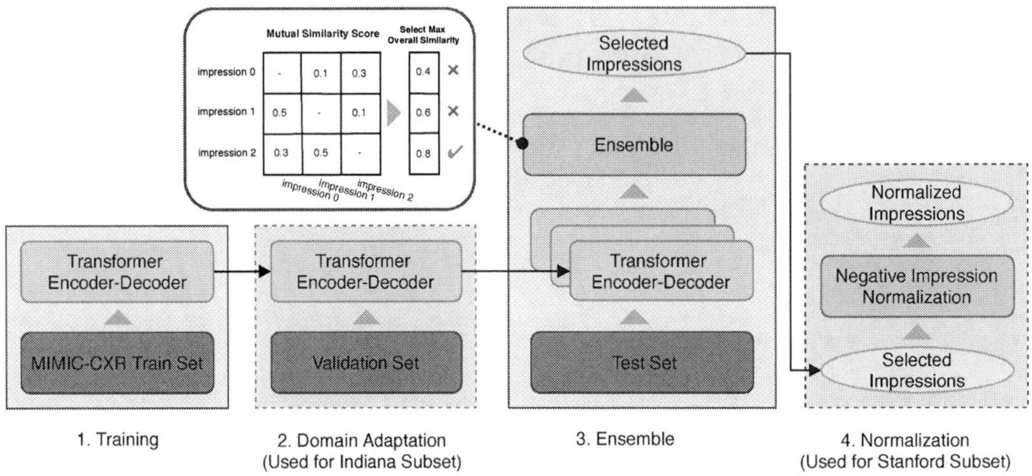

Figure 2: An overview of our system, which consists of (1) a Transformer encoder-decoder tuning module, (2) a domain adaptation module, (3) an ensemble module, (4) a negative impression normalization module. The domain adaptation module is activated only for test instances in the Indiana subset, and the final normalization module is activated only for test instances in the Stanford subset.

et al., 2020). These prior studies, however, are all based on traditional sequence-to-sequence models like RNN, BiLSTM, as well as pointer-generator network (See et al., 2017), and none of them actually touches the generalization or transfer issue.

In the past few years, pre-training Transformer-based encoder-decoder architectures from large-scale text corpora has been proposed and quickly received massive attention (Radford et al., 2018; Dong et al., 2019; Xiao et al., 2020). Quite a number of such pre-trained models, e.g., MASS (Song et al., 2019), BART (Lewis et al., 2020), and T5 (Raffel et al., 2020), have been devised and proved extremely effective in various language generation tasks. Against this background, we choose PEGASUS (Zhang et al., 2020), a pre-trained model that reports state-of-the-art performance on abstractive text summarization, as the backbone of our system. Since radiology report summarization is a special form of abstractive text summarization, we expect this choice to yield optimal performance. Besides, we employ a simple yet effective domain adaptation strategy, by further fine-tuning on a small amount of in-domain data to improve generalization and transfer abilities. We also use model ensemble and negative impression normalization strategies to further enhance the performance. Figure 2 provides an overview of our system.

With all these strategies, our system achieves an overall ROUGE-2 score of 0.436 on the whole test set, ranked at the 1st place among all participating systems. We will discuss later in the experimental section the performance of different pre-trained models and the effect of each individual strategy.

2 Task Description

This section gives a formal definition of the radiology report summarization task, and introduces data and evaluation metrics used for the task.

2.1 Task Definition

The MEDIQA 2021 Radiology Report Summarization task aims to automatically summarize radiology findings into natural language impression statements. Figure 1 provides an example of a standard radiology report, which consists of a *background*, *findings*, and *impression* section, detailed as below:

- **Background:** This section provides general information about the patient and exam, e.g., clinical history of the patient, type of the exam, and examination techniques. This kind of information helps diagnose diseases when combined with specific findings.

- **Findings:** This section records notable details in each part of the body observed in the exam, after reading an X-ray image. It describes the normality and abnormality a radiologist found in each part of the body. If a specific part was examined but not mentioned, there is probably no obvious abnormality found in that part.

- **Impression:** This section is a concise summarization of the findings written by a radiologist. It lists the patient's symptoms and sometimes with suggested diagnoses. This section is the most crucial part of a radiology report, providing valuable information for doctors to make clinical decisions.

Radiology Report Summarization is to generate the impression given the background and findings. Formally, given a passage of findings represented as a sequence of tokens $\mathbf{x} = \{x_1, x_2, \cdots, x_L\}$ along with the background represented as a sequence of tokens $\mathbf{y} = \{y_1, y_2, \cdots, y_M\}$, the goal is to generate another sequence of tokens $\mathbf{z} = \{z_1, z_2, \cdots, z_N\}$ that best summarizes salient and clinically significant findings in $\mathbf{x}$. Here, L, M, N are the lengths of the findings, the background, and the impression, respectively.

2.2 Official Data

The official data consists of a training split, two validation splits, and two test splits collected from different sources, detailed as follows:

- **Training split:** The training split is composed of 91,544 chest radiology reports picked from MIMIC-CXR database (Johnson et al., 2019). These reports are collected from patients presenting to the Beth Israel Deaconess Medical Center Emergency Department between 2011 and 2016.

- **Validation split I:** The first validation split consists of 2,000 chest radiology reports sampled also from MIMIC-CXR. It therefore has the same distribution with the training split.

- **Validation split II:** The second validation split consists of 2,000 radiology reports sampled from the Indiana chest X-ray collection (Demner-Fushman et al., 2016). These reports are collected from the Indiana Network for Patient Care, thus bearing a risk of inconsistency with the training split.

- **Test split I:** The first test split is also extracted from the Indiana chest X-ray collection, composed of 300 radiology reports in total.

- **Test split II:** The second test split comprises another 300 chest radiology reports collected from the picture archiving and communication system at the Stanford Hospital.

Split	# Reports	Source
Training	91,544	MIMIC-CXR database
Validation I	2,000	MIMIC-CXR database
Validation II	2,000	Indiana collection
Test I	300	Indiana collection
Test II	300	Stanford collection

Table 1: Statistics and sources of the official data.

The statistics and sources of the data splits are summarized in Table 1. As we can see, both test splits come from different sources with the training split. This poses significant challenges to the generalization and transfer ability of participating systems.

2.3 Evaluation Metrics

The task uses ROUGE (Lin, 2004) to evaluate the performance of participating systems. F1 scores for ROUGE-1, ROUGE-2 and ROUGE-L are reported on the whole test set, and also on the Indiana and Stanford splits. The metrics measure the word-level unigram-overlap, bigram-overlap and the longest common sequence between reference summaries and system predicted summaries respectively. The overall **ROUGE-2** on the whole test set is selected as the primary metric to rank participating systems.

3 Our Approach

We employ a Transformer-based encoder-decoder architecture for radiology report summarization. Our system, as illustrated in Figure 2, consists of four consecutive modules:

- a *Transformer encoder-decoder training* module that fine-tunes a pre-trained language generation model, e.g., PEGASUS (Zhang et al., 2020), on the training split;

- a *domain adaptation* module that further fine-tunes the model on a small amount of validation data coming from the same source with the test split, designed specifically to enhance generalization and transfer ability to unseen data;

- an *ensemble* module that combines diverse predictions from multiple models to generate robust summarization;

- a final *normalization* module that normalizes system predicted negative impressions into a specific form.

Our system is simple yet highly effective, ranked at the 1st place among all participating systems. In the rest of this section, we detail key modules of the system.

3.1 Transformer Encoder-Decoder Training

Transformer-based encoder-decoder architectures pre-trained from large-scale text corpora have recently stood out as the most promising techniques for natural language generation, outperforming the traditional RNN- or LSTM-based opponents in a wide range of language generation tasks (Radford et al., 2018; Raffel et al., 2020). We thereby choose a pre-trained Transformer encoder-decoder model as the backbone of our system, and fine-tunes the model on the training split.

During the fine-tuning process, for each training radiology report, we concatenate the findings $\mathbf{x}$ and background $\mathbf{y}$ into a single sequence, and pair that sequence with the impression $\mathbf{z}$, i.e.,

- Source: $x_1, x_2, \cdots, x_L, [\text{SEP}], y_1, y_2, \cdots, y_M$
- Target: $z_1, z_2, \cdots, z_N$

where [SEP] is a special token separating the findings and the background. The source sequence is fed into the encoder, and the decoder autoregressively decodes the next token conditioned on the encoder output and previous tokens.

We are free to use any pre-trained Transformer encoder-decoder models. We investigate three representatives: BART, ERNIE-GEN, and PEGASUS, detailed as below.

- **BART** (Lewis et al., 2020) is a denoising autoencoder for sequence-to-sequence learning. It is trained by corrupting text with a noising function, and learning a model to reconstruct the original text. It achieves promising results on a range of abstractive dialogue, question answering, and summarization tasks.

- **ERNIE-GEN** (Xiao et al., 2020) is a multi-flow sequence-to-sequence model that mitigates exposure bias with an infilling generation mechanism and a noise-aware generation method. It achieves comparable results with a smaller number of parameters on several abstractive summarization, question generation, and dialogue response generation tasks.

Model	# Parameters	Corpus Size
BART	400M	160GB
ERNIE-GEN	340M	430GB
PEGASUS	568M	3.8TB + 750GB

Table 2: Number of parameters and size of pre-training corpus of the three models.

- **PEGASUS** (Zhang et al., 2020) is a Transformer encoder-decoder model specifically designed for abstractive text summarization. It is trained by masking out important sentences from an input document and generating the masked sentences together from the remaining sentences, similar to an extractive summary. It achieves state-of-the-art performance on 12 summarization tasks spanning across news, science, stories, instructions, emails, patents, and legislative bills.

Table 2 compares number of parameters and size of pre-training corpus of the three models. PEGASUS gets the largest number of parameters and is trained on the largest amount of data.

3.2 Domain Adaptation

As the test splits (Indiana and Stanford) are collected from different sources with the training split (MIMIC-CXR), participating systems need to address the generalization and transfer issue. Inspired by (Gururangan et al., 2020), we employ a domain adaptation strategy. Specifically, after fine-tuning a pre-trained model on the MIMIC-CXR training set, we further fine-tune the model on a small amount of data similar to the test splits. In this way, we can effectively adapt the model trained from MIMIC-CXR to target test domains.

For the Indiana test split where there is a validation split sampled from the same source, we simply use this validation split for further fine-tuning. After a few epochs over the Indiana validation split, we use the resultant model to make predictions for reports in this test split. As we will show later in the experiments, this adaptation strategy, though conceptually simple, is highly effective, leading to a remarkable boost in ROUGE-2 on this test split.

For the Stanford test split, there is no validation split sampled from the same source. Therefore we construct a subset from the training split to conduct domain adaptation. For each case in this test split (a radiology report without impression), we exploit

Negative Impression	Indiana Freq.	MIMIC-CXR Freq.	Overall Freq.
No acute cardiopulmonary abnormality.	14.2%	4.9%	9.6%
No acute cardiopulmonary process.	3.0%	15.0%	9.0%
No acute cardiopulmonary findings.	6.0%	0.1%	3.1%
No acute cardiopulmonary disease.	0.2%	4.9%	2.6%
No acute cardiopulmonary abnormalities.	4.9%	0.1%	2.5%

Table 3: Top 5 frequent negative impressions and their frequencies on the validation splits.

ElasticSearch[1] to retrieve the top 10 reports from the MIMIC-CXR training split that share the most similar findings. We obtain 2,618 such radiology reports in total after removing duplicates. Then we conduct further fine-tuning on these reports, which, however, downgrades the performance. So we just use the model trained from training split to predict for reports in this test split.

3.3 Model Ensemble

We further employ ensemble that combines diverse predictions from multiple models for robust summarization. Suppose we have T candidate models, e.g., multiple runs with different seeds, each producing a predicted impression $\hat{\mathbf{z}}^i$ $(1 \le i \le T)$ for the given findings along with the background. We first compute the mutual similarity score $\mathrm{Sim}(\hat{\mathbf{z}}^i, \hat{\mathbf{z}}^j)$ between each pair of predictions, and aggregate these scores to measure the overall similarity of a specific prediction against all the other predictions:

$$s(\hat{\mathbf{x}}^i) = \sum_{j \neq i} \mathrm{Sim}(\hat{\mathbf{z}}^i, \hat{\mathbf{z}}^j), \quad i = 1, \cdots, T.$$

Then we select the prediction $\hat{\mathbf{z}}^i$ with the highest overall similarity $s(\hat{\mathbf{x}}^i)$ as our final prediction. Figure 2 visualizes this ensemble process. We have tried various similarity scoring functions $\mathrm{Sim}(\cdot, \cdot)$, e.g., ROUGE-1, ROUGE-2, ROUGE-L, and token-level F1, but observed no significant differences between their performance. We finally use ROUGE-1 as the similarity scoring function.

3.4 Negative Impression Normalization

The final normalization module normalizes system predicted negative impressions into a specific form. Roughly speaking, the impression of a radiology report can be divided into two categories: *positive* or *negative*. A positive impression typically reveals symptoms observed during the exam, e.g., "*Mild pulmonary edema and tracebilateral pleural effusions*", whereas a negative impression indicates no symptoms at all, e.g., "*No acute cardiopulmonary abnormality*". Unlike positive impressions which vary drastically w.r.t. input findings, negative impressions tend to be expressed in specific forms. Table 3 presents the top 5 frequent negative impressions and their frequencies on the validation splits. Though expressed in different forms, these negative impressions are all of the same meaning. The choice of a particular form is just a matter of writing style. As the writing style usually varies across organizations, predicting negative impressions by a complex model trained from another organization is prone to over-fitting and may not work well. In contrast, simple heuristics based on basic statistics may lead to less over-fitting and perform better.

Based on this observation, we introduce a heuristic strategy, i.e., for any negative prediction starting with "No acute", we normalize it into "No acute cardiopulmonary abnormality", which is the most frequent negative impression in the validation sets. This normalization process is carried out only for the Stanford test split, for which there is no training or validation set from same organization.

4 Experiments and Results

This section presents experiments and results of our system on the official data.

4.1 Experimental Setups

Our system is built upon a pre-trained Transformer encoder-decoder architecture, PEGASUS (Zhang et al., 2020). The maximum lengths of source and target sequences are restricted to 512 and 128 respectively, covering 99% of the cases in the training and validation splits. Throughout all experiments, we employ a decoding process with beam size of 5, length penalty of 0.8, and early stopping.

Fine-tuning Setup We first fine-tune PEGASUS-large[2] on the MIMIC-CXR training split. We tune

[1] `https://www.elastic.co`

[2] `https://huggingface.co/google/pegasus-large`

Rank	Team	All Test Set ROUGE-1/-2/-L			Indiana Test Set ROUGE-1/-2/-L			Stanford Test Set ROUGE-1/-2/-L		
1	**BDKG (Ours)**	**.5573**	**.4362**	**.5366**	**.6834**	**.5956**	**.6717**	**.4312**	.2769	**.4014**
2	IBMResearch	.5328	.4082	.5134	.6772	.5881	.6657	.3884	.2284	.3611
3	optumize	.5186	.3918	.4957	.6188	.5182	.6050	.4183	.2655	.3864
4	JB	.4955	.3778	.4794	.5895	.5039	.5824	.4015	.2517	.3763
5	low_rank_AI	.4716	.3311	.4487	.5129	.3846	.5026	.4302	**.2777**	.3948
6	med_qa_group	.4642	.3265	.4440	.5051	.3774	.4965	.4233	.2757	.3916
7	ChicHealth	.4606	.3236	.4411	.5070	.3782	.4984	.4143	.2690	.3838
8	hEALTHai	.4481	.3084	.4273	.4845	.3527	.4752	.4118	.2641	.3794
9	DAMO_ali	.4330	.2763	.4116	.4371	.2839	.4278	.4289	.2687	.3954
10	I_have_no_flash	.4303	.2743	.4092	.4351	.2826	.4258	.4256	.2661	.3926

Table 4: Official results of top 10 systems on the test splits. Systems ranked by ROUGE-2 on the whole test set.

the initial learning rate $\in \{1e\text{–}5, 3e\text{–}5, 6e\text{–}5, 1e\text{–}4\}$, batch size $\in \{8, 16, 32\}$, and number of epochs $\in \{5, 10, 15, 25\}$. Other hyper-parameters are fixed to their default values. The optimal configuration is determined by ROUGE-2 on the whole validation set (a combination of the MIMIC-CXR and Indiana splits), which is learning rate $= 6e-5$, batch size $= 8$, and number of epochs $= 15$.

Domain Adaptation Setup We further fine-tune the model derived above on the Indiana validation split, so as to adapt the model from MIMIC-CXR to our target test domain. Specifically, we split the Indiana validation set into 1700 : 300 subsets. We tune the model with initial learning rate $\in \{1e-4, 2e-4, 4e-4\}$, batch size $\in \{8, 16\}$, and number of epochs $\in \{10, 20, 50, 100\}$ on the former, and determine the optimal configuration on the latter (by ROUGE-2). The optimal configuration is initial learning rate $= 2e-4$, batch size $= 8$, and number of epochs $= 100$, with other hyper-parameters set, again, to their default values. After determining the optimal configuration, we re-tune the model on the whole Indiana validation set.

Ensemble Setup We ensemble 16 models further fine-tuned with in-domain data for the Indiana test split. These models are obtained with the same optimal configuration determined during domain adaptation, but different random seeds. We ensemble another 15 models trained from MIMIC-CXR training split for the Stanford test split. These models are obtained, again, with the same configuration but different seeds.

4.2 MEDIQA 2021 Official Results

Table 4 shows the official results of top 10 participating systems on the test splits, where systems are ranked by ROUGE-2 score on the whole test set. Our system, though conceptually simple, is highly effective, ranked the 1st place among participating systems. Notably, it consistently outperforms the other systems across all three test splits and almost in all metrics.

4.3 Further Analyses

This section provides in-depth analyses to show the effect of each individual module in our system.

Effect of Pre-trained Models We first examine the effect of different pre-trained models. Specifically, besides PEGASUS-large, we consider other pre-trained models including BART[3], DistilBART[4], ERNIE-GEN[5], and PEGASUS-xsum[6], all in the "large" setting. We tune their hyper-parameters in the same ranges as in PEGASUS-large, and report optimal results on the validation splits. The results are summarized in Table 5, where (S) scores denote results for single models averaged over five runs. Among these models, the two PEGASUS variants (-large and -xsum), which are designed specifically for abstractive text summarization, consistently perform better. And the -large variant performs even better than the -xsum one. The reason may be that the -xsum variant has been further tuned on XSum

[3]https://huggingface.co/facebook/bart-large

[4]https://huggingface.co/sshleifer/distilbart-xsum-12-6

[5]https://github.com/PaddlePaddle/ERNIE/tree/repro/ernie-gen

[6]https://huggingface.co/google/pegasus-xsum

Model	All Valid Set ROUGE-1/-2/-L			MIMC-CXR Valid Set ROUGE-1/-2/-L			Indiana Valid Set ROUGE-1/-2/-L		
BART (S)	.5352	.3871	.5103	.6209	.4902	.5865	.4495	.2840	.4340
BART (E)	.5535	.4057	.5284	.6425	.5125	.6077	.4644	.2989	.4491
DistilBART (S)	.5456	.3987	.5214	.6385	.5109	.6055	.4526	.2865	.4372
DistilBART (E)	.5604	.4144	.5360	.6516	.5244	.6189	.4691	.3043	.4531
ERNIE-GEN (S)	.5385	.3951	.5167	.6237	.4996	.5926	.4532	.2905	.4409
ERNIE-GEN (E)	.5476	.4035	.5229	.6313	.5070	.6002	.4638	.3000	.4515
PEGASUS-xsum (S)	.5506	.4107	.5303	.6413	.5233	.6117	.4600	.2981	.4489
PEGASUS-xsum (E)	.5566	.4172	.5361	.6441	.5266	.6141	.4691	.3078	.4581
PEGASUS-large (S)	.5559	.4129	.5330	.6511	.5290	.6188	.4608	.2968	.4471
PEGASUS-large (E)	**.5649**	**.4224**	**.5413**	**.6572**	**.5329**	**.6235**	**.4725**	**.3088**	**.4591**

Table 5: Results of different pre-trained models on validation splits. We run each model five times with different seeds under its optimal configuration. (S)/(E) respectively denotes the averaged/ensemble results of the five runs.

Ablation	All Test Set ROUGE-1/-2/-L			Indiana Test Set ROUGE-1/-2/-L			Stanford Test Set ROUGE-1/-2/-L		
Full Model	**.5573**	**.4362**	**.5366**	**.6834**	**.5956**	**.6717**	**.4312**	**.2769**	**.4014**
– Domain Adaptation	.4539	.2916	.4333	.4766	.3062	.4652	**.4312**	**.2769**	**.4014**
– Normalization	.5487	.4221	.5281	**.6834**	**.5956**	**.6717**	.4139	.2486	.3844

Table 6: Ablation results of domain adaptation and negative impression normalization on test splits.

(Narayan et al., 2018), which consists of articles from the British Broadcasting Corporation and exhibits drastic distinctions from radiology reports. This thereby may result in catastrophic forgetting.

Effect of Ensemble We further investigate the effect of model ensemble. To this end, for each of the pre-trained models considered above, we run the model five times with its optimal configuration but different seeds. We then compare performance of the single model (S) and the ensemble (E) on the validation splits, and report the results in Table 5. We can see that ensemble is a generally effective strategy, leading to about 1% to 2% gains across all data splits and metrics, not matter which pre-trained model is used.

Effect of Domain Adaptation We then evaluate the effect of our domain adaptation module, which is applied solely to the Indiana test split. We consider an ablation that uses the model trained from MIMIC-CXR to predict on both Indiana and Stanford test splits, without further fine-tuning on the in-domain Indiana validation split. Table 6 reports the performance of this ablation on the test splits, and makes comparisons to the full model. We can see that the adaptation module, though conceptually simple, is extremely useful, pushing the ROUGE-2 score drastically from 0.3062 to 0.5956 on Indiana test split.

Effect of Normalization We finally evaluate the effect of negative impression normalization, which is applied solely to the Stanford test split. Table 6 compares performance with and without this final normalization strategy on the test splits. We can see that this simple strategy brings meaningful gains, pushing the ROUGE-2 score from 0.2486 to 0.2769 on Stanford test split.

5 Conclusion

This paper presents our winning system at the Radiology Report Summarization track of the MEDIQA 2021 shared task. Participating systems in this track are required to summarize radiology findings into natural language impressions, and be able to generalize or transfer to reports collected from previously unseen hospitals. We build our system on the basis of a pre-trained Transformer encoder-decoder architecture, namely PEGASUS. We further employ a domain adaptation module to enhance generalization and transfer ability. Heuristics such as ensemble and negative impression normalization are also used. Our system finally achieves a ROUGE-2 score of 0.436 on the test set, ranked the 1st place among all participating systems.

Acknowledgements

We would like to thank the MEDIQA 2021 organizers for the challenging and exciting tasks. This work is supported by the National Key Research and Development Program of China No.2020AAA0109400.

References

Asma Ben Abacha, Yassine Mrabet, Yuhao Zhang, Chaitanya Shivade, Curtis Langlotz, and Dina Demner-Fushman. 2021. Overview of the MEDIQA 2021 shared task on summarization in the medical domain. In *Proceedings of the 20th SIGBioMed Workshop on Biomedical Language Processing, NAACL-BioNLP 2021*.

Dina Demner-Fushman, Marc D Kohli, Marc B Rosenman, Sonya E Shooshan, Laritza Rodriguez, Sameer Antani, George R Thoma, and Clement J McDonald. 2016. Preparing a collection of radiology examinations for distribution and retrieval. *Journal of the American Medical Informatics Association*, 23(2):304–310.

Li Dong, Nan Yang, Wenhui Wang, Furu Wei, Xiaodong Liu, Yu Wang, Jianfeng Gao, Ming Zhou, and Hsiao-Wuen Hon. 2019. Unified language model pre-training for natural language understanding and generation. *arXiv preprint arXiv:1905.03197*.

Esteban F Gershanik, Ronilda Lacson, and Ramin Khorasani. 2011. Critical finding capture in the impression section of radiology reports. In *AMIA Annual Symposium Proceedings*, pages 465–469.

Sajad Sotudeh Gharebagh, Nazli Goharian, and Ross Filice. 2020. Attend to medical ontologies: Content selection for clinical abstractive summarization. In *Proceedings of the 58th Annual Meeting of the Association for Computational Linguistics*, pages 1899–1905.

Suchin Gururangan, Ana Marasović, Swabha Swayamdipta, Kyle Lo, Iz Beltagy, Doug Downey, and Noah A. Smith. 2020. Don't stop pretraining: Adapt language models to domains and tasks. In *Proceedings of ACL*.

Alistair EW Johnson, Tom J Pollard, Seth J Berkowitz, Nathaniel R Greenbaum, Matthew P Lungren, Chih-ying Deng, Roger G Mark, and Steven Horng. 2019. MIMIC-CXR, a de-identified publicly available database of chest radiographs with free-text reports. *Scientific Data*, 6(1):1–8.

Charles E Kahn Jr, Curtis P Langlotz, Elizabeth S Burnside, John A Carrino, David S Channin, David M Hovsepian, and Daniel L Rubin. 2009. Toward best practices in radiology reporting. *Radiology*, 252(3):852–856.

Mike Lewis, Yinhan Liu, Naman Goyal, Marjan Ghazvininejad, Abdelrahman Mohamed, Omer Levy, Veselin Stoyanov, and Luke Zettlemoyer. 2020. BART: Denoising sequence-to-sequence pre-training for natural language generation, translation, and comprehension. In *Proceedings of the 58th Annual Meeting of the Association for Computational Linguistics*, pages 7871–7880.

Chin-Yew Lin. 2004. ROUGE: A package for automatic evaluation of summaries. In *Text summarization branches out*, pages 74–81.

Sean MacAvaney, Sajad Sotudeh, Arman Cohan, Nazli Goharian, Ish Talati, and Ross W Filice. 2019. Ontology-aware clinical abstractive summarization. In *Proceedings of the 42nd International ACM SIGIR Conference on Research and Development in Information Retrieval*, pages 1013–1016.

Shashi Narayan, Shay B Cohen, and Mirella Lapata. 2018. Don't give me the details, just the summary! topic-aware convolutional neural networks for extreme summarization. In *EMNLP*, pages 1797–1807.

Alec Radford, Karthik Narasimhan, Tim Salimans, and Ilya Sutskever. 2018. Improving language understanding by generative pre-training. Technical report, Technical report, OpenAI.

Colin Raffel, Noam Shazeer, Adam Roberts, Katherine Lee, Sharan Narang, Michael Matena, Yanqi Zhou, Wei Li, and Peter J Liu. 2020. Exploring the limits of transfer learning with a unified text-to-text transformer. *Journal of Machine Learning Research*, 21(140):1–67.

Abigail See, Peter J Liu, and Christopher D Manning. 2017. Get to the point: Summarization with pointer-generator networks. In *Proceedings of the 55th Annual Meeting of the Association for Computational Linguistics*, pages 1073–1083.

Kaitao Song, Xu Tan, Tao Qin, Jianfeng Lu, and Tie-Yan Liu. 2019. MASS: Masked sequence to sequence pre-training for language generation. In *Proceedings of the 36th International Conference on Machine Learning*, pages 5926–5936.

Dongling Xiao, Han Zhang, Yukun Li, Yu Sun, Hao Tian, Hua Wu, and Haifeng Wang. 2020. ERNIE-GEN: An enhanced multi-flow pre-training and fine-tuning framework for natural language generation. In *Proceedings of the Twenty-Ninth International Joint Conference on Artificial Intelligence*, pages 3997–4003.

Jingqing Zhang, Yao Zhao, Mohammad Saleh, and Peter Liu. 2020. PEGASUS: Pre-training with extracted gap-sentences for abstractive summarization. In *Proceedings of the 37th International Conference on Machine Learning*, pages 11328–11339.

Yuhao Zhang, Daisy Yi Ding, Tianpei Qian, Christopher D Manning, and Curtis P Langlotz. 2018. Learning to summarize radiology findings. In *Proceedings of the Ninth International Workshop on Health Text Mining and Information Analysis*, pages 204–213.

Yuhao Zhang, Derek Merck, Emily Bao Tsai, Christopher D Manning, and Curtis P Langlotz. 2019. Optimizing the factual correctness of a summary: A study of summarizing radiology reports. *arXiv preprint arXiv:1911.02541*.

damo_nlp at MEDIQA 2021: Knowledge-based Preprocessing and Coverage-oriented Reranking for Medical Question Summarization

Yifan He and **Mosha Chen** and **Songfang Huang**
Alibaba Group
{y.he, chenmosha.cms, songfang.hsf}@alibaba-inc.com

Abstract

Medical question summarization is an important but difficult task, where the input is often complex and erroneous while annotated data is expensive to acquire.

We report our participation in the MEDIQA 2021 question summarization task in which we are required to address these challenges. We start from pre-trained conditional generative language models, use knowledge bases to help correct input errors, and rerank single system outputs to boost coverage. Experimental results show significant improvement in string-based metrics.

1 Introduction

Question summarization for medical forum is important for medical knowledge discovery and retrieval and facilitates downstream tasks such as biomedical question answering (Jin et al., 2021). Medical questions are often complex, scattered with non-medical information, and can sometimes be erroneous because forum users are not domain experts (Ben Abacha and Demner-Fushman, 2019). In addition, annotation in the medical domain is harder to acquire than in the general domain. These challenges make medical question summarization an important and difficult task where annotation is often scarce.

The MEDIQA 2021 shared task 1 (Ben Abacha et al., 2021), medical question summarization, requires participants to build summarization systems for noisy medical forum texts with limited annotation data. The official training set of the task is the MeQSum dataset (Ben Abacha and Demner-Fushman, 2019), which is composed of 1,000 medical questions and their corresponding summaries. The validation and test sets consist of 50 and 100 questions respectively and topic words are sometimes misspelled.

Scarcity of data, noisy input, and complexity and redundancy of text all pose challenges for question summarization systems. We try to address these challenges using a combination of knowledge-based error correction, pre-trained generative language models, and output reranking.

Knowledge-based error correction leverages multiple levels of lexical resources and a high coverage knowledge base to correct errors in input. Our experiments show that knowledge-based error correction helps downstream summarization performance according to the Rouge metric.

Pre-trained generative language models are transformer-based language models trained with loss functions that facilitate sequence to sequence generation. Models such as Pegasus (Zhang et al., 2020a), BART (Lewis et al., 2020), and T5 (Raffel et al., 2020) achieve state-of-the-art performance on various text generation tasks and are shown to perform well on few-shot generation scenarios (Goodwin et al., 2020). We finetune pre-trained language models to obtain baseline systems with limited amount of training data.

Output reranking picks the best output among multiple systems. The availability of different language models offers a diverse set of summaries to choose from. We observe difference in summarization styles between the training and the validation set and devise a simple heuristic to pick the best output based on this observation.

In the rest of the paper, we describe these components and report evaluation results on the validation and the test set.

2 Task and Architecture Overview

The MEDIQA question summarization task requires participants to summarize user generated medical queries into shorter, more focused questions. We present an example from the MEDIQA 2021 task 1 validation set in Figure 1 (a). We note that the name of the disease "folliculitis" is spelled

Proceedings of the BioNLP 2021 workshop, pages 112–118
June 11, 2021.

Question Hi, Please can you help - I am writing from South Africa. My daughter suffers with acute folliculitus, and has been since the age of 13. She is now 20 and is in so much distress as nothing seems to alleviate the itching and soreness... I am writing to you for any help you could give me to try and assist her. Could you recommend a specialist and someone who could help us with research? Please could you point us in the right direction? I am happy to send through her lab tests - please let me know. Thanks

Summary How can we find a specialist or clinical trial for chronic folliculitis?

(a) Example from MEDIQA 2021 Task 1 Validation Set

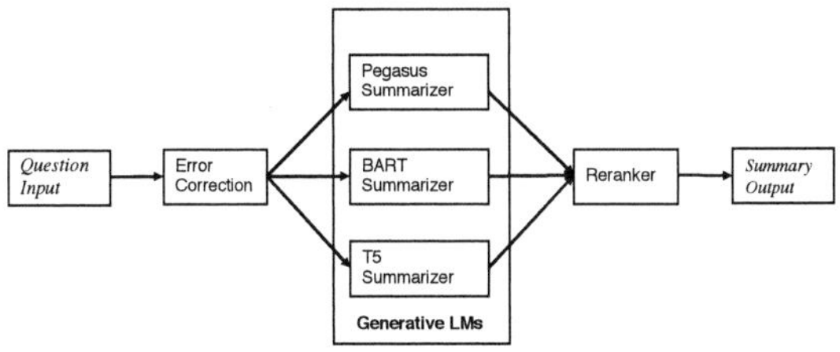

(b) Architecture of our submission

Figure 1: Question-summarization example and system architecture

incorrectly in the input question and the question contains a lot of irrelevant information. We attempt to correct misspellings with a dedicated module in our system. As useful information is often scattered in different sentences in the input, abstractive summarization suits this task better than extractive summarization. We perform abstractive summarization with pre-trained language models.

We illustrate the architecture of our submission in Figure 1 (b): we first try to correct spell errors in the input; then summarize each question with three generative LMs: Pegasus, BART, and T5; finally, for each question, we pick the best output with a feature-based reranker and the best output is chosen as the summarization of the question.

3 Knowledge-based Error Correction

Misspellings are prevalent in medical forums, where non-expert users discuss highly specialized medical topics. Uncorrected misspellings can lead to mismatch between the source text and the summary during training and cause errors if copied verbatim during prediction. These errors are penalized heavily by string matching-based metrics like Rouge as they break n-grams.

In this shared task, we conservatively correct misspelled words in input by reusing a cascade of candidate generation modules from an entity linking system. Entity linking is the task to link entity mentions in text to entities in a knowledge base (KB). Candidate generation is an intermediate step in entity linking to generate candidate KB entities from potentially abbreviated, misspelled, or alias text mentions (see e.g. (Charton et al., 2014)). Our method is also comparable to previous work on Levenshtein distance-based (Levenshtein, 1966) medical query correction (Soualmia et al., 2012), but we augment that approach with cascaded knowledge sources and an alias table.

Error correction can be implemented easier and with possibly higher quality if search suggestions from online search engines (Zhou et al., 2015) are utilized. We use in-house error correction to keep the submission offline.

3.1 Resources

The error correction module relies on the following resources:

- **Medical word list**. We collect tokens from the English side of ~20K bilingual medical phrases collected from dictionaries and drug names.
- **Wikipedia dump**. We use a 20210101 dump of the English Wikipedia as the knowledge base and alias table.
- **High frequency word list**. We use the top 10,000 words in the Google 1T corpus [1].

We use Wikipedia instead of a medical KB because of its broad coverage. Edges (redirects, links etc.) in the Wikipedia KB can be used as an alias table to capture common misspellings and aliases.

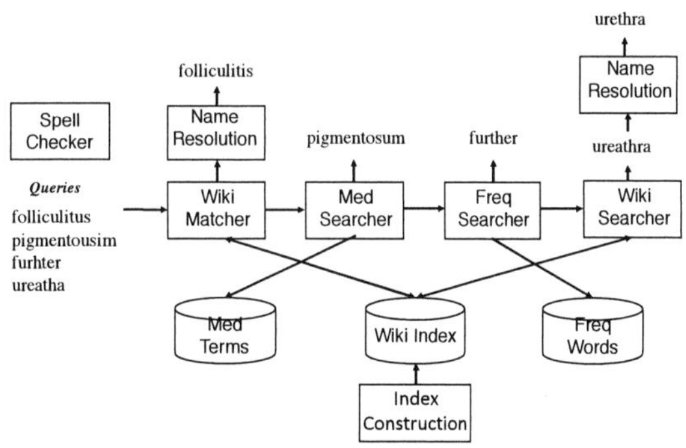

Figure 2: Example of error correction

3.2 Error correction steps

During error correction, we handle tokens composed entirely of alphabetical characters and allow at most 2 edits in similarity searches. We only consider tokens that share 3-prefix or 3-suffix with the query to limit search space.

[1] https://books.google.com/ngrams/info

Error correction consists of the following steps:

- **Index construction**. We build a token index of Wikipedia. We only index titles with no more than two tokens and tokens more than 5 characters long. We use the first token to represent the title. When a token can map to more than one titles, we map it to the title with the lowest id.

- **Spell checking**. We pass the text through a spell checker with medical terms[2] to detect potential errors. The flagged tokens are the *query* words for the error correction pipeline.

- **Wikipedia match**. If the query has an exact match in the Wikipedia token index, we link the query to the token and its corresponding Wikipedia title. Note that a title can either be an entity or an alias, which we resolve later in the name resolution step.

- **Medical word search**. We search the medical word list to find medical terms that spell similarly to the query. We choose the medical term if a result is found.

- **Frequent word search**. We search the high frequency word list to recall common English words that spell similarly to the query. We choose the word if a result is found.

- **Wikipedia search**. We search the Wikipedia token index for queries longer than 5. To further constrain search space, we only consider tokens that share 5-prefix, 5-suffix, or all consonants with the query. We choose the token with the highest sequence matching ratio[3].

- **Name resolution**. For corrected tokens retrieved from the medical word list and the Wikipedia, we search the Wikipedia dump to check if it is an alias of another entity and maps it to its canonical form.

Consider the example in Figure 2. Input queries of the error correction pipeline are the misspelled words identified by the spell checker. Wikipedia match catches the common misspelling **folliculitus* and recovers its canonical form *folliculitis*. Medical word search recovers *pigmentosum* from the medical dictionary. Frequent word search recovers misspellings of popular words, avoid sending them to the noisy Wikipedia search. Finally, Wikipedia search first map **ureatha* to its closest alias *ureathra* in Wikipedia and then maps *ureathra* to the canonical form *urethra*.

On the validation set, the process is unable to recover the word **preagnet* (pregnant). We are able to recover most other errors on the validation set. Impact of error correction is evaluated in Section 6.2.1.

4 Summarization with Pre-trained Conditional Generative Language Models

Pre-trained conditional generative language models have become the dominating paradigm for text generation and especially summarization, with recent models such as Pegasus (Zhang et al., 2020a), BART (Lewis et al., 2020), T5 (Raffel et al., 2020), and PALM (Bi et al., 2020) achieving state-of-the-art results on standard benchmarks CNN-Dailymail (See et al., 2017) and XSUM (Narayan et al., 2018). Recent work has also shown that these models achieve good performance in few-shot medical summarization settings (Goodwin et al., 2020).

Following (Goodwin et al., 2020), we use Pegasus, BART, and T5 single systems as our baselines.

- **Pegasus** (Zhang et al., 2020a) is a conditional language model designed specifically for abstractive summarization and is pre-trained with a self-supervised gap-sentence-generation objective, where the model is pre-trained to predict entire masked sentences from the document.

- **BART** (Lewis et al., 2020) is a model combining bi-directional and auto-regressive transformers, trained to both denoise and reconstruct corrupted texts.

- **T5** (Raffel et al., 2020) is pre-trained on multiple objectives, including masking, translation, classification, machine reading comprehension (MRC) and summarization, all formulated as conditional generation tasks.

We use `Pegasus-large`, `BART-large`, and `T5-base` respectively in our experiments.

[2]`https://github.com/glutanimate/hunspell-en-med-glut`

[3]`https://docs.python.org/3/library/difflib.html`

5 Output Reranking

Following previous work on reranking generative LM outputs (Mi et al., 2021), we pick the best summary for each question using the following linear model from outputs of three heterogeneous generative LMs,

$$T^* = \operatorname*{argmax}_{T'} \sum_i \psi_i(\mathbf{T}, T', S) w_i \quad (1)$$

where T' is output of a single system, $\mathbf{T}$ is the set of outputs of all single systems, and S is the input text. T^* is the ensemble output, which is picked from single system outputs by highest score.

The feature function $\psi(\mathbf{T}, T', S)$ is a function to estimate the quality of T' using information from $\mathbf{T}$ and S. w_i is a weight of $\psi(\mathbf{T}, T', S)$. In sequence generation tasks such as machine translation (Kumar and Byrne, 2004), ψ is usually a combination of consensus and linguistic features and w_i can be tuned by optimization algorithms such as MERT (Och, 2003) towards an automatic evaluation metric.

Our approach. We use a simple and coverage-oriented approach for reranking, based on the size and characteristics of the validation data. We notice that the writing style of the validation set is different from the MeQSum data set which we use for training: in MeQSum 18.5% sentences start with "*What are the treatments for*", 14.6% start with "*Where can I find*", and 2.5% start with "*What are the causes of*". A model trained on MeQSum tends to generate these topic-based boilerplates that are not mentioned in the source text. But in the validation set, summaries do not have these boilerplate texts and resemble the content of the source text more closely, which inspires us to pick the output with high coverage of the source.

We consider the validation set (50 sentences) too small for automatic tuning, so we design a minimal set of features and set the weights w_i manually.

Features. We use fidelity, length, consensus and wellformedness features:

- **Fidelity** (w_f). We calculate the Rouge-2 score between the input and the prediction. A higher score indicate a high-coverage summary.
- **Length** (w_l). The length ratio between the prediction and the input.

	Rouge-2	Rouge-L
Pegasus	0.187	0.333
Pegasus EC	0.206	0.344
BART	0.220	0.342
BART EC	0.227	0.342
T5	0.213	0.353
T5 EC	0.208	0.354

Table 1: Single system results on validation set. EC: Input error correction

	Rouge-2	Rouge-L
Best Single	0.220	0.342
Reranked	0.217	0.361
Best Single EC	0.227	0.342
Reranked EC	0.230	0.364

Table 2: Reranking results on validation set. EC: Input error correction

- **Consensus** (w_c). 1 if T' shares any bigram with $\mathbf{T} - T'$, 0 otherwise.
- **Wellformedness** (w_w). 1 if T' has less than three subsentences and starts with one question marker, 0 otherwise.

For our experiments on the validation set and Rouge-2 experiments on the test set, we set $w_f = 1$, $w_l = 0.01$, $w_c = 10$, $w_w = 10$. The idea is to select the summary that has highest coverage of the source that is a one sentence question, with at least one bi-gram in common with other summaries.

The choice to favor high coverage summary is based on this particular pair of training and validation data, rather than general ensemble principles for text generation. We switch the weights for w_f and w_l for length reranking experiments on the test set. Impact of reranking is evaluated in Section 6.2.2.

6 Experiments

6.1 Experimental settings

Our systems are based on the Transformers (Wolf et al., 2020) package. We finetune baseline models on the MeQSum (Ben Abacha and Demner-Fushman, 2019) dataset for 50 epochs, with batch size 8 and learning rate 2e-5 with the AdamW optimizer on Nvidia P100 GPUs. Finetuning is indispensable for this task: without finetuning, `BART-large` scores 0.06 Rouge-2 and 0.15

	ID	R1	R2 P	R2 R	R2 F1	R-L	HOLMS	BERTScore
Single Systems								
1	T5	0.296	0.122	0.109	0.107	0.267	0.541	0.673
2	BART	0.286	0.120	0.090	0.098	0.258	0.550	0.667
3	Pegasus	0.312	0.130	0.123	0.118	0.281	0.547	0.684
Length rerank								
4	3 Sys	0.342	0.149	0.166	0.148	0.299	0.561	0.689
5	3 Sys EC	0.351	0.157	0.175	0.155	0.307	0.566	0.688
6	4 Sys EC	0.358	0.160	0.181	0.159	0.310	0.565	0.689
Coverage rerank								
7	3' Sys EC	0.350	0.177	0.169	0.161	0.313	0.571	0.691
8	4 Sys EC	**0.351**	0.173	**0.173**	**0.161**	0.313	0.568	0.689
-	Best team	0.351	0.185	0.173	0.161	0.315	0.579	0.703

Table 3: Results on the test set. EC: Input error correction; R1/2/L: Rouge-1/2/L; P: Precision, R: Recall; Best team: Best score among all teams; Scores in bold when our system achieves the best score.

Rouge-L on the validation set in preliminary experiments.

For experiments on the test set, models for ensemble are further finetuned for 50 epochs on the validation set. Models for error-corrected input are finetuned on an automatically corrected version of the validation set.

6.2 Validation set experiments

We report single and reranking system performance in Tables 1 and 2 respectively. Results are evaluated by Rouge (Lin, 2004), which is based on n-gram or longest common sequence (LCS) matching of strings.

6.2.1 Single systems and error correction

Among the pre-trained LMs in Table 1, BART performs the best on the validation set. Comparing error-corrected (Pegasus/BART/T5 EC) and original (Pegasus/BART/T5) inputs, we note that error-corrected input significantly boosts the performance of Pegasus. In addition to corrected entity names, the fixed input also leads Pegasus to generate 5% longer output and results in a much higher Rouge-2 score in this small dataset. This trend is less significant on BART and T5, but adding error correction has a positive impact in general.

6.2.2 Reranking

We compare the reranked systems against baselines, with or without error-corrected input in Table 2. In both cases, reranking does not have significant effect on Rouge-2, but helps Rouge-L significantly. We suspect that reranking does improve word and style choice, but the room for increasing 2-gram matches is small on the validation set.

6.3 Test set experiments

We run three sets of experiments on the test set and report results in Table 3: single systems are the same systems tested on the validation set and ensembles are reranked outputs from systems further finetuned on the validation set.

In addition to string-based Rouge (Lin, 2004), test set results are also evaluated by pre-trained language model-based BERTScore (Zhang et al., 2020b) and HOLMS (Mrabet and Demner-Fushman, 2020) metrics:

- **BERTScore** (Zhang et al., 2020b) leverages the pre-trained contextual embeddings from BERT and matches words in candidate and reference sentences by cosine similarity, where matching is performed greedily for each word by choosing the most similar word in the other sentence.

- **HOLMS** (Mrabet and Demner-Fushman, 2020) combines soft matching of contextual embeddings derived from pre-trained LMs and a string-based metric (Rouge-1 recall in practice).

String-based and pre-trained language model-based metrics rank summaries differently. We discuss the impact of the choice of metrics in Section 6.4.

We run two other experiments validating post-processing and the UniLM language model (Dong et al., 2019), they perform inferior to their respective baselines and are not reported in Table 3.

We notice in single system experiments that the characteristics of the test set is still different from the validation set: all systems suffer from low recall, which leads us to perform more aggressive length-based reranking.

Length reranking. We experiment with a baseline approach that explicitly picks the longest output sentence by switching the weight of length and fidelity features in (1). The 3 systems in runs 4 and 5 are Pegasus and T5 finetuned on the validation set and the Pegasus system in run 3. Run 6 adds BART finetuned on the validation set.

We observe that this simple heuristic, together with further finetuning on the validation set, leads to significantly higher Rouge scores between runs 3 and 4 in Table 3. This change also improves HOLMS and BERTScore, suggesting that recall / coverage-based sentence selection does correlate to summarization quality in this scenario. Rouge is further improved by adding BART to the combination between runs 5 and 6.

Correcting input errors between runs 4 and 5 also helps Rouge significantly. BERTScore, which is based on word matching and utilizes BERT embeddings, is much less sensitive to small spelling errors and changes negatively. HOLMS changes positively as it has a Rouge component.

The negative change of BERTScore also suggests that we should be more cautious applying input error correction to summarization: mistakes in error correction might not hurt string-based metrics (the word is often misspelled already), but they can change the meaning of the sentence and degrade summarization quality.

Coverage reranking. In runs 7 and 8, we experiment with the the same setting as in Table 2. 3 systems are Pegasus, BART, and T5 finetuned on the validation set. These runs achieve balanced Rouge precision and recall, and the highest Rouge-2 score across all runs. There are small improvement on all metrics, which is expected, as Rouge-2 is a better indicator of summarization coverage than length.

According to BERT-based metrics, coverage-based reranking also leads to more steady improvement than length-based reranking. The overall improvement in all metrics suggests that coverage-based reranking does improve summarization quality in this task.

6.4 Lessons learned

In this shared task, we experimented with knowledge-based input error correction and coverage-oriented system reranking. These methods are effective in boosting string matching between the prediction and the reference summaries. According to Rouge metrics, our submissiong ranks first according to Rouge-1/2 metrics and ranks second according to the Rouge-L metric.

According to BERT-based metrics, however, reranking has a smaller impact on summarization quality and error correction has little to no effect: we are about 1 point below the best submission according to BERTScore and HOLMS, which are shown to often have higher correlation with human judgement (Zhang et al., 2020b; Mrabet and Demner-Fushman, 2020).

The discrepancy between the string-based and LM-based metrics makes the real improvement of summarization quality hard to measure. It is arguable that by focusing on misspellings and using coverage as surrogate for summarization quality, we might be optimizing more for the writing style and spelling, rather than the content of the summary. This shows the need of an efficient, optimizable summarization evaluation metric with high correlation with human judgement that our field agrees upon. We plan to look more into the choice of metric and optimization objectives for summarzation tasks in future work.

7 Conclusion

We reported our experiments in MEDIQA 2021 shared task 1. We used knowledge-based error correction and coverage-oriented reranking improve summarization. Our system performed well on string-based Rouge metrics, but less so on BERT-based semantic metrics. We plan to investigate methods that improve summarization according to human judgement.

References

Asma Ben Abacha and Dina Demner-Fushman. 2019. On the summarization of consumer health questions. In *Proceedings of the 57th Annual Meeting of the Association for Computational Linguistics*, pages 2228–2234, Florence, Italy.

Asma Ben Abacha, Yassine Mrabet, Yuhao Zhang, Chaitanya Shivade, Curtis Langlotz, and Dina Demner-Fushman. 2021. Overview of the mediqa

2021 shared task on summarization in the medical domain. In *Proceedings of the 20th SIG-BioMed Workshop on Biomedical Language Processing, NAACL-BioNLP 2021*.

Bin Bi, Chenliang Li, Chen Wu, Ming Yan, Wei Wang, Songfang Huang, Fei Huang, and Luo Si. 2020. PALM: Pre-training an autoencoding&autoregressive language model for context-conditioned generation. In *Proceedings of the 2020 Conference on Empirical Methods in Natural Language Processing (EMNLP)*, pages 8681–8691.

Eric Charton, Marie-Jean Meurs, Ludovic Jean-Louis, and Michel Gagnon. 2014. Improving entity linking using surface form refinement. In *Proceedings of the Ninth International Conference on Language Resources and Evaluation (LREC'14)*, pages 4609–4615, Reykjavik, Iceland.

Li Dong, Nan Yang, Wenhui Wang, Furu Wei, Xiaodong Liu, Yu Wang, Jianfeng Gao, Ming Zhou, and Hsiao-Wuen Hon. 2019. Unified language model pre-training for natural language understanding and generation. In *33rd Conference on Neural Information Processing Systems (NeurIPS 2019)*.

Travis Goodwin, Max Savery, and Dina Demner-Fushman. 2020. Flight of the PEGASUS? comparing transformers on few-shot and zero-shot multi-document abstractive summarization. In *Proceedings of the 28th International Conference on Computational Linguistics*, pages 5640–5646, Barcelona, Spain (Online). International Committee on Computational Linguistics.

Qiao Jin, Zheng Yuan, Guangzhi Xiong, Qianlan Yu, Chuanqi Tan, Mosha Chen, Songfang Huang, Xiaozhong Liu, and Sheng Yu. 2021. Biomedical question answering: A comprehensive review. *arXiv preprint arXiv:2102.05281*.

Shankar Kumar and William Byrne. 2004. Minimum bayes-risk decoding for statistical machine translation. In *Proceedings of HLT-NAACL*.

Vladimir I Levenshtein. 1966. Binary codes capable of correcting deletions, insertions, and reversals. In *Soviet physics doklady*, volume 10, pages 707–710.

Mike Lewis, Yinhan Liu, Naman Goyal, Marjan Ghazvininejad, Abdelrahman Mohamed, Omer Levy, Veselin Stoyanov, and Luke Zettlemoyer. 2020. BART: Denoising sequence-to-sequence pre-training for natural language generation, translation, and comprehension. In *Proceedings of the 58th Annual Meeting of the Association for Computational Linguistics*, pages 7871–7880, Online.

Chin-Yew Lin. 2004. ROUGE: A package for automatic evaluation of summaries. In *Text Summarization Branches Out*, pages 74–81, Barcelona, Spain.

Haitao Mi, Qiyu Ren, Yinpei Dai, Yifan He, Jian Sun, Yongbin Li, Jing Zheng, and Peng Xu. 2021. Towards generalized models for beyond domain api task-oriented dialogue. In *Proceedings of the 9th Dialog System Technology Challenge*.

Yassine Mrabet and Dina Demner-Fushman. 2020. HOLMS: Alternative summary evaluation with large language models. In *Proceedings of the 28th International Conference on Computational Linguistics*, pages 5679–5688, Barcelona, Spain (Online).

Shashi Narayan, Shay B. Cohen, and Mirella Lapata. 2018. Don't give me the details, just the summary! Topic-aware convolutional neural networks for extreme summarization. In *Proceedings of the 2018 Conference on Empirical Methods in Natural Language Processing*, Brussels, Belgium.

Franz Josef Och. 2003. Minimum error rate training in statistical machine translation. In *Proceedings of the 41st Annual Meeting of the Association for Computational Linguistics*, pages 160–167, Sapporo, Japan.

Colin Raffel, Noam Shazeer, Adam Roberts, Katherine Lee, Sharan Narang, Michael Matena, Yanqi Zhou, Wei Li, and Peter J. Liu. 2020. Exploring the limits of transfer learning with a unified text-to-text transformer. *Journal of Machine Learning Research*, 21(140):1–67.

Abigail See, Peter J. Liu, and Christopher D. Manning. 2017. Get to the point: Summarization with pointer-generator networks. *CoRR*, abs/1704.04368.

Lina F Soualmia, Elise Prieur-Gaston, Zied Moalla, Thierry Lecroq, and Stéfan J Darmoni. 2012. Matching health information seekers' queries to medical terms. *BMC bioinformatics*, 13(14):1–15.

Thomas Wolf, Lysandre Debut, Victor Sanh, Julien Chaumond, Clement Delangue, Anthony Moi, Pierric Cistac, Tim Rault, Remi Louf, Morgan Funtowicz, Joe Davison, Sam Shleifer, Patrick von Platen, Clara Ma, Yacine Jernite, Julien Plu, Canwen Xu, Teven Le Scao, Sylvain Gugger, Mariama Drame, Quentin Lhoest, and Alexander Rush. 2020. Transformers: State-of-the-art natural language processing. In *Proceedings of the 2020 Conference on Empirical Methods in Natural Language Processing: System Demonstrations*, pages 38–45, Online.

Jingqing Zhang, Yao Zhao, Mohammad Saleh, and Peter Liu. 2020a. PEGASUS: Pre-training with extracted gap-sentences for abstractive summarization. In *Proceedings of the 37th International Conference on Machine Learning*, pages 11328–11339.

Tianyi Zhang, Varsha Kishore, Felix Wu, Kilian Q. Weinberger, and Yoav Artzi. 2020b. Bertscore: Evaluating text generation with bert. In *International Conference on Learning Representations*.

X. Zhou, An Zheng, Jiaheng Yin, R. Chen, Xianyang Zhao, Wei Xu, Wenqing Cheng, T. Xia, and S. Lin. 2015. Context-sensitive spelling correction of consumer-generated content on health care. *JMIR Medical Informatics*.

Stress Test Evaluation of Biomedical Word Embeddings

Vladimir Araujo[1,2], Andrés Carvallo[1,2], Carlos Aspillaga[1], Camilo Thorne[3], Denis Parra[1,2]
[1]Pontificia Universidad Católica de Chile
[2] Millennium Institute for Foundational Research on Data (IMFD)
[3]Elsevier
{vgaraujo,afcarvallo,cjaspill}@uc.cl
c.thorne.1@elsevier.com
dparra@ing.puc.cl

Abstract

The success of pretrained word embeddings has motivated their use in the biomedical domain, with contextualized embeddings yielding remarkable results in several biomedical NLP tasks. However, there is a lack of research on quantifying their behavior under severe "stress" scenarios. In this work, we systematically evaluate three language models with adversarial examples – automatically constructed tests that allow us to examine how robust the models are. We propose two types of stress scenarios focused on the biomedical named entity recognition (NER) task, one inspired by spelling errors and another based on the use of synonyms for medical terms. Our experiments with three benchmarks show that the performance of the original models decreases considerably, in addition to revealing their weaknesses and strengths. Finally, we show that adversarial training causes the models to improve their robustness and even to exceed the original performance in some cases.

1 Introduction

Biomedical NLP (BioNLP) is the field concerned with developing NLP tools and methods for the life sciences domain. Some applications of these techniques include e.g., discovery of gene-disease interactions (Pletscher-Frankild et al., 2015), development of new drugs (Tari et al., 2010), or automatic screening of biomedical documents (Carvallo et al., 2020). With the exponential growth of digital biomedical literature, the importance of BioNLP has become especially relevant as a tool to extract relevant knowledge for making decisions in clinical settings as well as in public health. In order to encourage the development of this area, public datasets and challenges have been shared with the community to solve these tasks, such as BioSSES (Soğancıoğlu et al., 2017), HOC (Hanahan and Weinberg, 2000), ChemProt (Kringelum et al., 2016) and BC5CDR (Li et al., 2016), among others. At the same time, neural language models have shown significant progress since the introduction of models such as W2V (Mikolov et al., 2013), and more recent models like ELMo (Peters et al., 2018) and BERT (Devlin et al., 2019). These models, trained over large corpora (MEDLINE and PubMed in the biomedical domain) have obtained remarkable results in most NLP tasks, including BioNLP benchmarks (Peng et al., 2019). However, they have not been systematically evaluated under severe stress conditions to test their robustness to specific linguistic phenomena. For this reason, the objective of this paper is to evaluate three well-known neural language models under stress conditions. As a case study, we evaluate NER benchmarks since it a key BioNLP information extraction task.

Our stress test evaluation is inspired by the work of Naik et al. (2018), which proposes the use of adversarial evaluation for natural language inference by adding distractions in sentences, and evaluating models on this test set. We propose an adversarial evaluation black-box methodology, which does not require access to the inner workings of the models in order to generate adversarial examples (Zhang et al., 2019). Specifically, we make perturbations to the input data, also known as edit adversaries, that could cause the models to fall into erroneous predictions. Additionally, we train the models with the proposed adversarial examples, which is a methodology used in previous works (Belinkov and Bisk, 2018; Jia and Liang, 2017) to strengthen the neural language models during the training process. We hope that our work will motivate the development and use of adversarial examples to evaluate models and obtain more robust biomedical embeddings.

2 Related Work

Adversarial Evaluation of NLP Models One way to test NLP models is by using adversarial tests, which consist of applying intentional distur-

Proceedings of the BioNLP 2021 workshop, pages 119–125
June 11, 2021.

Original (O)	Linoleic acid autoxidation inhibitions on all fractions were higher than that on alpha-tocopherol.
Keyboard (K)	Linoleic avid autoxidatiob inh9bitions on all fractjons were higher than that on zlpha-toclpherol.
Swap (W)	Linoleic aicd autoxidtaion inhibtiions on all fractoins were higher than that on aplha-tocohperol.
Synonymy (S)	Linoleic acid autoxidation inhibitions on all fractions were higher than that on vitamin E.

Table 1: Examples of sentences of the stress tests.

bances to a gold standard, to test whether the attack leads the models into incorrect predictions. Previous works on adversarial attacks have demonstrated how dangerous it can be to use machine learning systems in real-world applications (Szegedy et al., 2014; Goodfellow et al., 2014). Indeed, it is known that even small amounts of noise can cause severe failures in neural computer vision models (Akhtar and Mian, 2018). However, such failures can be mitigated through adversarial training (Goodfellow et al., 2014). These properties have in turn motivated novel adversarial strategies designed for various NLP tasks (Zhang et al., 2019), as well as work on adversarial attacks focused on recurrent and transformer networks applied to *generic* NLP benchmarks (Aspillaga et al., 2020).

Evaluation of Biomedical Models Models used in BioNLP tasks elicit particular interest in this context because an erroneous prediction can potentially be very harmful in practice – e.g., put at risk the health of patients (Sun et al., 2018). Although adversarial attacks have been widely studied in tasks related to image analysis (Paschali et al., 2018; Finlayson et al., 2019; Ma et al., 2019), to the best of our knowledge, a gap still exists regarding BioNLP models and tasks (Araujo et al., 2020).

3 Methodology

We follow a black-box attack methodology (Zhang et al., 2019), which consists of making alterations in the input data to cause erroneous predictions in the models. The following subsections describe each of the adversarial sets, and their construction[1]. We show examples of the stress tests in Table 1.

Noise Adversaries These adversaries test the robustness of models to *spelling errors*. Inspired by (Belinkov and Bisk, 2018), we constructed adversarial examples that try to emulate spelling errors made by human beings. We used SpaCy models (Neumann et al., 2019) to retrieve the medical words of each corpus and add noise to them. We used two types of alterations: i) **Keyboard typo noise (K)** involves replacing a random character in each relevant word with an adjacent character on QWERTY English keyboards. This methodology could be adapted to keyboards with other designs or languages. ii) **Swap noise (W)** consists of selecting a random pair of consecutive characters in each relevant word and then swapping them.

Synonymy Adversaries (S) These adversaries test if a model can *understand synonymy relations*. Unlike the noise adversaries, this set focuses on modifying chemical and disease words (entities). We used PyMedTermino (Jean-Baptiste et al., 2015), which uses the vocabulary of UMLS (Bodenreider, 2004), to find the most similar or related term (synonym) to a certain word. If a synonym is retrieved, the original word is replaced; otherwise, it remains the same. In some cases, this method changes a simple entity (one word) to a composite one (multiple words), so the gold labels are also adjusted to avoid a mismatch in the dataset.

Task and Datasets Biomedical NER is the task that aims at detecting biomedical entities of interest such as proteins, cell types, chemicals, or diseases in biomedical documents. We conducted our evaluation on three biomedical NER benchmarks using the IOB2 tag format (Ramshaw and Marcus, 1999). The **BC5CDR** corpus (Li et al., 2016) is composed of mentions of chemicals and diseases found in 1,500 PubMed articles. The **BC4CHEMD** corpus (Krallinger et al., 2015) contains mentions of chemicals and drugs from 10,000 MEDLINE abstracts. The **NCBI-Disease** corpus (Doğan et al., 2014) consists of 793 PubMed abstracts annotated with disease mentions. Table 2 lists the datasets used in this work along with their most relevant statistics.

Embeddings and NER Models We evaluated both word (W2V) and contextualized embeddings. On the one hand, we assessed BioMedical W2V (Pyysalo et al., 2013) and ChemPatent W2V (Zhai et al., 2019). The ChemPatent embeddings were trained on a 1.1 billion word corpus of chemical patents from 7 patent offices, whereas all the other embeddings were trained on the PubMed corpus. On the other hand, we evaluated BioBERT v1.1 (Lee et al., 2019) and BlueBERT (P) (Peng et al., 2019), both in their base version for convenience.

[1] All stress tests available at https://github.com/ialab-puc/BioNLP-StressTest.

Train / Test	Entity	# of sentences (annotated)	# of tokens	% K	% W	% S
BC5CDR	Chemical	4560 (1609) / 4797 (1706)	122730 /129547	36.3 / 36.1	33.7 / 33.2	6.8 / 6.5
BC5CDR	Disease	4560 (1902) / 4797 (1955)	122730 /129547	36.3 / 36.1	33.7 / 33.2	10.6 / 9.9
BC4CHEMD	Chemical	30681 (16175) / 26363 (13935)	922609 / 792369	37.8 / 37.6	33.9 / 33.9	5.2 / 5.3
NCBI-Disease	Disease	5423 (2501) / 939 (401)	141092 / 25397	37.4 / 37.5	33.4 / 33.3	9.2 / 8.6

Table 2: Details of the datasets used. The last three columns present the percentage of tokens modified for each of the adversarial datasets. The slash separates the values belonging to the training and the test set.

Model	BC5CDR-Chemical				BC5CDR-Disease				BC4CHEMD				NCBI-Disease			
	O	K	W	S	O	K	W	S	O	K	W	S	O	K	W	S
BioBERT	.937 ±.004	.745 ±.006	.635 ±.008	.770 ±.011	.863 ±.004	.407 ±.008	.473 ±.010	.366 ±.007	.919 ±.004	.585 ±.005	.675 ±.007	.678 ±.009	.887 ±.004	.483 ±.007	.628 ±.011	.683 ±.006
BlueBERT	.901 ±.003	.583 ±.005	.708 ±.008	.739 ±.010	.838 ±.004	.368 ±.007	.441 ±.011	.362 ±.007	.820 ±.003	.472 ±.004	.570 ±.009	.607 ±.010	.773 ±.003	.332 ±.006	.438 ±.009	.615 ±.006
BERT	.887 ±.004	.563 ±.007	.684 ±.010	.738 ±.015	.816 ±.006	.356 ±.009	.431 ±.013	.336 ±.008	.808 ±.004	.443 ±.006	.509 ±.008	.598 ±.013	.771 ±.005	.305 ±.008	.433 ±.014	.583 ±.007
BioELMo	.923 ±.001	.838 ±.003	.726 ±.010	.757 ±.032	.845 ±.002	.656 ±.018	.482 ±.025	.408 ±.013	.915 ±.001	.770 ±.003	.634 ±.004	.668 ±.004	.869 ±.005	.711 ±.017	.543 ±.026	.677 ±.012
ChemPatent ELMo	.910 ±.001	.822 ±.004	.745 ±.005	.757 ±.016	.824 ±.001	.637 ±.013	.508 ±.013	.380 ±.017	.898 ±.001	.766 ±.003	.662 ±.005	.642 ±.005	.863 ±.004	.693 ±.018	.586 ±.020	.655 ±.009
ELMo	.879 ±.002	.702 ±.010	.637 ±.017	.720 ±.018	.800 ±.003	.461 ±.023	.373 ±.020	.378 ±.014	.866 ±.001	.612 ±.007	.507 ±.011	.611 ±.005	.848 ±.004	.575 ±.034	.495 ±.023	.643 ±.008
BioMedical W2V	.873 ±.004	.231 ±.012	.238 ±.021	.719 ±.016	.788 ±.008	.132 ±.009	.133 ±.011	.351 ±.015	.846 ±.005	.233 ±.008	.244 ±.013	.589 ±.012	.827 ±.005	.284 ±.014	.292 ±.019	.596 ±.021
ChemPatent W2V	.871 ±.003	.224 ±.011	.221 ±.012	.715 ±.015	.772 ±.007	.127 ±.005	.122 ±.009	.347 ±.016	.828 ±.007	.253 ±.009	.260 ±.010	.584 ±.012	.816 ±.007	.269 ±.021	.252 ±.019	.582 ±.013
W2V	.818 ±.004	.237 ±.013	.227 ±.013	.641 ±.017	.760 ±.003	.120 ±.008	.120 ±.009	.341 ±.013	.766 ±.007	.264 ±.011	.260 ±.012	.513 ±.008	.785 ±.005	.281 ±.022	.271 ±.019	.526 ±.009

Table 3: Stress test evaluation results in terms of terms F1-score for each model and dataset. We report means and standard deviations by training and evaluating ten times with different seeds.

BioBERT embeddings were trained on PubMed abstracts and full-text corpora consisting of 4.3 billion and 13.5 billion words each. BlueBERT was trained on 4 billion words from PubMed abstracts. We used the implementation provided by Peng et al. (2019) for NER with default hyperparameters.[2] Finally, we evaluate BioELMo (Jin et al., 2019) and ChemPatent ELMo (Zhai et al., 2019). As NER models we either (a) fine-tuned BERT as proposed by Peng et al. (2019) or (b) used AllenNLP's basic biLSTM-CRF implementation[3], with no hyperparameter tuning other than changing the initial embedding layer with one of the ELMo or W2V embeddings. For comparison purposes, we also include the "vanilla" version of the models mentioned above, which are pretrained with general corpora. We trained each model 10 times using different random seeds, for 15 epochs every time. We use CoNLL evaluation (Agirre and Soroa, 2007), reporting the F1 score for all datasets.

4 Experiments

In this section we report the results of our experiments. Note that all percentage drops or increases are expressed relative to the original score, not as percentage points.

Adversarial Evaluation Results Table 3 shows the evaluation results on the original (**O**) and adversarial test sets (**K**, **W**, and **S**). In general, the performance of models drops across all adversarial attacks. For BERT-based models, we observe that **K** attacks decrease performance by on average 43.1%, **W** by 34.3% and **S** by 30.8%. BioBERT has the smallest decrease in performance, 34.4%, followed by BlueBERT, with a 37.9% decrease. We hypothesize that BioBERT is more robust than BlueBERT since the former was trained on a larger and more varied corpus. Furthermore, when comparing the performance across all datasets, we see that **BC5CDR-Disease** is the most affected in all stress tests, with a 37.7% performance drop, and the least affected is **BC5CDR-Chemical**, with 16.1%.

The performance reduction of ELMo-based models is similar to those of BERT-based models. An exception is when subject to **W** and **S** noise, where they showed increased robustness with respect to BERT and W2V models (**W**: 55.3% better, **S**: 6.9% better). In almost all the tests, BioELMo performed better than ChemPatent ELMo, except under **W** noise, where ChemPatent ELMo performed con-

[2] https://github.com/ncbi-nlp/bluebert

[3] https://github.com/allenai/allennlp-models

Model	Training	BC5CDR-Chemical		BC5CDR-Disease		BC4CHEMD		NCBI-Disease	
BioBERT	O + K	.934 (O)	.888 (K)	.863 (O)	.755 (K)	.920 (O)	.874 (K)	.886 (O)	.820 (K)
	O + W	.931 (O)	.899 (W)	.865 (O)	.781 (W)	.922 (O)	.892 (W)	.872 (O)	.848 (W)
	O + S	.933 (O)	.910 (S)	.840 (O)	.819 (S)	.919 (O)	.923 (S)	.874 (O)	.875 (S)
BlueBERT	O + K	.898 (O)	.820 (K)	.844 (O)	.717 (K)	.819 (O)	.750 (K)	.789 (O)	.668 (K)
	O + W	.896 (O)	.656 (W)	.841 (O)	.759 (W)	.818 (O)	.785 (W)	.784 (O)	.729 (W)
	O + S	.900 (O)	.890 (S)	.818 (O)	.814 (S)	.820 (O)	.788 (S)	.773 (O)	.804 (S)
BioELMo	O + K	.923 (O)	.870 (K)	.833 (O)	.732 (K)	.912 (O)	.837 (K)	.864 (O)	.820 (K)
	O + W	.922 (O)	.825 (W)	.838 (O)	.654 (W)	.913 (O)	.820 (W)	.875 (O)	.777 (W)
	O + S	.919 (O)	.901 (S)	.826 (O)	.799 (S)	.912 (O)	.901 (S)	.871 (O)	.848 (S)
ChemPatent ELMo	O + K	.910 (O)	.859 (K)	.823 (O)	.713 (K)	.898 (O)	.828 (K)	.860 (O)	.793 (K)
	O + W	.907 (O)	.835 (W)	.813 (O)	.682 (W)	.899 (O)	.824 (W)	.863 (O)	.804 (W)
	O + S	.904 (O)	.895 (S)	.813 (O)	.757 (S)	.895 (O)	.874 (S)	.848 (O)	.819 (S)
BioMedical W2V	O + K	.888 (O)	.467 (K)	.773 (O)	.303 (K)	.832 (O)	.486 (K)	.820 (O)	.543 (K)
	O + W	.873 (O)	.598 (W)	.796 (O)	.482 (W)	.836 (O)	.609 (W)	.819 (O)	.639 (W)
	O + S	.867 (O)	.883 (S)	.781 (O)	.787 (S)	.837 (O)	.852 (S)	.836 (O)	.804 (S)
ChemPatent W2V	O + K	.867 (O)	.454 (K)	.768 (O)	.307 (K)	.817 (O)	.482 (K)	.822 (O)	.548 (K)
	O + W	.785 (O)	.619 (W)	.765 (O)	.477 (W)	.819 (O)	.626 (W)	.792 (O)	.663 (W)
	O + S	.868 (O)	.864 (S)	.738 (O)	.779 (S)	.818 (O)	.835 (S)	.797 (O)	.801 (S)

Table 4: Adversarial training results in terms of F1-score for each model and dataset. The training column shows the **O** set merged with **K**, **W**, or **S**. The test set is shown in parentheses for each scenario.

sistently better, by 5.1% on average. We hypothesize that these results are due to ELMo using a character-based input representation, which would allow handling of swap characters inside the words.

W2V-based models were the most brittle but showed similar patterns to the previous models. Adversaries examples produced performance drops ranging from 53.8% on **NCBI-Disease** to 74.1% on **BC5CDR-Disease**. In the case of **S** adversaries, W2V-based showed performance drops ranging from 17.8% on **BC5CDR-Chemical** to 55.3% on **BC5CDR-Disease**.

Regarding the "vanilla" models, we see that they are all the worst in the original dataset (**O**) compared to their biomedical counterparts. In the same way, they are more fragile to adversary attacks in the biomedical scenario. In average, BERT has a decrease in performance of 39.6%, ELMo of 34.4% and W2V of 59.6% across all datasets.

Even though the **BC5CDR** dataset covers both chemicals and diseases, the disease task is more affected by **S** adversaries. We believe this is due to the higher number of words affected by the attacks compared to the other benchmarks (Table 2). Another possible cause is the kind of synonyms used to replace the entities, which tend to be both superficially dissimilar and more extensive than their originals, e.g., *arrhythmia* is replaced by *heart conduction disorder*. By contrast, chemical synonyms often include terms derived from the original, e.g., *morphine* is changed to *morphine sulfate*.

Training on Adversarial Examples Additionally, we subjected the training sets to adversarial attacks, and evaluated the models both against the original test sets and their noisy counterparts. When training with **K** noise, we observed performance decreases by 21.2%, followed by **W**, 15.8%, and **S** with a slight decline of 0.8%, compared to 44.4%, 46.3% and 31.3% respectively in the Adversarial Evaluation setting. Besides, and interestingly, training with **S** improves performance in some cases, by up to 5.5% compared to the original **S** test set. We hypothesize that this is because the introduced adversarial samples work as a data augmentation mechanism. In terms of datasets, we see that **BC5CDR-Disease** is the most affected by adversaries, with an average 17.5% drop, and the least affected is **NCBI-Disease**, with an average 9.7% drop compared to the non-adversarial test set. When comparing the three architectures we see that BERT is affected by 6.3%, ELMo by 7.6% and W2V by 24.0% on average compared to the original test set. This result stands in line with findings on other NLP tasks, where BERT comes up first, followed by ELMo and W2V (Peng et al., 2019). This is because BERT uses recent methods and techniques like Transformer (Vaswani et al., 2017) and WordPiece tokenizer (Schuster and Nakajima, 2012) that allow it to learn better representations.

BioBERT Error Analysis This section seeks to understand how the most robust model – BioBERT – behaves under adversarial evaluation. To this end, we analyzed NER model confusions with respect to the original datasets, synonym (**S**), swap (**W**), and keyboard (**K**) perturbations on the BC5CDR chemical and disease dataset(s).

In the original dataset (Figure 1(a)), we see that

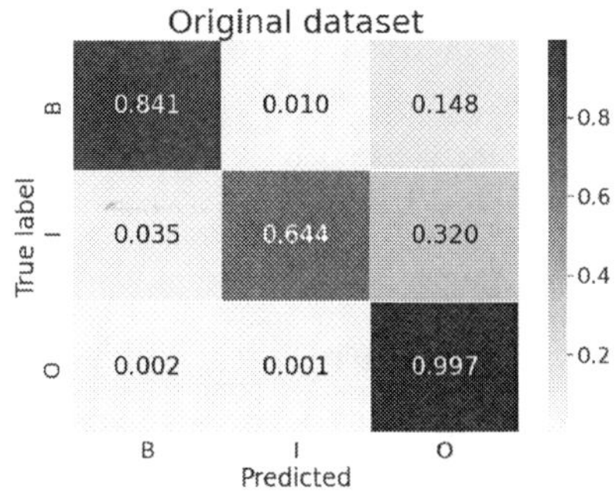

(a)

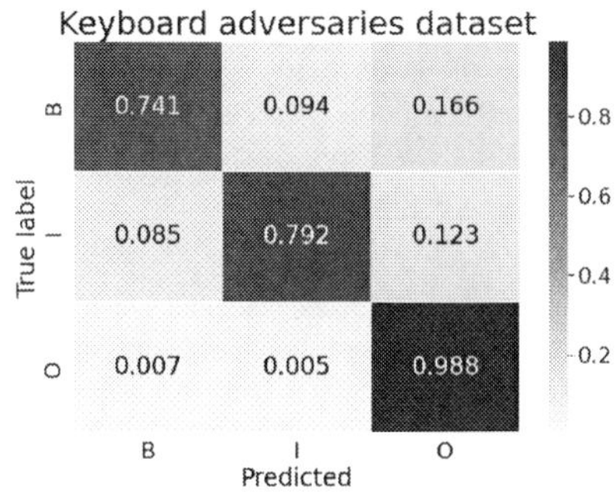

(b)

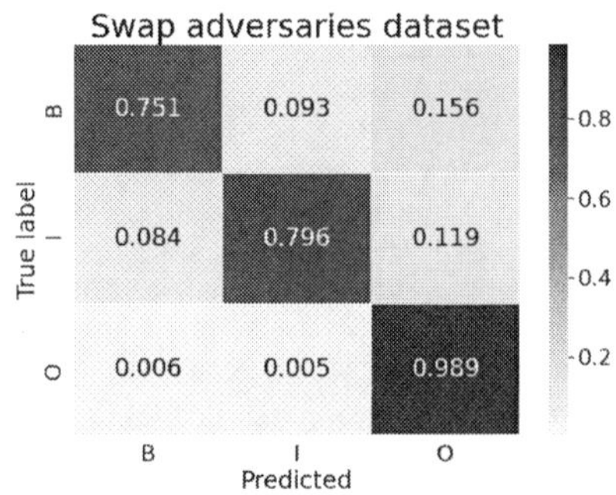

(c)

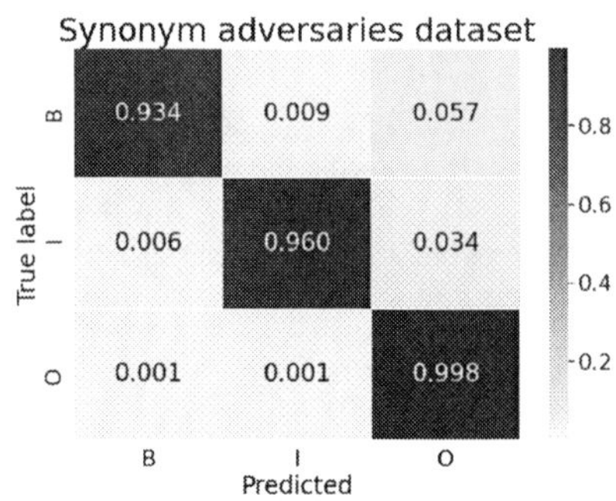

(d)

Figure 1: Normalized confusion matrices for test results with (a) original (**O**), (b) keyboard (**K**), (c) swap (**S**) and (d) synonym (**S**) BC5CDR-Disease and Chemical datasets on average.

most of the errors come from confusing I and O labels (32% of the cases). Under adversarial attacks, this type of error spreads to other IOB labels. For keyboard (**K**) errors (Figure 1(b)), the most frequent mistake is to confuse B with O, with 16.6% of these cases. The same goes for swap (**W**) perturbations (Figure 1(c)), where this error is repeated 15% of the time. When using synonyms (**S**) (Figure 1(d)), error rates become by contrast globally low compared to **K** and **W**. We believe that this happens because entities are converted into similar ones. For instance, "stomach neoplasm" gets transformed into "stomach tumor".

Lastly, regardless of the adversaries, there are confusions with numbers and special character sequences that the model classifies as I (i.e., lie inside an entity span) but whose ground truth label is O (i.e., lie outside an entity span).

5 Conclusions

In this work, we have investigated whether large scale biomedical word (W2V) and contextualized word embeddings (BERT and ELMo) are robust with respect to black-box adversarial attacks in the biomedical NER task. Our experimental results show different sensitivities of the models to misspellings and synonyms. Among the main findings, we show that BERT-based models are generally better prepared for adversarial attacks, but they are still fragile, leaving room for future improvement in the field. ELMo-based models show lower robustness in most cases but consistently outperformed BERT in some specific scenarios. W2V proves to be more brittle but shows similar patterns in terms of relative performance drops. We also demonstrate that by training with adversaries, we can considerably decrease the drop in performance and even improve the models' original performance when trained with synonyms, as they act as a form of regularization and augmentation of data.

Acknowledgements

We are grateful to the anonymous reviewers for their valuable feedback on earlier versions of this paper. This work was partially funded by ANID - Millennium Science Initiative Program - Code ICN17_002 and by ANID, FONDECYT grant 1191791, as well as supported by the TPU Research Cloud (TRC) program.

References

Eneko Agirre and Aitor Soroa. 2007. Semeval-2007 task 02: Evaluating word sense induction and discrimination systems. In *Proceedings of the fourth international workshop on semantic evaluations (semeval-2007)*, pages 7–12.

Naveed Akhtar and Ajmal Mian. 2018. Threat of adversarial attacks on deep learning in computer vision: A survey. *IEEE Access*, 6:14410–14430.

Vladimir Araujo, Andrés Carvallo, and Denis Parra. 2020. Adversarial evaluation of bert for biomedical named entity recognition. In *Proceedings of the The Fourth Widening Natural Language Processing Workshop*, Seattle, USA. Association for Computational Linguistics.

Carlos Aspillaga, Andrés Carvallo, and Vladimir Araujo. 2020. Stress test evaluation of transformer-based models in natural language understanding tasks. In *Proceedings of The 12th Language Resources and Evaluation Conference*, pages 1882–1894, Marseille, France. European Language Resources Association.

Yonatan Belinkov and Yonatan Bisk. 2018. Synthetic and natural noise both break neural machine translation. In *International Conference on Learning Representations*.

O. Bodenreider. 2004. The unified medical language system (UMLS): integrating biomedical terminology. *Nucleic Acids Research*, 32(90001):267D–270.

Andres Carvallo, Denis Parra, Hans Lobel, and Alvaro Soto. 2020. Automatic document screening of medical literature using word and text embeddings in an active learning setting. *Scientometrics*, 125(3):3047–3084.

Jacob Devlin, Ming-Wei Chang, Kenton Lee, and Kristina Toutanova. 2019. BERT: Pre-training of deep bidirectional transformers for language understanding. In *Proceedings of the 2019 Conference of the North American Chapter of the Association for Computational Linguistics: Human Language Technologies, Volume 1 (Long and Short Papers)*, pages 4171–4186, Minneapolis, Minnesota. Association for Computational Linguistics.

Rezarta Islamaj Doğan, Robert Leaman, and Zhiyong Lu. 2014. NCBI disease corpus: A resource for disease name recognition and concept normalization. *Journal of Biomedical Informatics*, 47:1–10.

Samuel G. Finlayson, John D. Bowers, Joichi Ito, Jonathan L. Zittrain, Andrew L. Beam, and Isaac S. Kohane. 2019. Adversarial attacks on medical machine learning. *Science*, 363(6433):1287–1289.

Ian J Goodfellow, Jonathon Shlens, and Christian Szegedy. 2014. Explaining and harnessing adversarial examples. *arXiv preprint arXiv:1412.6572*.

Douglas Hanahan and Robert A. Weinberg. 2000. The hallmarks of cancer. *Cell 100(1)*, pages 57–70.

Lamy Jean-Baptiste, Venot Alain, and Duclos Catherine. 2015. Pymedtermino: an open-source generic api for advanced terminology services. *Studies in Health Technology and Informatics*, 210(Digital Healthcare Empowering Europeans):924–928.

Robin Jia and Percy Liang. 2017. Adversarial examples for evaluating reading comprehension systems. In *Proceedings of the 2017 Conference on Empirical Methods in Natural Language Processing*, pages 2021–2031, Copenhagen, Denmark. Association for Computational Linguistics.

Qiao Jin, Bhuwan Dhingra, William Cohen, and Xinghua Lu. 2019. Probing biomedical embeddings from language models. In *Proceedings of the 3rd Workshop on Evaluating Vector Space Representations for NLP*, pages 82–89.

Martin Krallinger, Obdulia Rabal, Florian Leitner, Miguel Vazquez, David Salgado, Zhiyong Lu, Robert Leaman, Yanan Lu, Donghong Ji, Daniel M Lowe, Roger A Sayle, Riza Theresa Batista-Navarro, Rafal Rak, Torsten Huber, Tim Rocktäschel, Sérgio Matos, David Campos, Buzhou Tang, Hua Xu, Tsendsuren Munkhdalai, Keun Ho Ryu, SV Ramanan, Senthil Nathan, Slavko Žitnik, Marko Bajec, Lutz Weber, Matthias Irmer, Saber A Akhondi, Jan A Kors, Shuo Xu, Xin An, Utpal Kumar Sikdar, Asif Ekbal, Masaharu Yoshioka, Thaer M Dieb, Miji Choi, Karin Verspoor, Madian Khabsa, C Lee Giles, Hongfang Liu, Komandur Elayavilli Ravikumar, Andre Lamurias, Francisco M Couto, Hong-Jie Dai, Richard Tzong-Han Tsai, Caglar Ata, Tolga Can, Anabel Usié, Rui Alves, Isabel Segura-Bedmar, Paloma Martínez, Julen Oyarzabal, and Alfonso Valencia. 2015. The CHEMDNER corpus of chemicals and drugs and its annotation principles. *Journal of Cheminformatics*, 7(S1).

J. Kringelum, S. K. Kjaerulff, S. Brunak, O. Lund, T. I. Oprea, and O. Taboureau. 2016. Chemprot-3.0: a global chemical biology diseases mapping. *Database*.

Jinhyuk Lee, Wonjin Yoon, Sungdong Kim, Donghyeon Kim, Sunkyu Kim, Chan Ho So, and Jaewoo Kang. 2019. BioBERT: a pre-trained biomedical language representation model for biomedical text mining. *Bioinformatics*, 36(4):1234–1240.

Jiao Li, Yueping Sun, Robin J. Johnson, Daniela Sciaky, Chih-Hsuan Wei, Robert Leaman, Allan Peter Davis, Carolyn J. Mattingly, Thomas C. Wiegers, and Zhiyong Lu. 2016. BioCreative v CDR task corpus: a resource for chemical disease relation extraction. *Database*, 2016:baw068.

Xingjun Ma, Yuhao Niu, Lin Gu, Yisen Wang, Yitian Zhao, James Bailey, and Feng Lu. 2019. Understanding adversarial attacks on deep learning based medical image analysis systems.

Tomas Mikolov, Ilya Sutskever, Kai Chen, Greg S Corrado, and Jeff Dean. 2013. Distributed representations of words and phrases and their compositionality. In C. J. C. Burges, L. Bottou, M. Welling, Z. Ghahramani, and K. Q. Weinberger, editors, *Advances in Neural Information Processing Systems 26*, pages 3111–3119. Curran Associates, Inc.

Aakanksha Naik, Abhilasha Ravichander, Norman Sadeh, Carolyn Rose, and Graham Neubig. 2018. Stress test evaluation for natural language inference. In *Proceedings of the 27th International Conference on Computational Linguistics*, pages 2340–2353, Santa Fe, New Mexico, USA. Association for Computational Linguistics.

Mark Neumann, Daniel King, Iz Beltagy, and Waleed Ammar. 2019. Scispacy: Fast and robust models for biomedical natural language processing. In *SciSpacy:Fast and Robust Models for Biomedical Natural Language Processing*.

Magdalini Paschali, Sailesh Conjeti, Fernando Navarro, and Nassir Navab. 2018. Generalizability vs. robustness: Investigating medical imaging networks using adversarial examples. In *Medical Image Computing and Computer Assisted Intervention – MICCAI 2018*, pages 493–501. Springer International Publishing.

Yifan Peng, Shankai Yan, and Zhiyong Lu. 2019. Transfer learning in biomedical natural language processing: An evaluation of BERT and ELMo on ten benchmarking datasets. In *Proceedings of the 18th BioNLP Workshop and Shared Task*, pages 58–65, Florence, Italy. Association for Computational Linguistics.

Matthew Peters, Mark Neumann, Mohit Iyyer, Matt Gardner, Christopher Clark, Kenton Lee, and Luke Zettlemoyer. 2018. Deep contextualized word representations. In *Proceedings of the 2018 Conference of the North American Chapter of the Association for Computational Linguistics: Human Language Technologies, Volume 1 (Long Papers)*, pages 2227–2237, New Orleans, Louisiana. Association for Computational Linguistics.

Sune Pletscher-Frankild, Albert Pallejà, Kalliopi Tsafou, Janos X Binder, and Lars Juhl Jensen. 2015. Diseases: Text mining and data integration of disease–gene associations. *Methods*, 74:83–89.

S. Pyysalo, F. Ginter, H. Moen, T. Salakoski, and S. Ananiadou. 2013. Distributional semantics resources for biomedical text processing. In *Proceedings of LBM 2013*, pages 39–44.

Lance A Ramshaw and Mitchell P Marcus. 1999. Text chunking using transformation-based learning. In *Natural language processing using very large corpora*, pages 157–176. Springer.

M. Schuster and K. Nakajima. 2012. Japanese and korean voice search. In *2012 IEEE International Conference on Acoustics, Speech and Signal Processing (ICASSP)*, pages 5149–5152.

Gizem Soğancıoğlu, Hakime Öztürk, and Arzucan Özgür. 2017. Biosses: a semantic sentence similarity estimation system for the biomedical domain. *Bioinformatics*, 33(14):i49–i58.

Mengying Sun, Fengyi Tang, Jinfeng Yi, Fei Wang, and Jiayu Zhou. 2018. Identify susceptible locations in medical records via adversarial attacks on deep predictive models. In *Proceedings of the 24th ACM SIGKDD International Conference on Knowledge Discovery and Data Mining*, KDD '18, page 793–801, New York, NY, USA. Association for Computing Machinery.

Christian Szegedy, Wojciech Zaremba, Ilya Sutskever, Joan Bruna, Dumitru Erhan, Ian Goodfellow, and Rob Fergus. 2014. Intriguing properties of neural networks. In *International Conference on Learning Representations*.

Luis Tari, Saadat Anwar, Shanshan Liang, James Cai, and Chitta Baral. 2010. Discovering drug–drug interactions: a text-mining and reasoning approach based on properties of drug metabolism. *Bioinformatics*, 26(18):i547–i553.

Ashish Vaswani, Noam Shazeer, Niki Parmar, Jakob Uszkoreit, Llion Jones, Aidan N Gomez, Łukasz Kaiser, and Illia Polosukhin. 2017. Attention is all you need. In *Advances in neural information processing systems*, pages 5998–6008.

Zenan Zhai, Dat Quoc Nguyen, Saber Akhondi, Camilo Thorne, Christian Druckenbrodt, Trevor Cohn, Michelle Gregory, and Karin Verspoor. 2019. Improving chemical named entity recognition in patents with contextualized word embeddings. In *Proceedings of the 18th BioNLP Workshop and Shared Task*, pages 328–338, Florence, Italy. Association for Computational Linguistics.

Wei Emma Zhang, Quan Z. Sheng, Ahoud Alhazmi, and Chenliang Li. 2019. Adversarial attacks on deep learning models in natural language processing: A survey.

BLAR: Biomedical Local Acronym Resolver

William Hogan[1,2], Yoshiki Vazquez Baeza[2], Yannis Katsis[3], Tyler Baldwin[3], Ho-Cheol Kim[3], Chun-Nan Hsu[2]
[1]Department of Computer Science & Engineering,
[2]Center for Microbiome Innovation
University of California, San Diego, La Jolla, CA 92093
[3]IBM Research-Almaden, 650 Harry Road, San Jose, CA 95120
whogan@ucsd.edu

Abstract

NLP has emerged as an essential tool to extract knowledge from the exponentially increasing volumes of biomedical texts. Many NLP tasks, such as named entity recognition and named entity normalization, are especially challenging in the biomedical domain partly because of the prolific use of acronyms. Long names for diseases, bacteria, and chemicals are often replaced by acronyms. We propose Biomedical Local Acronym Resolver (BLAR), a high-performing acronym resolver that leverages state-of-the-art (SOTA) pre-trained language models to accurately resolve local acronyms in biomedical texts. We test BLAR on the Ab3P corpus and achieve state-of-the-art results compared to the current best-performing local acronym resolution algorithms and models.

1 Introduction

In the past decade, natural language processing (NLP) has greatly advanced in the biomedical domain. Given the troves of biomedical texts, NLP has emerged as a critical tool for knowledge extraction. NLP has been used to automatically analyze clinical notes, electronic medical records, biological literature, and other biomedical texts in the hopes of unearthing new knowledge and deeper insights.

Acronyms are especially common in science and even more so in biomedical publications, as authors regularly seek to shorten the long names for diseases, bacteria, and chemicals. Barnett and Doubleday (2020) documented acronym use in more than 24 million scientific article titles and 18 million scientific articles published between 1950 and 2019. They report that 19% of titles and 73% of abstracts contain acronyms. Of the more than one million unique acronyms in their data, 0.2% appeared regularly and most acronyms, 79%, appeared less than 10 times.

Acronym resolution (AR) can be performed by either leveraging acronym definitions found in the text (referred to as *local AR*) or by consulting external resources, such as ontologies (known as *disambiguation* or *global AR*). While a lot of progress has been recently done on the latter, local AR has seen surprisingly little recent work. In particular, the SOTA approaches in local AR are rule-based or simple machine learning approaches from more than a decade ago. As a result, this task has not benefited from recent advances in transformers (Vaswani et al., 2017). To address this issue, in this work we focus on local AR where we try to answer the question: Can transformers be leveraged to further improve traditional local AR approaches?

To answer this question, we present Biomedical Local Acronym Resolver (BLAR); a transformer-based model designed to resolve local acronyms in biomedical texts. In particular, this work makes the following contributions:

1. *Design of a novel transformer-based model for local acronym resolution*, which resolves acronyms through a combination of a two-step architecture and appropriate leveraging of pre-trained language models. To the best of our knowledge, this is the first transformer-based approach for local AR.

2. *Experimental evaluation of BLAR against SOTA local AR approaches*, showing that it outperforms the latter. In particular, evaluated on the Ab3P corpus (Sohn et al., 2008), BLAR reaches an F1 score of 0.966 compared to 0.899 of the best performing existing approach.

2 Background and Related Work

There are a few challenges inherent in acronym resolution that make a simple dictionary-lookup and

Proceedings of the BioNLP 2021 workshop, pages 126–130
June 11, 2021.

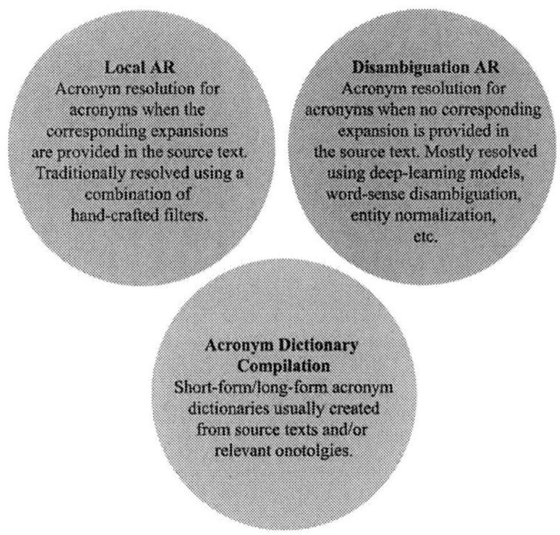

Figure 1: Sub-tasks of acronym resolution (AR). Our approach is applicable to both "Local AR" and "Acronym Dictionary Compilation."

other rule-based models less effective. First, short-form acronym representations are rarely unique. For instance, "CD" is an acronym for "Crohn's disease" and "Cowden Disease." A simple dictionary lookup of "CD" using an acronym disease dictionary will produce ambiguous results and requires additional steps of acronym disambiguation. Moreover, the number of letters in a short-form may not match the number of words in the corresponding long-form (e.g. the short-form of "systemic sclerosis" is "SSc"). Lastly, long-form entities can have complicated short-forms. For example, the short-form of "heparin-induced thrombocytopenia type II" is "HIT type II," a short-form that shortens the first three words of the long-form and leaves the last two words unmodified.

To address these challenges, approaches to acronym resolution have been developed and can be classified into three broad categories: *local* acronym resolution (Schwartz and Hearst, 2003; Sohn et al., 2008), *disambiguation* acronym resolution (also referred to as *non-local* or *global* acronym resolution) (Jin et al., 2019; Jacobs et al., 2020), and *acronym dictionary compilation* (Grossman et al., 2018). We refer to approaches that resolve acronyms by leveraging their definitions found in the containing text as *local* acronym resolution techniques. In contrast, *non-local* or *global* techniques resolve acronyms by using external resources. These typically target acronyms whose long-form is not contained within the text, which is common among more established acronyms, such as "mRNA" and "DNA." Finally, *acronym dictionary compilation* refers to the creation of an acronym dictionary based on the source text or external ontologies, or a combination of the two. These three sub-categories of AR approaches are depicted in Figure 1.

Our approach specifically targets local acronym resolution and acronym dictionary compilation. Local acronyms appear as a pair of entities featuring a short-form (SF) entity and a corresponding long-form (LF) entity. Historically, local acronym resolution has been handled by rule-based algorithms. From 2003 to 2009, Schwartz et al. (2003) and Sohn et al. (2008) demonstrated the best performance of local acronym resolution. They used a combination of hand-crafted filters to identify SF-LF pairs. Kuo et al. (2009) introduced the first local acronym resolution model that leveraged machine learning. It produced SOTA results with the help of four sets of hand-crafted features, including rule-based text filters. Yeganova et al. (2011) further improved upon local acronym resolution by introducing a hybrid machine learning and rule-base model that does not rely on labeled data. They extract potential SF-LF pairs from PubMed articles using rules similar to the rules developed by Sohn et al. and train a classifier to identify SF-LF pairs.

Our approach to local acronym resolution is simple in its architecture yet novel in its application. Our two-stage model leverages transfer learning from modern, SOTA pretrained transformers and is able to learn the features of short-form and long-form acronym pairs without the help of a predefined dictionary, hand-crafted features, filters, or rules. Our model processes batches of documents, such as abstracts from PubMed, and creates an acronym dictionary specific to each inputted document.

3 Method

The intuition behind local acronym resolution is that authors of scientific publications commonly define the acronyms that they employ later on in the document. This is typically done by defining acronyms within the text in the form of pairs of short-form (SF) and corresponding long-form (LF) entities. We can then use the identified SF-LF acronym pairs to either resolve the acronyms appearing in the input document or populate an SF-LF dictionary that can be used to accurately resolve future uses of the SF versions of the acronyms in the remainder of the text.

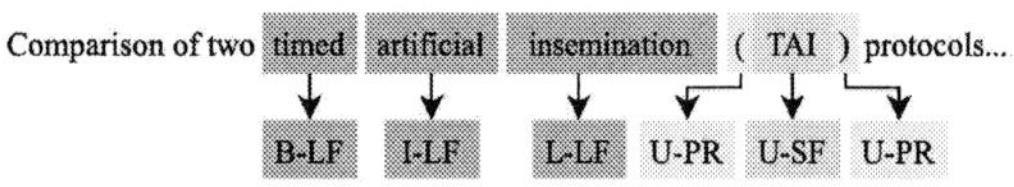

Figure 2: Sample output of *Step 2* showing the various tagged entities of a short and long-form acronym pair. We use a BILOU (Beginning, Inside, Last, Outside, Unit) tagging scheme (Ratinov and Roth, 2009) to identify long-form (LF) entities, short-form (SF) entities, and parenthesis (PR) enclosing a paired SF or LF entity.

Identifying the definitions of SF-LF pairs poses two major challenges: First, one has to identify the location in the text where the definition of an SF-LF pair is provided. Second, one has to identify the exact span (i.e., text) of both the short and long-form within the definition.

Two-step AR: Following the above structure, BLAR splits the problem into two separate subtasks:

- *Step 1: Sentence Classification.* Given the input text, identify sentences containing definitions of SF-LF pairs. This is modeled as a binary classification task.
- *Step 2: SF-LF Acronym Tagging:* Given a sentence predicted to contain a definition of an SF-LF pair, identify the exact form (i.e., text) of the SF and LF entities. This is modeled as a token classification task, where each token in the sentence is classified as being part of an acronym short-form, acronym long-form, or the parenthesis enclosing a paired entity. Token classification follows the BILOU (Beginning, Inside, Last, Outside, and Unit) encoding scheme (Ratinov and Roth, 2009), as shown in Figure 2 through a simple example.

Model architecture: The sentence classification model (*Step 1*) leverages transfer learning by fine-tuning the pretrained SciBERT model (Beltagy et al., 2019) for the specific task of sentence classification. The sentences that have been predicted as containing SF-LF pairs are given as input to the SF and LF tagging model (*Step 2*). The tagging model also leverages SciBERT by fine-tuning it on the SF and LF tagging task. To avoid exposure bias resulting from training on a set of perfect inputs (e.g. sentences containing acronym pairs as labeled in the dataset), we use the output from the sentence classification model from *Step 1* to train the tagging model in *Step 2*. The output of the tagging model is a dictionary that can then be used to replace all the short-form acronyms with their corresponding long-forms within a single source text.

Model training: We developed BLAR using the BioADI corpus (Kuo et al., 2009) and tested it on the Ab3P corpus (Sohn et al., 2008). BioADI includes 1,668 true SF-LF pairs from 1,200 annotated PubMed abstracts and Ab3P includes 1,221 true SF-LF pairs from 1,250 annotated PubMed abstracts. Both provide span-level data identifying short and long-form acronym pairs within PubMed abstracts and differ only in the articles selected for annotation. During development, we fine-tuned both our sentence and acronym token classifiers on the BioADI corpus randomly split into three subsets for training (80% of the corpus), validation (10% of the corpus), and testing (10% of the corpus). We use BioADI as a training dataset and Ab3P as a testing dataset to best compare our model's performance to existing SOTA benchmarks for local acronym resolution which use the same train/test splits. The BioADI and Ab3P corpora are described in Section 4. Since the models in both steps are fine-tuned versions of SciBERT, they are able to train fairly quick on CPUs. *Step 1* and *Step 2* converged within eight epochs, taking roughly 10 hours and 2 hours to complete, respectively, on two Intel Xeon CPUs (E5-2640 v3 @ 2.60GH) with 16GB of RAM.

Ablation study: To determine the importance of the 2-step architecture, we conduct an ablation study where we train a model to resolve acronyms without the help of a sentence classification step. This model is identical to the tagging model used in *Step 2*, only, it is trained on raw sentences that may or may not contain an acronym pair. This single-step architecture must simultaneously learn to detect and resolve an acronym pair. We refer to this model variation as "BLAR (single step)."

4 Datasets

BioADI: We use the BioADI (Kuo et al., 2009) corpus to train BLAR. It includes 1,668 true SF-LF pairs from 1,200 annotated PubMed abstracts.

Ab3P: We use the Ab3P (Sohn et al., 2008) corpus for testing. It includes 1,221 true SF-LF pairs from 1,250 annotated PubMed abstracts.

At the time of writing, both datasets are available for download on the BioC (Comeau et al., 2013) website.

5 Results and Discussion

To measure BLAR's performance, we first compare it against SOTA local AR approaches. As explained in the *Background and Previous Work* section, to the best of our knowledge, local acronym resolution has not seen significant advances since 2009. More recent acronym resolution works have focused instead on disambiguation acronym resolution, still relying on simpler rule-based algorithms for local acronym resolution (Jin et al., 2019; Jacobs et al., 2020). As a result, we compare BLAR to Kuo et al. (2009), Sohn et al. (2008), and Schwartz and Hearst (2003), which represent the SOTA in local acronym resolution.

Table 1 depicts the performance of BLAR against SOTA AR models. In this experiment, all models were trained on the BioADI dataset and tested on the Ab3P dataset. For each model, we evaluate Precision, Recall, and F1 score based on exact matches of long-form and short-form pairs. The results show that BLAR significantly outperforms all previous approaches, achieving an F1 score of 0.966 compared to 0.899 of the next best approach. We observe that, without a sentence classification step, the single-step BLAR model under-performs compared to the two-step architecture, highlighting the benefit of the sentence classification step in the full two-step architecture.

AR Model	P	R	F1
Schwartz et al. (2003)	0.950	0.788	0.861
Sohn et al. (2008)	**0.970**	0.836	0.898
Kuo et al. (2009)	0.959	0.846	0.899
Yeganova et al. (2011)	0.936	0.893	0.914
BLAR (single step)	0.950	0.957	0.953
BLAR (two step)	0.966	**0.966**	**0.966**

Table 1: Evaluation results of BLAR against SOTA local acronym resolution models. All models, save Yeganova et al., were trained on BioADI and tested on Ab3P. Yeganova et al. is trained on 1M automatically extracted potential SF-LF pairs from PubMed abstracts.

Model Output Analysis: Finally, to further understand the performance of BLAR, we perform an instance-level analysis of its output.

Analyzing the correct predictions, we see that the model successfully overcomes some of the complex challenges inherent in acronym resolution. For example, it correctly resolves the acronyms "SSc" to "systemic sclerosis" and "IUAG" to "intrauterine growth retardation." These examples show that BLAR learns to resolve short-forms that contain a different number of letters compared to the number of words in the corresponding long-form. In another example, BLAR correctly resolves "HIT type II" to "heparin-induced thrombocytopenia type II" which illustrates that the model was able to learn more complex acronyms that consist of a mix of short-form entities and complete words.

Moving to the incorrect predictions, we classify BLAR's errors into three categories: missed acronyms (false negatives), added acronyms (false positives), and modified acronyms (i.e., acronyms where the model correctly identifies a short-form but either truncates or extends the corresponding long-form).

A majority of the errors come from modified acronyms. Analyzing the modified acronyms, we find that 63.7% of cases are long-forms expanded or truncated by a single word/token. We identify that many of the erroneously expanded long-forms add a word or words preceding the ground truth long-form. For example, in the text "... heat stroke by reducing iNOS-dependent nitric oxide (NO)... ", BLAR identified "iNOS-dependent nitric oxide" as the long-form expansion of the short-form "NO.", instead of the correct "nitric oxide."

Another common error within the modified acronyms category is a truncated long-form. For example, BLAR predicts the long-form of "FVC" to be "forced vital capacity" but the ground truth is "forced expiratory volume in 1 s vital capacity." Here, BLAR predicts a simple long-form when the ground truth long-form is actually more complex. We plan to explore these insights in future work to further improve the model.

6 Conclusion and Future Work

Local acronym resolution has seen limited progress in recent years and has not benefited from the recent advancements in machine learning approaches. To address this problem, we develop BLAR; a deep-learning model that leverages a two-step architecture on top of pre-trained language models to identify SF-LF pairs in input documents. Our experimental results show that BLAR outperforms other local acronym resolution approaches and achieves state-of-the-art performance. We release BLAR and its source code for public use. As part of our future work, we will be exploring two threads: first, further improving the model based on our error analysis, and second, exploring how BLAR (which in this case has been fine-tuned for the scientific domain) can be extended to cover acronyms found in other domains. We believe future work could also focus on a hybrid model that leverages both deep-learning and rule-based algorithms.

Acknowledgement

Thank you to the anonymous reviewers for their thoughtful comments and corrections. This work is supported by IBM Research AI through the AI Horizons Network.

References

Adrian Barnett and Zoe Doubleday. 2020. Meta-research: The growth of acronyms in the scientific literature. *eLife*, 9:e60080.

Iz Beltagy, Kyle Lo, and Arman Cohan. 2019. SciBERT: A pretrained language model for scientific text. In *Proceedings of the 2019 Conference on Empirical Methods in Natural Language Processing and the 9th International Joint Conference on Natural Language Processing (EMNLP-IJCNLP)*, pages 3615–3620, Hong Kong, China. Association for Computational Linguistics.

Donald C. Comeau, Rezarta Islamaj Doğan, Paolo Ciccarese, Kevin Bretonnel Cohen, Martin Krallinger, Florian Leitner, Zhiyong Lu, Yifan Peng, Fabio Rinaldi, Manabu Torii, Alfonso Valencia, Karin Verspoor, Thomas C. Wiegers, Cathy H. Wu, and W. John Wilbur. 2013. Bioc: a minimalist approach to interoperability for biomedical text processing. *Database : the journal of biological databases and curation*, 2013:bat064–bat064. 24048470[pmid].

Lisa V. Grossman, Elliot G. Mitchell, George Hripcsak, Chunhua Weng, and David K. Vawdrey. 2018. A method for harmonization of clinical abbreviation and acronym sense inventories. *Journal of Biomedical Informatics*, 88:62 – 69.

Kayla Jacobs, Alon Itai, and Shuly Wintner. 2020. Acronyms: identification, expansion and disambiguation. *Annals of Mathematics and Artificial Intelligence*, 88(5):517–532.

Qiao Jin, Jinling Liu, and Xinghua Lu. 2019. Deep contextualized biomedical abbreviation expansion. *Proceedings of the 18th BioNLP Workshop and Shared Task*.

Cheng-Ju Kuo, Maurice HT Ling, Kuan-Ting Lin, and Chun-Nan Hsu. 2009. Bioadi: a machine learning approach to identifying abbreviations and definitions in biological literature. *BMC Bioinformatics*, 10(15):S7.

Lev Ratinov and Dan Roth. 2009. Design challenges and misconceptions in named entity recognition. In *Proceedings of the Thirteenth Conference on Computational Natural Language Learning (CoNLL-2009)*, pages 147–155, Boulder, Colorado. Association for Computational Linguistics.

Ariel S. Schwartz and Marti A. Hearst. 2003. A simple algorithm for identifying abbreviation definitions in biomedical text. *Pacific Symposium on Biocomputing. Pacific Symposium on Biocomputing*, pages 451–62.

Sunghwan Sohn, Donald C. Comeau, Won Kim, and W. John Wilbur. 2008. Abbreviation definition identification based on automatic precision estimates. *BMC bioinformatics*, 9:402–402. PMC2576267[pmcid].

Ashish Vaswani, Noam Shazeer, Niki Parmar, Jakob Uszkoreit, Llion Jones, Aidan N Gomez, Ł ukasz Kaiser, and Illia Polosukhin. 2017. Attention is all you need. In *Advances in Neural Information Processing Systems*, volume 30. Curran Associates, Inc.

Lana Yeganova, Donald Comeau, and W. Wilbur. 2011. Machine learning with naturally labeled data for identifying abbreviation definitions. *BMC bioinformatics*, 12 Suppl 3:S6.

Claim Detection in Biomedical Twitter Posts

Amelie Wührl and **Roman Klinger**
Institut für Maschinelle Sprachverarbeitung, University of Stuttgart, Germany
{amelie.wuehrl,roman.klinger}@ims.uni-stuttgart.de

Abstract

Social media contains unfiltered and unique information, which is potentially of great value, but, in the case of misinformation, can also do great harm. With regards to biomedical topics, false information can be particularly dangerous. Methods of automatic fact-checking and fake news detection address this problem, but have not been applied to the biomedical domain in social media yet. We aim to fill this research gap and annotate a corpus of 1200 tweets for implicit and explicit biomedical claims (the latter also with span annotations for the claim phrase). With this corpus, which we sample to be related to COVID-19, measles, cystic fibrosis, and depression, we develop baseline models which detect tweets that contain a claim automatically. Our analyses reveal that biomedical tweets are densely populated with claims (45 % in a corpus sampled to contain 1200 tweets focused on the domains mentioned above). Baseline classification experiments with embedding-based classifiers and BERT-based transfer learning demonstrate that the detection is challenging, however, shows acceptable performance for the identification of explicit expressions of claims. Implicit claim tweets are more challenging to detect.

1 Introduction

Social media platforms like Twitter contain vast amounts of valuable and novel information, and biomedical aspects are no exception (Correia et al., 2020). Doctors share insights from their everyday life, patients report on their experiences with particular medical conditions and drugs, or they discuss and hypothesize about the potential value of a treatment for a particular disease. This information can be of great value – governmental administrations or pharmaceutical companies can for instance learn about unknown side effects or potentially beneficial off-label use of medications.

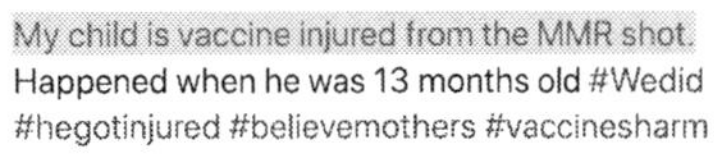

Figure 1: Tweet with a biomedical claim (highlighted).

At the same time, unproven claims or even intentionally spread misinformation might also do great harm. Therefore, contextualizing a social media message and investigating if a statement is debated or can actually be proven with a reference to a reliable resource is important. The task of detecting such claims is essential in argument mining and a prerequisite in further analysis for tasks like fact-checking or hypotheses generation. We show an example of a tweet with a claim in Figure 1.

Claims are widely considered the conclusive and therefore central part of an argument (Lippi and Torroni, 2015; Stab and Gurevych, 2017), consequently making it the most valuable information to extract. Argument mining and claim detection has been explored for texts like legal documents, Wikipedia articles, essays (Moens et al., 2007; Levy et al., 2014; Stab and Gurevych, 2017, i.a.), social media and web content (Goudas et al., 2014; Habernal and Gurevych, 2017; Bosc et al., 2016a; Dusmanu et al., 2017, i.a.). It has also been applied to scientific biomedical publications (Achakulvisut et al., 2019; Mayer et al., 2020, i.a.), but biomedical arguments as they occur on social media, and particularly Twitter, have not been analyzed yet.

With this paper, we fill this gap and explore claim detection for tweets discussing biomedical topics, particularly tweets about COVID-19, the measles, cystic fibrosis, and depression, to allow for drawing conclusions across different fields.

Our contributions to a better understanding of biomedical claims made on Twitter are, (1), to publish the first biomedical Twitter corpus manually labeled with claims (distinguished in explicit and implicit, and with span annotations for explicit claim phrases), and (2), baseline experiments to detect

Proceedings of the BioNLP 2021 workshop, pages 131–142
June 11, 2021. ©2021 Association for Computational Linguistics

(implicit and explicit) claim tweets in a classification setting. Further, (3), we find in a cross-corpus study that a generalization across domains is challenging and that biomedical tweets pose a particularly difficult environment for claim detection.

2 Related Work

Detecting biomedical claims on Twitter is a task rooted in both the argument mining field as well as the area of biomedical text mining.

2.1 Argumentation Mining

Argumentation mining covers a variety of different domains, text, and discourse types. This includes online content, for instance Wikipedia (Levy et al., 2014; Roitman et al., 2016; Lippi and Torroni, 2015), but also more interaction-driven platforms, like fora. As an example, Habernal and Gurevych (2017) extract argument structures from blogs and forum posts, including comments. Apart from that, Twitter is generally a popular text source (Bosc et al., 2016a; Dusmanu et al., 2017). Argument mining is also applied to professionally generated content, for instance news (Goudas et al., 2014; Sardianos et al., 2015) and legal or political documents (Moens et al., 2007; Palau and Moens, 2009; Mochales and Moens, 2011; Florou et al., 2013). Another domain of interest are persuasive essays, which we also use in a cross-domain study in this paper (Lippi and Torroni, 2015; Stab and Gurevych, 2017; Eger et al., 2017).

Existing approaches differ with regards to which tasks in the broader argument mining pipeline they address. While some focus on the detection of arguments (Moens et al., 2007; Florou et al., 2013; Levy et al., 2014; Bosc et al., 2016a; Dusmanu et al., 2017; Habernal and Gurevych, 2017), others analyze the relational aspects between argument components (Mochales and Moens, 2011; Stab and Gurevych, 2017; Eger et al., 2017).

While most approaches cater to a specific domain or text genre, Stab et al. (2018) argue that domain-focused, specialized systems do not generalize to broader applications such as argument search in texts. In line with that, Daxenberger et al. (2017) present a comparative study on cross-domain claim detection. They observe that diverse training data leads to a more robust model performance in unknown domains.

2.2 Claim Detection

Claim detection is a central task in argumentation mining. It can be framed as a classification (Does a document/sentence contain a claim?) or as sequence labeling (Which tokens make up the claim?). The setting as classification has been explored, inter alia, as a retrieval task of online comments made by public stakeholders on pending governmental regulations (Kwon et al., 2007), for sentence detection in essays, (Lippi and Torroni, 2015), and for Wikipedia (Roitman et al., 2016; Levy et al., 2017). The setting as a sequence labeling task has been tackled on Wikipedia (Levy et al., 2014), on Twitter, and on news articles (Goudas et al., 2014; Sardianos et al., 2015).

One common characteristic in most work on automatic claim detection is the focus on relatively formal text. Social media, like tweets, can be considered a more challenging text type, which despite this aspect, received considerable attention, also beyond classification or token sequence labeling. Bosc et al. (2016a) detect relations between arguments, Dusmanu et al. (2017) identify factual or opinionated tweets, and Addawood and Bashir (2016) further classify the type of premise which accompanies the claim. Ouertatani et al. (2020) combine aspects of sentiment detection, opinion, and argument mining in a pipeline to analyze argumentative tweets more comprehensively. Ma et al. (2018) specifically focus on the claim detection task in tweets, and present an approach to retrieve Twitter posts that contain argumentative claims about debatable political topics.

To the best of our knowledge, detecting biomedical claims in tweets has not been approached yet. Biomedical argument mining, also for other text types, is generally still limited. The work by Shi and Bei (2019) is one of the few exceptions that target this challenge and propose a pipeline to extract health-related claims from headlines of health-themed news articles. The majority of other argument mining approaches for the biomedical domain focus on research literature (Blake, 2010; Alamri and Stevenson, 2015; Alamri and Stevensony, 2015; Achakulvisut et al., 2019; Mayer et al., 2020).

2.3 Biomedical Text Mining

Biomedical natural language processing (BioNLP) is a field in computational linguistics which also receives substantial attention from the bioinformat-

Query category			
Disease Names	Topical Hashtags	Combinations	Drugs
COVID-19, #COVID-19	#socialdistancing, #chinesevirus	COVID-19 AND cured, COVID-19 AND vaccines	Hydroxychloroquine, Kaletra, Remdesivir
measles, #measles	#vaccineswork, #dontvaccinate	measles AND vaccine, measles AND therapize	M-M-R II, Priorix, ProQuad
cystic fibrosis, #cysticfibrosis	#livesavingdrugs4cf, #orkambinow	cystic fibrosis AND treated, cystic fibrosis AND heal	Orkambi, Trikafta, Tezacaftor
depression, #depression	#depressionisreal, #notjustsad	depression AND cure, depression AND treatment	Alprazolam, Buspirone, Xanax

Table 1: Examples of the four categories of search terms used to retrieve tweets about COVID-19, the measles, cystic fibrosis, and depression via the Twitter API.

ics community. One focus is on the automatic extraction of information from life science articles, including entity recognition, e.g., of diseases, drug names, protein and gene names (Habibi et al., 2017; Giorgi and Bader, 2018; Lee et al., 2019, i.a.) or relations between those (Lamurias et al., 2019; Sousa et al., 2021; Lin et al., 2019, i.a.).

Biomedical text mining methods have also been applied to social media texts and web content (Wegrzyn-Wolska et al., 2011; Yang et al., 2016; Sullivan et al., 2016, i.a.). One focus is on the analysis of Twitter with regards to pharmacovigilance. Other topics include the extraction of adverse drug reactions (Nikfarjam et al., 2015; Cocos et al., 2017), monitoring public health (Paul and Dredze, 2012; Choudhury et al., 2013; Sarker et al., 2016), and detecting personal health mentions (Yin et al., 2015; Karisani and Agichtein, 2018).

A small number of studies looked into the comparison of biomedical information in social media and scientific text: Thorne and Klinger (2018) analyze quantitatively how disease names are referred to across these domains. Seiffe et al. (2020) analyze laypersons' medical vocabulary.

3 Corpus Creation and Analysis

As the basis for our study, we collect a novel Twitter corpus in which we annotate which tweets contain biomedical claims, and (for all explicit claims) which tokens correspond to that claim.

3.1 Data Selection & Acquisition

The data for the corpus was collected in June/July 2020 using Twitter's API[1] which offers a keyword-based retrieval for tweets. Table 1 provides a sample of the search terms we used.[2] For each of the medical topics, we sample English tweets from keywords and phrases from four different query categories. This includes (1) the name of the disease as well as the respective hashtag for each topic, e.g., *depression* and *#depression*, (2) topical hashtags like *#vaccineswork*, (3) combinations of the disease name with words like *cure, treatment* or *therapy* as well as their respective verb forms, and (4) a list of medications, products, and product brand names from the pharmaceutical database DrugBank[3].

When querying the tweets, we exclude retweets by using the API's '-filter:retweets' option. From overall 902,524 collected tweets, we filter out those with URLs since those are likely to be advertisements (Cocos et al., 2017; Ma et al., 2018), and further remove duplicates based on the tweet IDs. From the resulting collection of 127,540 messages we draw a sample of 75 randomly selected tweets per topic (four biomedical topics) and search term category (four categories per topic). The final corpus to be annotated consists of 1200 tweets about four medical issues and their treatments: measles, depression, cystic fibrosis, and COVID-19.

3.2 Annotation

3.2.1 Conceptual Definition

While there are different schemes and models of argumentative structure varying in complexity as well as in their conceptualization of claims, the claim element is widely considered the core component of an argument (Daxenberger et al., 2017).

[1]https://developer.twitter.com/en/docs/twitter-api

[2]The full list of search terms (1771 queries in total) is available in the supplementary material.

[3]https://go.drugbank.com/. At the time of creating the search term list, COVID-19 was not included in DrugBank. Instead, medications which were under investigation at the time of compiling this list as outlined on the WHO website were included for Sars-CoV-2 in this category: https://www.who.int/emergencies/diseases/novel-coronavirus-2019/global-research-on-novel-coronavirus-2019-ncov/solidarity-clinical-trial-for-covid-19-treatments.

Aharoni et al. (2014) suggest a framework in which an argument consists of two main components: a claim and premises. We follow Stab and Gurevych (2017) and define the claim as the argumentative component in which the speaker or writer expresses the central, controversial conclusion of their argument. This claim is presented as if it were true even though objectively it can be true or false (Mochales and Ieven, 2009). The premise which is considered the second part of an argument includes all elements that are used either to substantiate or disprove the claim. Arguments can contain multiple premises to justify the claim. (Refer to Section 3.4 for examples and a detailed analysis of argumentative tweets in the dataset.)

For our corpus, we focus on the claim element and assign all tweets a binary label that indicates whether the document contains a claim. Claims can be either explicitly voiced or the claim property can be inferred from the text in cases in which they are expressed implicitly (Habernal and Gurevych, 2017). We therefore annotate explicitness or implicitness if a tweet is labeled as containing a claim. For explicit cases the claim sequence is additionally marked on the token level. For implicit cases, the claim which can be inferred from the implicit utterance is stated alongside the implicitness annotation.

3.2.2 Guideline Development

We define a preliminary set of annotation guidelines based on previous work (Mochales and Ieven, 2009; Aharoni et al., 2014; Bosc et al., 2016a; Daxenberger et al., 2017; Stab and Gurevych, 2017). To adapt those to our domain and topic, we go through four iterations of refinements. In each iteration, 20 tweets receive annotations by two annotators. Both annotators are female and aged 25–30. Annotator A1 has a background in linguistics and computational linguistics. A2 has a background in mathematics, computer science, and computational linguistics. The results are discussed based on the calculation of Cohen's κ (Cohen, 1960).

After Iteration 1, we did not make any substantial changes, but reinforced a common understanding of the existing guidelines in a joint discussion. After Iteration 2, we clarified the guidelines by adding the notion of an argumentative intention as a prerequisite for a claim: a claim is only to be annotated if the author actually appears to be intentionally argumentative as opposed to just sharing an opinion (Šnajder, 2016; Habernal and Gurevych, 2017). This is illustrated in the following example, which is not to be annotated as a claim, given this additional constraint:

> This popped up on my memories from two years ago, on Instagram, and honestly I'm so much healthier now it's quite unbelievable. A stone heavier, on week 11 of no IVs (back then it was every 9 weeks), and it's all thanks to #Trikafta and determination. I am stronger than I think.

We further clarified the guidelines with regards to the claim being the conclusive element in a Twitter document. This change encouraged the annotators to reflect specifically if the conclusive, main claim is conveyed explicitly or implicitly.

After Iteration 3, we did not introduce any changes, but went through an additional iteration to further establish the understanding of the annotation tasks.

	Cohen's κ		
	C/N	E/I/N	Span
Iteration 1	.31	.43	.32
Iteration 2	.34	.24	.12
Iteration 3	.61	.42	.42
Iteration 4	.60	.68	.41
Final corpus	.56	.48	.38

Table 2: Inter-annotator agreement during development of the annotation guidelines and for the final corpus. C/N: Claim/non-claim, E/I/N: Explicit/Implicit/Non-claim, Span: Token-level annotation of the explicit claim expression.

Table 2 shows the results of the agreement of the annotators in each iteration as well as the final κ-score for the corpus. We observe that the agreement substantially increased from Iteration 1 to 4. However, we also observe that obtaining a substantial agreement for the span annotation remains the most challenging task.

3.2.3 Annotation Procedure

The corpus annotation was carried out by the same annotators that conducted the preliminary annotations. A1 labeled 1000 tweets while A2 annotated 300 instances. From these both sets, 100 tweets were provided to both annotators, to track agreement (which remained stable, see Table 2). Annotating 100 tweets took approx. 3.3 hours. Overall, we observe that the agreement is generally moderate. Separating claim-tweets from non-claim tweets shows an acceptable κ=.56. Including the decision of explicitness/implicitness leads to κ=.48.

Class	# Instances	%	Length
non-claim	663	55.25	30.56
claim (I+E)	537	44.75	39.88
expl. claim	370	30.83	39.89
claim phrase			17.59
impl. claim	167	13.92	39.88
total	1200	100 %	34.73

Table 3: Distribution of the annotated classes and average instance lengths (in tokens).

	incompl.		blended		anecdotal		impl.	
M	8	.16	14	.28	9	.18	14	.28
C	17	.34	15	.30	8	.16	14	.28
CF	12	.24	10	.20	26	.52	18	.36
D	16	.32	9	.18	23	.46	11	.22
total	53	.27	48	.24	66	.33	57	.29

Table 4: Manual analysis of a subsample of 50 tweets/topic. Each column shows raw counts and percentage/topic.

id	Instance
1	*The French have had great success #hydroxycloroquine.*
2	Death is around 1/1000 in measles normally, same for encephalopathy, hospitalisation around 1/5. *With all the attendant costs, the vaccine saves money, not makes it.*
3	Latest: Kimberly isn't worried at all. *She takes #Hydroxychloroquine and feels awesome the next day.* Just think, it's more dangerous to drive a car than to catch corona
4	Lol exactly. It's not toxic to your body idk where he pulled this information out of. *Acid literally cured my depression/anxiety I had for 5 years in just 5 months (3 trips).* It literally reconnects parts of your brain that haven't had that connection in a long time.
5	Hopefully! The MMR toxin loaded vaccine I received many years ago seemed to work very well. More please!
6	Wow! Someone tell people with Cystic fibrosis and Huntington's that they can cure their genetics through Mormonism!

Table 5: Examples of explicit and implicit claim tweets from the corpus. Explicit claims are in italics.

The span-based annotation has limited agreement, with κ=.38 (which is why we do not consider this task further in this paper). These numbers are roughly in line with previous work. Achakulvisut et al. (2019) report an average κ=0.63 for labeling claims in biomedical research papers. According to Habernal and Gurevych (2017), explicit, intentional argumentation is easier to annotate than texts which are less explicit.

Our corpus is available with detailed annotation guidelines at http://www.ims.uni-stuttgart.de/data/bioclaim.

3.3 Corpus Statistics

Table 3 presents corpus statistics. Out of the 1200 documents in the corpus, 537 instances (44.75 %) contain a claim and 663 (55.25 %) do not. From all claim instances, 370 tweets are explicit (68 %). The claims are not equally distributed across topics (not shown in table): 61 % of measle-related tweets contain a claim, 49 % of those related to COVID-19, 40 % of cystic fibrosis tweets, and 29 % for depression.

The longest tweet in the corpus consists of 110 tokens[4], while the two shortest consist only of two tokens[5]. On average, a claim tweet has a length of ≈40 tokens. Both claim tweet types, explicit and implicit, have similar lengths (39.89 and 39.88 tokens respectively). In contrast to that, the average non-claim tweet is shorter, consisting of about 30 tokens. Roughly half of an explicit claim corresponds to the claim phrase.

We generally see that there is a connection between the length of a tweet and its class membership. Out of all tweets with up to 40 tokens, 453 instances are non-claims, while 243 contain a claim. For the instances that consist of 41 and more tokens, only 210 are non-claim tweets, whereas 294 are labeled as claims. The majority of the shorter tweets ($\leq$ 40 tokens) tend to be non-claim instances, while mid-range to longer tweets ($\geq$ 40 tokens) tend to be members of the claim class.

3.4 Qualitative Analysis

To obtain a better understanding of the corpus, we perform a qualitative analysis on a subsample of 50 claim-instances/topic. We manually analyze four claim properties: the tweet exhibits an incomplete argument structure, different argument components blend into each other, the text shows anecdotal evidence, and it describes the claim implicitly. Refer to Table 4 for an overview of the results.

In line with Šnajder (2016), we find that argument structures are often incomplete, e.g., in-

[4]The tweet includes 50 @-mentions followed by a measles-related claim: "Oh yay! I can do this too, since you're going to ignore the thousands of children who died in outbreaks last year from measles... Show me a proven death of a child from vaccines in the last decade. That's the time reference, now? So let's see a death certificate that says it, thx"

[5]"Xanax damage" and "Holy fuck".

stances only contain a stand-alone claim without any premise. This characteristic is most prevalent in the COVID-19-related tweets In Table 5, Ex. 1 is missing a premising element, Ex. 2 presents premise and claim.

Argument components (claim, premise) are not very clear cut and often blend together. Consequently they can be difficult to differentiate, for instance when authors use claim-like elements as a premise. This characteristic is again, most prevalent for COVID-19. In Ex. 3 in Table 5, the last sentence reads like a claim, especially when looked at in isolation, yet it is in fact used by the author to explain their claim.

Premise elements which substantiate and give reason for the claim (Bosc et al., 2016b) traditionally include references to studies or mentions of expert testimony, but occasionally also anecdotal evidence or concrete examples (Aharoni et al., 2014). We find the latter to be very common for our data set. This property is most frequent for cystic fibrosis and depression. Ex. 4 showcases how a personal experience is used to build an argument.

Implicitness in the form of irony, sarcasm or rhetoric questions are common features for these types of claims on Twitter. We observe claims related to cystic fibrosis are most often (in our sample) implicit. Ex. 5 and 6 show instances that use sarcasm or irony. The fact that implicitness is such a common feature in our dataset is in line with the observation that implicitness is a characteristic device not only in spoken language and everyday, informal argumentation (Lumer, 1990), but also in user-generated web content in general (Habernal and Gurevych, 2017).

4 Methods

In the following sections we describe the conceptual design of our experiments and introduce the models that we use to accomplish the claim detection task.

4.1 Classification Tasks

We model the task in a set of different model configurations.

Binary. A trained classifier distinguishes between claim and non-claim.

Multiclass. A trained classifier distinguishes between exlicit claim, implicit claim, and non-claim.

Multiclass Pipeline. A first classifier learns to discriminate between claims and non-claims (as in Binary). Each tweet that is classified as claim is further separated into implicit or explicit with another binary classifier. The secondary classifier is trained on gold data (not on predictions of the first model in the pipeline).

4.2 Model Architecture

For each of the classification tasks (binary/multiclass, steps in the pipeline), we use a set of standard text classification methods which we compare. The first three models (NB, LG, BiLSTM) use 50-dimensional FastText (Bojanowski et al., 2017) embeddings trained on the Common Crawl corpus (600 billion tokens) as input[6].

NB. We use a (Gaussian) naive Bayes with an average vector of the token embeddings as input.

LG. We use a logistic regression classifier with the same features as in NB.

BiLSTM. As a classifier which can consider contextual information and makes use of pretrained embeddings, we use a bidirectional long short-term memory network (Hochreiter and Schmidhuber, 1997) with 75 LSTM units followed by the output layer (sigmoid for binary classification, softmax for multiclass).

BERT. We use the pretrained BERT (Devlin et al., 2019) base model[7] and fine-tune it using the claim tweet corpus.

5 Experiments

5.1 Claim Detection

With the first experiment we explore how reliably we can detect claim tweets in our corpus and how well the two different claim types (*explicit* vs. *implicit claim tweets*) can be distinguished. We use each model mentioned in Section 4.2 in each setting described in Section 4.1. We evaluate each classifier in a binary or (where applicable) in a multi-class setting, to understand if splitting the claim category into its subcomponents improves the claim prediction overall.

[6]https://fasttext.cc/docs/en/english-vectors.html
[7]https://huggingface.co/bert-base-uncased

			NB			LG			LSTM			BERT		
Eval.	Task	Class	P	R	F_1	P	R	F_1	P	R	F_1	P	R	F_1
binary	binary	claim	.67	.65	.66	.66	.74	**.70**	.68	.48	.57	.66	.72	.69
		n-claim	.75	.77	**.76**	.79	.72	**.76**	.69	.84	.75	.78	.72	.75
	multiclass	claim	.66	.65	**.66**	.73	.53	.61	.75	.35	.48	.81	.49	.61
		n-claim	.74	.76	.75	.71	.85	.78	.66	.91	.76	.71	.91	**.80**
multi-class	multiclass	expl	.55	.45	.50	.63	.39	.48	.59	.27	.37	.62	.45	**.52**
		impl	.31	.44	**.36**	.33	.35	.34	.18	.09	.12	.29	.09	.13
		n-claim	.74	.76	.75	.71	.85	.78	.66	.91	.76	.71	.91	**.80**
	pipeline	expl	.56	.45	.50	.52	.55	.53	.50	.37	.43	.54	.65	**.59**
		impl	.31	.44	**.36**	.28	.35	.31	.07	.04	.05	.26	.22	.24
		n-claim	.75	.77	**.76**	.79	.72	**.76**	.69	.84	.75	.78	.72	.75

Table 6: Results for the claim detection experiments, separated into binary and multi-class evaluation. The best F_1 scores for each evaluation setting and class are printed in bold face.

5.1.1 Experimental Setting

From our corpus of 1200 tweets we use 800 instances for training, 200 as validation data to optimize hyperparameters and 200 as test data. We tokenize the documents and substitute all @-mentions by "@username". For the LG models, we use an l2 regularization. For the LSTM models, the hyper-parameters learning rate, dropout, number of epochs, and batch size were determined by a randomized search over a parameter grid and also use l2 regularization. For training, we use Adam (Kingma and Ba, 2015). For the BERT models, we experiment with combinations of the recommended fine-tuning hyper-parameters from Devlin et al. (2019) (batch size, learning rate, epochs), and use those with the best performance on the validation data. An overview of all hyper-parameters is provided in Table 9 in the Appendix. For the Bi-LSTM, we use the Keras API (Chollet et al., 2015) for TensorFlow (Abadi et al., 2015). For the BERT model, we use Simple Transformers (Rajapakse, 2019) and its wrapper for the Hugging Face transformers library (Wolf et al., 2020). Further, we oversample the minority class of implicit claims to achieve a balanced training set (the test set remains with the original distribution). To ensure comparability, we oversample in both the binary and the multi-class setting. For parameters that we do not explicitly mention, we use default values.

5.1.2 Results

Table 6 reports the results for the conducted experiments. The top half lists the results for the binary claim detection setting. The bottom half of the table presents the results for the multi-class claim classification.

For the binary evaluation setting, we observe that casting the problem as a ternary prediction task is not beneficial – the best F_1 score is obtained with the binary LG classifier (.70 F_1 for the class claim in contrast to .61 F_1 for the ternary LG). The BERT and NB approaches are slightly worse (1 pp and 4pp less for binary, respectively), while the LSTM shows substantially lower performance (13pp less).

In the ternary/multi-class evaluation, the scores are overall lower. The LSTM shows the lowest performance. The best result is obtained in the pipeline setting, in which separate classifiers can focus on distinguishing claim/non-claim and explicit/implicit – we see .59 F_1 for the explicit claim class. Implicit claim detection is substantially more challenging across all classification approaches.

We attribute the fact that the more complex models (LSTM, BERT) do not outperform the linear models across the board to the comparably small size of the dataset. This appears especially true for implicit claims in the multi-class setting. Here, those models struggle the most to predict implicit claims, indicating that they were not able to learn sufficiently from the training instances.

5.1.3 Error Analysis

From a manual introspection of the best performing model in the binary setting, we conclude that it is difficult to detect general patterns. We show two cases of false positives and two cases of false negatives in Table 7. The false positive instances show that the model struggles with cases that rely on judging the argumentative intention. Both Ex. 1 and 2 contain potential claims about depression and therapy, but they have not been annotated as such, because the authors' intention is motivational rather than argumentative. In addition, it appears

id	G	P	Text
1	n	c	#DepressionIsReal #MentalHealthAwareness #mentalhealth ruins lives. #depression destroys people. Be there when someone needs you. It could change a life. It may even save one.
2	n	c	The reason I stepped away from twitch and gaming with friends is because iv been slowly healing from a super abusive relationship. Going to therapy and hearing you have ptsd isnt easy. But look how far iv come, lost some depression weight and found some confidence:)plz stay safe
3	c	n	Not sure who knows more about #COVID19, my sister or #DrFauci. My money is on Stephanie.
4	c	n	How does giving the entire world a #COVID19 #vaccine compare to letting everyone actually get #covid? What would you prefer? I'm on team @username #WHO #CDC #math #VaccinesWork #Science

Table 7: Examples of incorrect predictions by the LG model in the binary setting (G:Gold, P:Predictions; n: no claim; c: claim).

that the model struggles to detect implicit claims that are expressed using irony (Ex. 3) or a rhetorical question (Ex. 4).

5.2 Cross-domain Experiment

We see that the models show acceptable performance in a binary classification setting. In the following, we analyze if this observation holds across domains or if information from another out-of-domain corpus can help.

As the binary LG model achieved the best results during the previous experiment, we use this classifier for the cross-domain experiments. We work with paragraphs of persuasive essays (Stab and Gurevych, 2017) as a comparative corpus. The motivation to use this resource is that while they are a distinctly different text type and usually linguistically much more formal than tweets, they are also opinionated documents.[8] We use the resulting essay model for making an in-domain as well as a cross-domain prediction and vice versa for the Twitter model. We further experiment with combining the training portions of both datasets and evaluate its performance for both target domains.

5.2.1 Experimental Setting

The comparative corpus contains persuasive essays with annotated argument structure (Stab and Gurevych, 2017). Eger et al. (2017) used this corpus subsequently and provide the data in CONLL-format, split into paragraphs, and predivided into train, development and test set.[9] We use their version of the corpus. The annotations for the essay corpus distinguish between major claims and claims. However, since there is no such hierarchical differentiation in the Twitter annotations, we consider both types as equivalent. We choose to use paragraphs instead of whole essays as the individual input documents for the classification and assign a claim label to every paragraph that contains a claim. This leaves us with 1587 essay paragraphs as training data, and 199 and 449 paragraphs respectively for validation and testing.

We follow the same setup as for the binary setting in the first experiment.

[8]An essay is defined as *"a short piece of writing on a particular subject, often expressing personal views"* (https://dictionary.cambridge.org/dictionary/english/essay).

Train	Test	P	R	F_1
Twitter	Twitter	.66	.74	.70
Essay	Twitter	.51	.69	.59
Twitter+Essay	Twitter	.58	.75	.66
Essay	Essay	.96	1.0	.98
Twitter	Essay	.94	.74	.83
Twitter+Essay	Essay	.95	1.0	.97

Table 8: Results of cross-domain experiments using the binary LG model on the Twitter and the essay corpus. We report precision, recall and F_1 for the claim tweet class.

5.2.2 Results

In Table 8, we summarize the results of the cross-domain experiments with the persuasive essay corpus. We see that the essay model is successful for classifying claim documents (.98 F_1) in the in-domain experiment. Compared to the in-domain setting for tweets all evaluation scores measure substantially higher.

When we compare the two cross-domain experiments, we observe that the performance measures decrease in both settings when we use the out-of-domain model to make predictions (11pp in F_1 for tweets, 15pp for essays). Combining the training portions of both data sets does not lead to an improvement over in-domain experiments. This shows the challenge of building domain-generic models that perform well across different data sets.

6 Discussion and Future Work

In this paper, we presented the first data set for biomedical claim detection in social media. In our

[9]https://github.com/UKPLab/acl2017-neural_end2end_am/tree/master/data/conll/Paragraph_Level

first experiment, we showed that we can achieve an acceptable performance to detect claims when the distinction between explicit or implicit claims is not considered. In the cross-domain experiment, we see that text formality, which is one of the main distinguishing feature between the two corpora, might be an important factor that influences the level of difficulty in accomplishing the claim detection task.

Our hypothesis in this work was that biomedical information on Twitter exhibits a challenging setting for claim detection. Both our experiments indicate that this is true. Future work needs to investigate what might be reasons for that. We hypothesize that our Twitter dataset contains particular aspects that are specific to the medical domain, but it might also be that other latent variables lead to confounders (e.g., the time span that has been used for crawling). It is important to better understand these properties.

We suggest future work on claim detection models optimize those to work well across domains. To enable such research, this paper contributed a novel resource. This resource could further be improved. One way of addressing the moderate agreement between the annotators could be to include annotators with medical expertise to see if this ultimately facilitates claim annotation. Additionally, a detailed introspection of the topics covered in the tweets for each disease would be interesting for future work since this might shed some light on which topical categories of claims are particularly difficult to label.

The COVID-19 pandemic has sparked recent research with regards to detecting misinformation and fact-checking claims (e.g., Hossain et al. (2020) or Wadden et al. (2020)). Exploring how a claim-detection-based fact-checking approach rooted in argument mining compares to other approaches is up to future research.

Acknowledgments

This research has been conducted as part of the FIBISS project[10] which is funded by the German Research Council (DFG, project number: KL 2869/5-1). We thank Laura Ana Maria Oberländer for her support and the anonymous reviewers for their valuable comments.

[10]Automatic Fact Checking for Biomedical Information in Social Media and Scientific Literature, https://www.ims.uni-stuttgart.de/en/research/projects/fibiss/

References

Martín Abadi, Ashish Agarwal, Paul Barham, Eugene Brevdo, Zhifeng Chen, Craig Citro, Greg S. Corrado, Andy Davis, Jeffrey Dean, Matthieu Devin, Sanjay Ghemawat, Ian Goodfellow, Andrew Harp, Geoffrey Irving, Michael Isard, Yangqing Jia, Rafal Jozefowicz, Lukasz Kaiser, Manjunath Kudlur, Josh Levenberg, Dandelion Mané, Rajat Monga, Sherry Moore, Derek Murray, Chris Olah, Mike Schuster, Jonathon Shlens, Benoit Steiner, Ilya Sutskever, Kunal Talwar, Paul Tucker, Vincent Vanhoucke, Vijay Vasudevan, Fernanda Viégas, Oriol Vinyals, Pete Warden, Martin Wattenberg, Martin Wicke, Yuan Yu, and Xiaoqiang Zheng. 2015. TensorFlow: Large-scale machine learning on heterogeneous systems. Software available from tensorflow.org.

Titipat Achakulvisut, Chandra Bhagavatula, Daniel Acuna, and Konrad Kording. 2019. Claim extraction in biomedical publications using deep discourse model and transfer learning. *arXiv preprint arXiv:1907.00962*.

Aseel Addawood and Masooda Bashir. 2016. "What is your evidence?" A study of controversial topics on social media. In *Proceedings of the Third Workshop on Argument Mining (ArgMining2016)*, pages 1–11, Berlin, Germany. Association for Computational Linguistics.

Ehud Aharoni, Anatoly Polnarov, Tamar Lavee, Daniel Hershcovich, Ran Levy, Ruty Rinott, Dan Gutfreund, and Noam Slonim. 2014. A benchmark dataset for automatic detection of claims and evidence in the context of controversial topics. In *Proceedings of the First Workshop on Argumentation Mining*, pages 64–68, Baltimore, Maryland. Association for Computational Linguistics.

Abdulaziz Alamri and Mark Stevenson. 2015. Automatic detection of answers to research questions from Medline abstracts. In *Proceedings of BioNLP 15*, pages 141–146, Beijing, China. Association for Computational Linguistics.

Abdulaziz Alamri and Mark Stevensony. 2015. Automatic identification of potentially contradictory claims to support systematic reviews. In *Proceedings of the 2015 IEEE International Conference on Bioinformatics and Biomedicine (BIBM)*, BIBM '15, page 930–937, USA. IEEE Computer Society.

Catherine Blake. 2010. Beyond genes, proteins, and abstracts: Identifying scientific claims from full-text biomedical articles. *Journal of biomedical informatics*, 43(2):173–189.

Piotr Bojanowski, Edouard Grave, Armand Joulin, and Tomas Mikolov. 2017. Enriching word vectors with subword information. *Transactions of the Association for Computational Linguistics*, 5:135–146.

Tom Bosc, Elena Cabrio, and Serena Villata. 2016a. DART: a dataset of arguments and their relations on Twitter. In *Proceedings of the Tenth International Conference on Language Resources and Evaluation (LREC'16)*, pages 1258–1263, Portorož, Slovenia. European Language Resources Association (ELRA).

Tom Bosc, Elena Cabrio, and Serena Villata. 2016b. Tweeties squabbling: Positive and negative results in applying argument mining on social media. *COMMA*, 2016:21–32.

François Chollet et al. 2015. Keras. https://keras.io.

Munmun De Choudhury, Scott Counts, and Eric Horvitz. 2013. Social media as a measurement tool of depression in populations. In *In Proceedings of the 5th ACM International Conference on Web Science (Paris, France, May 2-May 4, 2013). WebSci 2013.*

Anne Cocos, Alexander G Fiks, and Aaron J Masino. 2017. Deep learning for pharmacovigilance: recurrent neural network architectures for labeling adverse drug reactions in twitter posts. *Journal of the American Medical Informatics Association*, 24(4):813–821.

Jacob Cohen. 1960. A coefficient of agreement for nominal scales. *Educational and psychological measurement*, 20(1):37–46.

Rion Brattig Correia, Ian B. Wood, Johan Bollen, and Luis M. Rocha. 2020. Mining Social Media Data for Biomedical Signals and Health-Related Behavior. *Annual Review of Biomedical Data Science*, 3(1):433–458. _eprint: https://doi.org/10.1146/annurev-biodatasci-030320-040844.

Johannes Daxenberger, Steffen Eger, Ivan Habernal, Christian Stab, and Iryna Gurevych. 2017. What is the essence of a claim? Cross-domain claim identification. In *Proceedings of the 2017 Conference on Empirical Methods in Natural Language Processing*, pages 2055–2066, Copenhagen, Denmark. Association for Computational Linguistics.

Jacob Devlin, Ming-Wei Chang, Kenton Lee, and Kristina Toutanova. 2019. BERT: Pre-training of deep bidirectional transformers for language understanding. In *Proceedings of the 2019 Conference of the North American Chapter of the Association for Computational Linguistics: Human Language Technologies, Volume 1 (Long and Short Papers)*, pages 4171–4186, Minneapolis, Minnesota. Association for Computational Linguistics.

Mihai Dusmanu, Elena Cabrio, and Serena Villata. 2017. Argument mining on Twitter: Arguments, facts and sources. In *Proceedings of the 2017 Conference on Empirical Methods in Natural Language Processing*, pages 2317–2322, Copenhagen, Denmark. Association for Computational Linguistics.

Steffen Eger, Johannes Daxenberger, and Iryna Gurevych. 2017. Neural end-to-end learning for computational argumentation mining. In *Proceedings of the 55th Annual Meeting of the Association for Computational Linguistics (Volume 1: Long Papers)*, pages 11–22, Vancouver, Canada. Association for Computational Linguistics.

Eirini Florou, Stasinos Konstantopoulos, Antonis Koukourikos, and Pythagoras Karampiperis. 2013. Argument extraction for supporting public policy formulation. In *Proceedings of the 7th Workshop on Language Technology for Cultural Heritage, Social Sciences, and Humanities*, pages 49–54, Sofia, Bulgaria. Association for Computational Linguistics.

John M Giorgi and Gary D Bader. 2018. Transfer learning for biomedical named entity recognition with neural networks. *Bioinformatics*, 34(23):4087–4094.

Theodosis Goudas, Christos Louizos, Georgios Petasis, and Vangelis Karkaletsis. 2014. Argument extraction from news, blogs, and social media. In *Artificial Intelligence: Methods and Applications*, pages 287–299, Cham. Springer International Publishing.

Ivan Habernal and Iryna Gurevych. 2017. Argumentation mining in user-generated web discourse. *Computational Linguistics*, 43(1):125–179.

Maryam Habibi, Leon Weber, Mariana Neves, David Luis Wiegandt, and Ulf Leser. 2017. Deep learning with word embeddings improves biomedical named entity recognition. *Bioinformatics*, 33(14):i37–i48.

Sepp Hochreiter and Jürgen Schmidhuber. 1997. Long short-term memory. *Neural computation*, 9(8):1735–1780.

Tamanna Hossain, Robert L. Logan IV, Arjuna Ugarte, Yoshitomo Matsubara, Sean Young, and Sameer Singh. 2020. COVIDLies: Detecting COVID-19 Misinformation on Social Media. In *Proceedings of the 1st Workshop on NLP for COVID-19 (Part 2) at EMNLP 2020*, Online. Association for Computational Linguistics.

Payam Karisani and Eugene Agichtein. 2018. Did you really just have a heart attack? Towards robust detection of personal health mentions in social media. In *Proceedings of the 2018 World Wide Web Conference*, page 137–146, Republic and Canton of Geneva, CHE.

Diederik P. Kingma and Jimmy Ba. 2015. Adam: A method for stochastic optimization. In *International Conference on Learning Representations (ICLR)*.

Namhee Kwon, Liang Zhou, Eduard Hovy, and Stuart W. Shulman. 2007. Identifying and classifying subjective claims. In *Proceedings of the 8th Annual International Conference on Digital Government Research: Bridging Disciplines & Domains*, dg.o '07, page 76–81. Digital Government Society of North America.

Andre Lamurias, Diana Sousa, Luka A. Clarke, and Francisco M. Couto. 2019. Bo-lstm: classifying relations via long short-term memory networks along biomedical ontologies. *BMC Bioinformatics*, 20(1):10.

Jinhyuk Lee, Wonjin Yoon, Sungdong Kim, Donghyeon Kim, Sunkyu Kim, Chan Ho So, and Jaewoo Kang. 2019. Biobert: a pre-trained biomedical language representation model for biomedical text mining. *Bioinformatics*.

Ran Levy, Yonatan Bilu, Daniel Hershcovich, Ehud Aharoni, and Noam Slonim. 2014. Context dependent claim detection. In *Proceedings of COLING 2014, the 25th International Conference on Computational Linguistics: Technical Papers*, pages 1489–1500, Dublin, Ireland. Dublin City University and Association for Computational Linguistics.

Ran Levy, Shai Gretz, Benjamin Sznajder, Shay Hummel, Ranit Aharonov, and Noam Slonim. 2017. Unsupervised corpus–wide claim detection. In *Proceedings of the 4th Workshop on Argument Mining*, pages 79–84, Copenhagen, Denmark. Association for Computational Linguistics.

Chen Lin, Timothy Miller, Dmitriy Dligach, Steven Bethard, and Guergana Savova. 2019. A BERT-based universal model for both within- and cross-sentence clinical temporal relation extraction. In *Proceedings of the 2nd Clinical Natural Language Processing Workshop*, pages 65–71, Minneapolis, Minnesota, USA. Association for Computational Linguistics, Association for Computational Linguistics.

Marco Lippi and Paolo Torroni. 2015. Context-independent claim detection for argument mining. In *Proceedings of the 24th International Conference on Artificial Intelligence*, IJCAI'15, page 185–191. AAAI Press.

Christoph Lumer. 1990. *Praktische Argumentationstheorie: theoretische Grundlagen, praktische Begründung und Regeln wichtiger Argumentationsarten*. Hochschulschrift, University of Münster, Braunschweig.

Wenjia Ma, WenHan Chao, Zhunchen Luo, and Xin Jiang. 2018. CRST: a claim retrieval system in Twitter. In *Proceedings of the 27th International Conference on Computational Linguistics: System Demonstrations*, pages 43–47, Santa Fe, New Mexico. Association for Computational Linguistics.

Tobias Mayer, Elena Cabrio, and Serena Villata. 2020. Transformer-based Argument Mining for Healthcare Applications. In *24th European Conference on Artificial Intelligence (ECAI2020)*, Santiago de Compostela, Spain.

Raquel Mochales and Aagje Ieven. 2009. Creating an argumentation corpus: Do theories apply to real arguments? a case study on the legal argumentation of the echr. In *Proceedings of the 12th International Conference on Artificial Intelligence and Law*, ICAIL '09, page 21–30, New York, NY, USA. Association for Computing Machinery.

Raquel Mochales and Marie-Francine Moens. 2011. Argumentation mining. *Artificial Intelligence and Law*, 19(1):1–22.

Marie-Francine Moens, Erik Boiy, Raquel Mochales Palau, and Chris Reed. 2007. Automatic detection of arguments in legal texts. In *Proceedings of the 11th International Conference on Artificial Intelligence and Law*, ICAIL '07, page 225–230, New York, NY, USA. Association for Computing Machinery.

Azadeh Nikfarjam, Abeed Sarker, Karen O'Connor, Rachel Ginn, and Graciela Gonzalez. 2015. Pharmacovigilance from social media: mining adverse drug reaction mentions using sequence labeling with word embedding cluster features. *Journal of the American Medical Informatics Association*, 22(3):671–681.

Asma Ouertatani, Ghada Gasmi, and Chiraz Latiri. 2020. Parsing argued opinion structure in Twitter content. *Journal of Intelligent Information Systems*, pages 1–27.

Raquel Mochales Palau and Marie-Francine Moens. 2009. Argumentation mining: The detection, classification and structure of arguments in text. In *Proceedings of the 12th International Conference on Artificial Intelligence and Law*, ICAIL '09, page 98–107, New York, NY, USA. Association for Computing Machinery.

Michael J. Paul and Mark Dredze. 2012. A model for mining public health topics from Twitter. *Health*, 11(1).

Thilina Rajapakse. 2019. Simple transformers. https://github.com/ThilinaRajapakse/simpletransformers.

Haggai Roitman, Shay Hummel, Ella Rabinovich, Benjamin Sznajder, Noam Slonim, and Ehud Aharoni. 2016. On the retrieval of wikipedia articles containing claims on controversial topics. In *Proceedings of the 25th International Conference Companion on World Wide Web*, WWW '16 Companion, page 991–996, Republic and Canton of Geneva, CHE. International World Wide Web Conferences Steering Committee.

Christos Sardianos, Ioannis Manousos Katakis, Georgios Petasis, and Vangelis Karkaletsis. 2015. Argument extraction from news. In *Proceedings of the 2nd Workshop on Argumentation Mining*, pages 56–66, Denver, CO. Association for Computational Linguistics.

Abeed Sarker, Karen O'Connor, Rachel Ginn, Matthew Scotch, Karen Smith, Dan Malone, and Graciela

Gonzalez. 2016. Social media mining for toxicovigilance: automatic monitoring of prescription medication abuse from Twitter. *Drug safety*, 39(3):231–240.

Laura Seiffe, Oliver Marten, Michael Mikhailov, Sven Schmeier, Sebastian Möller, and Roland Roller. 2020. From witch's shot to music making bones - resources for medical laymen to technical language and vice versa. In *Proceedings of the 12th Language Resources and Evaluation Conference*, pages 6185–6192, Marseille, France. European Language Resources Association.

Yuan Shi and Yu Bei. 2019. HClaimE: A tool for identifying health claims in health news headlines. *Information Processing & Management*, 56(4):1220–1233.

Jan Šnajder. 2016. Social media argumentation mining: the quest for deliberateness in raucousness. *arXiv preprint arXiv:1701.00168*.

Diana Sousa, Andre Lamurias, and Francisco M. Couto. 2021. *Using Neural Networks for Relation Extraction from Biomedical Literature*, pages 289–305. Springer US, New York, NY.

Christian Stab and Iryna Gurevych. 2017. Parsing argumentation structures in persuasive essays. *Computational Linguistics*, 43(3):619–659.

Christian Stab, Tristan Miller, Benjamin Schiller, Pranav Rai, and Iryna Gurevych. 2018. Cross-topic argument mining from heterogeneous sources. In *Proceedings of the 2018 Conference on Empirical Methods in Natural Language Processing*, pages 3664–3674, Brussels, Belgium. Association for Computational Linguistics.

Ryan Sullivan, Abeed Sarker, Karen O'Connor, Amanda Goodin, Mark Karlsrud, and Graciela Gonzalez. 2016. Finding potentially unsafe nutritional supplements from user reviews with topic modeling. In *Biocomputing 2016*, pages 528–539, Kohala Coast, Hawaii, USA.

Camilo Thorne and Roman Klinger. 2018. On the semantic similarity of disease mentions in medline and twitter. In *Natural Language Processing and Information Systems: 23rd International Conference on Applications of Natural Language to Information Systems, NLDB 2018, Paris, France, June 13-15, 2018, Proceedings*, Cham. Springer International Publishing.

David Wadden, Shanchuan Lin, Kyle Lo, Lucy Lu Wang, Madeleine van Zuylen, Arman Cohan, and Hannaneh Hajishirzi. 2020. Fact or Fiction: Verifying Scientific Claims. In *Proceedings of the 2020 Conference on Empirical Methods in Natural Language Processing (EMNLP)*, pages 7534–7550, Online. Association for Computational Linguistics.

Parameter	LSTM2	LSTM3
Embedding	fastText	fastText
Emb. Dim.	50	50
# LSTM units	75	75
Training epochs	60	70
Training batch size	10	30
Loss function	Binary CE	Categ. CE
Optimizer	Adam	Adam
Learning rate	1e-3	1e-3
L2 regularization	1e-3	1e-3
dropout	0.5	0.6

(a) Overview of architectural choices and hyper-parameter settings for the binary (LSTM2) and multi-class (LSTM3) LSTM-based models used in our experiments.

Parameter	BERT2	BERT3
Training epochs	4	4
Training batch size	16	16
Learning rate	2e-5	3e-5

(b) Overview of fine-tuning hyper-parameters for the binary (BERT2) and multi-class (BERT3) models used in our experiments.

Table 9: Overview of model hyper-parameters.

Katarzyna Wegrzyn-Wolska, Lamine Bougueroua, and Grzegorz Dziczkowski. 2011. Social media analysis for e-health and medical purposes. In *2011 International Conference on Computational Aspects of Social Networks (CASoN)*, pages 278–283.

Thomas Wolf, Lysandre Debut, Victor Sanh, Julien Chaumond, Clement Delangue, Anthony Moi, Pierric Cistac, Tim Rault, Rémi Louf, Morgan Funtowicz, Joe Davison, Sam Shleifer, Patrick von Platen, Clara Ma, Yacine Jernite, Julien Plu, Canwen Xu, Teven Le Scao, Sylvain Gugger, Mariama Drame, Quentin Lhoest, and Alexander M. Rush. 2020. Transformers: State-of-the-art natural language processing. In *Proceedings of the 2020 Conference on Empirical Methods in Natural Language Processing: System Demonstrations*, pages 38–45, Online. Association for Computational Linguistics.

Fu-Chen Yang, Anthony J.T. Lee, and Sz-Chen Kuo. 2016. Mining health social media with sentiment analysis. *Journal of Medical Systems*, 40(11):236.

Zhijun Yin, Daniel Fabbri, S Trent Rosenbloom, and Bradley Malin. 2015. A scalable framework to detect personal health mentions on Twitter. *Journal of Medical Internet Research*, 17(6):e138.

BioELECTRA:Pretrained Biomedical text Encoder using Discriminators

Kamal Raj Kanakarajan and **Bhuvana Kundumani** and **Malaikannan Sankarasubbu**
SAAMA AI Research Lab, Chennai, India
{kamal.raj, bhuvana.kundumani, malaikannan.sankarasubbu}@saama.com

Abstract

Recent advancements in pretraining strategies in NLP have shown a significant improvement in the performance of models on various text mining tasks. In this paper, we introduce BioELECTRA, a biomedical domain-specific language encoder model that adapts ELECTRA (Clark et al., 2020) for the Biomedical domain. BioELECTRA outperforms the previous models and achieves state of the art (SOTA) on all the 13 datasets in BLURB benchmark and on all the 4 Clinical datasets from BLUE Benchmark across 7 NLP tasks. BioELECTRA pretrained on PubMed and PMC full text articles performs very well on Clinical datasets as well. BioELECTRA achieves new SOTA 86.34%(1.39% accuracy improvement) on MedNLI and 64% (2.98% accuracy improvement) on PubMedQA dataset.

1 Introduction

Following the success of BERT (Devlin et al., 2018) (Bidirectional Encoder Representations from Transformers) in the general domain, the pretrain-and-finetune approach has been used in the Biomedical domain. With large scale free text available from PubMed and PubMed central (millions of articles), biomedical domain has large unlabelled domain-specific corpus. However, the biomedical domain has labelled datasets that are very small compared to the general domain. Thus the transfer learning approach is well suited for Biomedical domain.

In the biomedical domain, BioBERT (Lee et al., 2020), BlueBERT (Peng et al., 2019) and ClinicalBERT (Alsentzer et al., 2019) are the initial models based on BERT. These models follow continual pretraining approach where the model weights are initialised with weights from BERT trained on Wikipedia and Book Corpus and uses the same vocabulary. Recent models SciBERT (Beltagy et al., 2019), PubMedBERT (Gu et al., 2020) and Biolm (Lewis et al., 2020) have shown that pretraining from scratch using domain specific corpora along with domain specific vocabulary improves the model performance significantly.

In this work, we adapt ELECTRA (Clark et al., 2020), a recent and powerful general domain model for the biomedical domain and we release BioELECTRA - a biomedical domain specific language encoder model. We follow the domain specific pretraining approach where the ELECTRA model is pretrained on PubMed and PubMed Central (PMC) full text articles. ELECTRA outperforms BERT, ALBERT (Lan et al., 2019), XLNet (Yang et al., 2020) and RoBERTa (Liu et al., 2019) on the GLUE (Wang et al., 2019) Benchmark and SQuAD (Rajpurkar et al., 2016a).

In particular, we make the following contributions.

1. We release BioELECTRA(P), BioELECTRA(P + F), BioELECTRA(P + F) LT(Longer Training of additional 1M steps) and BioELECTRA(W + P) pretrained from scratch using Biomedical domain text. Pretrained weights for all these models are publicly released through huggingface transformers(Wolf et al., 2020) model hub.
2. We evaluate our BioELECTRA models on all the 13 datasets in the BLURB (Gu et al., 2020) benchmark and on all the 4 clinical datasets from BLUE (Peng et al., 2019) benchmark across 7 NLP tasks.
3. BioELECTRA model achieves state-of-the-art (SOTA) results on all the 13 datasets in BLURB benchmark and achieves SOTA on all the Clinical datasets from BLUE Benchmark.
4. We publicly release the code[1] and parameters to reproduce our research results.

[1]The code and models are available at https://github.com/kamalkraj/BioELECTRA

Proceedings of the BioNLP 2021 workshop, pages 143–154
June 11, 2021.

2 Related work

Pretrained word embeddings (Mikolov et al., 2013), (Pennington et al., 2014) and contextualised word embeddings (Peters et al., 2018) have helped the deep learning algorithms to improve their performance in NLP tasks. ULMFiT (Howard and Ruder, 2018), introduces the transfer learning approach to Natural language processing and OpenAI GPT (Radford et al., 2018), pretrains a transformer (Vaswani et al., 2017) for learning general language representations. Similar to ULMFiT and OpenAI GPT, BERT (Devlin et al., 2018) follows this fine tuning approach and introduces a powerful bidirectional language representation model using the transformer based model architecture. BERT achieves SOTA on most NLP tasks without any heavily-engineered task specific architectures. Following the success of BERT, XLNet (Yang et al., 2020) with generalized autoregressive pretraining and RoBERTa (Liu et al., 2019) with robust pretraining techniques experiment with different pretraining objectives. ALBERT (Lan et al., 2019) uses weight sharing and embedding factorisation to reduce memory consumption and increase the training speed. ELECTRA (Clark et al., 2020) introduces sample-efficient 'replaced token detection' pretraining technique. ELECTRA-small, trained with very little compute outperforms GPT and performs comparably with larger models like RoBERTa and XLNet.

Recent works adapt BERT to scientific, biomedical and clinical domains. BioBERT (Lee et al., 2020) pretrains BERT with data from PubMed and PubMed Central (PMC) articles. BlueBERT (Peng et al., 2019) pretrains BERT on PubMed, PMC and MIMIC III (Johnson et al., 2016) data. ClinicalBERT (Alsentzer et al., 2019) initialises with BioBERT weights and pretrains on data from MIMIC III. SciBERT (Beltagy et al., 2019), PubMedBERT (Gu et al., 2020) and Bio-lm (Lewis et al., 2020) pretrain BERT based models from scratch with domain specific data. SciBERT pretrains on 1.14M papers from Semantic Scholar (Ammar et al., 2018), PubMedBERT on PubMed and PMC data and Bio-lm (Lewis et al., 2020) on data from PubMed, PMC and MIMIC III. Benchmarks in biomedical NLP - BLUE (Biomedical Language Understanding Evaluation) and BLURB (Biomedical Language Understanding & Reasoning Benchmark) are released by BlueBERT and PubMedBERT respectively.

3 Methods

3.1 Pretraining from scratch using domain specific corpora

The pioneers in applying transfer learning to NLP, pretrain Language Model(LM) on unlabelled large corpora in the general domain like Wikipedia articles, Web Text, Books corpus, Gigaword, web crawl etc. Biomedical literature has specific concepts and terms that are not part of the general domain. To enable the models to learn these features very specific to the biomedical domain, BioNLP models, BioBERT (Lee et al., 2020) and BlueBERT (Peng et al., 2019) use the mixed-domain pretraining approach (Gu et al., 2020). In mixed-domain approach, the model initialises with BERT weights and vocabulary trained on general domain text and the model is pretrained on the biomedical text.

Biomedical domain with its publicly available literature which is growing exponentially by the year makes it well suited for domain specific pretraining from scratch. Using a general domain vocabulary for biomedical text results in complex and specific terms being split into numerous subwords, as they do not exist in the general domain vocabulary. Hence a model trained on these word pieces might not generalise well for the domain specific downstream tasks. Recent work PubMedBERT (Gu et al., 2020) and Bio-lm (Lewis et al., 2020) pretrain a language model from scratch on PubMed abstracts and use the vocabulary that is generated from PubMed abstracts. These models outperform the BioBERT and BlueBERT models on biomedical and clinical NLP tasks .

3.2 Data

We use data very similar to PubMedBERT for fair comparison.

PubMed Abstracts We use text from 22 million PubMed abstracts downloaded as of January 2021. 27 GB of cleaned text with approximately 4.2 billion words are used.

PubMed Central (PMC) We obtained full text from 3.2 million PubMed Central (PMC) [2] articles as of January 2021. After cleaning the data, we use 57GB of text with approximately 9.6 billion words.

Preprocessing We used pubmed_parser parser[3] for

[2]https://www.ncbi.nlm.nih.gov/pmc/

[3]https://github.com/titipata/pubmed_parser

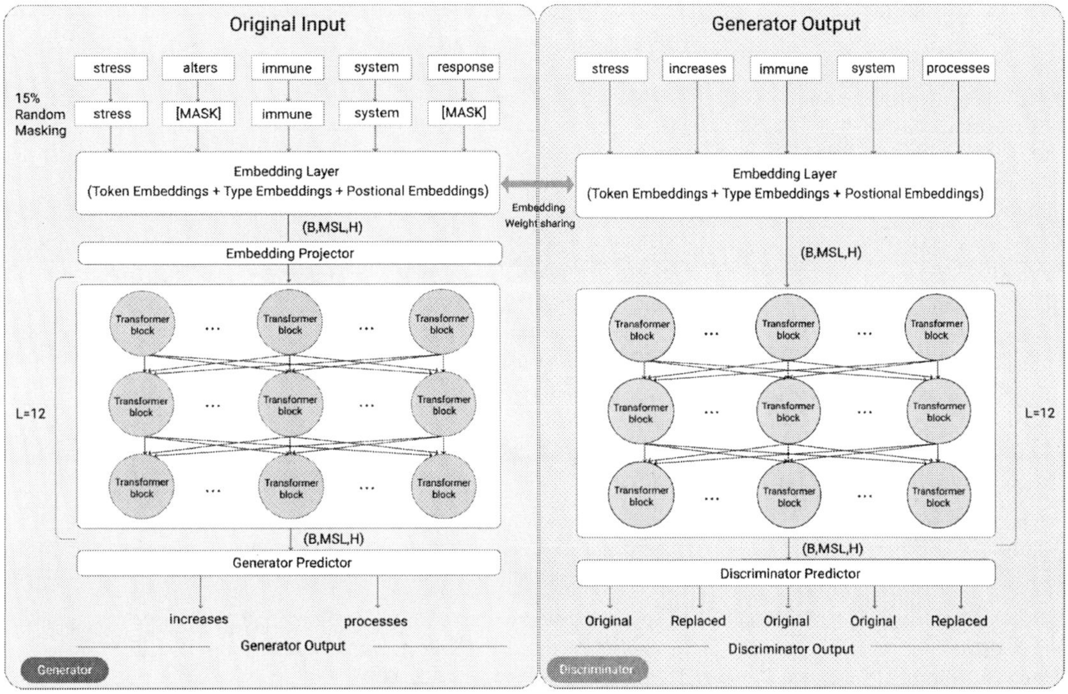

Figure 1: Overview of ELECTRA-Base model Pretraining. Output shapes are mentioned in parenthesis after each block.(B=Batch Size, MSL=Maximum Sequence Length, H=Hidden size)

extracting the abstracts and full text articles. We used SciSpacy(Neumann et al., 2019) for sentence tokenization.

3.3 ELECTRA

Architecture ELECTRA (Clark et al., 2020) pretraining architecture consists of a Generator and a Discriminator network. Each of them consists of Encoder blocks of the transformer (Vaswani et al., 2017) architecture. The generator size is chosen smaller than the Discriminator to make ELECTRA computationally efficient. The size of the Hidden dimension (H) of the transformer encoder in Generator is reduced to $1/3$ the size of the Discriminator. The Generator and Discriminator share the weights of the Embedding layer, which is composed of token embeddings, position embeddings and type embeddings. An embedding projector is added to Generator after the Embedding layer to project the embedding dimension H to H/3. Figure 1 shows pretraining configuration of ELECTRA-Base model. The Generator is trained with maximum likelihood as in ELECTRA paper and Generator is not given a noise input vector as in General Adversarial Networks (GANs). The Discriminator is trained very similar to a classifier with cross entropy loss. After pretraining only the Discriminator is used for all the finetuning.

Input/Output representations ELECTRA follows the Input/Output representations of BERT (Devlin et al., 2018). The first token is always the [CLS] token whose final hidden state is used for finetuning sentence level tasks. For single sentence tasks, the tokenized input sequence should follow the [CLS] token and end with [SEP]. For sentence pair tasks, the tokenized input sentences should be separated by a [SEP] token. Type and Position embeddings which indicate the sentence that it belongs to (sentence1/sentence2) are added to the input token embeddings. Final input representation of a given token is the summation of its token, position and type embeddings which are learnt during the training.

Pretraining Task ELECTRA introduces ***replaced token prediction*** pretraining task where the model is trained to distinguish real input tokens from synthetically generated tokens. Random words are selected in the input text and replaced with tokens generated by a small Generator network. The Discriminator network then predicts whether the input token is original or replaced. This novel approach ensures that the model learns from all the input tokens and not just from 15% of the

Dataset	Task	Train	Dev	Test	Evaluation Metrics
BC5-chem (Li et al., 2016)	NER	5203	5347	5385	F1 entity-level
BC5-disease (Li et al., 2016)	NER	4182	4244	4424	F1 entity-level
NCBI-disease (Doğan et al., 2014)	NER	5134	787	960	F1 entity-level
BC2GM (Smith et al., 2008)	NER	15197	3061	6325	F1 entity-level
JNLPBA (Collier and Kim, 2004)	NER	46750	4551	8662	F1 entity-level
ShARe/CLEFE* (Suominen et al., 2013)	NER	4628	1075	5195	F1 entity-level
EBM PICO(Nye et al., 2018)	PICO	339167	85321	16364	Macro F1 word-level
ChemProt (Krallinger et al., 2017)	Relation Extraction	18035	11268	15745	Micro F1
DDI (Herrero-Zazo et al., 2013)	Relation Extraction	25296	2496	5716	Micro F1
GAD (Bravo et al., 2015)	Relation Extraction	4261	535	534	Micro F1
i2b2-2010* (Uzuner et al., 2011)	Relation Extraction	3110	11	6293	Micro F1
BIOSSES (Soğancıoğlu et al., 2017)	Sentence Similarity	64	16	20	Pearson
ClinicalSTS** (Wang et al., 2020)	Sentence Similarity	1312	329	412	Pearson
HoC (Baker et al., 2015)	Document Classification	1295	186	371	Micro F1
MedNLI* (Romanov and Shivade, 2018)	Inference	11232	1395	1422	Accuracy
PubMedQA (Jin et al., 2019)	Question Answering	450	50	500	Accuracy
BioASQ (Nentidis et al., 2019)	Question Answering	670	75	140	Accuracy

Table 1: Datasets from BLURB and BLUE benchmark. Number of instances in train, dev, and test set along with the evaluation metrics used for each of the datasets is listed. * Clinical domain dataset from BLUE. ** Instead of MedSTS from BLUE we used ClinicalSTS released by (Wang et al., 2020)

tokens in the input text as in BERT. This makes the pretraining task computationally effective. As recent work (Liu et al., 2019) (Yang et al., 2020) suggests that using 'next sentence prediction' does not show consistent improvement in the scores, ELECTRA does not use any such 'contrastive learning' techniques for pretraining. Since ELECTRA does not have a contrastive learning technique, there is no pooling projection layer in ELECTRA.

4 Experiments

4.1 BioELECTRA pretraining

We pretrain ELECTRA from scratch with PubMed abstracts and PMC full text articles mentioned in Section 3.2. PubMedBERT (Gu et al., 2020) and BioBERT (Lee et al., 2020) pretrained BERT-Base models with biomedical domain specific corpus. In this paper, we experiment only with ELECTRA-Base architecture to ensure a fair comparison with these models. Four ELECTRA-Base models are trained - BioELECTRA (P) on PubMed abstracts, BioELECTRA (P+F) on PubMed abstracts and PMC full text articles, BioELECTRA (P+F) with longer training (2M steps) and BioELECTRA (W+P) on Wikipedia and PubMed abstracts. BioELECTRA(P) and BioELECTRA(P+F) models are trained with 1M steps with a batch size of 512. The number of training steps are chosen to make our work comparable with BioBERT[4] and PubMedBERT.[5] BioELECTRA(P+F) LT is trained like BioELECTRA(P+F) with an additional 1M steps. For BioELECTRA(W+F), a continual training approach is adopted where the model is initialised with ELECTRA-BASE general domain weights. It is pretrained further with PubMed abstracts for 100k, 200k and 400k steps. We publish our results of BioELECTRA(W+F) pretrained with 200k steps as these results were comparable with PubMedBERT BLURB (Gu et al., 2020) score.

SciBERT (Beltagy et al., 2019) shows that models trained on uncased vocabularies perform slightly better than the cased models in biomedical domain even for NER tasks. Hence we use the uncased biomedical domain-specific vocabularies from PubMedBERT for all our experiments. The optimization techniques and parameters from ELECTRA paper are followed. All our models are trained on Tensor Processing Unit(TPU) v3-8 instances. Refer Appendix A for complete model and optimizer details.

4.2 Datasets

We finetune our ELECTRA-Base models on 17 NLP datasets - 13 biomedical datasets from the

[4]BioBERT was trained with a batch size of 256 with 1M steps in pretraining and 1M steps in continual pretraining.

[5]PubMedBERT was trained with a batch size of 8,192 for 62,500 steps.

BLURB (Gu et al., 2020) benchmark and 4 clinical datasets from the BLUE (Peng et al., 2019) benchmark. We group our datasets based on the NLP tasks. We do not discuss the datasets in detail due to space constraints. Details on train, dev, test split, benchmark they belong to, evaluation metric used can be found in Table 1. Detailed description of the datasets are available in the BLURB(Gu et al., 2020) and BLUE(Peng et al., 2019) paper.

4.2.1 Named Entity Recognition (NER)

NER task aims at recognizing and predicting the entities e.g (chemicals, diseases, genes, proteins) in the given text. We use *BC5-Chemical, BC5-Disease, NCBI-Disease, BC2GM, JNLPBA* biomedical datasets from the BLURB benchmark. These datasets have the same train, dev and test split as released by (Crichton et al., 2017). In addition to these, *ShARe/CLEFE* clinical dataset used by BLUE benchmark which uses the train, dev and test split released by (Suominen et al., 2013) is used for NER task.

4.2.2 PICO extraction (PICO)

PICO task is very similar to NER, where the model aims to predict the Participants, Interventions, Comparisons and Outcomes entities in the given text. *EBM PICO* (Nye et al., 2020) dataset from the BLURB benchmark which has the same train, test and dev split as the original dataset is used for this task.

4.2.3 Relation Extraction (RE)

Relation Extraction task predicts relations and their types between the two entities mentioned in the given sentences (e.g, gene–disease relations, protein–chemical relations). We use *DDI, ChemProt and GAD* datasets from the BLURB benchmark and *i2b2-2010* clinical dataset in the BLUE benchmark. GAD dataset in BLURB benchmark uses train, dev and test split created by (Lee et al., 2020). For DDI, BLURB uses the original dataset by (Herrero-Zazo et al., 2013) and release their own train, dev and test datasets. BLURB uses the train, dev and test split from the original dataset (Krallinger et al., 2017) for ChemProt. BLUE uses the train, dev and test split released by (Uzuner et al., 2011)

4.2.4 Sentence Similarity

Sentence Similarity task predicts the similarity score based on how similar are the given pair of sentences. *BIOSSES* dataset from BLURB benchmark and *ClinicalSTS* dataset instead of the MedSTS dataset is chosen from BLUE benchmark. BLURB uses the train, dev and split created by (Peng et al., 2019). *ClinicalSTS* dataset is chosen as that is the latest version provided by n2c2 2019 challenge(Wang et al., 2020). It has added 574 more samples for training and a new test set of 412 samples. As this dataset doesn't have a public train and dev split, we have split it into 80% train and 20% dev set and we use the original test set for evaluation.

4.2.5 Document classification

Document classification task aims to predict the multiple labels for the given text. Evaluation for Document classification task is done at the document level where we aggregate the labels over all the sentences in a document. We use *HoC* dataset from BLURB benchmark which uses the original dataset by (Baker et al., 2015) to create their own train, dev and test split.

4.2.6 Natural Language Inference (NLI)

Natural Language Inference task predicts whether the relation between two sentences are entailment, contradiction or neutrality. *MedNLI* (Romanov and Shivade, 2018) dataset from the BLUE benchmark which uses the original train, dev and test split is used for this task.

4.2.7 Question Answering (QA)

Question Answering task aims to predict the answers in the context when a question text is given as the first sentence. The answers are either two-way (yes/ no) or three-way (yes/ maybe/ no). *PubMedQA and BioASQ* datasets from BLURB benchmark are used for our experiments. For both PubMedQA (Jin et al., 2019) and BioASQ (Nentidis et al., 2019), BLURB uses the original train, dev and test split.

4.3 Fine tuning

ELECTRA (Clark et al., 2020) applies very minimal architectural changes for finetuning downstream tasks. We follow the same approach as ELECTRA for finetuning BioELECTRA on the various downstream tasks. BIO encoding scheme is adopted for the NER tasks where B stands for Beginning, I stands for Inside and O stands for Outside. All the NER datasets in BLURB benchmark and *ShARe/CLEFE* in BLUE benchmark have

	BioBERT cased (P)	SciBERT uncased (CS+F)	ClinicalBERT cased (W+P+M)	BlueBERT cased (W+P+M)	PubMedBERT uncased (P)	BioELECTRA uncased (P)
BC5-chem	92.85	92.49	90.80	91.19	93.33	**93.60**
BC5-disease.	84.70	84.54	83.04	83.69	85.62	**85.84**
NCBI-disease	89.13	88.10	86.32	88.04	87.82	**89.38**
BC2GM	83.82	83.36	81.71	81.87	84.52	**84.69**
JNLPBA	79.35	79.45	78.59	78.68	80.06	**80.17**
EBM PICO	73.18	73.12	72.06	72.54	73.38	**74.26**
ChemProt	76.14	75.24	72.04	71.46	77.24	**78.20**
DDI	80.88	81.06	78.20	77.78	82.36	**82.76**
GAD	80.94	80.90	78.40	77.24	82.34	**83.70**
BIOSSES	89.52	86.25	91.23	85.38	92.30	**92.49**
HoC	81.54	80.66	80.74	80.48	82.32	**83.50**
PubMedQA	60.24	57.38	49.08	48.44	55.84	**64.02**
BioASQ	84.14	78.86	68.50	68.71	87.56	**88.57**
BLURB score	80.29	78.80	77.19	76.19	81.10	**82.47**

Table 2: Comparison of pretrained BioNLP models on the BLURB (Gu et al., 2020) benchmark. The BLURB score is the macro average of mean test results for each of the six tasks (NER, PICO, Relation Extraction, Sentence Similarity, Document Classification, Question Answering). Refer Table 1 for the evaluation metric used for each task. (P - PubMed abstracts, CS - Computer Science, F - PubMed Central full text articles, W - Wikipedia, M - MIMIC III (Johnson et al., 2016))

a single entity. (e.g. Disease in BC5-disease). PICO, a sequential tagging task is solved using the NER task approach and Document classification task for *HoC* dataset is solved as multi label classification task. The datasets in NER, PICO and Document classification tasks follow the single sentence representation. As mentioned in section 3.3, each tokenized input sequence follows the [CLS] token and ends with the [SEP] token. Sentence Similarity, Question Answering and Natural Language Inference tasks all have sentence pairs in their inputs. We process the sentence pairs as [CLS]sentence1[SEP]sentence2[SEP] very similar to BERT. In the Question Answering task, 'question' is treated as sentence1 and 'context' is treated as sentence2.

ELECTRA (Clark et al., 2020) uses the vector representation of the [CLS] token to generate the output for all the given NLP tasks except NER and PICO. For NER and PICO, representations for each token is used to classify the entities. A simple linear layer is added to the output of ELECTRA for finetuning. ELECTRA does not use LSTM (Hochreiter and Schmidhuber, 1997), CRF (Lafferty et al., 2001) layers for NER tasks. Figure 2 in appendix B illustrates the finetuning architecture for the NLP tasks. Mean-square error is used for regression tasks and cross entropy loss is used for classification tasks. Similar to BERT finetuning, all the layers are fine-tuned together along with task specific prediction layer. We use 'discriminative finetuning' similar to ELECTRA, where only the final layer is trained with the original learning rate and all other layers use a learning rate with a decay factor. For finetuning, Adam (Kingma and Ba, 2017) optimizer with a slanted triangular learning rate scheduler which linearly warms up (10% of steps) followed by linear decay (90% of steps) is used. We also use a dropout probability of 10%. We experiment with the following hyper parameters: learning rate [3e-5, 5e-5, 1e-4, 1.5e-4, 2e-4], batch size [16, 32], layer-wise learning-rate decay out of [0.9, 0.8, 0.7] and epochs [3,5]. BIOSSES (Soğancıoğlu et al., 2017), PubMedQA (Jin et al., 2019), BioASQ (Nentidis et al., 2019) and ClinicalSTS (Wang et al., 2020) are finetuned for longer epochs. For more details on the hyper parameters, refer Appendix B. We ran 10 fine tuning runs on BIOSSES, BioASQ and PubMedQA since the datasets are relatively smaller and 5 runs on all the other datasets. The average score is reported as the final score for the evaluation metric.

	BioBERT cased (P)	ClinicalBERT cased (W+P+M)	BlueBERT cased (P)	BlueBERT cased (P+M)	PubMedBERT uncased (P)	PubMedBERT uncased (P+F)	BioELECTRA uncased (P)	BioELECTRA uncased (P+F)
MedNLI	82.63	82.70	82.2	84	83.82	84.17	86.27	**86.34**
i2b2-2010	72.81	74.82	74.4	76.4	75.14	73.93	**76.50**	75.73
ShARe/CLEFE	80.73	82.15	75.4	77.1	74.45	74.77	**83.71**	83.15
ClinicalSTS	85.91	85.63	86.03	84.57	86.72	86.16	**89.07**	88.34

Table 3: Comparison of pretrained language models on the BLUE (Peng et al., 2019) benchmark. (P - PubMed abstracts, F - PubMed Central full text articles, W - Wikipedia, M - MIMIC III (Johnson et al., 2016))

5 Results

We finetune all of the four BioELECTRA models mentioned in 4.1 for seven biomedical text mining tasks (NER, PICO, Relation Extraction, Sentence Similarity, Document Classification, Question Answering and Natural Language Inference) that are part of the BLURB (Gu et al., 2020) and BLUE (Peng et al., 2019) benchmark.

BLURB benchmark Out of the four BioELECTRA models, BioELECTRA (P) model pretrained from scratch on PubMed abstracts alone along with biomedical domain specific vocabulary (from PubMedBERT (Gu et al., 2020)) achieves new State-of-the-Art (SOTA) results on all of the datasets in BLURB benchmark. Our results on BioELECTRA (P) along with the scores for BioBERT (Lee et al., 2020), SciBERT (Beltagy et al., 2019), ClinicalBERT (Alsentzer et al., 2019) , BlueBERT (Peng et al., 2019) and PubMedBERT (Gu et al., 2020) for all the tasks in the BLURB benchmark are shown in table 2. The scores on these datasets for all these models are taken from the BLURB benchmark. As we do not have details on train, test and dev split of datasets used by Bio-lm (Lewis et al., 2020) paper, we are not able to compare our results with their results. For NCBI-Disease, where the train, test and dev split is publicly available, our model (89.38%) performs better than the Bio-lm Base (PM + Voc) model (88.2%). ELECTRA performs significantly better than all other BERT based models on the SQuAD (Rajpurkar et al., 2016b) benchmark in the general domain. Similarly, BioELECTRA (P) model has significantly higher scores on the Question Answering tasks. It achieves new SOTA of 64.02% (3.78% increase over the previous SOTA) on PubMedQA and with a new SOTA of 88.57% (1.01 % increase over the previous SOTA) on BioASQ. Our overall BLURB score (macro average of the average metric for each of the six tasks) is 82.40% which is 1.3% higher than PubMedBERT BLURB score of 81.10%.

BLUE benchmark We present results of BioELECTRA (P) pretrained on PubMed abstracts alone and BioELECTRA (P+F) pretrained on both PubMed abstracts and PubMed full text articles on four of the clinical datasets in the BLUE benchmark in table3. We compare the performance of our models with the results of BioBERT, ClinicalBERT, BlueBERT and PubMedBERT. Since the scores on the train, dev and test split of these clinical datasets by BioBERT, ClinicalBERT, BlueBERT and PubMedBERT are not available, we used their pretrained weights on these datasets and documented the results. We do not have the results of SciBERT model as it was trained on mixed domain data. Out of the four datasets in the BLUE benchmark, we have results of Biolm for i2b2-2010 and MedNLI. Since we do not have the train, dev and test split used by Biolm for i2b2-2010, we compare our results only for the MedNLI dataset. Score of our BioELECTRA (P+F) model 86.34% is significantly higher than Biolm Base model (PM + Voc) score of 83.2%. We also note that BioELECTRA performs better than BERT based models trained on MIMIC data. BioELECTRA (P) achieves new SOTA on three of the datasets - i2b2-2010, ShARe/CLEFE and ClinicalSTS. BioELECTRA (P+F)'s score of 86.34% on MedNLI task is marginally (0.07%) higher than the score of BioELECTRA (P)'s score of 86.27% and this is the new SOTA for MedNLI dataset for models trained on PubMed abstracts and PubMed Central full text articles.

Our models pretrained on domain specific text along with domain specific vocabulary have consistently shown that the pretraining from scratch with domain specific data enables the model to capture the contextual representations of the language better.

Vocab	BioELECTRA P PubMed	BioELECTRA P+F PubMed	BioELECTRA P+F (LT) PubMed	BioELECTRA W+P General
BC5-chem	93.60	93.51	**93.75**	93.03
BC5-disease	**85.84**	85.55	85.32	84.66
NCBI-disease	**89.38**	88.43	88.73	88.45
BC2GM	**84.69**	84.61	84.68	83.90
JNLPBA	**80.17**	79.98	80.10	79.63
EBM PICO	**74.26**	73.88	73.86	73.33
ChemProt	**78.20**	77.76	76.76	77.06
DDI	82.76	**83.53**	82.34	79.68
GAD	83.70	84.18	**85.67**	83.16
BIOSSES	92.49	**93.80**	91.45	88.65
HoC	**83.50**	82.79	83.20	82.30
PubMedQA	**64.02**	63.80	62.21	61.20
BioASQ	88.57	91.42	**91.50**	90.01
BLURB Score	82.47	**82.72**	82.24	80.96
MedNLI	86.27	**86.34**	85.36	83.53
i2b2-2010	**76.50**	75.73	76.17	75.48
ShARe/CLEFE	**83.71**	83.15	83.52	83.02
ClinicalSTS	**89.07**	88.34	89.02	88.46

Table 4: Comparison of BioELECTRA models on BLURB (Gu et al., 2020) and BLUE (Peng et al., 2019) benchmark. (P - PubMed abstracts, F - PubMed Central full text articles, W - Wikipedia, LT - Longer Training)

Comparison of BioELECTRA models Table 4 shows the comparison of results of our models BioELECTRA(P), BioELECTRA (P+F) and BioELECTRA (P+F) LT with longer training of additional 1 million steps and BioELECTRA (W+P). BioELECTRA (W+P) is pretrained from scratch on Wikipedia and PubMed abstracts along with a general domain vocabulary (BERT (Devlin et al., 2018) uncased vocabulary). We observe that BioELECTRA (P+F) LT with longer training of 2 million steps does not give substantial improvements on all of the tasks. BioELECTRA (P+F) LT model's result is slightly better than BioELECTRA (P) on BC5-chem dataset. BioELECTRA (P+F) LT model's result on GAD and BioASQ datasets are marginally better than BioELECTRA (P+F). BioELECTRA (P+F) performs slightly better than BioELECTRA (P) on DDI and BIOSSES datasets.

The results clearly show that all BioELECTRA models pretrained from scratch with biomedical domain text and domain specific vocabulary perform better than the model pretrained on both general and biomedical domain text with general domain vocabulary. However it is interesting to note that BioELECTRA (W+P) model has significantly better results for i2b2-2010, ShARe/CLEFE and ClinicalSTS datasets than PubMedBERT. BioELECTRA (W+P)'s score for MedNLI is comparable to that of PubMedBERT (Gu et al., 2020).

6 Conclusion and Future Work

We release BioELECTRA-base models pretrained from scratch on biomedical domain specific text and evaluate the performance on seven different biomedical NLP tasks with 17 datasets. We achieve SOTA on all the datasets in the BLURB (Gu et al., 2020) benchmark and all four clinical datasets in the BLUE (Peng et al., 2019) benchmark. Our results show that pretraining from scratch with biomedical domain text helps the model to learn better contextual representations. We release the pretrained weights for all our models and the code for reproducibility.

We plan to explore and experiment with our domain specific pretraining approach on ELECTRA-LARGE models. We also intend to train ELECTRA-BASE and ELECTRA-LARGE models on MIMIC III (Johnson et al., 2016) clinical notes and evaluate the performance of the models on biomedical NLP tasks. As ELECTRA shows a significant improvement on SQuAD (Rajpurkar et al., 2016b), we want to focus on Biomedical QA tasks (span prediction) and evaluate domain specific pretrained ELECTRA models performance.

Acknowledgements

This research was supported by Google's TensorFlow Research Cloud (TFRC). We also extend our thanks to Samuel Gurudas for his assistance in creating the diagrams in this research paper.

References

Emily Alsentzer, John Murphy, William Boag, Wei-Hung Weng, Di Jindi, Tristan Naumann, and Matthew McDermott. 2019. Publicly available clinical BERT embeddings. In *Proceedings of the 2nd Clinical Natural Language Processing Workshop*, pages 72–78, Minneapolis, Minnesota, USA. Association for Computational Linguistics.

Waleed Ammar, Dirk Groeneveld, Chandra Bhagavatula, Iz Beltagy, Miles Crawford, Doug Downey, Jason Dunkelberger, Ahmed Elgohary, Sergey Feldman, Vu Ha, Rodney Kinney, Sebastian Kohlmeier, Kyle Lo, Tyler Murray, Hsu-Han Ooi, Matthew Peters, Joanna Power, Sam Skjonsberg, Lucy Wang, Chris Wilhelm, Zheng Yuan, Madeleine van Zuylen, and Oren Etzioni. 2018. Construction of the literature graph in semantic scholar. In *Proceedings of the 2018 Conference of the North American Chapter of the Association for Computational Linguistics: Human Language Technologies, Volume 3 (Industry Papers)*, pages 84–91, New Orleans - Louisiana. Association for Computational Linguistics.

Simon Baker, Ilona Silins, Yufan Guo, Imran Ali, Johan Högberg, Ulla Stenius, and Anna Korhonen. 2015. Automatic semantic classification of scientific literature according to the hallmarks of cancer. *Bioinformatics*, 32(3):432–440.

Iz Beltagy, Kyle Lo, and Arman Cohan. 2019. SciBERT: A pretrained language model for scientific text. In *Proceedings of the 2019 Conference on Empirical Methods in Natural Language Processing and the 9th International Joint Conference on Natural Language Processing (EMNLP-IJCNLP)*, pages 3615–3620, Hong Kong, China. Association for Computational Linguistics.

Àlex Bravo, Janet Piñero, Núria Queralt-Rosinach, Michael Rautschka, and Laura I Furlong. 2015. Extraction of relations between genes and diseases from text and large-scale data analysis: implications for translational research. *BMC bioinformatics*, 16(1):55.

Kevin Clark, Minh-Thang Luong, Quoc V Le, and Christopher D Manning. 2020. Electra: Pre-training text encoders as discriminators rather than generators. *arXiv preprint arXiv:2003.10555*.

Nigel Collier and Jin-Dong Kim. 2004. Introduction to the bio-entity recognition task at JNLPBA. In *Proceedings of the International Joint Workshop on Natural Language Processing in Biomedicine and its Applications (NLPBA/BioNLP)*, pages 73–78, Geneva, Switzerland. COLING.

Gamal Crichton, Sampo Pyysalo, Billy Chiu, and Anna Korhonen. 2017. A neural network multi-task learning approach to biomedical named entity recognition. *BMC Bioinformatics*, 18.

Jacob Devlin, Ming-Wei Chang, Kenton Lee, and Kristina Toutanova. 2018. Bert: Pre-training of deep bidirectional transformers for language understanding. *arXiv preprint arXiv:1810.04805*.

Rezarta Islamaj Doğan, Robert Leaman, and Zhiyong Lu. 2014. Ncbi disease corpus: a resource for disease name recognition and concept normalization. *Journal of biomedical informatics*, 47:1–10.

Yu Gu, Robert Tinn, Hao Cheng, Michael Lucas, Naoto Usuyama, Xiaodong Liu, Tristan Naumann, Jianfeng Gao, and Hoifung Poon. 2020. Domain-specific language model pretraining for biomedical natural language processing.

María Herrero-Zazo, Isabel Segura-Bedmar, Paloma Martínez, and Thierry Declerck. 2013. The ddi corpus: An annotated corpus with pharmacological substances and drug–drug interactions. *Journal of biomedical informatics*, 46(5):914–920.

Sepp Hochreiter and Jürgen Schmidhuber. 1997. Long short-term memory. *Neural Computation*, 9(8):1735–1780.

Jeremy Howard and Sebastian Ruder. 2018. Universal language model fine-tuning for text classification. In *Proceedings of the 56th Annual Meeting of the Association for Computational Linguistics (Volume 1: Long Papers)*, pages 328–339, Melbourne, Australia. Association for Computational Linguistics.

Qiao Jin, Bhuwan Dhingra, Zhengping Liu, William Cohen, and Xinghua Lu. 2019. PubMedQA: A dataset for biomedical research question answering. In *Proceedings of the 2019 Conference on Empirical Methods in Natural Language Processing and the 9th International Joint Conference on Natural Language Processing (EMNLP-IJCNLP)*, pages 2567–2577, Hong Kong, China. Association for Computational Linguistics.

Alistair EW Johnson, Tom J Pollard, Lu Shen, Li-wei H Lehman, Mengling Feng, Mohammad Ghassemi, Benjamin Moody, Peter Szolovits, Leo Anthony Celi, and Roger G Mark. 2016. Mimic-iii, a freely accessible critical care database. *Scientific data*, 3:160035.

Diederik P. Kingma and Jimmy Ba. 2017. Adam: A method for stochastic optimization.

Martin Krallinger, Obdulia Rabal, Saber A Akhondi, Martın Pérez Pérez, Jesús Santamaría, GP Rodríguez, et al. 2017. Overview of the biocreative vi chemical-protein interaction track. In *Proceedings of the sixth BioCreative challenge evaluation workshop*, volume 1, pages 141–146.

John D. Lafferty, Andrew McCallum, and Fernando C. N. Pereira. 2001. Conditional random fields: Probabilistic models for segmenting and labeling sequence data. In *Proceedings of the Eighteenth International Conference on Machine Learning*, ICML '01, pages 282–289, San Francisco, CA, USA. Morgan Kaufmann Publishers Inc.

Zhenzhong Lan, Mingda Chen, Sebastian Goodman, Kevin Gimpel, Piyush Sharma, and Radu Soricut. 2019. Albert: A lite bert for self-supervised learning of language representations. *arXiv preprint arXiv:1909.11942*.

Jinhyuk Lee, Wonjin Yoon, Sungdong Kim, Donghyeon Kim, Sunkyu Kim, Chan Ho So, and Jaewoo Kang. 2020. Biobert: a pre-trained biomedical language representation model for biomedical text mining. *Bioinformatics*, 36(4):1234–1240.

Patrick Lewis, Myle Ott, Jingfei Du, and Veselin Stoyanov. 2020. Pretrained language models for biomedical and clinical tasks: Understanding and extending the state-of-the-art. In *Proceedings of the 3rd Clinical Natural Language Processing Workshop*, pages 146–157, Online. Association for Computational Linguistics.

Jiao Li, Yueping Sun, Robin J Johnson, Daniela Sciaky, Chih-Hsuan Wei, Robert Leaman, Allan Peter Davis, Carolyn J Mattingly, Thomas C Wiegers, and Zhiyong Lu. 2016. Biocreative v cdr task corpus: a resource for chemical disease relation extraction. *Database*, 2016.

Yinhan Liu, Myle Ott, Naman Goyal, Jingfei Du, Mandar Joshi, Danqi Chen, Omer Levy, Mike Lewis, Luke Zettlemoyer, and Veselin Stoyanov. 2019. Roberta: A robustly optimized bert pretraining approach. *arXiv preprint arXiv:1907.11692*.

Tomas Mikolov, Ilya Sutskever, Kai Chen, Greg Corrado, and Jeffrey Dean. 2013. Distributed representations of words and phrases and their compositionality.

Anastasios Nentidis, Konstantinos Bougiatiotis, Anastasia Krithara, and Georgios Paliouras. 2019. Results of the seventh edition of the bioasq challenge. In *Joint European Conference on Machine Learning and Knowledge Discovery in Databases*, pages 553–568. Springer.

Mark Neumann, Daniel King, Iz Beltagy, and Waleed Ammar. 2019. ScispaCy: Fast and robust models for biomedical natural language processing. In *Proceedings of the 18th BioNLP Workshop and Shared Task*, pages 319–327, Florence, Italy. Association for Computational Linguistics.

Benjamin Nye, Junyi Jessy Li, Roma Patel, Yinfei Yang, Iain J Marshall, Ani Nenkova, and Byron C Wallace. 2018. A corpus with multi-level annotations of patients, interventions and outcomes to support language processing for medical literature. In *Proceedings of the conference. Association for Computational Linguistics. Meeting*, volume 2018, page 197. NIH Public Access.

Benjamin Nye, Ani Nenkova, Iain Marshall, and Byron C. Wallace. 2020. Trialstreamer: Mapping and browsing medical evidence in real-time. In *Proceedings of the 58th Annual Meeting of the Association for Computational Linguistics: System Demonstrations*, pages 63–69, Online. Association for Computational Linguistics.

Yifan Peng, Shankai Yan, and Zhiyong Lu. 2019. Transfer learning in biomedical natural language processing: An evaluation of BERT and ELMo on ten benchmarking datasets. In *Proceedings of the 18th BioNLP Workshop and Shared Task*, pages 58–65, Florence, Italy. Association for Computational Linguistics.

Jeffrey Pennington, Richard Socher, and Christopher Manning. 2014. GloVe: Global vectors for word representation. In *Proceedings of the 2014 Conference on Empirical Methods in Natural Language Processing (EMNLP)*, pages 1532–1543, Doha, Qatar. Association for Computational Linguistics.

Matthew E Peters, Mark Neumann, Mohit Iyyer, Matt Gardner, Christopher Clark, Kenton Lee, and Luke Zettlemoyer. 2018. Deep contextualized word representations. *arXiv preprint arXiv:1802.05365*.

Alec Radford, Karthik Narasimhan, Tim Salimans, and Ilya Sutskever. 2018. Improving language understanding by generative pre-training.

Pranav Rajpurkar, Jian Zhang, Konstantin Lopyrev, and Percy Liang. 2016a. SQuAD: 100,000+ questions for machine comprehension of text. In *Proceedings of the 2016 Conference on Empirical Methods in Natural Language Processing*, pages 2383–2392, Austin, Texas. Association for Computational Linguistics.

Pranav Rajpurkar, Jian Zhang, Konstantin Lopyrev, and Percy Liang. 2016b. Squad: 100,000+ questions for machine comprehension of text.

Alexey Romanov and Chaitanya Shivade. 2018. Lessons from natural language inference in the clinical domain.

Larry Smith, Lorraine K Tanabe, Rie Johnson nee Ando, Cheng-Ju Kuo, I-Fang Chung, Chun-Nan Hsu, Yu-Shi Lin, Roman Klinger, Christoph M Friedrich, et al. 2008. Overview of biocreative ii gene mention recognition. *Genome biology*, 9(S2):S2.

Gizem Soğancıoğlu, Hakime Öztürk, and Arzucan Özgür. 2017. Biosses: a semantic sentence similarity estimation system for the biomedical domain. *Bioinformatics*, 33(14):i49–i58.

Hanna Suominen, Sanna Salanterä, Sumithra Velupillai, Wendy W. Chapman, Guergana Savova, Noemie Elhadad, Sameer Pradhan, Brett R. South, Danielle L. Mowery, Gareth J. F. Jones, Johannes Leveling, Liadh Kelly, Lorraine Goeuriot, David Martinez, and Guido Zuccon. 2013. Overview of the share/clef ehealth evaluation lab 2013. In *Information Access Evaluation. Multilinguality, Multimodality, and Visualization*, pages 212–231, Berlin, Heidelberg. Springer Berlin Heidelberg.

Özlem Uzuner, Brett R South, Shuying Shen, and Scott L DuVall. 2011. 2010 i2b2/VA challenge on concepts, assertions, and relations in clinical text. *Journal of the American Medical Informatics Association*, 18(5):552–556.

Ashish Vaswani, Noam Shazeer, Niki Parmar, Jakob Uszkoreit, Llion Jones, Aidan N Gomez, Łukasz Kaiser, and Illia Polosukhin. 2017. Attention is all you need. In *Advances in neural information processing systems*, pages 5998–6008.

Alex Wang, Yada Pruksachatkun, Nikita Nangia, Amanpreet Singh, Julian Michael, Felix Hill, Omer Levy, and Samuel Bowman. 2019. Superglue: A stickier benchmark for general-purpose language understanding systems. In *Advances in Neural Information Processing Systems*, pages 3266–3280.

Y. Wang, S. Fu, F. Shen, S. Henry, O. Uzuner, and H. Liu. 2020. The 2019 n2c2/OHNLP Track on Clinical Semantic Textual Similarity: Overview. *JMIR Med Inform*, 8(11):e23375.

Thomas Wolf, Lysandre Debut, Victor Sanh, Julien Chaumond, Clement Delangue, Anthony Moi, Pierric Cistac, Tim Rault, Remi Louf, Morgan Funtowicz, Joe Davison, Sam Shleifer, Patrick von Platen, Clara Ma, Yacine Jernite, Julien Plu, Canwen Xu, Teven Le Scao, Sylvain Gugger, Mariama Drame, Quentin Lhoest, and Alexander Rush. 2020. Transformers: State-of-the-art natural language processing. In *Proceedings of the 2020 Conference on Empirical Methods in Natural Language Processing: System Demonstrations*, pages 38–45, Online. Association for Computational Linguistics.

Zhilin Yang, Zihang Dai, Yiming Yang, Jaime Carbonell, Ruslan Salakhutdinov, and Quoc V. Le. 2020. Xlnet: Generalized autoregressive pretraining for language understanding.

A Pretraining

Hyperparameter	Discriminator/Generator
Number of layers	12
Hidden Size	768/256
FFN inner hidden size	3072/1024
Attention heads	12/4
Attention head size	64
Embedding Size	768
Mask percent	15
Learning Rate Decay	Linear
Warmup steps	10000
Learning Rate	2e-4
Adam ϵ	1e-6
Adam β_1	0.9
Adam β_2	0.999
Attention Dropout	0.1
Dropout	0.1
Weight Decay	0.01
Batch Size	512
Train Steps	1M

Table 5: Pre-train hyperparameters.

All the BioELECTRA models are trained on TPU v3-8 instances. Adopting *bfloat16*[6] training helped us in improving the training speed. Very similar to BERT, we train the model in 2 phases, 90% of steps with sequence length of 128 (phase1) and 10% of steps with sequence length of 512 (phase2) to learn the positional embeddings. Model training reached 1M steps in 5 days (phase1 - 4 days and phase2 - 1day). For pretraining, we use the original ELECTRA code[7] released by authors. Refer table 5 for details regarding all the parameters.

B Finetuning

Figure 2 shows different architecture schema of different models.

- Single Sentence Classification : ChemProt, DDI, GAD, i2b2-2010, HoC
- Entity Classification: BC5-chem, BC5-disease, NCBI-Disease, BC2GM, JNLPBA, ShARe/CLEFE, EBM PICO

[6]https://cloud.google.com/blog/products/ai-machine-learning/bfloat16-the-secret-to-high-performance-on-cloud-tpus

[7]https://github.com/google-research/electra

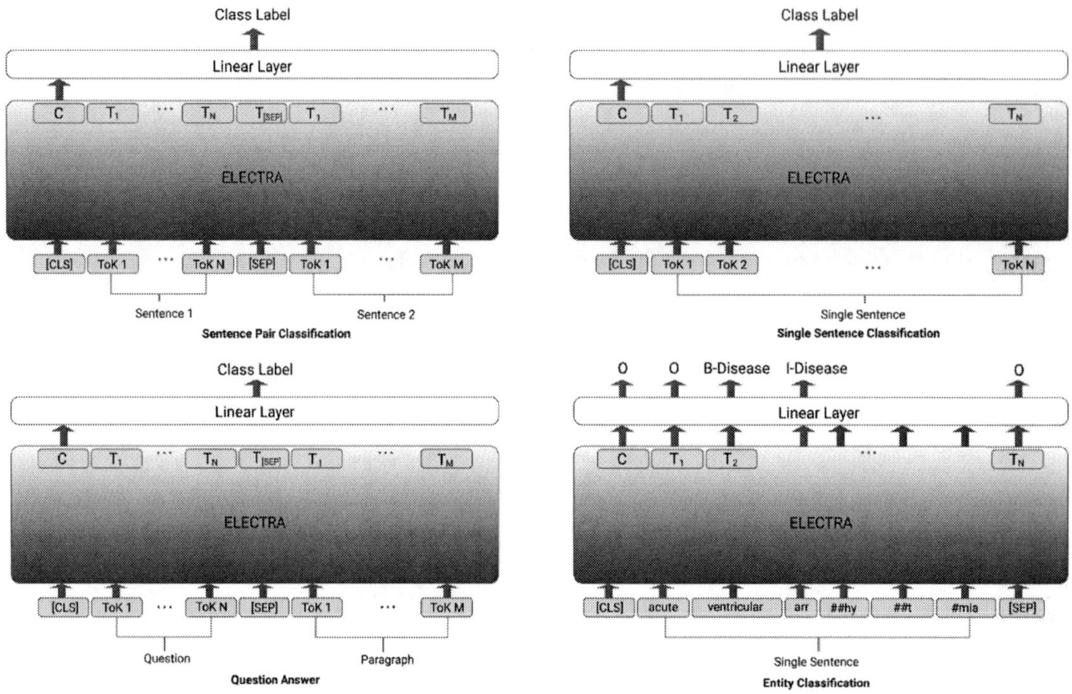

Figure 2: Overview of BioELECTRA model finetuning.

Hyperparameter	Value
Adam ϵ	1e-6
Adam β_1	0.9
Adam β_2	0.999
Layerwise LR decay	0.8
Learning rate decay	Linear
Warmup fraction	0.1
Attention Dropout	0.1
Dropout	0.1
Weight Decay	0

Table 6: Common hyperparamters across tasks

- Sentence Pair Classification: BIOSSES, ClinicalSTS
- Question Answering: PubMedQA, BioASQ

'Discriminative finetuning' is adopted where the learning rate varies across the layers. The learning rate decays across the layers from top to bottom with a factor of 0.8 for all the NLP tasks. The colour gradient in figure 2 represents this . For a learning rate of 1e-4 , only the task specific prediction layer (final layer) is finetuned at this rate. With a decay factor of 0.8, the embedding layer

Dataset	LR	BS	MSL	EPOCHS
BC5-chem	2e-4	16	256	5
BC5-disease	2e-4	16	256	5
NCBI-disease	2e-4	32	128	5
BC2GM	2e-4	32	256	5
JNLPBA	2e-4	16	256	3
ShARe/CLEFE	2e-4	32	512	5
EBM PICO	2e-4	32	256	3
ChemProt	1e-4	32	256	5
DDI	2e-4	32	256	3
GAD	2e-4	32	128	5
i2b2-2010	2e-4	32	128	5
BIOSSES	1.5e-4	16	128	60
ClinicalSTS	5e-5	32	128	10
HoC	2e-4	32	128	5
MedNLI	1e-4	32	128	5
PubMedQA	2e-4	32	512	20
BioASQ	2e-4	32	512	20

Table 7: LR : Learning Rate, BS : Batch Size, MSL : Maximum Sequence Length

for that particular task is finetuned at a learning rate of 5.5e-6. Table 6 shows the common hyperparameters used across tasks, and table 7 shows task specific hyperparameters.

Word centrality constrained representation for keyphrase extraction

Zelalem Gero and **Joyce C Ho**
Emory University
{zgero,joyce.c.ho}@emory.edu

Abstract

To keep pace with the increased generation and digitization of documents, automated methods that can improve search, discovery and mining of the vast body of literature are essential. Keyphrases provide a concise representation by identifying salient concepts in a document. Various supervised approaches model keyphrase extraction using local context to predict the label for each token and perform much better than the unsupervised counterparts. Unfortunately, this method fails for short documents where the context is unclear. Moreover, keyphrases, which are usually the gist of a document, need to be the central theme. We propose a new extraction model that introduces a centrality constraint to enrich the word representation of a Bidirectional long short-term memory. Performance evaluation on two publicly available datasets demonstrate our model outperforms existing state-of-the art approaches. Our model is publicly available at `https://github.com/ZHgero/keyphrases_centrality.git`

1 Introduction

Keyphrase extraction is an important information extraction task that identifies single or multi-word linguistic units that concisely represent a document. They can also serve to provide a brief summary of the document content. Keyphrases are widely used in variety of natural language processing tasks such as document summarization (Bharti and Babu, 2017; Sarkar, 2014), query formulation (Jones and Staveley, 1999), text classification (Coenen et al., 2007), clustering (Hammouda et al., 2005), and recommendation systems (Naw and Hlaing, 2013). Keyphrases have become increasingly important for biomedical documents as there has been an exponential growth with over 32 million articles indexed by PubMed (NLM). Figure 1 shows a PubMed document with the author-specified keyphrases highlighted in blue.

Existing keyphrase extraction methods mainly fall either under a supervised or unsupervised approach. Common unsupervised approaches use word co-occurrence statistics to build graph-based ranking algorithms. Each word is mapped to a node and edges connect words that co-occur within a specified window size. Even though unsupervised approaches are desirable for datasets which do not have manually-labeled ground truth values, most such methods perform worse compared to the supervised counterparts.

The supervised approaches use classification to label every token as being part of a keyphrase or not by using features such as part-of speech tags, term-frequency inverse document frequency (tf-idf), and the position of the token in the document. Recently, supervised methods based on deep learning have been employed for keyphrase extraction. In Thomaidou and Vazirgiannis (2011) and Gollapalli et al. (2017), the authors posed the problem as a sequence labeling task and applied a Long Short-Term Memory network (LSTM) and conditional random fields (CRF) to tag each token in document as positive (i.e., part of a keyphrase) or negative. While these approaches achieve much better performance, they still suffer from a major limitation when applied on biomedical literature. The task of labelling each token does not consider how central the token is to the document contents. For Figure 1, the main theme of the keyphrases are genes associated with breast cancer. Thus, the document theme can be used as additional information to improve the keyphrase extraction performance.

To this end, we propose to address the problem of keyphrase extraction as a sequence labelling task with *an additional component to capture the centrality of each token.* We design a centrality layer built on top of a bidirectional LSTM (BiLSTM) layer to constrain each token with regards to the central theme of the document. The output dependencies are then modeled using a CRF layer. The

Proceedings of the BioNLP 2021 workshop, pages 155–161
June 11, 2021.

Protein tyrosine kinase (PTK) is one of the major signaling enzymes in the process of cell signal transduction, which catalyzes the transfer of ATP-γ-phosphate to the tyrosine residues of the substrate protein, making it phosphorylation, regulating cell growth, differentiation, death and a series of physiological and biochemical processes. Abnormal expression of PTK usually leads to cell proliferation disorders, and is closely related to tumor invasion, metastasis and tumor angiogenesis. At present, a variety of PTKs have been used as targets in the screening of anti-tumor drugs. Tyrosine kinase inhibitors (TKIs) compete with ATP for the ATP binding site of PTK and reduce tyrosine kinase phosphorylation, thereby inhibiting cancer cell proliferation. TKI has made great progress in the treatment of cancer, but the attendant acquired resistance is still inevitable, restricting the treatment of cancer. In this paper, we summarize the role of PTK in cancer, TKI treatment of tumor pathways and TKI acquired resistance mechanisms, which provide some reference for further research on TKI treatment of tumors.

Figure 1: An example document from PubMed with author-provided keyphrases in blue.

contributions of our work are:

- Introducing a centrality constraint layer to better capture the main theme of the document and how strongly each token is related to the main theme.
- Thorough evaluation of the centrality layer using an ablation study on biomedical and general domain abstracts.

The next section presents a brief description of the related work. The proposed keyphrase extraction method is introduced in Section 3. Sections 4, and 5 present experimental results and conclusion respectively.

2 Related Work

Keyphrase extraction methods mainly take either supervised or unsupervised approach. Unsupervised approaches generate candidates and rank using features such as tf-idf and topic proportions (Barker and Cornacchia, 2000; Liu et al., 2009b), graph-based centrality measures (Grineva et al., 2009; Wan and Xiao, 2008), topic modeling (Liu et al., 2009a; Teneva and Cheng, 2017), and document's citation network (Gollapalli and Caragea, 2014). Unsupervised, graph-based methods build a graph from the input document where all the candidate keyphrases are nodes and the connection between each candidate is represented by edges. A graph-based ranking method then determines the weights for each node based on the relatedness between the candidates. Alternatively, topic-based approaches cluster candidate keyphrases into topics in the document so that all the topics in the input document are represented by the selected keyphrases. Recently (Sun et al., 2020) proposed a sentence embedding model named SIFRank that uses autoregressive pre-trained language model to extract keyphrases from short documents. Yet unsupervised methods often fail to achieve state-of-the-art performance.

Under the supervised approach, the keyphrase extraction problem is treated as a binary classification task (Alzaidy et al., 2019; Turney, 2000, 2002), where learning algorithms such as support vector machines (Witten et al., 2005; Jiang et al., 2009) and maximum entropy (Kim and Kan, 2009; Yih et al., 2006) are used. Supervised keyphrase extraction can also be posed as a ranking problem between candidates (Witten et al., 2005). The candidates keys are extracted using statistical features (tf-idf, number of occurrences, first occurrence of the key) and structural features (part of speech tags).

Deep learning based models have also been used for keyphrase extraction. Word embeddings are used to measure the relatedness between words in graph-based models (Wang et al., 2014). Zhang et al. (2016) used a Recurrent Neural Network (RNN) based approach to identify keyphrases in Twitter data. The model addresses the problem as sequence labeling for very short text, where a joint-layer RNN is used to capture the semantic dependencies in the input sequence. Alzaidy et al. (2019) employed a LSTM-CRF architecture to model keyphrase extraction as a sequence labelling task to learn the labels of the entire input sequence. Santosh et al. (2020) extended the LSTM-CRF to utilize BiLSTM and incorporated an attention mechanism to retrieve additional information from other sentences within the same document. Sahrawat et al. (2020) evaluated the effect of various pre-trained word embeddings for the BiLSTM-CRF architecture in extracting keyphrases from benchmark datasets and found contextual embeddings offered better performance. While these models offer better performance, they fail to capture the centrality of the keyphrases which represent a salient feature of the document.

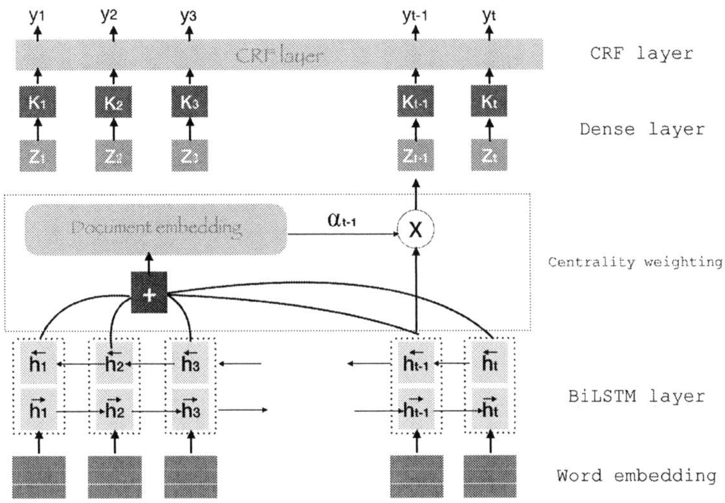

Figure 2: Our model architecture with the BiLSTM, centrality weighting, and CRF layer.

3 Methodology

The keyphrase extraction task is formulated as a sequence labelling task. Given a document $X = w_1, w_2, \cdots, w_t$ where w_i is the ith word and t is the number of words in the document, we predict the labels $y = y_1, y_2, \cdots, y_t$ where each label y_i is whether word w_i is a keyphrase or not.

3.1 Word Embedding Layer

Each word in the document is represented by pre-trained low-dimensional vector representations. Any pre-trained vector representation can be used, and we experiment with various pre-trained embeddings such as GloVe (Pennington et al., 2014), BERT (Devlin et al., 2019) and BioBERT (Lee et al., 2020). The impact of each embedding type is discussed in the experiments section.

3.2 BiLSTM Layer

This layer is used to encode each document to obtain the local contextual representation. A forward and backward LSTMs are used to read the input sequence from left to right, $\overrightarrow{h_1}, \overrightarrow{h_2}, \cdots, \overrightarrow{h_t}$, and right to left, $\overleftarrow{h_1}, \overleftarrow{h_2}, \cdots, \overleftarrow{h_t}$, respectively. The outputs from the two directions are concatenated and summed for the final hidden state representation of the document, $H = [\sum_{i=1}^{t} \overrightarrow{h_i}, \sum_{i=1}^{t} \overleftarrow{h_i}]$.

3.3 Centrality Weighting Layer

Sequence labelling is commonly used for other token encoding tasks such as Named Entity Recognition (NER) where the task is to determine whether a token is a named entity or not. However, keyphrase extraction is different from other sequence labelling tasks (for example NER) in that the tokens should capture the main gist of the document. This is in contrast to NER where the importance of the token is irrelevant as long as it is a named entity. To incorporate the idea of centrality, we use the similarity between each token and the document embedding, H, to bias the model towards tokens which are central (i.e., similar) to the document.

For words $\{w_1, w_2, \cdots, w_t\}$ in a document D, we compute the centrality weight for each word $\alpha_1, \alpha_2, \cdots, \alpha_t$. Each α_i is calculated as the cosine similarity between the document vector (H) and each word (w_i). This is then used to weight the document vector when concatenating with each word's representation from the BiLSTM.

The output representation, z_i for each word is then the centrality weight, α_i multiplied by the output of the biLSTM, $z_i = [\alpha_i \overrightarrow{h_i}, \alpha_i \overleftarrow{h_i}]$. A dense layer is then used to transform the output representation, $k_i = f(z_i)$.

3.4 Conditional Random Fields (CRF)

The obtained contextual representations of each word, k_i are given as input sequence to a CRF layer. CRFs are widely used to model sequence labeling tasks (Lafferty et al., 2001). Given the input document as sequence of tokens, CRF produces a probability distribution over the output label sequence using the dependencies among the labels of the entire input sequence. This formulation considers the correlations between neighboring labels and allows joint decoding for the best sequence of

Table 1: Datasets used for experiments

Dataset	PubMed	INSPEC
Tot. documents	2532	500
Tot. # of tokens	654389	67200
Tot. # of keyphrases	31871	4912
Avg. # of keyphrases	12.5	9.8

labels for the input sequence, rather than decoding each label independently. Moreover, by utilizing two different labels for the keyphrase to denote the beginning (t_B) and intermediate part (t_I) of the keyphrase, the model can learn a multi-token keyphrase. As an example, given a sentence with five tokens ($t_1, t_2, t_3, t_4 t_5$) of which two (t_2, t_3) are part of a keyphrase, the label would be represented as (t_O, t_B, t_I, t_O, t_O). Figure 2 illustrates our model architecture with the various layers.

4 Experiments

Datasets. We ran our experiment on 2 publicly available keyphrase datasets: PubMed (Gero and Ho, 2019) and INSPEC (Hulth, 2003). PubMed consists of 2532 articles from PubMed Central Open Access Subset with at least 5 author-provided keyphrases while INSPEC contains 200 abstracts of scientific journal papers from Computer Science collected between the years 1998 and 2002. Each document in INSPEC has two sets of keywords assigned: the controlled keywords, which are manually controlled assigned keywords that appear in the Inspec thesaurus but may not appear in the document, and the uncontrolled keywords which are freely assigned by the editors. The union of both sets is considered as the ground-truth in this work. Summary statistics for the datasets are shown in Table 1.

Since we use a sequence labeling formulation of the keyphrase extraction problem, the abstract/keyphrases data pairs are prepared such that each document is a sequence of word tokens, each with positive labels if it occurs in a keyphrase (k_B, k_I), or with a negative label (k_O).

Experiment Settings. As baseline models, we train BiLSTM and BiLSTM-CRF with 100-dimension Glove pre-trained embedding vectors (Pennington et al., 2014). We also train BiLSTM-CRF with two 768-dimension contextual embeddings, BERT (Devlin et al., 2019) and BioBERT (Lee et al., 2020). DAKE (Santosh et al., 2020), a state-of-the art baseline, uses a sentence enriching process from all the documents using sentence embedding. To replicate their work, we used the BERT model to extract sentence embeddings for each document and enrich the representation. Finally, our model is trained using BERT word embeddings for the INSPEC dataset and BioBERT embeddings for the PubMed dataset.

Table 2: Model performance on different datasets

Model	PubMed	INSPEC
BiLSTM (GloVe)	0.543	0.427
BiLSTM-CRF (GloVe)	0.554	0.453
BiLSTM-CRF (BERT)	0.604	0.581
BiLSTM-CRF (BioBERT)	0.622	0.464
DAKE	0.623	0.463
Ours	**0.644**	**0.586**

The results reported are from three runs using 80/20/20 split for train/val/test sets respectively. The BiLSTM, and BiLSTM-CRF are optimized during training using stochastic gradient descent with the learning rate 0.0001. Gradient clipping and drop-out are used to prevent overflow and overfitting. We select the model with the best F1 score on the validation set over three runs. The final test scores reported are the averages running the best model on the test sets.

The code was implemented in Tensorflow 2.4.1 and the code is available at `https://github.com/ZHgero/keyphrases_centrality.git`.

5 Results

The performance comparisons between the baselines and our model are shown in Table 2. Our model performs significantly better on the PubMed dataset compared to the existing baselines. In particular, the results show the impact of the centrality layer as it provides a boost in AUC of 0.02 from BiLSTM-CRF (BioBERT) to our model. The improvement gained from our model is not as large on the INSPEC dataset. We hypothesize that for the centrality constraint to be effective, the input sequence should be relatively longer. The sentences in the INSPEC dataset are much shorter hence the difficulty in learning the central theme.

We also compared our models with several state-of-the-art unsupervised approaches including SingleRank (Litvak and Last, 2008), PositionRank (Florescu and Caragea, 2017), TopicRank (Bougouin et al., 2013), and SIFRank (Sun et al.,

Table 3: Ranking comparison on the PubMed dataset

Model	F1@5	F1@10	F1@15
SingleRank	15.2	16.3	19.2
PositionRank	18.3	18.3	20.9
TopicRank	26.4	28.7	29.2
SIFRank	32.3	48.4	56.2
Ours	**34.8**	**53.1**	**62.6**

2020). Table 3 presents the comparison on the PubMed dataset. Since the unsupervised methods are ranking-based methods, the performances are evaluated in terms of F1-measure when a fixed number of keyphrases are extracted. To convert our model into a ranking model, we compute the probability for the predicted keyphrases by using an independence assumption after calculating the marginal probabilities from the CRF layer. The results illustrate that our model outperforms previous unsupervised methods by a significant margin.

In Figure 3, we compare keyphrases tagged by the BioBERT model and our model on a sample abstract. The true positives are colored blue while false negatives are in red. We observe that the BioBERT model fails to identify 'chronic thromboembolic pulmonary hypertension' as an important keyphrase whereas our model correctly identifies it. This may be due to the single occurrence of 'pulmonary hypertension' in the input text. Meanwhile our model leverages the document embedding to 'understand' that pulmonary hypertension is semantically relevant in the context of the entire abstract. We also observe a similar pattern with the keyphrase 'duration of anticoagulation'. Even though both models fail to capture the entire phrase, our model identifies 'anticoagulation' as a strong candidate because of its semantic meaning in the context of the whole abstract.

The figure also illustrates the limitation of the models as both struggle with common words such as 'post' and 'high' that are attached as prefixes to important keywords. 'High risk', 'duration of' and 'post-' are considered unimportant by both models. This can be explained by the fact that such words usually occur outside a keyphrase boundary and get overlooked even when they appear with important words. False positives by both models are important terms as the phrases are very relevant in the context of abstract but were not selected by the authors.

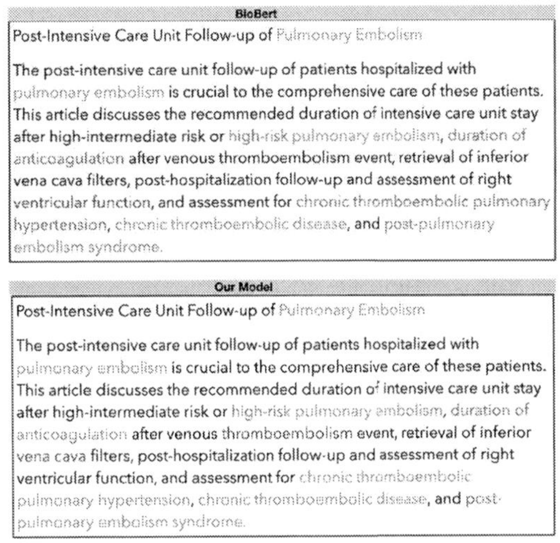

Figure 3: Comparison of keyphrases tagged by two models. True positives are colored blue while false negatives are in red.Purple represents keys that are false positive.

6 Conclusion

In this paper, we proposed a keyphrase extraction method that focuses on identifying words which are central to the document semantics. The problem of keyphrase extraction is posed as a sequence labeling task where each token is tagged as either a keyphrase or not. In addition to our novel centrality constraint layer, we have used Bi-LSTM layers to capture the long term dependencies among the input sequences. Finally, we have a CRF layer which is well suited to capture the dependencies from the output labels. Empirical results on two datasets show that our method gains significant improvement in the PubMed dataset while performing slightly better on the INSPEC dataset.

Acknowledgements

This work is supported by funding from the National Institute of Health awards #5K01LM012924 and #1R01LM013323.

References

Pubmed search engine. https://pubmed.ncbi.nlm.nih.gov/. Accessed: 2021-03-20.

Rabah Alzaidy, Cornelia Caragea, and C. Lee Giles. 2019. Bi-lstm-crf sequence labeling for keyphrase extraction from scholarly documents. In *The World Wide Web Conference, WWW 2019, San Francisco, CA, USA, May 13-17, 2019*, pages 2551–2557.

Ken Barker and Nadia Cornacchia. 2000. Using noun phrase heads to extract document keyphrases. In *conference of the canadian society for computational studies of intelligence*, pages 40–52. Springer.

Santosh Kumar Bharti and Korra Sathya Babu. 2017. Automatic keyword extraction for text summarization: A survey. *arXiv preprint arXiv:1704.03242*.

Adrien Bougouin, Florian Boudin, and Béatrice Daille. 2013. TopicRank: Graph-based topic ranking for keyphrase extraction. In *Proceedings of the Sixth International Joint Conference on Natural Language Processing*, pages 543–551.

Frans Coenen, Paul Leng, Robert Sanderson, and Yanbo J Wang. 2007. Statistical identification of key phrases for text classification. In *International Workshop on Machine Learning and Data Mining in Pattern Recognition*, pages 838–853. Springer.

Jacob Devlin, Ming-Wei Chang, Kenton Lee, and Kristina Toutanova. 2019. BERT: Pre-training of deep bidirectional transformers for language understanding. In *Proceedings of the 2019 Conference of the North American Chapter of the Association for Computational Linguistics: Human Language Technologies, Volume 1 (Long and Short Papers)*, pages 4171–4186.

Corina Florescu and Cornelia Caragea. 2017. PositionRank: An unsupervised approach to keyphrase extraction from scholarly documents. In *Proceedings of the 55th Annual Meeting of the Association for Computational Linguistics (Volume 1: Long Papers)*, pages 1105–1115.

Zelalem Gero and Joyce C Ho. 2019. Namedkeys: Unsupervised keyphrase extraction for biomedical documents. In *Proceedings of the 10th ACM International Conference on Bioinformatics, Computational Biology and Health Informatics*, pages 328–337.

Sujatha Das Gollapalli and Cornelia Caragea. 2014. Extracting keyphrases from research papers using citation networks. In *Proceedings of the Twenty-Eighth AAAI Conference on Artificial Intelligence, July 27 -31, 2014, Québec City, Québec, Canada*, pages 1629–1635.

Sujatha Das Gollapalli, Xiaoli Li, and Peng Yang. 2017. Incorporating expert knowledge into keyphrase extraction. In *Proceedings of the Thirty-First AAAI Conference on Artificial Intelligence, February 4-9, 2017, San Francisco, California, USA*, pages 3180–3187.

Maria P. Grineva, Maxim N. Grinev, and Dmitry Lizorkin. 2009. Extracting key terms from noisy and multitheme documents. In *Proceedings of the 18th International Conference on World Wide Web, WWW 2009, Madrid, Spain, April 20-24, 2009*, pages 661–670.

Khaled M Hammouda, Diego N Matute, and Mohamed S Kamel. 2005. Corephrase: Keyphrase extraction for document clustering. In *International workshop on machine learning and data mining in pattern recognition*, pages 265–274. Springer.

Anette Hulth. 2003. Improved automatic keyword extraction given more linguistic knowledge. In *Proceedings of the 2003 Conference on Empirical Methods in Natural Language Processing*, pages 216–223.

Xin Jiang, Yunhua Hu, and Hang Li. 2009. A ranking approach to keyphrase extraction. In *Proceedings of the 32nd international ACM SIGIR conference on Research and development in information retrieval*, pages 756–757.

Steve Jones and Mark S Staveley. 1999. Phrasier: a system for interactive document retrieval using keyphrases. In *Proceedings of the 22nd annual international ACM SIGIR conference on Research and development in information retrieval*, pages 160–167.

Su Nam Kim and Min-Yen Kan. 2009. Re-examining automatic keyphrase extraction approaches in scientific articles. In *Proceedings of the Workshop on Multiword Expressions: Identification, Interpretation, Disambiguation and Applications (MWE 2009)*, pages 9–16.

John D. Lafferty, Andrew McCallum, and Fernando C. N. Pereira. 2001. Conditional random fields: Probabilistic models for segmenting and labeling sequence data. In *Proceedings of the Eighteenth International Conference on Machine Learning (ICML 2001), Williams College, Williamstown, MA, USA, June 28 - July 1, 2001*, pages 282–289.

Jinhyuk Lee, Wonjin Yoon, Sungdong Kim, Donghyeon Kim, Sunkyu Kim, Chan Ho So, and Jaewoo Kang. 2020. Biobert: a pre-trained biomedical language representation model for biomedical text mining. *Bioinformatics*, 36(4):1234–1240.

Marina Litvak and Mark Last. 2008. Graph-based keyword extraction for single-document summarization. In *Coling 2008: Proceedings of the workshop Multi-source Multilingual Information Extraction and Summarization*, pages 17–24.

Feifan Liu, Deana Pennell, Fei Liu, and Yang Liu. 2009a. Unsupervised approaches for automatic keyword extraction using meeting transcripts. In *Proceedings of Human Language Technologies: The*

2009 Annual Conference of the North American Chapter of the Association for Computational Linguistics, pages 620–628.

Zhiyuan Liu, Peng Li, Yabin Zheng, and Maosong Sun. 2009b. Clustering to find exemplar terms for keyphrase extraction. In *Proceedings of the 2009 Conference on Empirical Methods in Natural Language Processing*, pages 257–266.

Naw Naw and Ei Ei Hlaing. 2013. Relevant words extraction method for recommendation system. *Bulletin of Electrical Engineering and Informatics*, 2(3):169–176.

Jeffrey Pennington, Richard Socher, and Christopher Manning. 2014. GloVe: Global vectors for word representation. In *Proceedings of the 2014 Conference on Empirical Methods in Natural Language Processing (EMNLP)*, pages 1532–1543.

Dhruva Sahrawat, Debanjan Mahata, Haimin Zhang, Mayank Kulkarni, Agniv Sharma, Rakesh Gosangi, Amanda Stent, Yaman Kumar, Rajiv Ratn Shah, and Roger Zimmermann. 2020. Keyphrase extraction as sequence labeling using contextualized embeddings. In *European Conference on Information Retrieval*, pages 328–335. Springer.

Tokala Yaswanth Sri Sai Santosh, Debarshi Kumar Sanyal, Plaban Kumar Bhowmick, and Partha Pratim Das. 2020. Dake: Document-level attention for keyphrase extraction. In *European Conference on Information Retrieval*, pages 392–401. Springer.

Kamal Sarkar. 2014. A keyphrase-based approach to text summarization for english and bengali documents. *International Journal of Technology Diffusion (IJTD)*, 5(2):28–38.

Yi Sun, Hangping Qiu, Yu Zheng, Zhongwei Wang, and Chaoran Zhang. 2020. Sifrank: A new baseline for unsupervised keyphrase extraction based on pre-trained language model. *IEEE Access*, 8:10896–10906.

Nedelina Teneva and Weiwei Cheng. 2017. Salience rank: Efficient keyphrase extraction with topic modeling. In *Proceedings of the 55th Annual Meeting of the Association for Computational Linguistics (Volume 2: Short Papers)*, pages 530–535.

Stamatina Thomaidou and Michalis Vazirgiannis. 2011. Multiword keyword recommendation system for online advertising. In *International Conference on Advances in Social Networks Analysis and Mining, ASONAM 2011, Kaohsiung, Taiwan, 25-27 July 2011*, pages 423–427.

Peter D Turney. 2000. Learning algorithms for keyphrase extraction. *Information retrieval*, 2(4):303–336.

Peter D Turney. 2002. Learning to extract keyphrases from text. *arXiv preprint cs/0212013*.

Xiaojun Wan and Jianguo Xiao. 2008. Single document keyphrase extraction using neighborhood knowledge. In *AAAI*, volume 8, pages 855–860.

Rui Wang, Wei Liu, and Chris McDonald. 2014. Corpus-independent generic keyphrase extraction using word embedding vectors. In *Software Engineering Research Conference*, volume 39, pages 1–8.

Ian H Witten, Gordon W Paynter, Eibe Frank, Carl Gutwin, and Craig G Nevill-Manning. 2005. Kea: Practical automated keyphrase extraction. In *Design and Usability of Digital Libraries: Case Studies in the Asia Pacific*, pages 129–152.

Wen-tau Yih, Joshua Goodman, and Vitor R. Carvalho. 2006. Finding advertising keywords on web pages. In *Proceedings of the 15th international conference on World Wide Web, WWW 2006, Edinburgh, Scotland, UK, May 23-26, 2006*, pages 213–222.

Qi Zhang, Yang Wang, Yeyun Gong, and Xuanjing Huang. 2016. Keyphrase extraction using deep recurrent neural networks on Twitter. In *Proceedings of the 2016 Conference on Empirical Methods in Natural Language Processing*, pages 836–845.

End-to-end Biomedical Entity Linking with Span-based Dictionary Matching

Shogo Ujiie[♠] **Hayate Iso**[♡*]
Shuntaro Yada[♠] **Shoko Wakamiya**[♠] **Eiji Aramaki**[♠]
[♠]Nara Institute of Science and Technology [♡]Megagon Labs
{ujiie, yada-s, wakamiya, aramaki}@is.naist.jp
hayate@magagon.ai

Abstract

Disease name recognition and normalization , which is generally called biomedical entity linking, is a fundamental process in biomedical text mining. Recently, neural joint learning of both tasks has been proposed to utilize the mutual benefits. While this approach achieves high performance, disease concepts that do not appear in the training dataset cannot be accurately predicted. This study introduces a novel end-to-end approach that combines span representations with dictionary-matching features to address this problem. Our model handles unseen concepts by referring to a dictionary while maintaining the performance of neural network-based models, in an end-to-end fashion. Experiments using two major datasets demonstrate that our model achieved competitive results with strong baselines, especially for unseen concepts during training.

1 Introduction

Identifying disease names , which is generally called biomedical entity linking, is the fundamental process of biomedical natural language processing, and it can be utilized in applications such as a literature search system (Lee et al., 2016) and a biomedical relation extraction (Xu et al., 2016). The usual system to identify disease names consists of two modules: named entity recognition (NER) and named entity normalization (NEN). NER is the task that recognizes the span of a disease name, from the start position to the end position. NEN is the post-processing of NER, normalizing a disease name into a controlled vocabulary, such as a MeSH or Online Mendelian Inheritance in Man (OMIM).

Although most previous studies have developed pipeline systems, in which the NER model first recognizs disease mentions (Lee et al., 2020; Weber et al., 2020) and the NEN model normalizes the recognized mention (Leaman et al., 2013; Ferré et al., 2020; Xu et al., 2020; Vashishth et al., 2020), a few approaches employ a joint learning architecture for these tasks (Leaman and Lu, 2016; Lou et al., 2017). These joint approaches simultaneously recognize and normalize disease names utilizing their mutual benefits. For example, Leaman et al. (2013) demonstrated that dictionary-matching features, which are commonly used for NEN, are also effective for NER. While these joint learning models achieve high performance for both NER and NEN, they predominately rely on hand-crafted features, which are difficult to construct because of the domain knowledge requirement.

Recently, a neural network (NN)-based model that does not require any hand-crafted features was applied to the joint learning of NER and NEN (Zhao et al., 2019). NER and NEN were defined as two token-level classification tasks, i.e., their model classified each token into IOB2 tags and concepts, respectively. Although their model achieved the state-of-the-art performance for both NER and NEN, a concept that does not appear in training data (i.e., zero-shot situation) can not be predicted properly.

One possible approach to handle this zero-shot situation is utilizing the dictionary-matching features. Suppose that an input sentence "Classic *polyarteritis nodosa* is a systemic vasculitis" is given, where "*polyarteritis nodosa*" is the target entity. Even if it does not appear in the training data, it can be recognized and normalized by referring to a controlled vocabulary that contains "*Polyarteritis Nodosa* (MeSH: D010488)." Combining such looking-up mechanisms with NN-based models, however, is not a trivial task; dictionary matching must be performed at the *entity*-level, whereas standard NN-based NER and NEN tasks are performed at the *token*-level (for example, Zhao et al., 2019).

To overcome this problem, we propose a novel end-to-end approach for NER and NEN that com-

[*]Work done while at Nara Institute of Science and Technology.

Proceedings of the BioNLP 2021 workshop, pages 162–167
June 11, 2021. ©2021 Association for Computational Linguistics

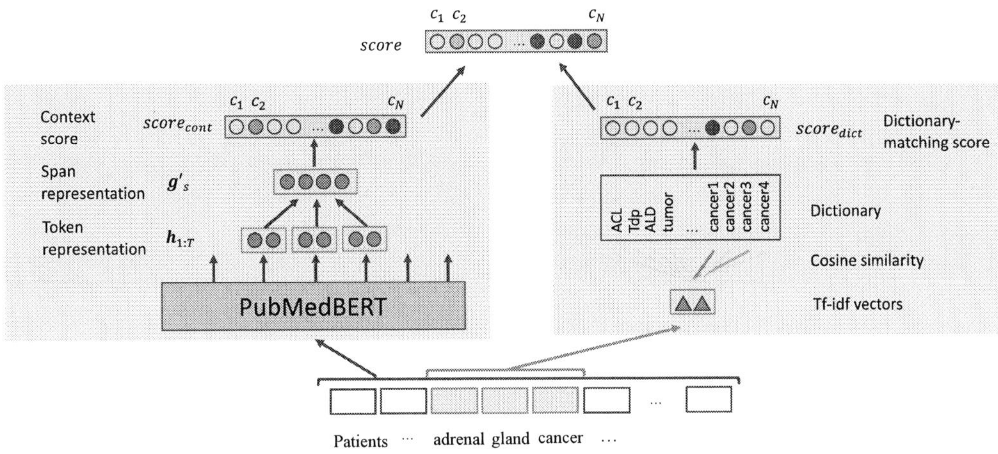

Figure 1: The overview of our model. It combines the dictionary-matching scores with the context score obtained from PubMedBERT. The red boxes are the target span and "ci" in the figure is the "i"-th concept in the dictionary.

bines dictionary-matching features with NN-based models. Based on the span-based model introduced by Lee et al. (2017), our model first computes span representations for all possible spans of the input sentence and then combines the dictionary-matching features with the span representations. Using the score obtained from both features, it directly classifies the disease concept. Thus, our model can handle the zero-shot problem by using dictionary-matching features while maintaining the performance of the NN-based models.

Our model is also effective in situations other than the zero-shot condition. Consider the following input sentence: "We report the case of a patient who developed acute *hepatitis*," where "*hepatitis*" is the target entity that should be normalized to "drug-induced hepatitis." While the longer span "acute hepatitis" also appears plausible for standalone NER models, our end-to-end architecture assigns a higher score to the correct shorter span "hepatitis" due to the existence of the normalized term ("drug-induced hepatitis") in the dictionary.

Through the experiments using two major NER and NEN corpora, we demonstrate that our model achieves competitive results for both corpora. Further analysis illustrates that the dictionary-matching features improve the performance of NEN in the zero-shot and other situations.

Our main contributions are twofold: (i) We propose a novel end-to-end model for disease name recognition and normalization that utilizes both NN-based features and dictionary-matching features; (ii) We demonstrate that combining dictionary-matching features with an NN-based model is highly effective for normalization, especially in the zero-shot situations.

2 Methods

2.1 Task Definition

Given an input sentence, which is a sequence of words $\boldsymbol{x} = \{x_1, x_2, \cdots, x_{|\boldsymbol{X}|}\}$ in the biomedical literature, let us define $\mathcal{S}$ as a set of all possible spans, and $\mathcal{L}$ as a set of concepts that contains the special label *Null* for a non-disease span. Our goal is to predict a set of labeled spans $\boldsymbol{y} = \{\langle i, j, d\rangle_k\}_{k=1}^{|\boldsymbol{Y}|}$, where $(i, j) \in \mathcal{S}$ is the word index in the sentence, and $d \in \mathcal{L}$ is the concept of diseases.

2.2 Model Architecture

Our model predicts the concepts for each span based on the score, which is represented by the weighted sum of two factors: the context score $score_{cont}$ obtained from span representations and the dictionary-matching score $score_{dict}$. Figure 1 illustrates the overall architecture of our model. We denote the score of the span s as follows:

$$score(s, c) = score_{cont}(s, c) + \lambda score_{dict}(s, c)$$

where $c \in \mathcal{L}$ is the candidate concept and λ is the hyperparameter that balances the scores. For the concept prediction, the scores of all possible spans and concepts are calculated, and then the concept with the highest score is selected as the predicted concept for each span as follows:

$$y = \arg\max_{c \in \mathcal{L}} score(s, c)$$

Context score The context score is computed in a similar way to that of Lee et al. (2017), which is based on the span representations. To compute the representations of each span, the input tokens are first encoded into the token embeddings. We used BioBERT (Lee et al., 2020) as the encoder, which is a variation of bidirectional encoder representations from transformers (BERT) that is trained on a large amount of biomedical text. Given an input sentence containing T words, we can obtain the contextualized embeddings of each token using BioBERT as follows:

$$\mathbf{h}_{1:T} = \mathrm{BERT}(x_1, x_2, \cdots, x_T)$$

where $\mathbf{h}_{1:T}$ is the input tokens embeddings.

Span representations are obtained by concatenating several features from the token embeddings:

$$\mathbf{g}_s = [\mathbf{h}_{start(s)}, \mathbf{h}_{end(s)}, \hat{\mathbf{h}}_s, \phi(s)]$$
$$\mathbf{g}'_s = \mathrm{GELU}(\mathrm{FFNN}(\mathbf{g}_s))$$

where $\mathbf{h}_{start(s)}$ and $\mathbf{h}_{end(s)}$ are the start and end token embeddings of the span, respectively; and $\hat{\mathbf{h}}_s$ is the weighted sum of the token embeddings in the span, which is obtained using an attention mechanism (Bahdanau et al., 2015). $\phi(i)$ is the size of span s. These representations $\mathbf{g}_s$ are then fed into a simple feed-forward NN, FFNN, and a nonlinear function, GELU (Hendrycks and Gimpel, 2016).

Given a particular span representation and a candidate concept as the inputs, we formulate the context score as follows:

$$score_{cont}(s, c) = \mathbf{g}_s \cdot \mathbf{W}_c$$

where $\mathbf{W} \in \mathbb{R}^{|\mathcal{L}| \times d^{\mathbf{g}}}$ is the weight matrix associated with each concept c, and $\mathbf{W}_c$ represents the weight vector for the concept c.

Dictionary-matching score We used the cosine similarity of the TF-IDF vectors as the dictionary-matching features. Because there are several synonyms for a concept, we calculated the cosine similarity for all synonyms of the concept and used the maximum cosine similarity as the score for each concept. The TF-IDF is calculated using the character-level n-gram statistics computed for all diseases appearing in the training dataset and controlled vocabulary. For example, given the span "breast cancer," synonyms with high cosine similarity are "breast cancer (1.0)" and "male breast cancer (0.829)."

3 Experiment

3.1 Datasets

To evaluate our model, we chose two major datasets used in disease name recognition and normalization against a popular controlled vocabulary, MEDIC (Davis et al., 2012). Both datasets, the National Center for Biotechnology Information Disease (NCBID) corpus (Doğan et al., 2014) and the BioCreative V Chemical Disease Relation (BC5CDR) task corpus (Li et al., 2016), comprise of PubMed titles and abstracts annotated with disease names and their corresponding normalized term IDs (CUIs). NCBID provides 593 training, 100 development, and 100 test data splits, while BC5CDR evenly divides 1500 data into the three sets. We adopted the same version of MEDIC as TaggerOne (Leaman and Lu, 2016) used, and that we dismissed non-disease entity annotations contained in BC5CDR.

3.2 Baseline Models

We compared several baselines to evaluate our model. DNorm (Leaman et al., 2013) and NormCo (Wright et al., 2019) were used as pipeline models due to their high performance. In addition, we used the pipeline systems consisting of state-of-the-art models: BioBERT (Lee et al., 2020) for NER and BioSyn (Sung et al., 2020) for NEN.

TaggerOne (Leaman and Lu, 2016) and Transition-based model (Lou et al., 2017) are used as joint-learning models. These models outperformed the pipeline models in NCBID and BC5CDR. For the model introduced by Zhao et al. (2019), we cannot reproduce the performance reported by them. Instead, we report the performance of the simple token-level joint learning model based on the BioBERT, which referred as "joint (token)".

3.3 Implementation

We performed several preprocessing steps: splitting the text into sentences using the NLTK toolkit (Bird et al., 2009), removing punctuations, and resolving abbreviations using Ab3P (Sohn et al., 2008), a common abbreviation resolution module. We also merged disease names in each training set into a controlled vocabulary, following the methods of Lou et al. (2017).

For training, we set the learning rate to 5e-5, and mini-batch size to 32. λ was set to 0.9 using the development sets. For BC5CDR, we trained the model using both the training and development sets

Models	NCBID NER	NCBID NEN	BC5CDR NER	BC5CDR NEN
TaggerOne	0.829	0.807	0.826	0.837
Transition-based model	0.821	0.826	0.862	**0.876**
NormCo	0.829	0.840	0.826	0.830
pipeline	0.874	0.841	0.865	0.818
joint (token)	0.864	0.765	0.855	0.817
Ours without dictionary	0.884	0.781	0.864	0.808
Ours	**0.891**	**0.854**	**0.867**	0.851

Table 1: F1 scores of NER and NEN in NCBID and BC5CDR. Bold font represents the highest score.

following Leaman and Lu (2016). For computational efficiency, we only consider spans with up to 10 words.

3.4 Evaluation Metrics

We evaluated the recognition performance of our model using micro-F1 at the entity level. We consider the predicted spans as true positive when their spans are identical. Following the previous work (Wright et al., 2019; Leaman and Lu, 2016), the performance of NEN was evaluated using micro-F1 at the abstract level. If a predicted concept was found within the gold standard concepts in the abstract, regardless of its location, it was considered as a true positive.

4 Results & Discussions

Table 1 illustrates that our model mostly achieved the highest F1-scores in both NER and NEN, except for the NEN in BC5CDR, in which the transition-based model displays its strength as a baseline. The proposed model outperformed the pipeline model of the state-of-the-art models for both tasks, which demonstrates that the improvement is attributed not to the strength of BioBERT but the model architecture, including the end-to-end approach and combinations of dictionary-matching features.

Comparing the model variation results, adding dictionary-matching features improved the performance in NEN. The results clearly suggest that dictionary-matching features are effective for NN-based NEN models.

4.1 Contribution of Dictionary-Matching

To analyze the behavior of our model in the zero-shot situation, we investigated the NEN performance on two subsets of both corpora: disease names with concepts that appear in the training data (i.e., standard situation), and disease names with concepts that do not appear in the training data (i.e., the zero-shot situation). Table 2 shows the number of mentions and concepts in each situation. Table 3 displays the results of the zero-shot and standard situation. The proposed model with dictionary-matching features can classify disease concepts in the zero-shot situation, whereas the NN-based classification model cannot normalize the disease names.

dataset	standard mention	standard concept	zero-shot mention	zero-shot concept
NCBID	781	135	179	56
BC5CDR	4031	461	391	179

Table 2: Number of mentions and concepts in standard and zero-shot situations.

	Methods	NCBID	BC5CDR
zero-shot	Ours without dictionary	0	0
	Ours	**0.704**	**0.597**
standard	Ours without dictionary	0.854	0.846
	Ours	**0.905**	**0.877**

Table 3: F1 scores for NEN of NCBID and BC5CDR subsets for zero-shot situation where disease concepts do not appear in training data and the standard situation where they do appear in training data.

The results of the standard situation demonstrate that combining dictionary-matching features also improves the performance even when target concepts appear in the training data. This finding implies that an NN-based model can benefit from dictionary-matching features, even if the models can learn from many training data.

4.2 Case study

We examined 100 randomly sampled sentences to determine the contributions of dictionary-matching features. There are 32 samples in which the models predicted concepts correctly by adding dictionary-matching features. Most of these samples are disease concepts that do not appear in the training set but appear in the dictionary. For example, "*pure red cell aplasis* (MeSH: D012010)" is not in the BC5CDR training set while the MEDIC contains "Pure Red-Cell Aplasias" for "D012010". In this case, a high dictionary-matching score clearly leads to a correct prediction in the zero-shot situation.

In contrast, there are 32 samples in which the dictionary-matching features cause errors. The sources of this error type are typically general disease names in the MEDIC. For example, "Death (MeSH:D003643)" is incorrectly predicted as a disease concept in NER. Because these words are also used in the general context, our model overestimated their dictionary-matching scores.

Furthermore, in the remaining samples, our model predicted the code properly and the span incorrectly. For example, although "thoracic hematomyelia" is labeled as "MeSH: D020758" in the BC5CDR test set, our model recognized this as "hematomyelia." In this case, our model mostly relied on the dictionary-matching features and misclassifies the span because 'hematomyelia" is in the MEDIC but not in the training data.

4.3 Limitations

Our model is inferior to the transition-based model for BC5CDR. One possible reason is that the transition-based model utilizes normalized terms that co-occur within a sentence, whereas our model does not. Certain disease names that co-occur within a sentence are strongly useful for normalizing disease names. Although BERT implicitly considers the interaction between disease names via the attention mechanism, a more explicit method is preferable for normalizing diseases.

Another limitation is that our model treats the dictionary entries equally. Because certain terms in the dictionary may also be used for non-disease concepts, such as gene names, we must consider the relative importance of each concept.

5 Conclusion

We proposed a end-to-end model for disease name recognition and normalization that combines the NN-based model with the dictionary-matching features. Our model achieved highly competitive results for the NCBI disease corpus and BC5CDR corpus, demonstrating that incorporating dictionary-matching features into an NN-based model can improve its performance. Further experiments exhibited that dictionary-matching features enable our model to accurately predict the concepts in the zero-shot situation, and they are also beneficial in the other situation. While the results illustrate the effectiveness of our model, we found several areas for improvement, such as the general terms in the dictionary and the interaction between disease names within a sentence. A possible future direction to deal with general terms is to jointly train the parameters representing the importance of each synonyms.

References

Dzmitry Bahdanau, Kyunghyun Cho, and Yoshua Bengio. 2015. Neural machine translation by jointly learning to align and translate. In *ICLR*.

Steven Bird, Ewan Klein, and Edward Loper. 2009. *Natural language processing with Python: Analyzing text with the natural language toolkit*. O'Reilly Media, Inc.

Allan Peter Davis, Thomas C Wiegers, Michael C Rosenstein, and Carolyn J Mattingly. 2012. MEDIC: A practical disease vocabulary used at the comparative toxicogenomics database. *Database*, 2012:bar065.

Rezarta Islamaj Doğan, Robert Leaman, and Zhiyong Lu. 2014. NCBI disease corpus: A resource for disease name recognition and concept normalization. *J. Biomed. Inform.*, 47:1–10.

Arnaud Ferré, Louise Deléger, Robert Bossy, Pierre Zweigenbaum, and Claire Nédellec. 2020. C-Norm: a neural approach to few-shot entity normalization. *BMC Bioinformatics*, 21(Suppl 23):579.

Dan Hendrycks and Kevin Gimpel. 2016. Gaussian error linear units (GELUs). *arXiv preprint arXiv:1606.08415*.

Robert Leaman, Rezarta Islamaj Dogan, and Zhiyong Lu. 2013. DNorm: Disease name normalization with pairwise learning to rank. *Bioinformatics*, 29(22):2909–2917.

Robert Leaman and Zhiyong Lu. 2016. TaggerOne: Joint named entity recognition and normalization with semi-markov models. *Bioinformatics*, 32(18):2839–2846.

Jinhyuk Lee, Wonjin Yoon, Sungdong Kim, Donghyeon Kim, Sunkyu Kim, Chan Ho So, and Jaewoo Kang. 2020. BioBERT: a pre-trained biomedical language representation model for biomedical text mining. *Bioinformatics*, 36(4):1234–1240.

Kenton Lee, Luheng He, Mike Lewis, and Luke Zettlemoyer. 2017. End-to-end neural coreference resolution. In *EMNLP*, pages 188–197.

Sunwon Lee, Donghyeon Kim, Kyubum Lee, Jaehoon Choi, Seongsoon Kim, Minji Jeon, Sangrak Lim, Donghee Choi, Sunkyu Kim, Aik-Choon Tan, and Jaewoo Kang. 2016. BEST: Next-Generation biomedical entity search tool for knowledge discovery from biomedical literature. *PLoS One*, 11(10):e0164680.

Jiao Li, Yueping Sun, Robin J Johnson, Daniela Sciaky, Chih-Hsuan Wei, Robert Leaman, Allan Peter Davis, Carolyn J Mattingly, Thomas C Wiegers, and Zhiyong Lu. 2016. BioCreative V CDR task corpus: A resource for chemical disease relation extraction. *Database*, 2016:baw068.

Yinxia Lou, Yue Zhang, Tao Qian, Fei Li, Shufeng Xiong, and Donghong Ji. 2017. A transition-based joint model for disease named entity recognition and normalization. *Bioinformatics*, 33(15):2363–2371.

Sunghwan Sohn, Donald C Comeau, Won Kim, and W John Wilbur. 2008. Abbreviation definition identification based on automatic precision estimates. *BMC Bioinformatics*, 9:402.

Mujeen Sung, Hwisang Jeon, Jinhyuk Lee, and Jaewoo Kang. 2020. Biomedical entity representations with synonym marginalization. In *ACL*, pages 3641–3650.

Shikhar Vashishth, Rishabh Joshi, Denis Newman-Griffis, Ritam Dutt, and Carolyn Rose. 2020. MedType: Improving Medical Entity Linking with Semantic Type Prediction. *arXiv preprint arXiv:2005.00460*.

Leon Weber, Jannes Münchmeyer, Tim Rocktäschel, Maryam Habibi, and Ulf Leser. 2020. HUNER: improving biomedical NER with pretraining. *Bioinformatics*, 36(1):295–302.

Dustin Wright, Yannis Katsis, Raghav Mehta, and Chun-Nan Hsu. 2019. NormCo: Deep disease normalization for biomedical knowledge base construction. In *AKBC*.

Dongfang Xu, Zeyu Zhang, and Steven Bethard. 2020. A Generate-and-Rank framework with semantic type regularization for biomedical concept normalization. In *ACL*, pages 8452–8464.

Jun Xu, Yonghui Wu, Yaoyun Zhang, Jingqi Wang, Hee-Jin Lee, and Hua Xu. 2016. CD-REST: A system for extracting chemical-induced disease relation in literature. *Database*, 2016:baw036.

Sendong Zhao, Ting Liu, Sicheng Zhao, and Fei Wang. 2019. A neural multi-task learning framework to jointly model medical named entity recognition and normalization. In *AAAI*, pages 817–824.

Word-Level Alignment of Paper Documents with their Electronic Full-Text Counterparts

Mark-Christoph Müller, Sucheta Ghosh, Ulrike Wittig, and Maja Rey
Heidelberg Institute for Theoretical Studies gGmbH, Heidelberg, Germany
{mark-christoph.mueller,sucheta.ghosh,
ulrike.wittig,maja.rey}@h-its.org

Abstract

We describe a simple procedure for the automatic creation of word-level alignments between printed documents and their respective full-text versions. The procedure is unsupervised, uses standard, off-the-shelf components only, and reaches an F-score of 85.01 in the basic setup and up to 86.63 when using pre- and post-processing. Potential areas of application are manual database curation (incl. document *triage*) and biomedical expression OCR.

1 Introduction

Even though most research literature in the life sciences is *born-digital* nowadays, manual data curation (International Society for Biocuration, 2018) from these documents still often involves paper. For curation steps that require close reading and markup of relevant sections, curators frequently rely on paper printouts and highlighter pens (Venkatesan et al., 2019). Figure 1a shows a page of a typical document used for manual curation. The potential reasons for this can be as varied as merely sticking to a habit, ergonomic issues related to reading from and interacting with a device, and functional limitations of that device (Buchanan and Loizides, 2007; Köpper et al., 2016; Clinton, 2019).

Whatever the *reason*, the *consequence* is a two-fold media break in many manual curation workflows: first from electronic format (either PDF or full-text XML) to paper, and then back from paper to the electronic format of the curation database. Given the above arguments in favor of paper-based curation, removing the first media break from the curation workflow does not seem feasible. Instead, we propose to bridge the gap between paper and electronic media by automatically creating an alignment between the words on the printed document pages and their counterparts in an electronic full-text version of the same document.

Our approach works as follows: We automatically create machine-readable versions of printed paper documents (which might or might not contain markup) by scanning them, applying optical character recognition (OCR), and converting the resulting semi-structured OCR output text into a flexible XML format for further processing. For this, we use the multilevel XML format of the annotation tool MMAX2[1] (Müller and Strube, 2006). We retrieve electronic full-text counterparts of the scanned paper documents from PubMedCentral® in `.nxml` format[2], and also convert them into MMAX2 format. By using a shared XML format for the two heterogeneous text sources, we can capture their content and structural information in a way that provides a *compatible*, though often not identical, word-level tokenization. Finally, using a sequence alignment algorithm from bioinformatics and some pre- and post-processing, we create a word-level alignment of both documents.

Aligning words from OCR and full-text documents is challenging for several reasons. The OCR output contains various types of **recognition errors**, many of which involve special symbols, Greek letters like μ or sub- and superscript characters and numbers, which are particularly frequent in chemical names, formulae, and measurement units, and which are notoriously difficult for OCR (Ohyama et al., 2019).

If the printed paper document is based on PDF, it usually has an **explicit page layout**, which is different from the way the corresponding full-text XML document is displayed in a web browser. Differences include double- vs. single-column layout, but also the way in which tables and figures are rendered and positioned.

Finally, printed papers might contain **additional**

[1] `https://github.com/nlpAThits/MMAX2`

[2] While PubMedCentral® is an obvious choice here, other resources with different full-text data formats exist and can also be used. All that needs to be modified is the conversion step (see Section 2.2).

Proceedings of the BioNLP 2021 workshop, pages 168–179
June 11, 2021.

content in headers or footers (like e.g. download timestamps). Also, while the references/bibliography section is an integral part of a printed paper and will be covered by OCR, in XML documents it is often structurally kept apart from the actual document text.

Given these challenges, attempting data extraction from document *images* if the documents are available in PDF or even full-text format may seem unreasonable. We see, however, the following useful applications:

1. Manual Database Curation As mentioned above, manual database curation requires the extraction, normalization, and database insertion of scientific content, often from *paper* documents. Given a paper document in which a human expert curator has manually marked a word or sequence of words for insertion into the database, having a *link* from these words to their electronic counterparts can eliminate or at least reduce error-prone and time-consuming steps like manual re-keying. Also, already existing annotations of the electronic full-text[3] would also be accessible and could be used to inform the curation decision or to supplement the database entry.

2. Automatic PDF Highlighting for Manual *Triage* Database curation candidate papers are identified by a process called document *triage* (Buchanan and Loizides, 2007; Hirschman et al., 2012) which, despite some attempts towards automation (e.g. Wang et al. (2020)), remains a mostly manual process. In a nut shell, triage normally involves querying a literature database (like PubMed[4]) for specific terms, skimming the list of search results, selecting and skim-reading some papers, and finally downloading and printing the PDF versions of the most promising ones for curation (Venkatesan et al., 2019). Here, the switch from *searching* in the electronic full-text (or abstract) to *printing* the PDF brings about a loss of information, because the terms that caused the paper to be retrieved will have to be located again in the print-out. A word-level alignment between the full-text and the PDF version would allow to create an enhanced version of the PDF with highlighted search term occurrences *before* printing.

3. Biomedical Expression OCR Current state-of-the-art OCR systems are very accurate at recognizing standard text using Latin script and baseline typography, but, as already mentioned, they are less reliable for more typographically complex expressions like chemical formulae. In order to develop specialized OCR systems for these types of expressions, ground-truth data is required in which image regions containing these expressions are labelled with the correct characters and their positional information (see also Section 5). If aligned documents are available, this type of data can easily be created at a large scale.

The remainder of this paper is structured as follows. In Section 2, we describe our data set and how it was converted into the shared XML format. Section 3 deals with the actual alignment procedure, including a description of the optional pre- and post-processing measures. In Section 4, we present experiments in which we evaluate the performance of the implemented procedure, including an ablation of the effects of the individual pre- and post-processing measures. Quantitative evaluation alone, however, does not convey a realistic idea of the actual usefulness of the procedure, which ultimately needs to be evaluated in the context of real applications including, but not limited to, database curation. Section 4.2, therefore, briefly presents examples of the alignment and highlighting detection functionality and the biomedical expression OCR use case mentioned above. Section 5 discusses relevant related work, and Section 6 summarizes and concludes the paper with some future work.

All the tools and libraries we use are freely available. In addition, our implementation can be found at `https://github.com/nlpAThits/BioNLP2021`.

2 Data

For the alignment of a paper document with its electronic full-text counterpart, what is minimally required is an image of every page of the document, and a full-text XML file of the same document. The document images can either be created by scanning or by directly converting the corresponding PDF into an image. The latter method will probably yield images of a better quality, because it completely avoids the physical printing and subsequent scanning step, while the output of the former method will be more realistic. We experiment with both types of images (see Section 2.1). We identify a document by its DOI, and refer to the different versions as DOI_{xml} (from the full-text XML), DOI_{conv}, and DOI_{scan}. Whenever

[3] Like `https://europepmc.org/Annotations`

[4] `https://pubmed.ncbi.nlm.nih.gov/`

a distinction between DOI_{conv} and DOI_{scan} is not required, we refer to these versions collectively as DOI_{ocr}.

Printable PDF documents and their associated `.nxml` files are readily available at PMC-OAI.[5] In our case, however, printed paper versions were already available, as we have access to a collection of more than 6.000 printed scientific papers (approx. 30.000 pages in total) that were created in the SABIO-RK[6] Biochemical Reaction Kinetics Database project (Wittig et al., 2017, 2018). These papers contain manual highlighter markup at different levels of granularity, including the word, line, and section level. Transferring this type of markup from printed paper to the electronic medium is one of the key applications of our alignment procedure. Our paper collection spans many publication years and venues. For our experiments, however, it was required that each document was freely available both as PubMedCentral® full-text XML and as PDF. While this leaves only a fraction of (currently) 68 papers, the data situation is still sufficient to demonstrate the feasibility of our procedure. Even more importantly, the procedure is unsupervised, i.e. it does not involve learning and does not require any training data.

2.1 Document Image to Multilevel XML

Since we want to compare downstream effects of input images of different quality, we created both a converted and a scanned image version for every document in our data set. For the DOI_{conv} version, we used `pdftocairo` to create a high-resolution (600 DPI) PNG file for every PDF page. Figure 1c shows an example. The DOI_{scan} versions, on the other hand, were extracted from 'sandwich' PDFs which had been created earlier by a professional scanning service provider. The choice of a service provider for this task was only motivated by the large number of pages to process, and not by expected quality or other considerations. A sandwich PDF contains, among other data, the document plain text (as recognized by the provider's OCR software) and a background image for each page. This background image is a by-product of the OCR process in which pixels that were recognized as parts of a character are inpainted, i.e. removed by being overwritten with colors of neighbouring regions. Figure 1b shows the background image corresponding to the page in Figure 1a. Note how the image retains the highlighting. We used `pdfimages` to extract the background images (72 DPI) from the sandwich PDF for use in highlighting extraction (see Section 2.1.1 below). We refer to these versions as DOI_{scan_bg}. For the actual DOI_{scan} versions, we again used `pdftocairo` to create a high-resolution (600 DPI) PNG file for every scanned page.

OCR was then performed on the DOI_{conv} and the DOI_{scan} versions with tesseract 4.1.1[7], using default recognition settings (`-oem 3 -psm 3`) and specifying hOCR[8] with character-level bounding boxes as output format. In order to maximize recognition accuracy (at the expense of processing speed), the default language models for English were replaced with optimized LSTM models[9]. No other modification or re-training of tesseract was performed. In a final step, the hOCR output from both image versions was converted into the MMAX2 (Müller and Strube, 2006) multilevel XML annotation format, using *words* as tokenization granularity, and storing word- and character-level confidence scores and bounding boxes as MMAX2 attributes.[10]

2.1.1 Highlighting Detection

Highlighting detection and subsequent extraction can be performed if the scanned paper documents contain manual markup. In its current state, the detection procedure described in the following *requires* inpainted OCR background images which, in our case, were produced by the third-party OCR software used by the scanning service provider. tesseract, on the other hand, does *not* produce these images. While it would be desirable to employ free software only, this fact does not severely limit the usefulness of our procedure, because 1) other software (either free or commercial) with the same functionality might exist, and 2) even for document collections of medium size, employing an external service provider might be the most economical solution even in academic / research settings, anyway. What is more, inpainted backgrounds are only required if highlighting detection is desired: For text-only alignment, plain scans are sufficient.

[5] `https://www.ncbi.nlm.nih.gov/pmc/tools/oai/`

[6] `http://sabio.h-its.org/`

[7] `https://github.com/tesseract-ocr/tesseract`

[8] `http://kba.cloud/hocr-spec/1.2/`

[9] `https://github.com/tesseract-ocr/tessdata_best`

[10] See the lower part of Figure A.1 in the Appendix.

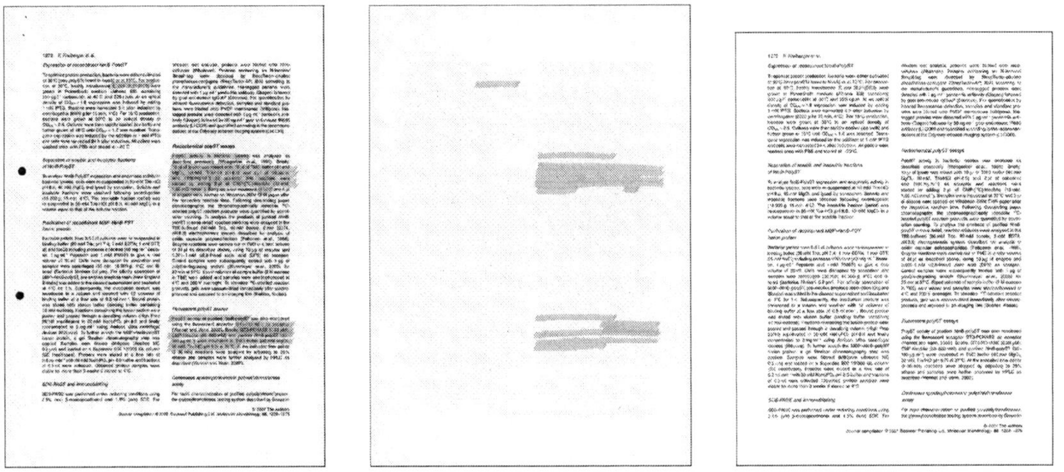

(a) DOI_{scan} (b) DOI_{scan_bg} (c) DOI_{conv}

Figure 1: Three image renderings of the same document page: Scanned print-out w/ manual markup (a), background with markup only (b), and original PDF (c).

The actual highlighting extraction works as follows (see Müller et al. (2020) for details): Since document highlighting comes mostly in strong colors, which are characterized by large differences among their three component values in the RGB color model, we create a binarized version of each page by going over all pixels in the background image and setting each pixel to 1 if the pairwise differences between the R, G, and B components are above a certain threshold (50), and to 0 otherwise. This yields an image with regions of higher and lower density of black pixels. In the final step, we iterate over the word-level tokens created from the hOCR output and converted into MMAX2 format earlier, compute for each word its degree of highlighting as the percentage of black pixels in the word's bounding box, and store that percentage value as another MMAX2 attribute if it is at least 50%. An example result will be presented in Section 4.2.

2.2 PMC® `.nxml` to Multilevel XML

The `.nxml` format employed for PubMedCentral® full-text documents uses the JATS scheme[11] which supports a rich meta data model, only a fraction of which is of interest for the current task. In principle, however, all information contained in JATS-conformant documents can also be represented in the multilevel XML format of MMAX2. The `.nxml` data provides precise information about both the textual content (including correctly encoded special characters) and its word- and section-level layout. At present, we only extract content from the **`<article-meta>`** section (`<article-title>`, `<surname>`, `<given-names>`, `<xref>`, `<email>`, `<aff>`, and `<abstract>`), and from the **`<body>`** (`<sec>`, `<p>`, `<tr>`, `<td>`, `<label>`, `<caption>`, and `<title>`). These sections cover the entire textual content of the document. We also extract the formatting tags `<italic>`, `<bold>`, `<underline>`, and in particular `<sup>` and `<sub>`. The latter two play a crucial role in the chemical formulae and other domain-specific expressions. Converting the `.nxml` data to our MMAX2 format is straightforward.[12] In some cases, the `.nxml` files contain embedded LaTex code in `<tex-math>` tags. If this tag is encountered, its content is processed as follows: LaTex Math encodings for sub- and superscript, _{} and ^{}, are removed, their content is extracted and re-inserted with JATS-conformant `<sub>...</sub>` and `<sup>...</sup>` elements. Then, the resulting LaTex-like string is sent through the `detex` tool to remove any other markup. While this obviously cannot handle layouts like e.g. fractions, it still preserves many simpler expressions that would otherwise be lost in the conversion.

[11] `https://jats.nlm.nih.gov/archiving/`

[12] See the upper part of Figure A.1 in the Appendix.

3 Outline of the Alignment Procedure

The actual word-level alignment of the DOI_{xml} version with the DOI_{ocr} version of a document operates on lists of $< token, id >$ tuples which are created from each version's MMAX2 annotation. These lists are characterized by longer and shorter stretches of tuples with matching tokens, which just happen to start and end at different list indices. These stretches are interrupted at times by (usually shorter) sequences of tuples with non-matching tokens, which mostly exist as the result of OCR errors (see below). *Larger* distances between stretches of tuples with matching tokens, on the other hand, can be caused by structural differences between the DOI_{xml} and the DOI_{ocr} version, which can reflect actual layout differences, but which can also result from OCR errors like incorrectly joining two adjacent lines from two columns.

The task of the alignment is to find the correct mapping on the token level for as many tuples as possible. We use the `align.globalxx` method from the `Bio.pairwise2` module of Biopython (Cock et al., 2009), which provides pairwise sequence alignment using a dynamic programming algorithm (Needleman and Wunsch, 1970). While this library supports the definition of custom similarity functions for minimizing the alignment cost, we use the most simple version which just applies a binary (=identity) matching scheme, i.e. full matches are scored as 1, all others as 0. This way, we keep full control of the alignment, and can identify and locally fix non-matching sequences during post-processing (cf. Section 3.2 below). The result of the alignment (after optional pre- and post-processing) is an n-to-m mapping between $< token, id >$ tuples from the DOI_{xml} and the DOI_{ocr} version of the same document.[13]

3.1 Pre-Processing

The main difference between pre- and post-processing is that the former operates on two still *unrelated* tuple lists of different lengths, while for the latter the tuple lists have the same length due to padding entries («`GAP`») that were inserted by the alignment algorithm in order to bridge sequences of non-alignable tokens. Pre-processing aims to smooth out trivial mismatches and thus to help alignment. Both pre- and post-processing, however, only modify the tokens in DOI_{ocr}, but never those in DOI_{xml}, which are considered as gold-standard.

[13]See also the central part of Figure A.1 in the Appendix.

Pre-compress matching sequences [`pre_compress=p`] The space complexity of the Needleman-Wunsch algorithm is $O(mn)$, where m and n are the numbers of tuples in each document. Given the length of some documents, the memory consumption of the alignment can quickly become critical. In order to reduce the number of tuples to be compared, we apply a simple pre-compression step which first identifies sequences of p tuples (we use $p = 20$ in all experiments) with *perfectly identical* tokens in both documents, and then replaces them with single tuples where the token and id part consist of concatenations of the individual tokens and ids. After the alignment, these compressed tuples are expanded again.

While pre-compression was always performed, the pre- and post-processing measures described in the following are optional, and their individual effects on the alignment will be evaluated in Section 4.1.

De-hyphenate DOI_{ocr} tokens [`dehyp`] Sometimes, words in the DOI_{ocr} versions are hyphenated due to layout requirements which, in principle, do not exist in the DOI_{xml} versions. These words appear as three consecutive tuples with either the '-' or '¬' token in the center tuple. For de-hyphenation, we search the tokens in the tuple list for DOI_{ocr} for single hyphen characters and reconstruct the potential un-hyphenated word by concatenating the tokens immediately before and after the hyphen. If this word exists *anywhere* in the list of DOI_{xml} tokens, we simply substitute the three original $< token, id >$ DOI_{ocr} tuples with one merged tuple. De-hyphenation (like all other pre- and post-processing measures) is completely lexicon-free, because the decision whether the un-hyphenated word exists is only based on the content of the DOI_{xml} document.

Diverging tokenizations in the DOI_{xml} and DOI_{ocr} document versions are a common cause of local mismatches. Assuming the tokenization in DOI_{xml} to be correct, tokenizations can be fixed by either joining or splitting tokens in DOI_{ocr}.

Join incorrectly split DOI_{ocr} tokens [`pre_join`] We apply a simple rule to detect and join tokens that were incorrectly split in DOI_{ocr}. We move a window of size 2 over the list of DOI_{ocr} tuples and concatenate the two tokens. We then iterate over all tokens in the DOI_{xml} version. If we find the reconstructed word in a matching context (one immediately preceeding and following token), we replace, in the DOI_{ocr} version, the first original tuple with the concatenated one, assigning the concatenated ID as new ID, and remove the second tuple from the list. Consider the following example.

```
< phen ,    word_3084 >n
< yl ,      word_3085 >n+1
==>
< phenyl ,  word_3084+word_3085 >n
```

This process (and the following one) is repeated until no more modifications can be performed.

Split incorrectly joined DOI_{ocr} tokens [`pre_split`] In a similar fashion, we identify and split incorrectly joined tokens. We move a window of size 2 over the list of DOI_{xml} tuples, concatenate the two tokens, and try to locate a corresponding single token, in a matching context, in the list of DOI_{ocr} tuples. If found, we replace the respective tuple in that list with two new tuples, one with the first token from the DOI_{xml} tuple and one with the second one. Both tuples retain the ID from the original DOI_{ocr} tuple. In the following example, the correct tokenization separates the trailing number 3 from the rest of the expression, because it needs to be typeset in subscript in order for the formula to be rendered correctly.

```
< KHSO3,   word_3228 >n
==>
< KHSO,    word_3228 >n
< 3,       word_3228 >n+1
```

3.2 Alignment Post-Processing

Force-align [`post_force_align`] The most frequent post-processing involves cases where single tokens of the same length and occurring in the same context are not aligned automatically. In the following, the left column contains the DOI_{ocr} and the right the DOI_{xml} tuples. In the first example, the β was not correctly recognized and substituted with a B. We identify force-align candidates like these by looking for sequences of s consecutive tuples with a «GAP» token in one list, followed by a similar sequence of the same length in the other. Then, if both the context and the number of characters matches, we force-align the two sequences.

```
<metallo , word_853>      <metallo , word_546>
<-, word_854>             <-, word_547>
<B, word_855>             <<<GAP>>, ->
<<<GAP>>, ->              <β, word_548>
<-, word_856>             <-, word_549>
<lactamase , word_857>    <lactamase , word_550>
==>
...
<B, word_855>             <β, word_548>
...
```

For $s = 2$, force-align will also fix the following.

```
<acid , word_1643>        <acid , word_997>
<,, word_1644>            <,, word_998>
<It , word_1645>          <<<GAP>>, ->
<Is , word_1646>          <<<GAP>>, ->
<<<GAP>>, ->              <it , word_999>
<<<GAP>>, ->              <is , word_1000>
<purified , word_1647>    <purified , word_1001>
<using , word_1648>       <using , word_1002>
==>
...
<It , word_1645>          <it , word_999>
<Is , word_1646>          <is , word_1000>
...
```

4 Experiments

4.1 Quantitative Evaluation

We evaluate the system on our 68 DOI_{xml} – DOI_{ocr} document pair data set by computing P, R, and F for the task of aligning tokens from DOI_{xml} (the gold-standard) to tokens in DOI_{ocr}. By defining the evaluation task in this manner, we take into account that the DOI_{ocr} version usually contains more tokens, mostly because it includes the bibliography, which is generally *not* included in the DOI_{xml} version. Thus, an alignment is perfect if every token in DOI_{xml} is correctly aligned to a token in DOI_{ocr}, regardless of there being *additional* tokens in DOI_{ocr}. In order to compute P and R, the number of *correct* alignments (=TP) among all alignments needs to be determined. Rather than inspecting and checking all alignments manually, we employ a simple heuristic: Given a pair of automatically aligned tokens, we create two KWIC string representations, KWIC_{xml} and KWIC_{ocr}, with a left and right context of 10 tokens each. Then, we compute the normalized Levenshtein similarity $lsim$ between each pair $ct1$ and $ct2$ of left and right contexts, respectively, as

$$1 - levdist(ct1, ct2)/max(len(ct1), len(ct2))$$

We count the alignment as correct (=TP) if $lsim$ of **both** the two left **and** the two right contexts is $>=$ .50, and as incorrect (=FP) otherwise.[14] The number of missed alignments (=FN) can be computed by substracting the number of TP from the number of all tokens in DOI_{xml}. Based on these counts, we compute precision (P), recall (R), and F-score (F) in the standard way. Results are provided in Table 1. For each parameter setting (first column), there are two result columns with P, R, and F each. The column **DOI_{xml} – DOI_{conv}** contains alignment results for which OCR was

[14] Figure A.2 in the Appendix provides an example.

Pre-/Post-Processing	$\mathbf{DOI}_{xml}$ – $\mathbf{DOI}_{conv}$			$\mathbf{DOI}_{xml}$ – $\mathbf{DOI}_{scan}$		
	P	R	F	P	R	F
–	**95.04**	76.90	85.01	**93.59**	75.29	83.45
`dehyp`	94.91	77.47	85.31	93.48	75.96	83.81
`pre`	**95.04**	77.40	85.32	93.57	75.83	83.77
`dehyp + pre`	94.90	77.97	85.61	93.47	76.52	84.15
`post_force_align`	95.03	78.57	86.02	93.57	76.99	84.48
`dehyp + post_force_align`	94.91	79.17	86.32	93.47	77.69	84.86
`pre + post_force_align`	95.02	79.08	86.32	93.56	77.55	84.81
`dehyp + pre + post_force_align`	94.90	**79.68**	**86.63**	93.47	**78.27**	**85.20**

Table 1: Alignment Scores (micro-averaged, n=68). All results using `pre_compress`=20. Max. values in bold.

performed on the converted PDF pages, while results in column $\mathbf{DOI}_{xml}$ – $\mathbf{DOI}_{scan}$ are based on scanned print-outs. Differences between these two sets of results are due to the inferior quality of the images used in the latter. The top row in Table 1 contains the result of using only the alignment without any pre- or post-processing. Subsequent rows show results for all possible combinations of pre- and post-processing measures (cf. Section 3.1). Note that `pre_split` and `pre_join` are not evaluated separately and appear combined as `pre`. The first observation is that, for $\mathbf{DOI}_{xml}$ – $\mathbf{DOI}_{conv}$ and $\mathbf{DOI}_{xml}$ – $\mathbf{DOI}_{scan}$, precision is very high, with max. values of 95.04 and 93.59, respectively. This is a result of the rather strict alignment method which will align two tokens only if they are *identical* (rather than merely *similar*). At the same time, precision is very stable across experiments, i.e. indifferent to changes in pre- and post-processing. This is because, as described in Section 3.1, pre- and post-processing exclusively aim to improve recall by either smoothing out trivial mismatches before alignment, or adding missing alignments afterwards. In fact, pre- and post-processing actually *introduce* precision errors, since they relax this alignment condition somewhat: This is evident in the fact that the two top precision scores result from the setup with no pre- or post-processing at all, and even though the differences across experiments are extremly small, the pattern is still clear. Table 1 also shows the intended positive effect of the different pre- and post-processing measures on recall. Without going into much detail, we can state the following: For $\mathbf{DOI}_{xml}$ – $\mathbf{DOI}_{conv}$ and $\mathbf{DOI}_{xml}$ – $\mathbf{DOI}_{scan}$, the lowest recall results from the setup without pre- or post-processing. When pre- and post-processing measures are added, recall increases constantly, at the expense of small drops in precision. However, the positive effect consistently outweighs the negative, causing the F-score to increase to a max. score of 86.63 and 85.20, respectively, when all pre- and post-processing measures are used. Finally, as expected, the inferior quality of the data in DOI_{scan} as compared to DOI_{conv} is nicely reflected in consistently lower scores across all measurements. The absolute differences, however, are very small, amounting to only about 1.5 points. This might be taken to indicate that converted (rather than printed and scanned) PDF documents can be functionally equivalent as input for tasks like OCR ground-truth data generation.

4.2 Qualitative Evaluation and Examples

This section complements the quantitative evaluation with some illustrative examples. Figure 2 shows two screenshots in which DOI_{scan} (left) and DOI_{xml} (right) are displayed in the MMAX2 annotation tool. The left image shows that the off-the-shelf text recognition accuracy of tesseract is very good for standard text, but lacking, as expected, when it comes to recognising special characters and subscripts (like μ, $ZnCl_2$, or k_{obs} in the example). For the highlighting detection, the yellow text background was chosen as visualization in MMAX2 in order to mimick the physical highlighting of the printed paper. Note that since the highlighting detection is based on layout position only (and not anchored to text), manually highlighted text is recognized as highlighted regardless of whether the actual underlying text is recognized correctly. The right image shows the rendering of the correct text extracted from the original PMC® full-text XML. The rendering of the title as bold and underlined is based on typographic information that was extracted at conversion time (cf. Section 2.2). The same is true for the subscripts, which are correctly rendered both in terms of the content and the position. Table 2 displays a different type of result, i.e. a small selection of a much larger set of OCR errors with their respective images and the correct recognition result. This data, automatically identi-

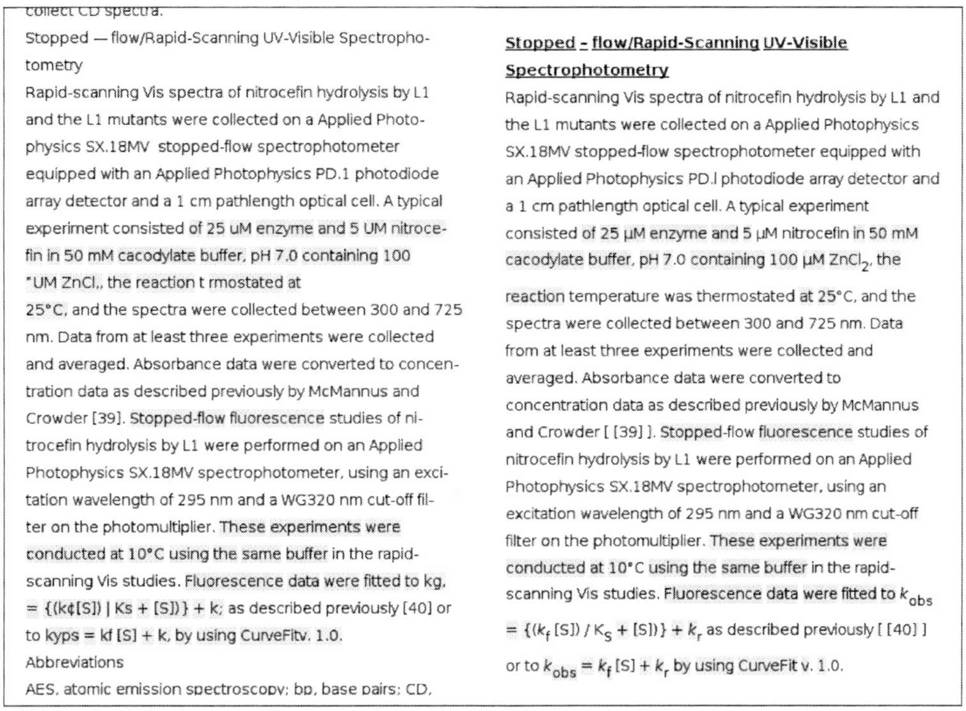

collect CD spectra.
Stopped — flow/Rapid-Scanning UV-Visible Spectrophotometry
Rapid-scanning Vis spectra of nitrocefin hydrolysis by L1 and the L1 mutants were collected on a Applied Photophysics SX.18MV stopped-flow spectrophotometer equipped with an Applied Photophysics PD.1 photodiode array detector and a 1 cm pathlength optical cell. A typical experiment consisted of 25 uM enzyme and 5 UM nitrocefin in 50 mM cacodylate buffer, pH 7.0 containing 100 "UM ZnCl,, the reaction t rmostated at 25°C, and the spectra were collected between 300 and 725 nm. Data from at least three experiments were collected and averaged. Absorbance data were converted to concentration data as described previously by McMannus and Crowder [39]. Stopped-flow fluorescence studies of nitrocefin hydrolysis by L1 were performed on an Applied Photophysics SX.18MV spectrophotometer, using an excitation wavelength of 295 nm and a WG320 nm cut-off filter on the photomultiplier. These experiments were conducted at 10°C using the same buffer in the rapid-scanning Vis studies. Fluorescence data were fitted to kg, = {(k¢[S]) | Ks + [S])} + k; as described previously [40] or to kyps = kf [S] + k, by using CurveFitv. 1.0.
Abbreviations
AES, atomic emission spectroscopy; bp, base pairs; CD,

Stopped - flow/Rapid-Scanning UV-Visible Spectrophotometry
Rapid-scanning Vis spectra of nitrocefin hydrolysis by L1 and the L1 mutants were collected on a Applied Photophysics SX.18MV stopped-flow spectrophotometer equipped with an Applied Photophysics PD.I photodiode array detector and a 1 cm pathlength optical cell. A typical experiment consisted of 25 μM enzyme and 5 μM nitrocefin in 50 mM cacodylate buffer, pH 7.0 containing 100 μM $ZnCl_2$, the reaction temperature was thermostated at 25°C, and the spectra were collected between 300 and 725 nm. Data from at least three experiments were collected and averaged. Absorbance data were converted to concentration data as described previously by McMannus and Crowder [[39]]. Stopped-flow fluorescence studies of nitrocefin hydrolysis by L1 were performed on an Applied Photophysics SX.18MV spectrophotometer, using an excitation wavelength of 295 nm and a WG320 nm cut-off filter on the photomultiplier. These experiments were conducted at 10°C using the same buffer in the rapid-scanning Vis studies. Fluorescence data were fitted to k_{obs} = $\{(k_f [S]) / K_S + [S])\} + k_r$ as described previously [[40]] or to $k_{obs} = k_f [S] + k_r$ by using CurveFit v. 1.0.

Figure 2: DOI_{scan} document with automatically detected, overlayed highlighting (left). DOI_{xml} with highlighting transferred from automatically aligned DOI_{scan} document (right).

fied by the alignment post-processing, is a valuable resource for the development of biomedical expression OCR systems.

Image	OCR	PMC®
$\Delta\varepsilon_{265}$	Agyg5	$\Delta\varepsilon_{265}$
μM	"UM	μM
k_{obs}	kyps	k_{obs}
7 ± 1	7 + 1	7 ± 1
DH5α	DH50	DH5α

Table 2: Examples of image snippets (left) with incorrect (middle) and correct (right) text representation.

5 Related Work

The work in this paper is obviously related to automatic text alignment, with the difference that what is mostly done there is the alignment of texts in different languages (i.e. **bi-lingual alignment**). Gale and Church (1993) align not words but entire sentences from two languages based on statistical properties. Even if words were aligned, alignment candidates in bi-lingual corpora are not identified on the basis of simple matching, with the exception of language-independent tokens like e.g. proper names.

Scanning and OCR is also often applied to historical documents, which are only available in paper (Hill and Hengchen, 2019; van Strien et al., 2020; Schaefer and Neudecker, 2020). Here, **OCR post-correction** attempts to map words with word- and character-level OCR errors (similar to those found in our DOI_{ocr} data) to their correct variants, but it does so by using general language models and dictionaries, and not an aligned correct version. Many of the above approaches have in common that they employ specialized OCR models and often ML/DL models of considerable complexity.

The idea of using an electronic and a paper version of **the same document** for creating a character-level alignment dates back at least to Kanungo and Haralick (1999), who worked on OCR ground-truth data generation. Like most later methods, the procedure of Kanungo and Haralick (1999) works on the *graphical* level, as opposed to the *textual* level. Kanungo and Haralick (1999) use LaTex to create what they call 'ideal document images' with controlled content. Print-outs of these images are created, which are then photocopied and scanned, yielding slightly noisy and skewed variants of the 'ideal' images. Then, corresponding feature points in both images are identified, and a projective transformation between these is computed. Finally, the actual ground-truth data is generated by applying this transformation for aligning the bounding boxes in the ideal images to their correspondences in the scanned images. Since Kanungo and Haralick (1999) have full control over the content of their 'ideal document images', extracting the ground-truth character data is trivial. The approach of van Beusekom et al. (2008) is similar to that of Kanungo and Haralick (1999), but the former use more sophisticated methods, including Canny edge detection (Canny, 1986) for finding corresponding sections in images of the original and the scanned document, and RAST (Breuel, 2001) for doing the actual alignment. Another difference is that van Beusekom et al. (2008) use pre-existing PDF documents as the source documents from which ground-truth data is to be extracted. Interestingly, however, their experiments only use synthetic ground-truth data from the UW3 data set[15], in which bounding boxes and the contained characters are explicitly encoded. In their conclusion, van Beusekom et al. (2008) concede that extracting ground-truth data from PDF is a non-trivial task in itself. Ahmed et al. (2016) work on automatic ground-truth data generation for *camera-captured* document images, which they claim pose different problems than document images created by scanning, like e.g. blur, perspective distortion, and varying lighting. Their procedure, however, is similar to that of van Beusekom et al. (2008). They also use pre-existing PDF documents and automatically rendered 300 DPI images of these documents.

6 Conclusions

In this paper, we described a completely unsupervised procedure for automatically aligning printed paper documents with their electronic full-text counterparts. Our point of departure and main motivation was the idea to alleviate the effect of the **paper-to-electronic media break** in manual biocuration, where printed paper is still very popular when it comes to close reading and manual markup. We also argued that the related task of document *triage* can benefit from the availability of alignments between electronic full-text documents (as retrieved from a literature database) and the corresponding PDF documents. Apart from this, we identified yet another field of application, biomedical expression OCR, which can benefit from ground-truth data which can automatically be generated with our procedure. Improvements in biomedical expression OCR, then, can feed back into the other use cases, by improving the OCR step and thus the alignment, thus potentially establishing a kind of bootstrapping development. Our implementation relies on *tried and tested* technology, including tesseract as off-the-shelf OCR component, Biopython for the alignment, and MMAX2 as visualization and data processing platform. The most computationally complex part is the actual sequence alignment with a dynamic programming algorithm from the Biopython library, which we keep tractable even for longer documents by using a simple pre-compression method. The main experimental finding of this paper is that our approach, although very simple, yields a level of performance that we consider suitable for practical applications. In quantitative terms, the procedure reaches a very good F-score of 86.63 on converted and 85.20 on printed and scanned PDF documents, with corresponding precision scores of 94.90 and 93.47, respectively. The negligible difference in results between the two types of images is interesting, as it seems to indicate that converted PDF documents, which are very easy to generate in large amounts, are almost equivalent to the more labour-intensive scans. In future work, we plan to implement solutions for the identified use cases, and to test them in actual biocuration settings. Also, we will start creating OCR ground-truth data at a larger scale, and apply that for the development of specialised tools for biomedical OCR. In the long run, procedures like the one presented in this paper might contribute to the development of systems that support curators to work in a more natural, practical, convenient, and efficient way.

Acknowledgements

This work was done as part of the project DeepCurate, which is funded by the German Federal Ministry of Education and Research (BMBF) (No. 031L0204) and the Klaus Tschira Foundation, Heidelberg, Germany. We thank the anonymous BioNLP reviewers for their helpful suggestions.

[15]http://tc11.cvc.uab.es/datasets/DFKI-TGT-2010_1

References

Sheraz Ahmed, Muhammad Imran Malik, Muhammad Zeshan Afzal, Koichi Kise, Masakazu Iwamura, Andreas Dengel, and Marcus Liwicki. 2016. A generic method for automatic ground truth generation of camera-captured documents. *CoRR*, abs/1605.01189.

Thomas M. Breuel. 2001. A practical, globally optimal algorithm for geometric matching under uncertainty. *Electron. Notes Theor. Comput. Sci.*, 46:188–202.

George Buchanan and Fernando Loizides. 2007. Investigating document triage on paper and electronic media. In *Research and Advanced Technology for Digital Libraries*, pages 416–427, Berlin, Heidelberg. Springer Berlin Heidelberg.

John F. Canny. 1986. A computational approach to edge detection. *IEEE Trans. Pattern Anal. Mach. Intell.*, 8(6):679–698.

Virginia Clinton. 2019. Reading from paper compared to screens: A systematic review and meta-analysis. *Journal of Research in Reading*, 42(2):288–325.

Peter J. A. Cock, Tiago Antao, Jeffrey T. Chang, Brad A. Chapman, Cymon J. Cox, Andrew Dalke, Iddo Friedberg, Thomas Hamelryck, Frank Kauff, Bartek Wilczynski, and Michiel J. L. de Hoon. 2009. Biopython: freely available python tools for computational molecular biology and bioinformatics. *Bioinform.*, 25(11):1422–1423.

William A. Gale and Kenneth W. Church. 1993. A program for aligning sentences in bilingual corpora. *Computational Linguistics*, 19(1):75–102.

Mark J. Hill and Simon Hengchen. 2019. Quantifying the impact of dirty OCR on historical text analysis: Eighteenth century collections online as a case study. *Digit. Scholarsh. Humanit.*, 34(4):825–843.

Lynette Hirschman, Gully Burns, Martin Krallinger, Cecilia Arighi, Kevin Cohen, Alfonso Valencia, Cathy Wu, Andrew Chatr-Aryamontri, Karen Dowell, Eva Huala, Anália Lourenco, Robert Nash, Anne-Lise Veuthey, Thomas Wiegers, and Andrew Winter. 2012. Text mining for the biocuration workflow. *Database : the journal of biological databases and curation*, 2012:bas020.

International Society for Biocuration. 2018. Biocuration: Distilling data into knowledge. *PLOS Biology*, 16(4):1–8.

Tapas Kanungo and Robert M. Haralick. 1999. An automatic closed-loop methodology for generating character groundtruth for scanned documents. *IEEE Trans. Pattern Anal. Mach. Intell.*, 21(2):179–183.

Maja Köpper, Susanne Mayr, and Axel Buchner. 2016. Reading from computer screen versus reading from paper: does it still make a difference? *Ergonomics*, 59(5):615–632. PMID: 26736059.

Christoph Müller and Michael Strube. 2006. Multi-level annotation of linguistic data with MMAX2. In Sabine Braun, Kurt Kohn, and Joybrato Mukherjee, editors, *Corpus Technology and Language Pedagogy: New Resources, New Tools, New Methods*, pages 197–214. Peter Lang, Frankfurt a.M., Germany.

Mark-Christoph Müller, Sucheta Ghosh, Maja Rey, Ulrike Wittig, Wolfgang Müller, and Michael Strube. 2020. Reconstructing manual information extraction with db-to-document backprojection: Experiments in the life science domain. In *Proceedings of the First Workshop on Scholarly Document Processing, SDP@EMNLP 2020, Online, November 19, 2020*, pages 81–90. Association for Computational Linguistics.

SB Needleman and CD Wunsch. 1970. A general method applicable to the search for similarities in the amino acid sequence of two proteins. *Journal of molecular biology*, 48(3):443—453.

Wataru Ohyama, Masakazu Suzuki, and Seiichi Uchida. 2019. Detecting mathematical expressions in scientific document images using a u-net trained on a diverse dataset. *IEEE Access*, 7:144030–144042.

Robin Schaefer and Clemens Neudecker. 2020. A two-step approach for automatic OCR post-correction. In *Proceedings of the The 4th Joint SIGHUM Workshop on Computational Linguistics for Cultural Heritage, Social Sciences, Humanities and Literature*, pages 52–57, Online. International Committee on Computational Linguistics.

Joost van Beusekom, Faisal Shafait, and Thomas M. Breuel. 2008. Automated OCR ground truth generation. In *The Eighth IAPR International Workshop on Document Analysis Systems, DAS 2008, September 16-19, 2008, Nara, Japan*, pages 111–117. IEEE Computer Society.

Daniel van Strien, Kaspar Beelen, Mariona Coll Ardanuy, Kasra Hosseini, Barbara McGillivray, and Giovanni Colavizza. 2020. Assessing the impact of OCR quality on downstream NLP tasks. In *Proceedings of the 12th International Conference on Agents and Artificial Intelligence, ICAART 2020, Volume 1, Valletta, Malta, February 22-24, 2020*, pages 484–496. SCITEPRESS.

A Venkatesan, N Karamanis, M Ide-Smith, J Hickford, and J McEntyre. 2019. Understanding life sciences data curation practices via user research [version 1; peer review: 1 approved, 1 approved with reservations]. *F1000Research*, 8(1622).

Jian Wang, Mengying Li, Qishuai Diao, Hongfei Lin, Zhihao Yang, and YiJia Zhang. 2020. Biomedical document triage using a hierarchical attention-based capsule network. *BMC Bioinformatics*, 21(380).

Ulrike Wittig, Maja Rey, Andreas Weidemann, Renate Kania, and Wolfgang Müller. 2018. SABIO-RK: an updated resource for manually curated biochemical reaction kinetics. *Nucleic Acids Research*, 46(D1):D656–D660.

Ulrike Wittig, Maja Rey, Andreas Weidemann, and Wolfgang Müller. 2017. Data management and data enrichment for systems biology projects. *Journal of biotechnology.*, 261:229–237.

Appendix

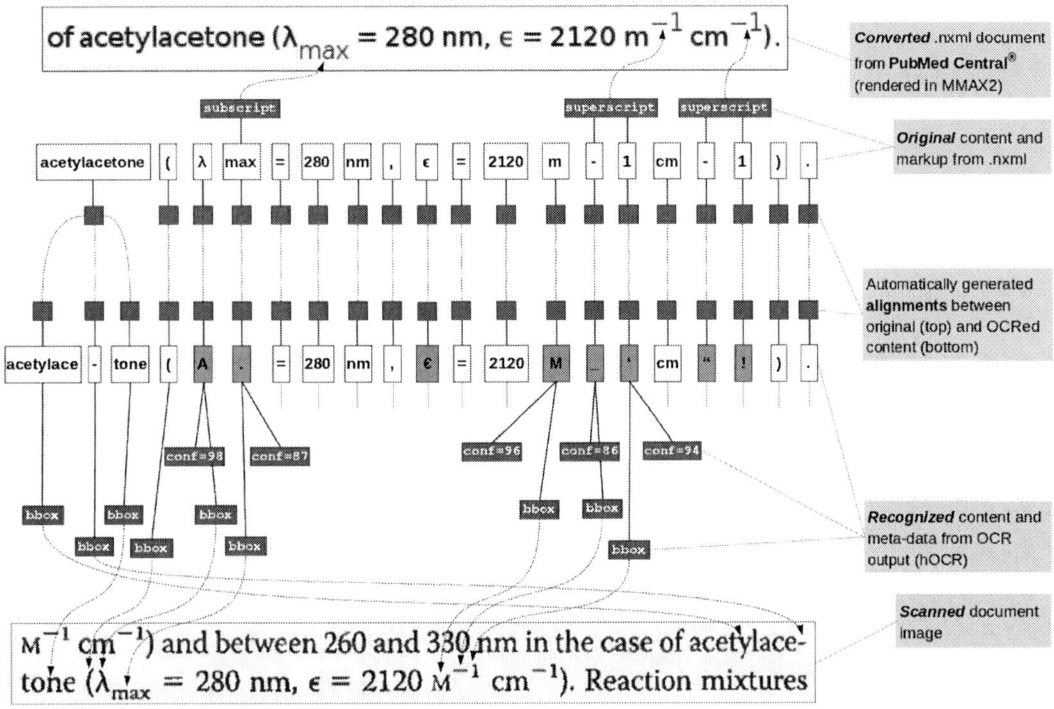

Figure A.1: Conversion and alignment data model. **Top**: Full-text and markup (`subscript`, `superscript`) is extracted from .nxml documents. Each content token is associated with an alignment token (solid blue boxes). **Bottom**: Text and meta-data is extracted from the OCR result of scanned document pages. Meta-data includes bounding boxes, which link the recognized text to image regions, and numerical recognition scores, which reflect the confidence with which the OCR system recognized the respective token. (Not all meta-data is given in the Figure to avoid clutter.)

```
                          MUT) DeficiencyFornyPatrick  1  2  † FroeseD >>  . <<    Sean  1  3  † SuormalaTerttu 1 YueWyatt W
                                .org 1,2% Communicated by Elizabeth F  >>  . <<    L Neufeld Received 6 June 2014; accepted revised manuscript
FP (0.136, 0.186)

                            Sean  1  3  † SuormalaTerttu 1 YueWyatt W  >>  . <<     3 BaumgartnerMatthias R.  1  21 Division for
                  6 June 2014; accepted revised manuscript 27 July 2014  >>  . <<    ABSTRACT: Methylmalonyl-CoA mutase (MUT) is
FP (0.208, 0.068)

  ZurichSwitzerland3 Structural Genomics ConsortiumUniversity of OxfordUK  >>  ABSTRACT << Methylmalonyl—CoA mutase (MUT) is
                     June 2014; accepted revised manuscript 27 July 2014.  >>  ABSTRACT << : Methylmalonyl-CoA mutase (MUT) is an
FP (0.169, 0.842)

Switzerland3 Structural Genomics ConsortiumUniversity of OxfordUKABSTRACT  >>  Methylmalonyl << —CoA mutase (MUT) is an
                    ; accepted revised manuscript 27 July 2014. ABSTRACT:  >>  Methylmalonyl << -CoA mutase (MUT) is an essential enzyme
FP (0.247, 0.550)

                        University of OxfordUKABSTRACTMethylmalonyl—CoA >>  mutase << (MUT) is an essential enzyme in propionate catabolism
                     manuscript 27 July 2014. ABSTRACT: Methylmalonyl-CoA  >>  mutase << (MUT) is an essential enzyme in propionate catabolism
TP (0.538, 1.000)

                            of OxfordUKABSTRACTMethylmalonyl—CoA mutase >>  ( <<   MUT) is an essential enzyme in propionate catabolism that
                        27 July 2014. ABSTRACT: Methylmalonyl-CoA mutase  >>  ( <<   MUT) is an essential enzyme in propionate catabolism that
TP (0.667, 1.000)

                             OxfordUKABSTRACTMethylmalonyl—CoA mutase ( >>  MUT << ) is an essential enzyme in propionate catabolism that requires
                         July 2014. ABSTRACT: Methylmalonyl-CoA mutase (  >>  MUT << ) is an essential enzyme in propionate catabolism that requires
TP (0.702, 1.000)
```

Figure A.2: Sample output of the KWIC-based alignment evaluation procedure (context size = 10 tokens (left and right)). In each pair of lines, the top line comes from DOI_{ocr}, the bottom line from DOI_{xml}. Pairs are labeled as TP (correctly aligned) if the normalized Levenshtein similarity of both the left and the right context strings (given in parentheses) is above 0.5.

Improving Biomedical Pretrained Language Models with Knowledge

Zheng Yuan[1*] **Yijia Liu**[2] **Chuanqi Tan**[2†] **Songfang Huang**[2] **Fei Huang**[2]
[1]Tsinghua University [2]Alibaba Group
yuanz17@mails.tsinghua.edu.cn
{yanshan.lyj,chuanqi.tcq,songfang.hsf,f.huang}@alibaba-inc.com

Abstract

Pretrained language models have shown success in many natural language processing tasks. Many works explore incorporating knowledge into language models. In the biomedical domain, experts have taken decades of effort on building large-scale knowledge bases. For example, the Unified Medical Language System (UMLS) contains millions of entities with their synonyms and defines hundreds of relations among entities. Leveraging this knowledge can benefit a variety of downstream tasks such as named entity recognition and relation extraction. To this end, we propose KeBioLM, a biomedical pretrained language model that explicitly leverages knowledge from the UMLS knowledge bases. Specifically, we extract entities from PubMed abstracts and link them to UMLS. We then train a knowledge-aware language model that firstly applies a text-only encoding layer to learn entity representation and applies a text-entity fusion encoding to aggregate entity representation. Besides, we add two training objectives as entity detection and entity linking. Experiments on the named entity recognition and relation extraction from the BLURB benchmark demonstrate the effectiveness of our approach. Further analysis on a collected probing dataset shows that our model has better ability to model medical knowledge.

1 Introduction

Large-scale pretrained language models (PLMs) are proved to be effective in many natural language processing (NLP) tasks (Peters et al., 2018; Devlin et al., 2019). However, there are still many works that explore multiple strategies to improve the PLMs. Firstly, in specialized domains (i.e biomedical domain), many works demonstrate that using in-domain text (i.e. PubMed and MIMIC for biomedical domain) can further improve downstream tasks over general-domain PLMs (Lee et al., 2020; Peng et al., 2019; Gu et al., 2020; Shin et al., 2020; Lewis et al., 2020; Beltagy et al., 2019; Alsentzer et al., 2019). Secondly, unlike training language models (LMs) with unlabeled text, many works explore training the model with structural knowledge (i.e. triplets and facts) for better language understanding (Zhang et al., 2019; Peters et al., 2019; Févry et al., 2020; Wang et al., 2019). In this work, we propose to combine the above two strategies for a better **K**nowledge **e**nhanced **Bio**medical pretrained **L**anguage **M**odel (KeBioLM).

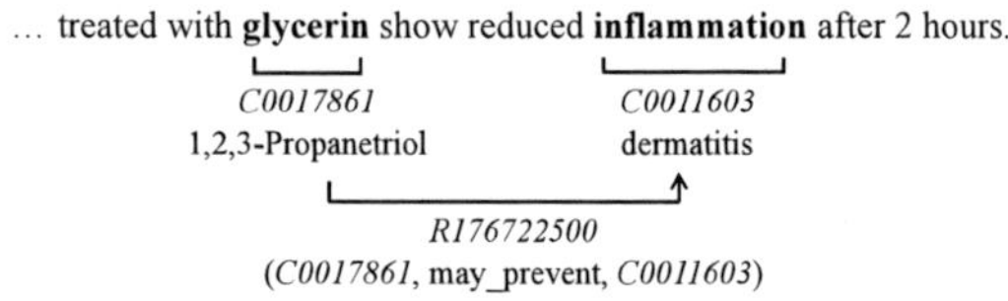

Figure 1: An example of the biomedical sentence. Two entities "glycerin" and "inflammation" are linked to *C0017861* (1,2,3-Propanetriol) and *C0011603* (dermatitis) respectively with a relation triplet (*C0017861, may_prevent, C0011603*) in UMLS.

As an applied discipline that needs a lot of facts and evidence, the biomedical and clinical fields have accumulated data and knowledge from a very early age (Ashburner et al., 2000; Stearns et al., 2001). One of the most representative work is Unified Medical Language System (UMLS) (Bodenreider, 2004) that contains more than 4M entities with their synonyms and defines over 900 kinds of relations. Figure 1 shows an example. There are two entities "glycerin" and "inflammation" that can be linked to *C0017861* (1,2,3-Propanetriol) and *C0011603* (dermatitis) respectively with a *may_prevent* relation in UMLS. As the most important facts in biomedical text, entities and relations can provide information for better text understanding (Xu et al., 2018; Yuan et al., 2020).

To this end, we propose to improve biomedical PLMs with explicit knowledge modeling. Firstly,

* Work done at Alibaba DAMO Academy.
† Corresponding author.

Proceedings of the BioNLP 2021 workshop, pages 180–190
June 11, 2021. ©2021 Association for Computational Linguistics

we process the PubMed text to link entities to the knowledge base. We apply an entity recognition and linking tool ScispaCy (Neumann et al., 2019) to annotate 660M entities in 3.5M documents. Secondly, we implement a knowledge enhanced language model based on Févry et al. (2020), which performs a text-only encoding and a text-entity fusion encoding. Text-only encoding is responsible for bridging text and entities. Text-entity fusion encoding fuses information from tokens and knowledge from entities. Finally, two objectives as entity extraction and linking are added to learn better entity representations. To be noticed, we initialize the entity embeddings with TransE (Bordes et al., 2013), which leverages not only entity but also relation information of the knowledge graph.

We conduct experiments on the named entity recognition (NER) and relation extraction (RE) tasks in the BLURB benchmark dataset. Results show that our KeBioLM outperforms the previous work with average scores of 87.1 and 81.2 on 5 NER datasets and 3 RE datasets respectively. Furthermore, our KeBioLM also achieves better performance in a probing task that requires models to fill the masked entity in UMLS triplets.

We summary our contributions as follows[1]:

- We propose KeBioLM, a biomedical pretrained language model that explicitly incorporates knowledge from UMLS.
- We conduct experiments on 5 NER datasets and 3 RE datasets. Results demonstrate that our KeBioLM achieves the best performance on both NER and RE tasks.
- We collect a cloze-style probing dataset from UMLS relation triplets. The probing results show that our KeBioLM absorbs more knowledge than other biomedical PLMs.

2 Related Work

2.1 Biomedical PLMs

Models like ELMo (Peters et al., 2018) and BERT (Devlin et al., 2019) show the effectiveness of the paradigm of first pre-training an LM on the unlabeled text then fine-tuning the model on the downstream NLP tasks. However, direct application of the LMs pre-trained on the encyclopedia and web text usually fails on the biomedical domain, because of the distinctive terminologies and idioms.

The gap between general and biomedical domains inspires the researchers to propose LMs specially tailored for the biomedical domain. BioBERT (Lee et al., 2020) is the most widely used biomedical PLM which is trained on PubMed abstracts and PMC articles. It outperforms vanilla BERT in named entity recognition, relation extraction, and question answering tasks. Jin et al. (2019) train BioELMo with PubMed abstracts, and find features extracted by BioELMo contain entity-type and relational information. Different training corpora have been used for enhancing performance of sub-domain tasks. ClinicalBERT (Alsentzer et al., 2019), BlueBERT (Peng et al., 2019) and bio-lm (Lewis et al., 2020) utilize clinical notes MIMIC to improve clinical-related downstream tasks. SciBERT (Beltagy et al., 2019) uses papers from the biomedical and computer science domain as training corpora with a new vocabulary. KeBioLM is trained on PubMed abstracts to adapt to PubMed-related downstream tasks.

To understand the factors in pretraining biomedical LMs, Gu et al. (2020) study pretraining techniques systematically and propose PubMedBERT pretrained from scratch with an in-domain vocabulary. Lewis et al. (2020) also find using an in-domain vocabulary enhances the downstream performances. This inspires us to utilize the in-domain vocabulary for KeBioLM.

2.2 Knowledge-enhanced LMs

LMs like ELMo and BERT are trained to predict correlation between tokens, ignoring the meanings behind them. To capture both the textual and conceptual information, several knowledge-enhanced PLMs are proposed.

Entities are used for bridging tokens and knowledge graphs. Zhang et al. (2019) align tokens and entities within sentences, and aggregate token and entity representations via two multi-head self-attentions. KnowBert (Peters et al., 2019) and Entity as Experts (EAE) (Févry et al., 2020) use the entity linker to perform entity disambiguation for candidate entity spans and enhance token representations using entity embeddings. Inspired by entity-enhanced PLMs, we follow the model of EAE to inject biomedical knowledge into KeBioLM by performing entity detection and linking.

Relation triplets provide intrinsic knowledge be-

[1]Our codes and model can be found at https://github.com/GanjinZero/KeBioLM.

tween entity pairs. KEPLER (Wang et al., 2019) learns the knowledge embeddings through relation triplets while pretraining. K-BERT (Liu et al., 2020) converts input sentences into sentence trees by relation triplets to infuse knowledge.

In the biomedical domain, He et al. (2020) inject disease knowledge to existing PLMs by predicting diseases names and aspects on Wikipedia passages. Michalopoulos et al. (2020) use UMLS synonyms to supervise masked language modeling. We propose KeBioLM to infuse various kinds of biomedical knowledge from UMLS including but not limited to diseases.

3 Approach

In this paper, we assume to access an entity set $\mathcal{E} = \{e_1, ..., e_t\}$. For a sentence $\mathbf{x} = \{x_1, ..., x_n\}$, we assume some spans $m = (x_i, ..., x_j)$ can be grounded to one or more entities in $\mathcal{E}$. We further assume the disjuncture of these spans. In this paper, we use UMLS to set the entity set.

3.1 Model Architecture

To explicitly model both the textual and conceptual information, we follow Févry et al. (2020) and use a multi-layer self-attention network to encode both the text and entities. The model can be viewed as building the links between text and entities in the lower layers and fusing the text and entity representation in the upper layers. The overall architecture is shown in Figure 2. To be more specific, we set the PubMedBERT (Gu et al., 2020) as our backbone. We split the layers of the backbone into two groups, performing a text-only encoding and text-entity fusion encoding respectively.

Text-only encoding. For the first group, which is closer to the input, we extract the final hidden states and perform a token-wise classification to identify if the token is at the beginning, inside, or outside of a *mention* (i.e., the BIO scheme). The probabilities of the B/I/O label $\{l_i\}$ are written as:

$$\mathbf{h}_1, ..., \mathbf{h}_n = \text{Transformers}^0(x_1, ..., x_n) \quad (1)$$

$$p(l_i \mid \mathbf{x}) = \text{softmax}(\mathbf{W}_l \mathbf{h}_i + \mathbf{b}_l) \quad (2)$$

After identifying the mention boundary, we maintain a function $\mathcal{M}(i) \rightarrow \mathcal{E} \cup \{\text{NIL}\}$, which returns the entity of the i-th token belongs.[2] We collect the mentions with a sentence $\mathbf{x}$. For a mention $m = (s, t)$, where s and t represents the starting and ending indexes of m, we encode it as the concatenation of hidden states of the boundary tokens $\mathbf{h}_m = [\mathbf{h}_s; \mathbf{h}_t]$.

For an entity $e_j \in \mathcal{E}$ in the KG, we denote its entity embedding as $\mathbf{e}_j$. For a mention m, we search the k nearest entities of its projected representation $\mathbf{h}'_m = \mathbf{W}_m \mathbf{h}_m + \mathbf{b}_m$ in the entity embedding space, obtaining a set of entities $\mathcal{E}'$. The normalized similarity between $\mathbf{h}'_m$ and $\mathbf{e}_j$ is calculated as

$$a_j = \frac{\exp(\mathbf{h}'_m \cdot \mathbf{e}_j)}{\sum_{e_k \in \mathcal{E}'} \exp(\mathbf{h}'_m \cdot \mathbf{e}_k)} \quad (3)$$

The additional entity representation $\mathbf{e}'_m$ of m is calculated as a weighted sum of the embeddings $\mathbf{e}'_m = \sum_{e_j \in \mathcal{E}'} a_j \cdot \mathbf{e}_j$.

Text-entity fusion encoding. After getting the mentions and entities, we fuse the entity embeddings with the text embedding by summation. For the i-th token, the entity-enhanced embedding is calculated as:

$$\mathbf{h}^*_i = \begin{cases} \mathbf{h}_i + (\mathbf{W}_e \mathbf{e}'_m + \mathbf{b}_e), & \exists m, \mathcal{M}(i) = m, \\ \mathbf{h}_i, & \text{otherwise.} \end{cases} \quad (4)$$

$\mathcal{M}(i) = m$ represents the i-th token belong to entity e_m. The sequence of $\mathbf{h}^*_1, ..., \mathbf{h}^*_n$ is then fed into the second group of transformer layers to generate text-entity representations. The final hidden states $\mathbf{h}^f_i$ are calculated as:

$$\mathbf{h}^f_1, ..., \mathbf{h}^f_n = \text{Transformers}^1(\mathbf{h}^*_1, ..., \mathbf{h}^*_n) \quad (5)$$

3.2 Pretraining Tasks

We have three pretraining tasks for KeBioLM. Masked language modeling is a cloze-style task for predicting masked tokens. Since the entities are the main focus of our model, we add two tasks as entity detection and linking respectively following Févry et al. (2020). Finally, we jointly minimize the following loss:

$$\mathcal{L} = \mathcal{L}_{MLM} + \mathcal{L}_{ED} + \mathcal{L}_{EL} \quad (6)$$

Masked Language Modeling Like BERT and other LMs, we predict the masked tokens $\{x_i\}$ in inputs using the final hidden representations $\{\mathbf{h}^f_i\}$. The loss $\mathcal{L}_{MLM}$ is calculated based on the cross-entropy of masked and predicted tokens:

$$p_M(x_i \mid \mathbf{x}) = \text{softmax}(\mathbf{W}_m \mathbf{h}^f_i + \mathbf{b}_m) \quad (7)$$

$$\mathcal{L}_{MLM} = \sum -\log p_M(x_i \mid \mathbf{x}) \quad (8)$$

[2]NIL is returned when there is no entity being matched.

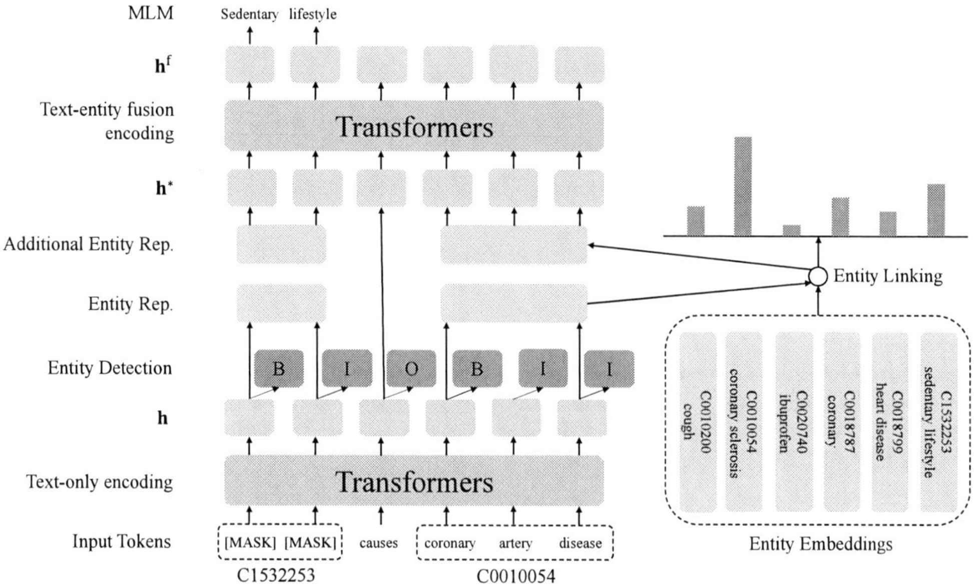

Figure 2: The overall architecture of KeBioLM.

Whole word masking is successful in training masked language models (Devlin et al., 2019; Cui et al., 2019). In the biomedical domain, entities are the semantic units of texts. Therefore, we extend this technique to whole entity masking. We mask all tokens within a word or entity span. KeBioLM replaces 12% of tokens to *[MASK]* and 1.5% tokens to random tokens. This is more difficult for models to recover tokens, which leads to learning better entity representations.

Entity Detection Entity detection is an important task in biomedical NLP to link the tokens to entities. Thus, We add an entity detection loss by calculating the cross-entropy for BIO labels:

$$\mathcal{L}_{ED} = \sum_{i=1}^{n} - \log p(l_i \mid \mathbf{x}) \tag{9}$$

Entity Linking One medical entity in different names linking to the same index permits the model to learn better text-entity representations. To link mention $\{m\}$ in texts with entities $\{e\}$ in entity set $\mathcal{E}$, we calculate the cross-entropy loss using similarities between $\{\mathbf{h}'_m\}$ and entities in $\mathcal{E}$:

$$\mathcal{L}_{EL} = \sum - \log \frac{\exp(\mathbf{h}'_m \cdot \mathbf{e})}{\sum_{e_j \in \mathcal{E}} \exp(\mathbf{h}'_m \cdot \mathbf{e}_j)} \tag{10}$$

3.3 Data Creation

Given a sentence S from PubMed content, we need to recognize entities and link them to the UMLS knowledge base. We use ScispaCy (Neumann et al., 2019), a robust biomedical NER and entity linking model, to annotate the sentence. Unlike previous work (Vashishth et al., 2020) that only retains recognized entities in a subset of Medical Subject Headings (MeSH) (Lipscomb, 2000), we relax the restriction to annotate all entities to UMLS 2020 AA release [3] whose linking scores are higher than a threshold of 0.85.

4 Experiments

In this section, we first introduce the pretraining details of KeBioLM. Then we introduce the BLURB datasets for evaluating our approach. Finally, we introduce a probing dataset based on UMLS triplets for evaluating knowledge modeling.

4.1 Pretraining Details

We use ScispaCy to acquire 477K CUIs and 660M entities among 3.5M PubMed documents[4] from PubMedDS dataset (Vashishth et al., 2020) as training corpora.

We initialize entity embeddings by TransE (Bordes et al., 2013) which learns embeddings from relation triplets. Relation triplets come from UMLS

[3] https://www.nlm.nih.gov/research/umls/licensedcontent/umlsarchives04.html#2020AA

[4] The count of documents in PubMedDS is based on https://arxiv.org/pdf/2005.00460v1.pdf.

	#Train	#Dev	#Test	#Ments	#Ments (UMLS)	#Ments (KeBioLM)
BC5chem	5,203	5,347	5,385	15,935	10,373	8,993
BC5dis	4,182	4,244	4,424	12,850	8,846	3,878
NCBI	5,137	787	960	6,884	1,985	1,091
BC2GM	15,197	3,061	6,325	24,583	2,808	2,423
JNLPBA	46,750	4,551	8,662	59,963	6,099	5,233
ChemProt	18,035	11,268	15,745	39,022	13,106	10,772
DDI	25,296	2,496	5,716	15,738	10,429	9,212
GAD	4,261	535	534	-	-	-

Table 1: The training instances (mentions for NER tasks and sentences with two entities for RE tasks) and the mention counts of NER and RE datasets preprocessed in BLURB benchmark respectively. The mention counts overlapping with UMLS 2020 AA release and KeBioLM are also listed. For the GAD dataset, annotated mentions do not appear in the BLURB preprocessed version.

2020 AA release. We train TransE with the L2-norm distance function and set embedding dim to 100. Adam (Kingma and Ba, 2014) is used as the optimizer with a learning rate of 1e-3, batch size of 2048, and train epoch of 30. Entity embeddings add 45.5M parameters to KeBioLM.

The parameters of transformers in KeBioLM are initialized from the checkpoint of PubMedBERT. We also use the vocabulary from PubMedBERT. AdamW (Loshchilov and Hutter, 2017) is used as the optimizer for KeBioLM with 10,000 steps warmup and linear decay. We use an 8-layer transformer for text-only encoding and a 4-layer transformer for text-entity fusion encoding. We set the learning rate to 5e-5, batch size to 512, max sequence length to 512, and training epochs to 2. For each input sequence, we limit the max entities count to 50 and the excessive entities will be truncated. To generate entity representation $\mathbf{e}'_m$, the most $k = 100$ similar entities are used. We train our model with 8 NVIDIA 16GB V100 GPUs.

4.2 Datasets

In this section, we evaluate KeBioLM on NER tasks and RE tasks of the BLURB benchmark[5] (Gu et al., 2020). For all tasks, we use the preprocessed version from BLURB. We measure the NER and RE datasets in terms of F1-score. Table 1 shows the counts of training instances in BLURB datasets (i.e., annotated mentions for NER datasets and sentences with two mentions for RE datasets). We also report the count of annotated mentions overlapping with the UMLS 2020 release and KeBioLM in each dataset. The percentage of mentions overlapping with KeBioLM ranges from 8.7% (NCBI-disease) to 58.5% (DDI) which indicates that KeBioLM learns entity knowledge related to downstream tasks.

4.2.1 Named Entity Recognition

BC5-chem & BC5-disease (Li et al., 2016) contain 1500 PubMed abstracts for extracting chemical and disease entities respectively.

NCBI-disease (Doğan et al., 2014) includes 793 PubMed abstracts to detect disease entities.

BC2GM (Smith et al., 2008) contains 20K PubMed sentences to extract gene entities.

JNLPBA (Collier and Kim, 2004) includes 2,000 PubMed abstracts to identify molecular biology-related entities. We ignore entity types in JNLPBA following Gu et al. (2020).

4.2.2 Relation Extraction

ChemProt (Krallinger et al., 2017) classifies the relation between chemicals and proteins within sentences from PubMed abstracts. Sentences are classified into 6 classes including a negative class.

DDI (Herrero-Zazo et al., 2013) is a RE dataset with sentence-level drug-drug relation on PubMed abstracts. There are four classes for relation: advice, effect, mechanism, and false.

GAD (Bravo et al., 2015) is a gene-disease relation binary classification dataset collected from PubMed sentences.

4.3 Fine-tuning Details

NER We follow Gu et al. (2020) to formulate NER tasks as sequential labeling tasks with the

[5] `https://microsoft.github.io/BLURB/`

	Bio-BERT	Sci-BERT	Clinical-BERT	Blue-BERT	disease-BERT	bio-lm†	PubMed-BERT	KeBio-LM
BC5chem	92.9	92.5	90.8	91.2	-	92.9	**93.3**	**93.3**$_{\pm 0.2}$
BC5dis	84.7	84.5	83.0	83.7	**86.5**	83.8	85.6	86.1$_{\pm 0.3}$*
NCBI	**89.1**	88.1	88.3	88.0	87.1	87.7	87.8	**89.1**$_{\pm 0.3}$*
BC2GM	83.8	83.4	81.7	81.9	-	87.0	84.5	**85.1**$_{\pm 1.6}$
JNLPBA	79.4	79.5	78.6	78.7	-	80.6	80.1	**82.0**$_{\pm 0.2}$*
NER	86.0	85.6	84.5	84.7	-	86.4	86.3	**87.1**$_{\pm 0.3}$*
ChemProt	76.1	75.2	72.0	71.5	-	75.4	77.2	**77.5**$_{\pm 0.3}$*
DDI	80.9	81.1	78.2	77.8	-	81.0	**82.4**	81.9$_{\pm 0.8}$
GAD	80.9	80.9	78.4	77.2	-	82.2	82.3	**84.3**$_{\pm 1.0}$*
RE	79.3	79.1	76.2	75.5	-	79.5	80.6	**81.2**$_{\pm 0.5}$*

Table 2: F1-scores on NER and RE tasks in BLURB benchmark. Standard deviations of KeBioLM are reported across five runs. Results of diseaseBERT-biobert and bio-lm come from their corresponded papers. Others are copied from BLURB. * indicates that $p \leq 0.05$ of one-sample t-test which compares whether the mean performance of KeBioLM is better than PubMedBERT. † Bio-lm applies different metrics with BLURB (micro F1 v.s. macro F1). Thus, we just list its results but do not directly compare with them.

BIO tagging scheme and ignore the entity types in NER datasets. We classify labels of tokens by a linear layer on top of the hidden representations.

RE We replace the entity mentions in RE datasets with entity indicators like @DISEASE$ or @GENE$ to avoid models classifying relations by memorizing entity names. We add these entity indicators into the vocabulary of LMs. We concatenate the representation of two concerned entities and feed it into a linear layer for relation classification.

Parameters We adopt AdamW as the optimizer with a 10% steps linear warmup and a linear decay. We search the hyperparameters of learning rate among 1e-5, 3e-5, and 5e-5. We fine-tune the model for 60 epochs. We evaluate the model at the end of each epoch and choose the best model according to the evaluation score on the development set. We set batch size as 16 when fine-tuning. The maximal input lengths are 512 for all NER datasets. We truncate ChemProt and DDI to 256 tokens, and GAD to 128 tokens. To perform a fair comparison, we fine-tune our model with 5 different seeds and report the average score.

4.4 Results

We compare KeBioLM with following base-size biomedical PLMs on the above-mentioned datasets: BioBERT (Lee et al., 2020), SciBERT (Beltagy et al., 2019), ClinicalBERT (Alsentzer et al., 2019), BlueBERT (Peng et al., 2019), bio-lm (Lewis et al., 2020), diseaseBERT (He et al., 2020), and PubMedBERT (Gu et al., 2020) [6].

Table 2 shows the main results on NER and RE datasets of the BLURB benchmark. In addition, we report the average scores for NER and RE tasks respectively. KeBioLM achieves state-of-the-art performance for NER and RE tasks. Compared with the strong baseline BioBERT, KeBioLM shows stable improvements in NER and RE datasets (+1.1 in NER, +1.9 in RE). Compared with our baseline model PubMedBERT, KeBioLM performs significantly better in BC5dis, NCBI, JNLPBA, ChemProt, and GAD ($p \leq 0.05$ based on one-sample t-test) and achieves better average scores (+0.8 in NER, +0.6 in RE). DiseaseBERT is a model carefully designed for predicting disease names and aspects, which leads to better performance in the BC5dis dataset (+0.4). They only report the promising results in disease-related tasks, however, our model obtains consistent promising performances across all kinds of biomedical tasks. In the BC2GM dataset, KeBioLM outperforms our baseline model PubMedBERT and other PLMs except for bio-lm, and the standard deviation of the BC2GM task is evidently larger than other tasks. Another exception is the DDI dataset, we observe a slight performance degradation compared to PubMedBERT (-0.5). The average performances demonstrate that fusing entity knowledge into the LM boosts the performances across the board.

[6] We use BioBERT v1.1, SciBERT-scivocab-uncased, Bio-ClinicalBERT, BlueBERT-pubmed-mimic, bio-lm(RoBERTa-base-PM-M3-Voc), diseaseBERT-biobert and PubMedBERT-abstract versions for comparison.

	KeBio-LM	-wem	+rand	+frz
BC5chem	**93.3**	92.8	92.8	92.3
BC5dis	**86.1**	85.9	85.5	85.5
NCBI	**89.1**	88.4	88.8	88.3
BC2GM	85.1	84.5	84.5	**85.7**
JNLPBA	**82.0**	81.5	81.9	81.8
NER	**87.1**	86.6	86.7	86.7
ChemProt	**77.5**	77.3	76.3	76.8
DDI	**81.9**	80.6	81.4	80.7
GAD	**84.3**	83.1	82.3	82.8
RE	**81.2**	80.3	80.0	80.1

Table 3: Ablation studies for KeBioLM architecture on the BLURB benchmark. We use -wem, +rand and +frz to represent pretraining setting (a), (b) and (c), respectively.

	$l_0 = 8$ $l_1 = 4$	$l_0 = 4$ $l_1 = 8$	$l_0 = 12$ $l_1 = 0$
BC5chem	**93.3**	93.1	93.2
BC5dis	**86.1**	85.7	86.0
NCBI	**89.1**	88.5	88.4
BC2GM	85.1	84.8	**86.8**
JNLPBA	**82.0**	81.7	78.8
NER	**87.1**	86.8	86.6
ChemProt	77.5	**77.7**	77.6
DDI	**81.9**	81.0	80.1
GAD	**84.3**	82.9	83.2
RE	**81.2**	80.5	80.3

Table 4: Ablation studies for transformer layers count in KeBioLM on the BLURB benchmark.

4.5 Ablation Test

We conduct ablation tests to validate the effectiveness of each part in KeBioLM. We pretrain the model with the following settings and reuse the same parameters described above: (a) Remove whole entity masking and retain whole word masking while pretraining (-wem); (b) Initialize entity embeddings randomly (+rand); (c) Initialize entity embeddings by TransE and freeze the entity embeddings while pretraining (+frz).

In Table 3, we observe the following results. Firstly, comparing KeBioLM with setting (a) shows that whole entity masking boosting the performances consistently in all datasets (+0.5 in NER, +0.9 in RE). Secondly, comparing KeBioLM with setting (b) indicates initializing the entity embeddings randomly degrades performances in NER tasks and RE tasks (-0.4 in NER, -1.2 in RE). Entity embeddings initialized by TransE utilize relation knowledge in UMLS and enhance the results. Thirdly, freezing the entity embeddings in setting (c) reduces the performances on all datasets compared to KeBioLM except BC2GM (-0.4 in NER, -1.1 in RE). This indicates that updating entity embedding while pretraining helps KeBioLM to have better text-entity representations, and this leads to better downstream performances.

To evaluate how the count of transformer layers affects our model, we pretrain KeBioLM with the different number of layers. For the convenience of notation, denote l_0 is the layer count of text-only encoding and l_1 is the layer count of text-entity fusion encoding. We have the following settings: (i) $l_0 = 8, l_1 = 4$ (our base model), (ii)$l_0 = 4, l_1 = 8$, (iii)$l_0 = 12, l_1 = 0$ (without the second group of transformer layers, $\{\mathbf{h}_i\}$ are used for token representations). Results are shown in Table 4. Our base model (i) has better performance than setting (ii) (+0.3 in NER, +0.7 in RE). Training setting (iii) is equal to a traditional BERT model with additional entity extraction and entity linking tasks. The comparison with (i) and (iii) indicates that text-entity representations have better performances than text-only representations (+0.5 in NER, +0.9 in RE) in the same amount of parameters.

4.6 UMLS Knowledge Probing

We establish a probing dataset based on UMLS triplets to evaluate how LMs understand medical knowledge via pretraining.

4.6.1 Probing Dataset

UMLS triplets are stored in the form of (s, r, o) where s and o are CUIs in UMLS and r is a relation type. We generate two queries for one triplet based on names of CUIs and relation type:

- Q_1: *[CLS]* s r *[MASK] [SEP]*
- Q_2: *[CLS] [MASK]* r o *[SEP]*

For example, we sample a triplet and terms of corresponded entities (*C0048038*:apraclonidine, *may_prevent*, *C0028840*:ocular hypertension). We remove the underscores of relation names and generate two queries (we omit *[CLS]* and *[SEP]*):

- Q_1: apraclonidine may prevent *[MASK]*.
- Q_2: *[MASK]* may prevent ocular hypertension.

#Queries	#Relations	#Avg. CUIs
143,771	922	2.39

Table 5: The number of generated UMLS relation probing dataset.

For relation names end with "of", "as" , and "by", we add "is" in front of relation names. For instance, *translation_of* is converted to *is translation of*, *classified_as* is converted to *is classified as*, and *used_by* is converted to *is used by*. Commonly, different relation triplets can generate same query since triplets may overlap $(s, r, -)$ or $(-, r, o)$ with each other. We deduplicate all repeat queries and randomly choose at most 200 queries from all relation types in UMLS. After deduplication, one query can have multiple CUIs as answers. For example:

- Q: *[MASK]* may treat essential tremor.
- A_1: *C0282321*: propranolol hydrochloride
- A_2: *C0033497*: propranolol

We summarize our generated UMLS relation probing dataset in Table 5. Unlike LAMA (Petroni et al., 2019) and X-FACTR (Jiang et al., 2020) that contain less than 50 kinds of relation, our probing task is a more difficult task requiring a model to decode entities over 900 kinds of relations.

4.6.2 Multi [MASK] Decoding

To probe PLMs using generated queries, we require models to recover the masked tokens. Since biomedical entities are usually formed by multiple words and each word can be tokenized into several wordpieces (Wu et al., 2016), models have to recover multiple *[MASK]* tokens. We limit the max length of one entity is 10 for decoding.

We decode the multi *[MASK]* tokens using the confidence-based method described in Jiang et al. (2020). We also implement a beam search for decoding. Unlike beam search in machine translation that decodes tokens from left to right, we decode tokens in an arbitrary order. For each step, we calculate the probabilities of all undecoded masked tokens based on original input and decoded tokens. We predict only one token within undecoded tokens with the top $B = 5$ accumulated log probabilities. Decoding will be accomplished after count of *[MASK]* times iterations and we keep the best $B = 5$ decoding results. We skip the refinement stage since it is time-consuming and does not significantly improve the results.

	Type 1	Type 2	Overall
SciBERT	13.92	1.01	2.75
ClinicalBERT	4.19	0.33	0.79
BlueBERT	4.67	0.39	1.02
KeBioLM	**14.01**	**1.48**	**3.26**

Table 6: Results of the probing test in terms of Recall@5.

4.6.3 Evaluation Metric

Since multiple correct CUIs exist for one query, we consider a model answering the query correctly if any decoded tokens in any *[MASK]* length hit any of the correct CUIs. We evaluate the probing results by the relation-level macro-recall@5.

4.6.4 Probing Results

We classify probing queries into two types based on their difficulties. Type 1: **answers within queries** (24,260 queries); Type 2: **answers not in queries** (119,511 queries). Here are examples of Type 1 (Q_1 and A_1) and Type 2 (Q_2 and A_2) queries:

- Q_1: *[MASK]* has form tacrolimus monohydrate.
- A_1: *C0085149*: tacrolimus
- Q_2: cosyntropin may diagnose *[MASK]*.
- A_2: *C0001614*: adrenal cortex disease

Table 6 summarizes the probing results of different PLMs according to query types. Checkpoints of BioBERT and PubMedBERT miss a cls/predictions layer and cannot perform the probe directly. Compared to other PLMs, KeBioLM achieves the best scores in both two types and obviously outperforms BlueBERT and ClincalBERT with a large margin, which indicates that KeBioLM learns more medical knowledge.

Table 7 lists some probing examples. SciBERT can decode medical entities for *[MASK]* tokens which may be unrelated. KeBioLM decodes relation correctly and is aware of the synonyms of hepatic. KeBioLM states that *Vaccination may prevent tetanus* which is a correct but not precise statement.

5 Conclusions

In this paper, we propose to improve biomedical pretrained language models with knowledge. We

Query & Answer CUI	SciBERT	KeBioLM
omalizumab may treat *[MASK]*	migraine	**asthma**
C0004096: asthma	the disease	severe allergic asthma
phentolamine may diagnose *[MASK]*	depression	**pheochromocytoma**
C0031511: phaeochromocytoma	the serotonin syndrome	renovascular hypertension
[MASK] is noun form of hepatic	it	**liver**
C0023884: liver	the form of hepatic	hepatic only
[MASK] may prevent tetanus	it	vaccination
C0305062: tetanus toxoid	bcg vaccination	prophylactic tetanus vaccination

Table 7: Probing examples of UMLS relation triplets. Queries and answer CUIs are listed. We only list one correct CUI for each query. For each model, one *[MASK]* token decoding result and an example of multi *[MASK]* decoding result are displayed. Bold text represents a term of the answer CUI.

propose KeBioLM which applies text-only encoding and text-entity fusion encoding and has two additional entity-related pretraining tasks: entity detection and entity linking. Extensive experiments have shown that KeBioLM outperforms other PLMs on NER and RE datasets of the BLURB benchmark. We further probe biomedical PLMs by querying UMLS relation triplets, which indicates KeBioLM absorbs more biomedical knowledge than others. In this work, we only leverage the relation information in TransE to initialize the entity embeddings. We will further investigate how to directly incorporate the relation information into LMs in the future.

Acknowledgements

We would like to thank the anonymous reviewers for their helpful comments and suggestions. This work is supported by Alibaba Group through Alibaba Research Intern Program.

References

Emily Alsentzer, John Murphy, William Boag, Wei-Hung Weng, Di Jindi, Tristan Naumann, and Matthew McDermott. 2019. Publicly available clinical BERT embeddings. In *Proceedings of the 2nd Clinical Natural Language Processing Workshop*, pages 72–78, Minneapolis, Minnesota, USA. Association for Computational Linguistics.

Michael Ashburner, Catherine A Ball, Judith A Blake, David Botstein, Heather Butler, J Michael Cherry, Allan P Davis, Kara Dolinski, Selina S Dwight, Janan T Eppig, et al. 2000. Gene ontology: tool for the unification of biology. *Nature genetics*, 25(1):25–29.

Iz Beltagy, Kyle Lo, and Arman Cohan. 2019. SciBERT: A pretrained language model for scientific text. In *Proceedings of the 2019 Conference on Empirical Methods in Natural Language Processing and the 9th International Joint Conference on Natural Language Processing (EMNLP-IJCNLP)*, pages 3615–3620, Hong Kong, China. Association for Computational Linguistics.

Olivier Bodenreider. 2004. The unified medical language system (umls): integrating biomedical terminology. *Nucleic acids research*, 32(suppl_1):D267–D270.

Antoine Bordes, Nicolas Usunier, Alberto Garcia-Duran, Jason Weston, and Oksana Yakhnenko. 2013. Translating embeddings for modeling multi-relational data. In *Neural Information Processing Systems (NIPS)*, pages 1–9.

Àlex Bravo, Janet Piñero, Núria Queralt-Rosinach, Michael Rautschka, and Laura I Furlong. 2015. Extraction of relations between genes and diseases from text and large-scale data analysis: implications for translational research. *BMC bioinformatics*, 16(1):1–17.

Nigel Collier and Jin-Dong Kim. 2004. Introduction to the bio-entity recognition task at JNLPBA. In *Proceedings of the International Joint Workshop on Natural Language Processing in Biomedicine and its Applications (NLPBA/BioNLP)*, pages 73–78, Geneva, Switzerland. COLING.

Yiming Cui, Wanxiang Che, Ting Liu, Bing Qin, Ziqing Yang, Shijin Wang, and Guoping Hu. 2019. Pre-training with whole word masking for chinese bert. *arXiv preprint arXiv:1906.08101*.

Jacob Devlin, Ming-Wei Chang, Kenton Lee, and Kristina Toutanova. 2019. BERT: Pre-training of deep bidirectional transformers for language understanding. In *Proceedings of the 2019 Conference of the North American Chapter of the Association for Computational Linguistics: Human Language Technologies, Volume 1 (Long and Short Papers)*, pages 4171–4186, Minneapolis, Minnesota. Association for Computational Linguistics.

Rezarta Islamaj Doğan, Robert Leaman, and Zhiyong Lu. 2014. Ncbi disease corpus: a resource for disease name recognition and concept normalization. *Journal of biomedical informatics*, 47:1–10.

Thibault Févry, Livio Baldini Soares, Nicholas FitzGerald, Eunsol Choi, and Tom Kwiatkowski. 2020. Entities as experts: Sparse memory access with entity supervision. In *Proceedings of the 2020 Conference on Empirical Methods in Natural Language Processing (EMNLP)*, pages 4937–4951, Online. Association for Computational Linguistics.

Yu Gu, Robert Tinn, Hao Cheng, Michael Lucas, Naoto Usuyama, Xiaodong Liu, Tristan Naumann, Jianfeng Gao, and Hoifung Poon. 2020. Domain-specific language model pretraining for biomedical natural language processing. *arXiv preprint arXiv:2007.15779*.

Yun He, Ziwei Zhu, Yin Zhang, Qin Chen, and James Caverlee. 2020. Infusing Disease Knowledge into BERT for Health Question Answering, Medical Inference and Disease Name Recognition. In *Proceedings of the 2020 Conference on Empirical Methods in Natural Language Processing (EMNLP)*, pages 4604–4614, Online. Association for Computational Linguistics.

María Herrero-Zazo, Isabel Segura-Bedmar, Paloma Martínez, and Thierry Declerck. 2013. The ddi corpus: An annotated corpus with pharmacological substances and drug–drug interactions. *Journal of biomedical informatics*, 46(5):914–920.

Zhengbao Jiang, Antonios Anastasopoulos, Jun Araki, Haibo Ding, and Graham Neubig. 2020. X-FACTR: Multilingual factual knowledge retrieval from pretrained language models. In *Proceedings of the 2020 Conference on Empirical Methods in Natural Language Processing (EMNLP)*, pages 5943–5959, Online. Association for Computational Linguistics.

Qiao Jin, Bhuwan Dhingra, William Cohen, and Xinghua Lu. 2019. Probing biomedical embeddings from language models. In *Proceedings of the 3rd Workshop on Evaluating Vector Space Representations for NLP*, pages 82–89.

Diederik P Kingma and Jimmy Ba. 2014. Adam: A method for stochastic optimization. *arXiv preprint arXiv:1412.6980*.

Martin Krallinger, Obdulia Rabal, Saber A Akhondi, Martın Pérez Pérez, Jesús Santamaría, Gael Pérez Rodríguez, et al. 2017. Overview of the biocreative vi chemical-protein interaction track. In *Proceedings of the sixth BioCreative challenge evaluation workshop*, volume 1, pages 141–146.

Jinhyuk Lee, Wonjin Yoon, Sungdong Kim, Donghyeon Kim, Sunkyu Kim, Chan Ho So, and Jaewoo Kang. 2020. Biobert: a pre-trained biomedical language representation model for biomedical text mining. *Bioinformatics*, 36(4):1234–1240.

Patrick Lewis, Myle Ott, Jingfei Du, and Veselin Stoyanov. 2020. Pretrained language models for biomedical and clinical tasks: Understanding and extending the state-of-the-art. In *Proceedings of the 3rd Clinical Natural Language Processing Workshop*, pages 146–157, Online. Association for Computational Linguistics.

Jiao Li, Yueping Sun, Robin J Johnson, Daniela Sciaky, Chih-Hsuan Wei, Robert Leaman, Allan Peter Davis, Carolyn J Mattingly, Thomas C Wiegers, and Zhiyong Lu. 2016. Biocreative v cdr task corpus: a resource for chemical disease relation extraction. *Database*, 2016.

Carolyn E Lipscomb. 2000. Medical subject headings (mesh). *Bulletin of the Medical Library Association*, 88(3):265.

Weijie Liu, Peng Zhou, Zhe Zhao, Zhiruo Wang, Qi Ju, Haotang Deng, and Ping Wang. 2020. K-bert: Enabling language representation with knowledge graph. *Proceedings of the AAAI Conference on Artificial Intelligence*, 34(03):2901–2908.

Ilya Loshchilov and Frank Hutter. 2017. Decoupled weight decay regularization. *arXiv preprint arXiv:1711.05101*.

George Michalopoulos, Yuanxin Wang, Hussam Kaka, Helen Chen, and Alex Wong. 2020. Umlsbert: Clinical domain knowledge augmentation of contextual embeddings using the unified medical language system metathesaurus.

Mark Neumann, Daniel King, Iz Beltagy, and Waleed Ammar. 2019. ScispaCy: Fast and robust models for biomedical natural language processing. In *Proceedings of the 18th BioNLP Workshop and Shared Task*, pages 319–327, Florence, Italy. Association for Computational Linguistics.

Yifan Peng, Shankai Yan, and Zhiyong Lu. 2019. Transfer learning in biomedical natural language processing: An evaluation of BERT and ELMo on ten benchmarking datasets. In *Proceedings of the 18th BioNLP Workshop and Shared Task*, pages 58–65, Florence, Italy. Association for Computational Linguistics.

Matthew Peters, Mark Neumann, Mohit Iyyer, Matt Gardner, Christopher Clark, Kenton Lee, and Luke Zettlemoyer. 2018. Deep contextualized word representations. In *Proceedings of the 2018 Conference of the North American Chapter of the Association for Computational Linguistics: Human Language Technologies, Volume 1 (Long Papers)*, pages 2227–2237, New Orleans, Louisiana. Association for Computational Linguistics.

Matthew E. Peters, Mark Neumann, Robert Logan, Roy Schwartz, Vidur Joshi, Sameer Singh, and Noah A. Smith. 2019. Knowledge enhanced contextual word representations. In *Proceedings of the 2019 Conference on Empirical Methods in Natural Language*

Processing and the 9th International Joint Conference on Natural Language Processing (EMNLP-IJCNLP), pages 43–54, Hong Kong, China. Association for Computational Linguistics.

Fabio Petroni, Tim Rocktäschel, Sebastian Riedel, Patrick Lewis, Anton Bakhtin, Yuxiang Wu, and Alexander Miller. 2019. Language models as knowledge bases? In *Proceedings of the 2019 Conference on Empirical Methods in Natural Language Processing and the 9th International Joint Conference on Natural Language Processing (EMNLP-IJCNLP)*, pages 2463–2473, Hong Kong, China. Association for Computational Linguistics.

Hoo-Chang Shin, Yang Zhang, Evelina Bakhturina, Raul Puri, Mostofa Patwary, Mohammad Shoeybi, and Raghav Mani. 2020. BioMegatron: Larger biomedical domain language model. In *Proceedings of the 2020 Conference on Empirical Methods in Natural Language Processing (EMNLP)*, pages 4700–4706, Online. Association for Computational Linguistics.

Larry Smith, Lorraine K Tanabe, Rie Johnson nee Ando, Cheng-Ju Kuo, I-Fang Chung, Chun-Nan Hsu, Yu-Shi Lin, Roman Klinger, Christoph M Friedrich, Kuzman Ganchev, et al. 2008. Overview of biocreative ii gene mention recognition. *Genome biology*, 9(2):1–19.

Michael Q Stearns, Colin Price, Kent A Spackman, and Amy Y Wang. 2001. Snomed clinical terms: overview of the development process and project status. In *Proceedings of the AMIA Symposium*, page 662. American Medical Informatics Association.

Shikhar Vashishth, Rishabh Joshi, Ritam Dutt, Denis Newman-Griffis, and Carolyn Rose. 2020. Medtype: Improving medical entity linking with semantic type prediction. *arXiv preprint arXiv:2005.00460*.

Xiaozhi Wang, Tianyu Gao, Zhaocheng Zhu, Zhiyuan Liu, Juanzi Li, and Jian Tang. 2019. Kepler: A unified model for knowledge embedding and pretrained language representation. *arXiv preprint arXiv:1911.06136*.

Yonghui Wu, Mike Schuster, Zhifeng Chen, Quoc V Le, Mohammad Norouzi, Wolfgang Macherey, Maxim Krikun, Yuan Cao, Qin Gao, Klaus Macherey, et al. 2016. Google's neural machine translation system: Bridging the gap between human and machine translation. *arXiv preprint arXiv:1609.08144*.

B. Xu, X. Shi, Z. Zhao, and W. Zheng. 2018. Leveraging biomedical resources in bi-lstm for drug-drug interaction extraction. *IEEE Access*, 6:33432–33439.

Zheng Yuan, Zhengyun Zhao, and Sheng Yu. 2020. Coder: Knowledge infused cross-lingual medical term embedding for term normalization. *arXiv preprint arXiv:2011.02947*.

Zhengyan Zhang, Xu Han, Zhiyuan Liu, Xin Jiang, Maosong Sun, and Qun Liu. 2019. ERNIE: Enhanced language representation with informative entities. In *Proceedings of the 57th Annual Meeting of the Association for Computational Linguistics*, pages 1441–1451, Florence, Italy. Association for Computational Linguistics.

EntityBERT: Entity-centric Masking Strategy for Model Pretraining for the Clinical Domain

Chen Lin[1], Timothy Miller[1], Dmitriy Dligach[2], Steven Bethard[3] and Guergana Savova[1]

[1]Boston Children's Hospital and Harvard Medical School
[2]Loyola University Chicago
[3]University of Arizona
[1]{first.last}@childrens.harvard.edu
[2]ddligach@luc.edu
[3]bethard@email.arizona.edu

Abstract

Transformer-based neural language models have led to breakthroughs for a variety of natural language processing (NLP) tasks. However, most models are pretrained on general domain data. We propose a methodology to produce a model focused on the clinical domain: continued pretraining of a model with a broad representation of biomedical terminology (PubMedBERT) on a clinical corpus along with a novel entity-centric masking strategy to infuse domain knowledge in the learning process. We show that such a model achieves superior results on clinical extraction tasks by comparing our entity-centric masking strategy with classic random masking on three clinical NLP tasks: cross-domain negation detection (Wu et al., 2014), document time relation (DocTimeRel) classification (Lin et al., 2020b), and temporal relation extraction (Wright-Bettner et al., 2020). We also evaluate our models on the PubMedQA(Jin et al., 2019) dataset to measure the models' performance on a non-entity-centric task in the biomedical domain. The language addressed in this work is English.

1 Introduction

Transformer-based neural language models, such as BERT (Devlin et al., 2018), have achieved state-of-the-art performance for a variety of natural language processing (NLP) tasks. Since most are pre-trained on large general domain corpora, many efforts have been made to continue pretaining general-domain language models on clinical/biomedical corpora to derive domain-specific language models (Lee et al., 2020; Alsentzer et al., 2019; Beltagy et al., 2019).

Yet, as Gu et al. (2020a) pointed out, in specialized domains such as biomedicine, continued pretraining from generic language models is inferior to domain-specific pretraining from scratch. Continued pre-training from a generic model would break down many of the domain specific terms into sub-words through the Byte-Pair Encoding (BPE) (Gage, 1994) or variants like WordPiece tokenization (Wu et al., 2016) because these specific terms are not in the vocabulary of the generic pretrained model. A clinical domain-specific pretraining from scratch would derive an in-domain vocabulary as many of the biomedical terms, such as diseases, signs/symptoms, medications, anatomical sites, procedures, would be represented in their original form. Such an improved word-level representation is expected to bring substantial performance gains in clinical domain tasks because the model would learn the characteristics of the term along with its surrounding context as one unit.

In our preliminary work on a clinical relation extraction task, we observed a performance gain with the PubMedBERT model (Gu et al., 2020a) which outperformed BioBERT (Lee et al., 2020), ClinicalBERT (Alsentzer et al., 2019), and even some larger general domain models like RoBERTa (Liu et al., 2019) and BART-large (Lewis et al., 2019). The performance gain was primarily attributed to PubMedBERT's in-domain vocabulary as we observed that PubMedBERT kept 30% more in-domain words in its vocabulary than BERT. When we swapped PubMedBERT tokenization with BERT or RoBERTa tokenization, the performance of PubMedBERT degraded.

Thus, PubMedBERT appears to provide a vocabulary that is helpful to the clinical domain. However, the language of biomedical literature is different from the language of the clinical documents found in electronic medical records (EMRs). In general, a clinical document is written by physicians who have very limited time to express the numerous details of a patient-physician encounter. Many nonstandard expressions, abbreviations, assumptions and domain knowledge are used in clinical notes which makes the text hard to understand outside of the clinical community and presents

Proceedings of the BioNLP 2021 workshop, pages 191–201
June 11, 2021.

challenges for automated systems. Pretraining a language model specific to the clinical domain requires large amounts of unlabeled clinical text on par with what the generic models are trained on. Unfortunately, such data are not available to the community. The only available such corpus is MIMIC III used to train ClinicalBERT (Alsentzer et al., 2019) and BlueBERT (Peng et al., 2019), but it is magnitudes smaller and represents one specialty in medicine – intensive care.

Pretraining is agnostic to downstream tasks: it learns representations for all words using a self-supervised data-rich task. Yet, not all words are important for downstream fine-tuning tasks. Numerous pretrained words are not even used in the fine-tuning step, while important words crucial for the downstream task are not well represented due to insufficient amounts of labeled data. Many clinical NLP tasks are centered around entities: clinical named entity recognition aims to detect clinical entities (Wu et al., 2017; Pradhan et al., 2014; Elhadad et al., 2015), clinical negation extraction decides if a certain clinical entity is negated (Chapman et al., 2001; Harkema et al., 2009; Mehrabi et al., 2015), clinical relation discovery extracts relations among clinical entities (Lv et al., 2016; Leeuwenberg and Moens, 2017), etc. Though various masking strategies have been employed during pretraining – masking contiguous spans of text (SpanBERT, Joshi et al., 2020; BART, Lewis et al., 2019), varying masking ratios (Raffel et al., 2019), building additional neural models to predict which words to mask (Gu et al., 2020b), incorporating knowledge graphs (Zhang et al., 2019), masking entities for a named entity recognition task (Ziyadi et al., 2020) – none of the masking techniques so far have investigated and focused on clinical entities.

Besides transformer-based models, there are other efforts (Beam et al., 2019; Chen et al., 2020) to characterize the biomedical/clinical entities at the word embedding level. There are also other statistical methods applied to the downstream tasks. We do not include these efforts in our discussion because the focus of our paper is the investigation of a novel entity-based masking strategy in a transformer-based setting.

In this paper, we propose a methodology to produce a model focused on clinical entities: continued pretraining of a model with a broad representation of biomedical terminology (the PubMedBERT model) on a clinical corpus, along with a novel entity-centric masking strategy to infuse domain knowledge in the learning process[1]. We show that such a model achieves superior results on clinical extraction tasks by comparing our entity-centric masking strategy with classic random masking on three clinical NLP tasks: cross-domain negation detetction (Wu et al., 2014), document time relation (DocTimeRel) classification (Lin et al., 2020b), and temporal relation extraction (Wright-Bettner et al., 2020).

The contributions of this paper are: (1) a continued pretraining methodology for clinical domain specific neural language models, (2) a novel entity-centric masking strategy to infuse domain specific knowledge, (3) evaluation of the proposed strategies on three clinical tasks: cross-domain negation detection, DocTimeRel classification, and temporal relation extraction, and (4) evaluation of our models on the PubMedQA (Jin et al., 2019) dataset to measure the models' performance on a non-entity-centric task in the biomedical domain.

2 Methods

In this section, we first describe our clinical text datasets and related NLP tasks, the details of our entity-centric masking strategy, and finally the settings we used for both pretraining and fine-tuning.

2.1 Transformer models

Transformer models learn a sequential contextual representation of the input sequence through a multi-layer, multi-head self-attention mechanism, which models long-range dependencies in texts through highly parallel computation. They are usually pretrained through a self-supervised masked language model (MLM) task i.e., predicting the randomly masked subset of the input tokens. Some transformer models also use next sentence prediction (NSP) as a self-supervision task i.e., predicting if two given sentences are adjacent in the original text. A language model can be continuously pretrained on new corpora to further expand its representative power especially for a target domain. For a task-specific application, a pretrained language model's parameters are usually refined through a fine-tuning process on the task-specific training data, and a special [CLS] token is usually used as

[1]Our pretrained models are submitted to PhysioNet(Goldberger et al., 2000). Once approved, they will be publicly available through PhysioNet Credentialed Health Data License 1.5.0.

Dataset	sentence#	word#	entities#/sentence
MIMIC-SMALL	4.6M	125.1M	1+
MIMIC-BIG	15.6M	728.6M	2+

Table 1: Two versions of curated MIMIC data.

the representation of the input instance for text-classification tasks.

2.2 Unlabeled Pre-training Data

MIMIC-III We use the freely-available MIMIC-III (Medical Information Mart for Intensive Care) Clinical dataset (Johnson et al., 2016) (version 1.4) for continued pretraining of the PubMedBERT model. This dataset comprises approximately 2M deidentified notes for over 40,000 patients who stayed in critical care units of the Beth Israel Deaconess Medical Center between 2001 and 2012.

We process the MIMIC-III corpus with the sentence detection, tokenization, and temporal modules of Apache cTAKES (Savova et al., 2010)[2] to identify all entities (events and time expressions) in the corpus. Events are recognized by cTAKES event annotator. Event types include diseases/disorders, signs/symptoms, medications, anatomical sites, and procedures. Time expressions are recognized by cTAKES timex annotator. Time classes includes: date, time, duration, quantifier, prepostesp, and set (Styler IV et al., 2014). Special XML tags (Dligach et al., 2017) are inserted into the text sequence to mark the position of identified entities. Time expressions are replaced by their time class (Lin et al., 2017, 2018) for better generalizability. All special XML-tags and time class tokens are added into the PubMedBERT vocabulary so that they can be recognized. The top line of Figure 1 shows a sample sentence from the MIMIC-III corpus. The entities of this sentence are identified by Apache cTAKES. The bottom line of Figure 1 shows the entities marked by XML tags and the temporal expression replaced by its class. We process the MIMIC corpus sentence by sentence, and discard sentences that have fewer than two entities. The resulting set (MIMIC-BIG) has 15.6 million sentences, 728.6 million words (the bottom row of Table 1). In another setting, from the pool of sentences with at least one entity, we sample a smaller set (MIMIC-SMALL), resulting in 4.6 million sentences and 125.1 million words (the top row of Table 1).

[2] http://ctakes.apache.org

The patient had [EVENT: fever], [EVENT: tachypnea], and elevated [EVENT: lactate] on [TIME: March 11, 2010].

⇓

The patient had **<e>** fever **</e>**, **<e>** tachypnea **</e>**, and elevated **<e>** lactate **</e>** on **<t>** date **</t>**.

Figure 1: MIMIC-III text with XML-tagged entities: <e> and </e> mark events; <t> and </t> mark time expressions.

#1: she is feeling reasonably well . she has not **<e>** noted **</e>** any new areas of pain and has had no fevers
#2: a **<e>** surgery **</e>** was scheduled on **<t>** date **</t>** .
#3: a **<e1>** surgery **</e1>** was **<e2>** scheduled **</e2>** on march 11th .
#4: she denies any **<e>** fevers **</e>** or chills .
#5: Inpatient versus outpatient management of neutropenic fever in gynecologic oncology patients: is risk stratification useful? **ANSWER:** Based on this pilot data, MASCC score appears promising in determining suitability for outpatient management of NF in gynecologic oncology patients. Prospective study is ongoing to confirm safety and determine impact on cost.

Figure 2: Sample instances for DocTimeRel(1), TLINK:event-time(2), TLINK:event-event(3), Negation (4), and PubMedQA (5).

2.3 Labeled Fine-tuning Data

The following sections describe the labeled datasets that are used as fine-tuning tasks. Figure 2 shows examples of how we format inputs for these tasks (more details below).

THYME The THYME corpus (Styler IV et al., 2014) is widely used (Bethard et al., 2015, 2016, 2017) for clinical temporal relation discovery. There are two types of temporal relations defined in it: (1) The document time relations (DocTimeRel), which link a clinical event (EVENT) to the document creation time (DCT) with possible values of *BEFORE, AFTER, OVERLAP, and BEFORE_OVERLAP*, and (2) pairwise temporal relations (TLINK) between two events (EVENT) or an event and a time expression (TIMEX3) using an extension of TimeML (Pustejovsky et al., 2003; Pustejovsky and Stubbs, 2011). Recently, the TLINK annotations of (2) were refined with values of *BEFORE, BEGINS-ON, CONTAINS, CON-SUB, ENDS-ON, NOTED-ON, OVERLAP*, with the revised corpus known as the THYME+ corpus (Wright-Bettner et al., 2020).

For the DocTimeRel task, we mark all events in THYME+ corpus with XML tags ("<e>", "</e>") and extract 10 tokens from each side of the event as the contextual information. The DocTimeRel labels are predicted using the special [CLS] embedding and a softmax function.

For the TLINK task, we use the THYME+ annotation and the same window-based processing (Lin et al., 2019; Wright-Bettner et al., 2020) for generating relational candidates. The two entities involved in a relation candidate are marked by XML tags following the style of Dligach et al. (2017). Time expressions are represented by their time classes. The TLINK labels are predicted using the special [CLS] embedding and a softmax function.

Cross-domain Negation We use the same corpora as Miller et al. (2017); Lin et al. (2020a): (1) 2010 i2b2/VA NLP Challenge Corpus (i2b2: Uzuner et al., 2011), (2) the Multi-source Integrated Platform for Answering Clinical Questions Corpus (MiPACQ: Albright et al., 2013), (3) the Strategic Health IT Advanced Research Projects (SHARP) Seed (Seed), and (4) SHARP Stratified (Strat). We use them for fine-tuning the pretrained models for the cross-domain negation task. The same XML tags as described above mark the entities for which the negation status is to be determined. The +1(negated) and -1(not negated) labels are predicted using the special [CLS] embedding and a softmax function.

PubMedQA PubMedQA (Jin et al., 2019) is a biomedical question answering (QA) dataset collected from PubMed abstracts. The task is to answer research questions with yes/no/maybe using the corresponding abstracts or the conclusion sections of the abstracts (i.e., the long answers). For simplicity, we only fine-tune pretrained models on the PubMedQA labeled (PQA-L) data of 1K expert annotations, with the original train/dev/test split with 450, 50, 500 questions, respectively. The unlabeled (PQA-U) and artificially generated QA instances (PQA-A) are not used. Pretrained models are fine-tuned on the PQA-L data in the reasoning-free setting (without reasoning the full abstracts as contexts) by concatenating the questions and related long answers. The question and the answer is separated by "ANSWER:" (as shown in the bottom case of fig. 2) instead of the special [SEP] token in order not to involve the Next Sentence Prediction (NSP). The yes/no/maybe labels are predicted using the special [CLS] embedding and a softmax function.

2.4 Entity-centric Masking

Conventional BERT-style Masked Language Model (MLM) randomly chooses 15% of the input tokens for corruption, among which 80% are replaced by a special token "[MASK]", 10% are left unchanged, and 10% are randomly replaced by a token from the vocabulary. The language model is trained to reconstruct the masked tokens.

We propose an entity-centric masking strategy (as shown in Figure 3). All entities in the input sequence are marked with XML tags, which are added into the vocabulary and mapped to unique IDs. Then 40% of entities and 12% of random words are chosen respectively within each sequence block for corruption, following the same 80%-10%-10% ratio for [MASK], unchanged, and random replacement. We refer to this masking strategy as entity-centric masking.

We did not use the Next Sentence Prediction (NSP) task in our pretraining experiments based on Liu et al. (2019).

The PubMedBERT base uncased version was pretrained from scratch using abstracts from PubMed and full-text articles from PubMedCentral. We applied continued pretraining on it with MIMIC-BIG and MIMIC-SMALL with entity-centric masking and random masking. We denote the model pretrained with entity-centric masking **EntityBERT**, and model pretrained with random masking **RandMask**. For both masking strategies, we use different random seeds.

The pretrained models are then fine-tuned for the three clinical tasks (TLINK temporal relation extraction, DocTimeRel classification, and negation detection) and one biomedical task (PubMedQA). Since the TLINK task has the most relation types and is the most complicated task among the three, we use it as the primary testing task. The best models derived on the TLINK task are then tested on the other tasks.

2.5 Settings

We used an NVIDIA Titan RTX GPU cluster of 7 nodes for pre-training and fine-tuning experiments through HuggingFace's Transformer API (Wolf et al., 2019) version 2.10.0.

For pretraining, we set the max steps to 200k to allow full model convergence, and set the block size to 100. For fine-tuning, the batch size is selected from (16, 32), the learning rate is selected from (1e-5, 2e-5, 3e-5, 4e-5, 5e-5).

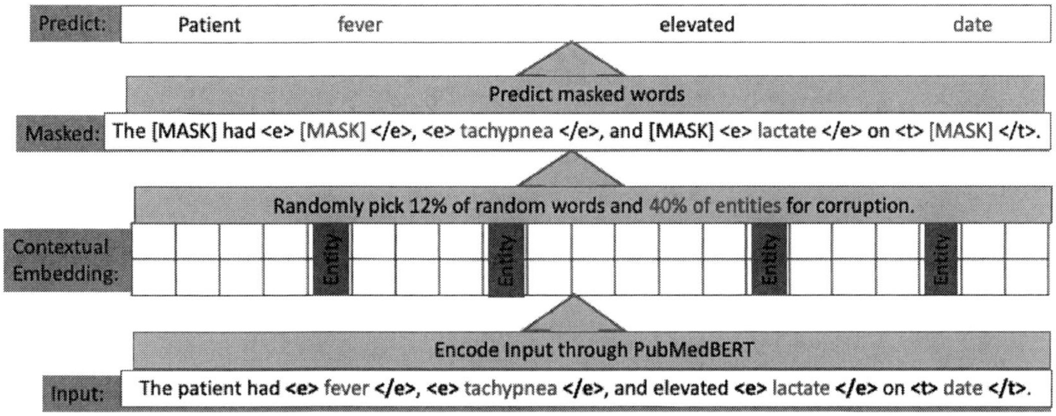

Figure 3: The architecture for continued pretraining of PubMedBERT with the entity-centric masking strategy.

For the TLINK task, the maximal sequence length is set to 100. The models are fine-tuned on the THYME colon cancer training set, parameters are optimized through the THYME colon development set, and tested on the THYME colon cancer test set. The performance is evaluated by the Clinical TempEval evaluation script (Bethard et al., 2017) modified to accommodate the refined temporal relations (Wright-Bettner et al., 2020).

For the DocTimeRel task, the maximal sequence length is set to 30. The models are fine-tuned on the THYME colon cancer training set, parameters are optimized on the THYME colon cancer development set, and tested on both the THYME colon cancer test set and the THYME brain cancer test set for portability evaluation.

For the negation task, the maximal sequence length is set to 64. We follow the same source-target setting as (Lin et al., 2020a) to carry out the cross-domain negation experiments.

For PubMedQA, the maximal sequence length is set to 100 to accommodate both the question and the long answer. The average PubMedQA question length is 14.4 tokens, while the average long answer length is 43.2 tokens (Jin et al., 2019).

Following (Reimers and Gurevych, 2017) in addition to reporting the best scores, we executed multiple runs with varied settings (e.g. random seeds, learning rates, etc.). We compared the distributions with two-sample t-test and report related p-values.

3 Results

Table 2 shows that on the test set of the TLINK task, the best rates for randomly masking entities and words are 40% and 12%, respectively. The table shows only the most successful rates; we considered more entity rates (20%, 40%, 60%, 80%, 100%) and word rates (0%, 8%, 10%, 12%, 14%, 16%). We found that (1) masking non-entity words in addition to masking entities is important as non-entity words capture semantic/syntactic information, and (2) masking too many tokens may make the reconstruction task too hard.

Entity-rate	Word-rate	Overall TLINK F1
30%	10%	0.631
30%	12%	0.644
30%	14%	0.642
40%	10%	0.640
40%	12%	**0.651**
40%	14%	0.642
40%	16%	0.639
50%	12%	0.643
50%	14%	0.641
60%	8%	0.638
60%	10%	0.634
60%	12%	0.631

Table 2: Effect of masking rates for entities (entity-rate) and random words (word-rate) when pretraining PubMedBERT on MIMIC-SMALL for temporal relation extraction. Performance is in terms of overall F1.

Table 3 shows that continuously pretraining PubMedBert (PM) with entity-centric masking (Entity) outperforms random masking (Rand) on both MIMIC-SMALL (p=0.034 with a two-sample t-test) and MIMIC-BIG (p=0.046). The best scores are marked in bold. MIMIC-BIG models have a lower inter-seed variance and slightly better average performance than MIMIC-SMALL. We also combined entity-centric masking with SpanBERT (Joshi et al., 2020) and continuously pre-

Mask	BERT	MIMIC	Random Seed 3	4	12	13	42	avg
Rand	PM	Small	0.628	0.641	0.632	0.628	0.641	0.634
Entity	PM	Small	0.643	0.641	0.640	0.634	**0.651**	0.642
Rand	PM	Big	0.634	0.637	0.641	0.634	0.635	0.636
Entity	PM	Big	0.641	0.642	0.640	0.643	**0.648**	0.643
Rand	Span	Small	0.632	0.630	0.632	0.641	0.636	0.634
Entity	Span	Small	0.638	0.635	0.637	**0.643**	**0.643**	0.639

Table 3: Effect of masking strategy (random or entity-centric) on continuously pretraining models (PubMed-BERT (PM) or SpanBERT) on MIMIC (BIG or SMALL) for the TLINK task, across different random seeds. Performance is in terms of overall F1.

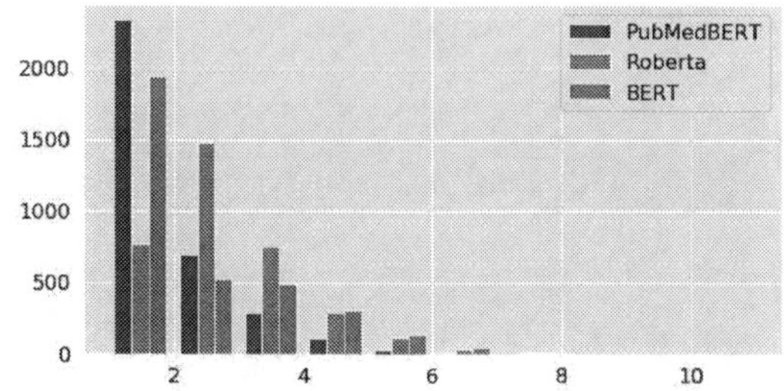

Figure 4: Histogram of token numbers after using different tokenization methods to process all single-token events in THYME Colon training set.

trained the model on MIMIC-SMALL with different random seeds. The last two rows of Table 3 show that entity-centric masking also helps SpanBERT on the TLINK task (p=0.004).

For our experiments on the downstream tasks, we choose the EntityBERT model continuously pretrained on MIMIC-SMALL with random seed 42 (0.651 F1) and the RandMask model continuously pretrained on MIMIC-BIG with random seed 12 (0.641 F1) because of their best performance. For RandMask models that all get 0.641 F1, we pick the one continuously pretrained on MIMIC-BIG. We fine-tuned them for the specific tasks. The detailed model performance on all TLINK categories is in the bottom two rows in Table 4. The top three rows of Table 4 show the previous best TLINK scores on the same THYME+ corpus by BioBERT and BART-large (Wright-Bettner et al., 2020) and the original PubMedBERT performance.

Table 5 shows that for cross-domain negation detection, out of 12 cross-domain pairs, the entity-centric masking is helpful for 9 pairs. EntityBERT's cross-domain negation average F1 is 0.781 while RandMask's average F1 is 0.773.

Table 6 shows that for DocTimeRel classification, EntityBERT improves over RandMask in the cross-domain setting. When trained and tested in the same colon cancer domain, EntityBERT gets the same overall F1 score as RandMask (0.92 F1). But when trained on colon cancer and tested on brain cancer, EntityBERT significantly improves the overall F1 from 0.69 F1 to 0.72 F1 (p=0.0007).

Table 7 shows PubMedBERT, RandMask and EntityBERT fine-tuning results on the PQA-L test set in the reasoning-free final-phase only setting. It is an extremely low resource setting where there are only 450 training instances used for fine-tuning the models. Results are reported in accuracy using the provided evaluation script. EntityBERT is on par with RandMask (p=0.307) even though these clinical-domain models are both out-of-domain for this biomedical-domain task.

4 Discussion

The benefit of an in-domain vocabulary. To study the in-domain vocabulary's contribution to a clinical task, we extract all 3,471 gold standard events in the THYME colon cancer training set and feed them into the PubMedBERT, RoBERTa, and BERT tokenizers. These events are all single-token events. Figure 4 shows the histogram of tokens per event after tokenization (x-axis shows the number of tokens each event is represented by). We see that PubMedBERT keeps the majority of the events (2,330) as one unit instead of breaking them into multiple sub-words. The BERT tokenizer keeps 1,729 events as one unit. The 601 events that PubMedBERT recognizes but BERT breaks into word pieces are of importance for the TLINK task. If we remove these 601 events from the PubMedBERT vocabulary – forcing them to be broken down into word pieces – the model performance on the TLINK task drops from 0.638 F1 (Table 4 row three) to 0.541 F1, which is the same performance we get if we replace PubMedBERT's tokenizer entirely with BERT's.

What makes a difference? By comparing the TLINK predictions (without applying temporal closure) produced by the best EntityBERT (0.651 F1) and by the best RandMask (0.641 F1), we found that EntityBERT predicted 4,924 correct TLINKs, while RandMask predicted 4,778 correct TLINKs (Table 8). By comparing the entities involved in those correct TLINKs, we found that EntityBERT recognized 131 more entities than Rand-

Model	BEFORE P	BEFORE R	BEFORE F1	BEGINS-ON P	BEGINS-ON R	BEGINS-ON F1	CONTAINS P	CONTAINS R	CONTAINS F1	ENDS-ON P	ENDS-ON R	ENDS-ON F1
BioBERT	0.278	0.458	0.346	0.423	0.175	0.248	0.793	0.708	0.748	0.112	0.210	0.146
BART-large	0.300	0.422	0.351	0.378	0.175	0.239	0.796	0.710	0.750	0.124	0.210	0.156
PubMedBERT	0.302	0.493	0.375	0.368	0.172	0.234	0.786	0.734	0.759	0.131	0.227	0.166
RandMask	0.309	0.460	0.370	0.376	0.165	0.229	0.804	0.726	0.763	0.131	0.160	0.144
EntityBERT	0.308	0.467	0.371	0.398	0.179	0.247	0.802	0.739	0.769	0.149	0.185	0.165

Model	NOTED-ON P	NOTED-ON R	NOTED-ON F1	OVERLAP P	OVERLAP R	OVERLAP F1	**OVERALL** P	**OVERALL** R	**OVERALL** F1
BioBERT	0.786	0.706	0.744	0.353	0.508	0.416	0.696	0.568	0.625
BART-large	0.786	0.707	0.744	0.404	0.470	0.435	0.718	0.558	0.628
PubMedBERT	0.791	0.728	0.758	0.403	0.489	0.442	0.704	0.583	0.638
RandMask	0.767	0.742	0.754	0.404	0.519	0.455	0.717	0.580	0.641
EntityBERT	0.783	0.758	0.770	0.408	0.534	0.462	**0.723**	**0.592**	**0.651**

Table 4: Performance of previous state-of-the-art and the proposed model (EntityBERT) on the TLINK task.

Source	Target	RandMask	EntityBERT
Seed	Strat	0.830	**0.834**
Seed	Mipacq	0.759	**0.761**
Seed	i2b2	0.827	**0.828**
Strat	Seed	0.722	**0.772**
Strat	Mipacq	0.758	0.754
Strat	i2b2	0.881	**0.886**
Mipacq	Seed	0.780	0.772
Mipacq	Strat	0.756	**0.785**
Mipacq	i2b2	0.878	0.871
i2b2	Seed	0.730	**0.732**
i2b2	Strat	0.662	**0.664**
i2b2	Mipacq	0.693	**0.713**
Overall		0.773	**0.781**

Table 5: Effect of masking strategy (Rand or Entity) on cross-domain negation detection. Performance is in terms of F1.

Model	Domain	after	before	bfr/ovlp	overlap	**overall**
RandMask	same	0.88	0.92	0.78	0.94	0.92
EntityBERT	same	0.88	0.92	0.79	0.94	0.92
RandMask	cross	0.65	0.65	0.34	0.74	0.69
EntityBERT	cross	0.64	0.66	0.40	0.77	**0.72**

Table 6: Effect of masking strategy (Rand or Entity) for in-domain (same) and cross-domain settings of the DocTimeRel task. Performance is in terms of F1.

Mask. Some entities only appear in EntityBERT-identified relations, e.g. *staging*, *hemoglobin*, *finding*, *consideration*, *consider*, *develops*, *request*, *treatment*, *neuropathy*, *carcinoma*, *metastasis*, *injection*, *resected*, and *staged* are involved in multiple relations. Entity-centric masking masks more entities than random masking so that those clinical entities can be better represented by the language model in terms of their semantic and syntactic usage. When the model is fine-tuned for an entity-centric task like the TLINK extraction task, these entities can be better utilized for reasoning relations which they are part of.

	PubMedBERT	RandMask	EntityBERT
Accuracy	**0.760**	0.738	0.750

Table 7: Performance of models on PubMedQA.

Model	within-sentence	cross-sentence	total
RandMask	4,021	757	4,778
EntityBERT	4,156	768	4,924

Table 8: Correctly predicted TLINK counts by EntityBERT and RandMask before temporal closure.

In Figure 5 we visualize with BertViz (Vig, 2019) the attention weights of head zero from the last layer of the fine-tuned RandMask and EntityBERT models on the TLINK task for a relation that EntityBERT correctly predicted but RandMask missed. The context is *he has had steroid <e> injection </e> <t> date </t>*. A plausible explanation is that because the key entities, *injection* and *date*, are not well represented in RandMask model, the [CLS] token of RandMask model (Figure 5 (a)) focuses on entity markers, *<e>*, *</e>*, *<t>*, and *</t>*. It may figure out this is an event-time relation but incorrectly infers its type. The [CLS] token of EntityBERT (Figure 5 (b)) bakes in representations of all tokens with knowledge that *injection* is related to *steroid* and *date* is related to *<e> injection </e>*, which shows the key entities are well represented.

Table 8 also shows that the EntityBERT model is most helpful for within-sentence relations (135 more correct within-sentence predictions vs. 11

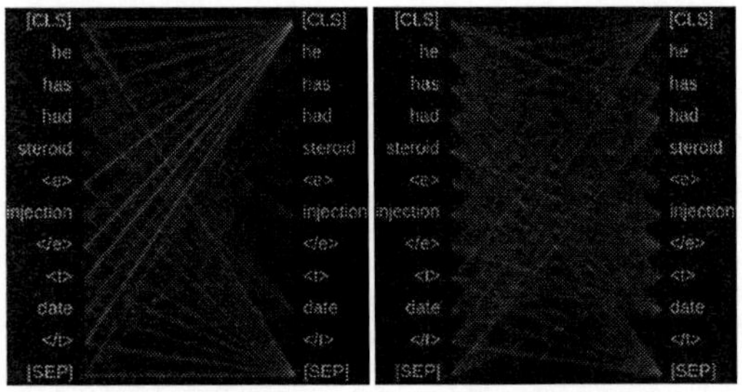

(a) RandMask (b) EntityBERT

Figure 5: Attention visualization of the last layer of RandMask (a) and EntityBERT (b).

more correct cross-sentence predictions). It could mean the better-learned entity representation is most helpful within a relatively close distance for the current model architecture. To help inferring longer-distanced relations, we may need enhanced model architectures, e.g., DeBERTa (He et al., 2020), that can represent the relative distance between two entities in a disentangled fashion.

Combining Entity-centric Masking with Another Masking Strategy. Some other pre-trained language models like BART (Lewis et al., 2019) and SpanBERT (Joshi et al., 2020) are not pretrained on clinical/biomedical corpora. Yet, they are suitable for clinical tasks in that they mask contiguous random spans instead of individual tokens/word pieces during pretraining – in the clinical domain there are a lot of events and entities that span multiple tokens (e.g., *ascending colon cancer*, *March 11, 2011*). Even without any continued pretraining on a clinical corpus, BART-large achieves 0.628 F1 on the TLINK task (Table 4, row 2), and with continued pretraining on MIMIC-SMALL, SpanBERT-base achieves 0.641 F1 (Table 3, row 5, seed 13). Interestingly, entity-centric masking can further increase SpanBERT performance in the continued pretraining setting (Table 3, last two rows, p=0.004). The reason could be that even though clinical entities could span multiple tokens, a contiguous random span may not be a clinical entity. So, specifically masking clinical entities still has its advantage during continued pretraining a contiguous-span-based language model. We may even see further improved performance if BART or SpanBERT can be pretained from scratch on large clinical/biomedical corpora (however, such a corpus is not available currently!) and then combined with entity-centric masking.

The Strength and Limitations of EntityBERT: EntityBERT assumes that clinical entities are important words, thus if a clinical language model can represent clinical entities better, it will benefit downstream clinical entity-centric tasks. Therefore, such a masking strategy increases the entity concentrations in the masked words during the model pretraining, but does not increase the overall computational loads either for pre-training or for fine-tuning since the overall total number of masked items is similar to random word masking. This is unlike building an additional neural network for selective masking Gu et al. (2020b) or incorporating knowledge graphs Zhang et al. (2019).

The better representation of clinical entities is not only beneficial in an in-domain setting, e.g., the TLINK task, but also effective in a cross-domain setting, e.g., the negation and DocTimeRel tasks. For the DocTimeRel task, both EntityBERT and RandMask achieve very good in-domain performance of 0.92 F1 (see Table 6). In its cross-domain setting, EntityBERT has a clear edge of 0.71 F1 over RandMask 0.69 F1 (see Table 6). Even though some of the improvements may not seem big, they are statistically significant.

We acknowledge some limitations of the current EntityBERT model. First, it is pretrained with a relatively small block size (100 tokens) which is sufficient for a sentence- or a short-paragraph-level reasoning tasks but may be not sufficient for document-level tasks. Second, EntityBERT aims to improve the performance of entity-centric clinical tasks. For tasks that may not directly leverage entities, such as question answering or document classification, entities may still play a supporting role but may not prove as effective. However, we hypothesize that even in those cases its performance would be on-par with RandMask because of its in-domain vocabulary and continued training on a clinical corpus.

Based on the results of Table 7 on PubMedQA, we can see that even though RandMask and EntityBERT models are continuously pretrained from the PubMedBERT model, the continued pretraining on a clinical corpus has made them diverge from its biomedical domain. For the PubMedQA biomedical domain task, the original PubMedBERT model was pretrained from scratch in this target domain, thus performs the best in this task. Yet, even for this non-entity-centric task, EntityBERT performs slightly (but not significantly) better than the RandMask model (0.750 vs. 0.738 in accuracy).

MIMIC-BIG vs. MIMIC-SMALL: RandMask and EntityBERT models pretrained on MIMIC-SMALL perform almost on par with models pretrained on the much bigger corpus, MIMIC-BIG (Table 3) for the TLINK task. The reason could be that even though clinical language varies, the crucial clinical entities are not that many. For example, for the TLINK task, there are only 3,471 unique gold standard events in the training set. Thus, although the size of the corpus is smaller, it could be sufficient for the model to learn representations of the important unique entities.

MIMIC-BIG was created by filtering sentences with fewer than two entities with the goal of capturing pair-wise interactions between events in the language model. One of the limitations of our architecture is its block size. Perhaps with a model that can effectively represent the relative distances, the interactions among entities can be represented better. In addition, by eliminating sentences that only have one or no entity, MIMIC-BIG misses some language phenomena. MIMIC-SMALL, despite its smaller size, thus may encounter more diverse language. This could be the explanation of why an EntityBERT model pretrained on MIMIC-SMALL gets the best TLINK performance (0.651 F1; Table 3 row 2 and Table 4 bottom row).

In the future, we will investigate combining entity-centric masking with DeBERTa (He et al., 2020) with the goal of developing a strategy for a deep neural model that combines entities and their relative position in an input sequence. We will experiment with more flavors of EntityBERT with different block sizes for a wider range of clinical applications. Further testing EntityBERT on a wider range of clinical and biomedical tasks would be helpful for understanding its capabilities.

Acknowledgments

The study was funded by R01LM10090, U24CA248010 and UG3CA243120 from the Unites States National Institutes of Health. The content is solely the responsibility of the authors and does not necessarily represent the official views of the National Institutes of Health.

The authors would like to thank the anonymous reviewers for their valuable suggestions and criticism.

The authors would also like to acknowledge Boston Children's Hospital's High-Performance Computing Resources BCH HPC Cluster Enkefalos 2 (E2) made available for conducting the research reported in this publication. Software used in the project was installed and configured by BioGrids (Morin et al., 2013).

References

Daniel Albright, Arrick Lanfranchi, Anwen Fredriksen, William F Styler IV, Colin Warner, Jena D Hwang, Jinho D Choi, Dmitriy Dligach, Rodney D Nielsen, James Martin, et al. 2013. Towards comprehensive syntactic and semantic annotations of the clinical narrative. *Journal of the American Medical Informatics Association*, 20(5):922–930.

Emily Alsentzer, John R Murphy, Willie Boag, Wei-Hung Weng, Di Jin, Tristan Naumann, and Matthew McDermott. 2019. Publicly available clinical bert embeddings. *arXiv preprint arXiv:1904.03323*.

Andrew L Beam, Benjamin Kompa, Allen Schmaltz, Inbar Fried, Griffin Weber, Nathan Palmer, Xu Shi, Tianxi Cai, and Isaac S Kohane. 2019. Clinical concept embeddings learned from massive sources of multimodal medical data. In *PACIFIC SYMPOSIUM ON BIOCOMPUTING 2020*, pages 295–306. World Scientific.

Iz Beltagy, Kyle Lo, and Arman Cohan. 2019. Scibert: A pretrained language model for scientific text. *arXiv preprint arXiv:1903.10676*.

Steven Bethard, Leon Derczynski, Guergana Savova, Guergana Savova, James Pustejovsky, and Marc Verhagen. 2015. Semeval-2015 task 6: Clinical tempeval. In *Proceedings of the 9th International Workshop on Semantic Evaluation (SemEval 2015)*, pages 806–814.

Steven Bethard, Guergana Savova, Wei-Te Chen, Leon Derczynski, James Pustejovsky, and Marc Verhagen. 2016. Semeval-2016 task 12: Clinical tempeval. *Proceedings of SemEval*, pages 1052–1062.

Steven Bethard, Guergana Savova, Martha Palmer, James Pustejovsky, and Marc Verhagen. 2017. Semeval-2017 task 12: Clinical tempeval. *Proceedings of the 11th International Workshop on Semantic Evaluation (SemEval-2017)*, pages 563–570.

Wendy W Chapman, Will Bridewell, Paul Hanbury, Gregory F Cooper, and Bruce G Buchanan. 2001. A simple algorithm for identifying negated findings and diseases in discharge summaries. *Journal of biomedical informatics*, 34(5):301–310.

Qingyu Chen, Kyubum Lee, Shankai Yan, Sun Kim, Chih-Hsuan Wei, and Zhiyong Lu. 2020. Bioconceptvec: creating and evaluating literature-based biomedical concept embeddings on a large scale. *PLoS computational biology*, 16(4):e1007617.

Jacob Devlin, Ming-Wei Chang, Kenton Lee, and Kristina Toutanova. 2018. Bert: Pre-training of deep bidirectional transformers for language understanding. *arXiv preprint arXiv:1810.04805*.

Dmitriy Dligach, Timothy Miller, Chen Lin, Steven Bethard, and Guergana Savova. 2017. Neural temporal relation extraction. In *Proceedings of the 15th Conference of the European Chapter of the Association for Computational Linguistics: Volume 2, Short Papers*, pages 746–751.

Noémie Elhadad, Sameer Pradhan, Sharon Gorman, Suresh Manandhar, Wendy Chapman, and Guergana Savova. 2015. SemEval-2015 task 14: Analysis of clinical text. In *Proceedings of the 9th International Workshop on Semantic Evaluation (SemEval 2015)*, pages 303–310, Denver, Colorado. Association for Computational Linguistics.

Philip Gage. 1994. A new algorithm for data compression. *C Users Journal*, 12(2):23–38.

Ary L Goldberger, Luis AN Amaral, Leon Glass, Jeffrey M Hausdorff, Plamen Ch Ivanov, Roger G Mark, Joseph E Mietus, George B Moody, Chung-Kang Peng, and H Eugene Stanley. 2000. Physiobank, physiotoolkit, and physionet: components of a new research resource for complex physiologic signals. *circulation*, 101(23):e215–e220.

Yu Gu, Robert Tinn, Hao Cheng, Michael Lucas, Naoto Usuyama, Xiaodong Liu, Tristan Naumann, Jianfeng Gao, and Hoifung Poon. 2020a. Domain-specific language model pretraining for biomedical natural language processing. *arXiv preprint arXiv:2007.15779*.

Yuxian Gu, Zhengyan Zhang, Xiaozhi Wang, Zhiyuan Liu, and Maosong Sun. 2020b. Train no evil: Selective masking for task-guided pre-training. *arXiv preprint arXiv:2004.09733*.

Henk Harkema, John N Dowling, Tyler Thornblade, and Wendy W Chapman. 2009. Context: an algorithm for determining negation, experiencer, and temporal status from clinical reports. *Journal of biomedical informatics*, 42(5):839–851.

Pengcheng He, Xiaodong Liu, Jianfeng Gao, and Weizhu Chen. 2020. Deberta: Decoding-enhanced bert with disentangled attention. *arXiv preprint arXiv:2006.03654*.

Qiao Jin, Bhuwan Dhingra, Zhengping Liu, William W Cohen, and Xinghua Lu. 2019. Pubmedqa: A dataset for biomedical research question answering. *arXiv preprint arXiv:1909.06146*.

Alistair EW Johnson, Tom J Pollard, Lu Shen, H Lehman Li-Wei, Mengling Feng, Mohammad Ghassemi, Benjamin Moody, Peter Szolovits, Leo Anthony Celi, and Roger G Mark. 2016. Mimic-iii, a freely accessible critical care database. *Scientific data*, 3(1):1–9.

Mandar Joshi, Danqi Chen, Yinhan Liu, Daniel S Weld, Luke Zettlemoyer, and Omer Levy. 2020. Spanbert: Improving pre-training by representing and predicting spans. *Transactions of the Association for Computational Linguistics*, 8:64–77.

Jinhyuk Lee, Wonjin Yoon, Sungdong Kim, Donghyeon Kim, Sunkyu Kim, Chan Ho So, and Jaewoo Kang. 2020. Biobert: a pre-trained biomedical language representation model for biomedical text mining. *Bioinformatics*, 36(4):1234–1240.

Artuur Leeuwenberg and Marie Francine Moens. 2017. Structured learning for temporal relation extraction from clinical records. In *Proceedings of the 15th Conference of the European Chapter of the Association for Computational Linguistics: Volume 1, Long Papers*, pages 1150–1158.

Mike Lewis, Yinhan Liu, Naman Goyal, Marjan Ghazvininejad, Abdelrahman Mohamed, Omer Levy, Ves Stoyanov, and Luke Zettlemoyer. 2019. Bart: Denoising sequence-to-sequence pre-training for natural language generation, translation, and comprehension. *arXiv preprint arXiv:1910.13461*.

Chen Lin, Steven Bethard, Dmitriy Dligach, Farig Sadeque, Guergana Savova, and Timothy A Miller. 2020a. Does bert need domain adaptation for clinical negation detection? *Journal of the American Medical Informatics Association*, 27(4):584–591.

Chen Lin, Timothy Miller, Dmitriy Dligach, Hadi Amiri, Steven Bethard, and Guergana Savova. 2018. Self-training improves recurrent neural networks performance for temporal relation extraction. In *Proceedings of the Ninth International Workshop on Health Text Mining and Information Analysis*, pages 165–176.

Chen Lin, Timothy Miller, Dmitriy Dligach, Steven Bethard, and Guergana Savova. 2017. Representations of time expressions for temporal relation extraction with convolutional neural networks. In *BioNLP 2017*, pages 322–327.

Chen Lin, Timothy Miller, Dmitriy Dligach, Steven Bethard, and Guergana Savova. 2019. A bert-based universal model for both within-and cross-sentence clinical temporal relation extraction. In *Proceedings of the 2nd Clinical Natural Language Processing Workshop*, pages 65–71.

Chen Lin, Timothy Miller, Dmitriy Dligach, Farig Sadeque, Steven Bethard, and Guergana Savova. 2020b. A bert-based one-pass multi-task model for clinical temporal relation extraction. In *Proceedings of the 19th SIGBioMed Workshop on Biomedical Language Processing*, pages 70–75.

Yinhan Liu, Myle Ott, Naman Goyal, Jingfei Du, Mandar Joshi, Danqi Chen, Omer Levy, Mike Lewis, Luke Zettlemoyer, and Veselin Stoyanov. 2019. Roberta: A robustly optimized bert pretraining approach. *arXiv preprint arXiv:1907.11692*.

Xinbo Lv, Yi Guan, Jinfeng Yang, and Jiawei Wu. 2016. Clinical relation extraction with deep learning. *International Journal of Hybrid Information Technology*, 9(7):237–248.

Saeed Mehrabi, Anand Krishnan, Sunghwan Sohn, Alexandra M Roch, Heidi Schmidt, Joe Kesterson, Chris Beesley, Paul Dexter, C Max Schmidt, Hongfang Liu, et al. 2015. Deepen: A negation detection system for clinical text incorporating dependency relation into negex. *Journal of biomedical informatics*, 54:213–219.

Timothy Miller, Steven Bethard, Hadi Amiri, and Guergana Savova. 2017. Unsupervised domain adaptation for clinical negation detection. In *BioNLP 2017*, pages 165–170.

Andrew Morin, Ben Eisenbraun, Jason Key, Paul C Sanschagrin, Michael A Timony, Michelle Ottaviano, and Piotr Sliz. 2013. Cutting edge: Collaboration gets the most out of software. *elife*, 2:e01456.

Yifan Peng, Shankai Yan, and Zhiyong Lu. 2019. Transfer learning in biomedical natural language processing: An evaluation of bert and elmo on ten benchmarking datasets.

Sameer Pradhan, Noémie Elhadad, Wendy Chapman, Suresh Manandhar, and Guergana Savova. 2014. SemEval-2014 task 7: Analysis of clinical text. In *Proceedings of the 8th International Workshop on Semantic Evaluation (SemEval 2014)*, pages 54–62, Dublin, Ireland. Association for Computational Linguistics.

James Pustejovsky, José M Castano, Robert Ingria, Roser Sauri, Robert J Gaizauskas, Andrea Setzer, Graham Katz, and Dragomir R Radev. 2003. Timeml: Robust specification of event and temporal expressions in text. *New directions in question answering*, 3:28–34.

James Pustejovsky and Amber Stubbs. 2011. Increasing informativeness in temporal annotation. In *Proceedings of the 5th Linguistic Annotation Workshop*, pages 152–160. Association for Computational Linguistics.

Colin Raffel, Noam Shazeer, Adam Roberts, Katherine Lee, Sharan Narang, Michael Matena, Yanqi Zhou, Wei Li, and Peter J Liu. 2019. Exploring the limits of transfer learning with a unified text-to-text transformer. *arXiv preprint arXiv:1910.10683*.

Nils Reimers and Iryna Gurevych. 2017. Reporting score distributions makes a difference: Performance study of lstm-networks for sequence tagging. *arXiv preprint arXiv:1707.09861*.

Guergana K Savova, James J Masanz, Philip V Ogren, Jiaping Zheng, Sunghwan Sohn, Karin C Kipper-Schuler, and Christopher G Chute. 2010. Mayo clinical text analysis and knowledge extraction system (ctakes): architecture, component evaluation and applications. *Journal of the American Medical Informatics Association*, 17(5):507–513.

William F Styler IV, Steven Bethard, Sean Finan, Martha Palmer, Sameer Pradhan, Piet C De Groen, Brad Erickson, Timothy Miller, Chen Lin, Guergana Savova, et al. 2014. Temporal annotation in the clinical domain. *Transactions of the association for computational linguistics*, 2:143–154.

Özlem Uzuner, Brett R South, Shuying Shen, and Scott L DuVall. 2011. 2010 i2b2/va challenge on concepts, assertions, and relations in clinical text. *Journal of the American Medical Informatics Association*, 18(5):552–556.

Jesse Vig. 2019. A multiscale visualization of attention in the transformer model. In *Proceedings of the 57th Annual Meeting of the Association for Computational Linguistics: System Demonstrations*, pages 37–42, Florence, Italy. Association for Computational Linguistics.

Thomas Wolf, Lysandre Debut, Victor Sanh, Julien Chaumond, Clement Delangue, Anthony Moi, Pierric Cistac, Tim Rault, Rémi Louf, Morgan Funtowicz, et al. 2019. Huggingface's transformers: State-of-the-art natural language processing. *arXiv preprint arXiv:1910.03771*.

Kristin Wright-Bettner, Chen Lin, Timothy Miller, Steven Bethard, Dmitriy Dligach, Martha Palmer, James H Martin, and Guergana Savova. 2020. Defining and learning refined temporal relations in the clinical narrative. In *Proceedings of the 11th International Workshop on Health Text Mining and Information Analysis*, pages 104–114.

Stephen Wu, Timothy Miller, James Masanz, Matt Coarr, Scott Halgrim, David Carrell, and Cheryl Clark. 2014. Negation's not solved: generalizability versus optimizability in clinical natural language processing. *PloS one*, 9(11):e112774.

Yonghui Wu, Min Jiang, Jun Xu, Degui Zhi, and Hua Xu. 2017. Clinical named entity recognition using deep learning models. In *AMIA Annual Symposium Proceedings*, volume 2017, page 1812. American Medical Informatics Association.

Yonghui Wu, Mike Schuster, Zhifeng Chen, Quoc V Le, Mohammad Norouzi, Wolfgang Macherey, Maxim Krikun, Yuan Cao, Qin Gao, Klaus Macherey, et al. 2016. Google's neural machine translation system: Bridging the gap between human and machine translation. *arXiv preprint arXiv:1609.08144*.

Zhengyan Zhang, Xu Han, Zhiyuan Liu, Xin Jiang, Maosong Sun, and Qun Liu. 2019. ERNIE: enhanced language representation with informative entities. *CoRR*, abs/1905.07129.

Morteza Ziyadi, Yuting Sun, Abhishek Goswami, Jade Huang, and Weizhu Chen. 2020. Example-based named entity recognition. *arXiv preprint arXiv:2008.10570*.

Contextual explanation rules for neural clinical classifiers

Madhumita Sushil[1] and **Simon Šuster**[2,*] and **Walter Daelemans**[1]
[1] Computational Linguistics and Psycholinguistics Research Center (CLiPS),
University of Antwerp, Belgium
firstname.lastname@uantwerpen.be

[2] Faculty of Engineering and Information Technology, University of Melbourne
simon.suster@unimelb.edu.au

Abstract

Several previous studies on explanation for recurrent neural networks focus on approaches that find the most important input segments for a network as its explanations. In that case, the manner in which these input segments combine with each other to form an explanatory pattern remains unknown. To overcome this, some previous work tries to find patterns (called rules) in the data that explain neural outputs. However, their explanations are often insensitive to model parameters, which limits the scalability of text explanations. To overcome these limitations, we propose a pipeline to explain RNNs by means of decision lists (also called rules) over skipgrams. For evaluation of explanations, we create a synthetic sepsis-identification dataset, as well as apply our technique on additional clinical and sentiment analysis datasets. We find that our technique persistently achieves high explanation fidelity and qualitatively interpretable rules.

1 Introduction

Understanding and explaining decisions of complex models such as neural networks has recently gained a lot of attention for engendering trust in these models, improving them, and understanding them better (Montavon et al., 2018; Alishahi et al., 2019; Belinkov and Glass, 2019). Several previous studies developing interpretability techniques provide a set of input features or segments that are the most salient for the model output. Approaches such as input perturbation and gradient computation are popular for this purpose (Ancona et al., 2018; Arras et al., 2019). A drawback of these approaches is the lack of information about interaction between different features. While heatmaps (Li et al., 2016b,a; Arras et al., 2017) and partial dependence plots (Lundberg and Lee, 2017) are popularly used, they only provide a qualitative view which quickly gets complex as the number of features increases. To overcome this limitation, rule induction for model interpretability has become popular, which accounts for interactions between multiple features and output classes (Lakkaraju et al., 2017; Puri et al., 2017; Ming et al., 2018; Ribeiro et al., 2018; Sushil et al., 2018; Evans et al., 2019; Pastor and Baralis, 2019). Most of these work treat the explained models as black boxes, and fit a separate interpretable model on the original input to find rules that mimic the output of the explained model. However, because the interpretable model does not have information about the parameters of the complex model, global explanation is expensive, and the explaining and explained models could fit different curves to arrive to the same output. Sushil et al. (2018) incorporates model gradients in the explanation process to overcome these challenges, but this technique cannot be used with current state-of-the-art models that use word embeddings due to their reliance on interpretable model input in the form of bag-of-words. Murdoch and Szlam (2017) explain long short term memory networks (LSTMs) (Hochreiter and Schmidhuber, 1997) by means of ngram rules, but their rules are limited to presence of single ngrams and do not capture interaction between ngrams in text. To learn explanation rules for RNNs while overcoming the limitations of the previous approaches, we have the following contributions in the paper:

1. We induce explanation rules over important skipgrams in text, while ensuring that these rules generalize to unseen data. To this end, we quantify skipgram importance in LSTMs by first pooling gradients across embedding dimensions to compute word importance, and thereby aggregating them into skipgram importance. Skipgrams incorporate word order in explanations and increase interpretability.

2. To overcome existing limitations with au-

*Research conducted while at CLiPS.

Proceedings of the BioNLP 2021 workshop, pages 202–212
June 11, 2021. ©2021 Association for Computational Linguistics

tomated explanation evaluation (Lertvittayakumjorn and Toni, 2019; Poerner et al., 2018), we provide a synthetic clinical text classification dataset for evaluating interpretability techniques. We construct this dataset according to existing medical knowledge and clinical corpus. We validate our explanation pipeline on this synthetic dataset by recovering the labeling rules of the dataset. We then apply our pipeline to two clinical datasets for sepsis classification, and one dataset for sentiment analysis. We confirm that the explanation results obtained on synthetic data are scalable to real corpora.

2 Explanation pipeline

We propose a method to find decision lists as explanation rules for RNNs with word embedding input. We quantify word importance in an RNN by comparing multiple pooling operations (qualitatively and quantitatively). After establishing a desired pooling technique, we move to finding importance of skipgrams, which provides larger context around words in explanations. We then find decision lists that associate the relative importance of multiple skipgrams in the RNN to an output class. This is an extension of our prior work (Sushil et al., 2018) where we find if-then-else rules for feedforward neural networks. However, the previous approach relies on using interpretable inputs independent of word order and is not scalable to the current state-of-the-art approaches that use word embeddings instead. Moreover, explanation of binary classifiers is not supported by that pipeline, and the explanation rules are not generalized to unseen examples. Furthermore, the previous explanation rules are hierarchical, and hence cannot be understood independently without parsing the entire rule hierarchy. In the proposed research, we address all these limitations and extend the explanations to binary cases, unseen data, and to sequential neural networks with word embedding input. Additionally, these explanation rules can be understood as an independent decision path. We present the complete pipeline for our approach, which we name UNRAVEL, in Figure 1. Code for the paper is available on `https://github.com/clips/rnn_expl_rules`.

2.1 Word importance computation

Saliency (importance) scores of input features are often computed as gradients of the predicted output node w.r.t. all the input nodes for all the instances (Simonyan et al., 2013; Adebayo et al., 2018). In neural architectures that have an embedding layer, interpretable input features are replaced by corresponding low-dimensional embeddings. Due to this, we obtain different saliency scores for different embedding dimensions of a word in a document. Because embedding dimensions are not interpretable, it is difficult to understand what these multiple saliency scores mean. To instead obtain a single score for a word by combining saliency values of all the dimensions, we consider the following commonly used pooling techniques:

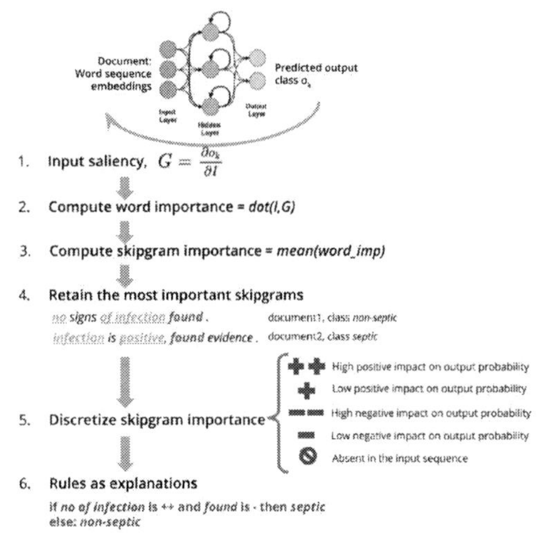

Figure 1: The complete UNRAVEL pipeline for gradient-informed rule induction in recurrent neural networks. The underlined terms in point 4 refer to different important skipgrams in the text.

- **L2 norm** of the gradient scores (Bansal et al., 2016; Hechtlinger, 2016; Poerner et al., 2018).

$$saliency_{L2} = \Sigma_{dim} grad^2$$

- **Sum** of gradients across all the dimensions.

$$saliency_{sum} = \Sigma_{dim} grad$$

- **Dot product** between the embeddings and the gradient scores (Denil et al., 2014; Montavon et al., 2018; Arras et al., 2019). This additionally accounts for the embedding value itself.

$$saliency_{dot} = \Sigma_{dim}(emb \odot grad)$$

We also experimented with max pooling, but we omit the discussion here because they have the same patterns as the L2 norm, albeit with higher magnitudes.

In Section 4.1, we analyze the importance scores obtained with these techniques qualitatively and quantitatively to identify the preferred one.

2.2 Skipgrams to incorporate context

One of the contributions of this work is to find explanation rules for sequential models such as RNNs. Conjunctive clauses of if-then-else rules are order independent although this order is critical for RNNs. To account for word order in input documents, some previous approaches find the most important ngrams instead of the top words only (Murdoch and Szlam, 2017; Jacovi et al., 2018). To incorporate word order also in explanation rules, we compute the importance of subsequences in the documents before combining different subsequences into conjunctive rules. We define importance of a subsequence as the mean saliency of all the tokens in that subsequence. We represent subsequences as skipgrams with length in the range [1,4] and with maximum two skip tokens[1]. After computing the scores, we retain the 50 most important skipgrams for every document (based on absolute importance scores). The number of unique skipgrams obtained in this manner is very high. To limit the complexity of explanations, we retain 5k skipgrams with the highest total absolute importance score across the entire training set and learn explanation rules over these. To this end, we create a bag-of-skipgram-importance representation of the documents, where the vocabulary corresponds to the 5k most important skipgrams across the training set. For ease of understanding, we discretize the importance scores of the skipgrams to represent five different levels of importance: $\{--, -, 0, +, ++\}$. Here $--$ and $++$ represent a high negative and positive importance, respectively, for the predicted output class, 0 means that the skipgram is absent in the document, and $-$ and $+$ indicate low negative and positive importance scores, respectively. This skipgram set, along with the output predictions of a model, is then input to a rule induction module to obtain decision lists as explanations.

[1]Length and skip values in skipgrams were manually decided to include sufficient context while limiting complexity. As these values increase further, the phrases become more sparse, resulting into a larger explanation vocabulary. Feature selection step hence selects a smaller proportion of phrases to retain the same computational complexity, which can limit the explanation coverage/recall.

2.3 Learning transferable explanations

In the prediction phase, a model merely applies the knowledge it has learned from the training data. Hence, an explanation technique should not require prior knowledge of the test set to find global explanations of a model. We hypothesize that explanation rules should be consistently accurate between the training data and the predictions on unseen data. In accordance to this hypothesis, instead of learning explanations directly from validation or test instances, which is common in interpretability research (Ribeiro et al., 2018; Sushil et al., 2018), we modify the explanation procedure to learn accurate, transferable explanations only from the training set. We first feed the training data to our neural network and record the corresponding output predictions. These output predictions, combined with the corresponding set of top discretized skipgrams, are used to fit the rule inducer. The hyperparameters of the rule inducer are optimized to best explain the validation set outputs. Finally, we report a score that quantifies how well the learned rules transfer to the test predictions. This training scheme ensures that the explanations are generalizable to unseen data, instead of overfitting the test set.

We obtain decision lists using PART (Frank and Witten, 1998), which finds simplified paths of partial C4.5 decision trees. These decision lists can be comprehended independent of the order, and support both binary and multi-class cases.

3 Datasets and Models

3.1 Synthetic dataset

A big challenge for interpretability research is the evaluation of the results (Lertvittayakumjorn and Toni, 2019). Human evaluation is not ideal because a model can learn correct classification patterns that are counter-intuitive for humans (Poerner et al., 2018). In complex domains like healthcare, such an evaluation is additionally infeasible. To overcome existing limitations with automated evaluation of explanations, we create a synthetic binary clinical document classification dataset. We base the dataset construction on the sepsis screening guidelines[2]. This is a critical task for preventing deaths in ICUs (Futoma et al., 2017) and new insights about the problem are important in the medical domain. The synthetic dataset includes a subset of sentences from the freely available clinical corpus

[2]`https://bit.ly/3575e3d`

MIMIC-III (Johnson et al., 2016). Dataset construction process is described here:

- From the MIMIC-III corpus, we sample 3–15 words long sentences that mention the keywords discussed in the screening guidelines, grouped into the following sets:
    1. I: Contains sentences that mention these infection-related keywords: {pneumonia *and*[3] empyema, meningitis, endocarditis, infection}.
    2. $Infl$: Contains sentences that mention these inflammation-related keywords: {hypothermia *or*[4] hyperthermia, leukocytosis *or* leukopenia, altered mental status, tachycardia, tachypnea, hyperglycemia}.
    3. $Others$: Sentences that do not mention any of the previously stated keywords: $Sentence \notin \{I \cup Infl\}$.
- We populate 50k documents with 17 sentences each by randomly sampling one sentence from set I, one sentence for each comma-separated term in set $Infl$, and 10 sentences from set $Others$. We additionally populate 20k documents with 17 sentences, all from set $Others$.
- We then run the CLAMP clinical NLP pipeline (Soysal et al., 2017) to identify if these keywords are negated in the documents.
- Next, we assign class labels to the documents using the following rule:

    if *the infection term sampled from set I is not negated* **and** *at least 2 responses sampled from set* $Infl$ *are not negated*
    $\implies$ Class label is *septic*,
    Class label is *non-septic* otherwise.

49% of the documents are thus labeled as *septic*.

Sampling sentences from the MIMIC-III corpus introduces language diversity through a large vocabulary and varied sentence structures. Use of an imperfect tool to identify negation for document labeling also adds noise to the dataset. These properties are desirable because they allow for controlled explanation evaluation while also simulating real world corpora and tasks, unlike several synthetic datasets used for explanation evaluation (Arras et al., 2019; Chrupala and Alishahi, 2019).

[3]Sentences mentioning both the keywords are sampled.

[4]Sentences mentioning either of the keywords are sampled.

3.1.1 Gold important terms

For every document, the set of words that are used to assign it a class label includes all the keyword terms about infection from set I that are mentioned in that document, keyword terms about inflammatory response from set $Infl$, and their corresponding negation markers as identified by the CLAMP pipeline. We mark these sets of terms, one set per document, as the gold set of important terms for this task. For example in the document:

> No signs of infection were found. Altered mental status exists. Patient is suffering from hypothermia,

the set of gold terms would include all the underlined words. Among these words, *infection*, *altered*, *mental*, *status*, and *hypothermia* are keyword terms, and *no*, *signs*, and *of* are terms corresponding to the negation scope.

3.1.2 Model:

We split the dataset into subsets of 80-10-10% as training-validation-test sets. We obtain a vocabulary of 47,015 tokens after lower-casing the documents without removing punctuation. We replace unknown words in validation and test sets with the ⟨unk⟩ token. We train LSTM classifiers to predict the document class from the hidden representation after the final timestep, which is obtained after processing the entire document as a sequence of tokens[5]. The classifiers use randomly initialized word embeddings and a single RNN layer without attention. The hidden state size and embedding dimension are set to either 50 or 100. We use the Adam optimizer (Kingma and Ba, 2014) with learning rate 0.001 and a batch size of 64 (without hyperparameter optimization). Classification performance is shown in Table 1.

3.2 Real clinical datasets

We additionally find explanation rules for sepsis classifiers on the MIMIC-III clinical corpus. We define sepsis label as all the cases where patients are assigned one of the following diagnostic codes:

- 995.91 (Sepsis): Two or more systemic inflammatory response criteria plus a known or suspected infection. 2% of the cases.

[5]We do not experiment with other types of classifiers because the focus of the work is to find and evaluate explanation rules for sequential models that use word embeddings as input, as opposed to comparing different classifiers.

- 995.92 (Severe Sepsis): Sepsis with acute organ dysfunction. 3% of the cases.

- 785.52 (Septic Shock): Form of severe sepsis where the organ dysfunction involves the cardiovascular system. 4% of the cases.

We analyze two different setups after removing blank notes and the notes marked as *error* in the MIMIC-III corpus:

1. We use the last discharge note for every patient to classify whether the patient has sepsis. Class distribution among 58,028 instances is 90-10% for non-septic and septic cases respectively, and the vocabulary size is 229,799. The task is easier in this setup because 70% of septic cases mention sepsis directly, whereas only 13% of non-septic cases mention sepsis.

2. We classify whether a patient has a sepsis diagnosis or not using the last note about a patient excluding the categories discharge notes, social work, rehab services and nutrition. We obtain 52,691 patients in this manner, out of which only 9% are septic. The vocabulary size is 87,753. In this setup, only 17% of septic cases mention sepsis, as opposed to 6% of non-septic cases mentioning sepsis.

3.2.1 Models:

We train 2-layer bidirectional LSTM classifiers with 100 dimensional randomly initialized word embeddings and 100 dimensional hidden layer. We train for 50 epochs with early stopping with patience 5. The remaining data processing and implementation details are the same as discussed for synthetic dataset. Macro F1 score of classification when using discharge notes is 0.68 (septic class F1 is 0.41), and without using discharge notes is 0.60 (septic class F1 is 0.27). Majority baseline is 0.5.

3.3 Sentiment analysis

Following Murdoch and Szlam (2017), we explain LSTM classifiers initialized with 300 dimensional Glove (Pennington et al., 2014) embeddings and 150 hidden nodes for binary sentiment classification on the Stanford sentiment analysis (SST2) dataset (Socher et al., 2013). We obtain 84.13% classification accuracy, and our vocabulary size is 13,983.

3.4 Baseline explanation rules

Several existing approaches for global rule-based interpretability (Lakkaraju et al., 2017; Puri et al., 2017) have one common aspect—they directly use the original input to find explanation rules for complex classifiers without making use of the parameters of the complex models. However, these approaches don't scale to NLP tasks due to combinatorial computational complexity in finding explanation rules. For comparison, as baseline rules, we induce explanations directly from the input data without using gradients of neural models. To this end, we create a bag-of-skipgrams by binarizing the most frequent skipgrams to represent whether they are present in a document. We then train rule induction classifiers on this binarized skipgram data to explain neural outputs.

We also compare to Anchors (Ribeiro et al., 2018) for SST2 explanations by implementing their submodular pick algorithm for obtaining global explanations. Anchors does not scale to longer documents used for sepsis classification.

3.5 Evaluation metrics

We record fidelity scores of the explanation rules on the test set, and the complexity of these explanations. Fidelity scores refer to how faithful the explanations are to the test output predictions of the explained neural network. Like our prior work (Sushil et al., 2018), we use macro F1-measure of explanations compared to original predictions to quantify it. We define explanation complexity as the number of rules in an explanation.

4 Evaluation

4.1 Comparing pooling techniques

To compare different pooling techniques described in Section 2.1, we evaluate sets of most important words obtained by different techniques against gold sets of important terms for the documents.

4.1.1 Qualitative analysis

In Figure 2, we compare word importance distribution for the pooling techniques for an instance in the validation set of the synthetic corpus. The L2 norm provides distributions over the positive values only and the importance scores are low because it squares the gradients. Sum pooling and dot product instead return a distribution over both positive and negative values, with dot product returning a more peaked distribution. However, as we can see, sum

Classification		Pooling		
Classifier	Acc.	L2	sum	dot
LSTM100, E100	96.5	17.8	13.7	**26.0**
LSTM100, E50	95.5	23.7	21.5	**35.4**
LSTM50, E100	92.0	38.2	33.5	**50.2**
LSTM50, E50	92.4	26.5	25.1	**36.1**

Table 1: Classification accuracy of different LSTM classifiers and the average accuracy for the top k words in documents in the **synthetic dataset** obtained with L2, sum and dot product pooling techniques. LSTMx, Ey refers to LSTM with x hidden nodes and y dimensional word embeddings.

and dot product often provide opposite importance signs for the same words. This is caused due to presence of word embeddings while computing dot product, which can take both positive and negative values. In this instance, both true and predicted classes are *non-septic*. Looking at Figure 2c, we find positive peaks over *negative* and *infection*, and negative peaks over *altered mental status* and *hyperglycemia*. This corresponds to the class labeling rule in the synthetic data, where *non-septic* class is assigned when infection terms are negated. These directions of influence are counter-intuitive for sum pooling in Figure 2b. Due to its intuitive, peaked importance distributions, dot product seems to be better than other techniques. However, we move to quantitative evaluation for a global perspective because this qualitative analysis is biased towards a specific instance and model.

4.1.2 Quantitative analysis

We find the top k tokens for test documents in the synthetic dataset by ranking absolute word importance scores, where k is the number of gold important terms used to label the document. We ignore the 20k documents that only consist of sentences that do not mention any keyword term, and hence have an empty gold set. We compute the accuracy of the set of most important words for every document compared to their corresponding gold set. Later, we take a mean across all the documents and report it in Table 1. We find that dot product consistently recovers more important tokens than other pooling techniques across all the classifiers, confirming the qualitative analysis earlier and the findings of Arras et al. (2019). Hence we use dot product for computing word importance before inducing explanation rules.

We additionally see that the mean accuracy is nearly twice for the classifier with 50 hidden nodes and 100 dimensional word embeddings as compared to the the larger classifier that uses 100 hidden units instead, although the latter classifier is nearly 5% more accurate. This suggests that the larger network obtains higher performance by focusing on tokens that are not incorporated within the gold keywords. The reason behind different tokens being considered important could be that our gold set of important terms is noisy:

- Some tokens such as punctuation symbols are missing from the gold set, although they are important for identifying the scope of negation, as seen in Figure 3.
- Some terms in the gold set are not required for correct classification. For example: 1. Too many words are included as negation triggers. For example, in the sentence *no signs of infection were found.*, 'no', 'signs', and 'of' are all added to the gold set as negation markers although the subset {'no', 'infection'} may be sufficient. 2. Similarly, the keyword *altered mental status* could already be recognized from a subset of these terms.

4.2 Explaining synthetic data classifiers

We obtain explanations of all the LSTM classifiers for the synthetic dataset. We record fidelity scores of explanations and the corresponding complexity in Table 2. We find that when we use the proposed pipeline UNRAVEL for learning gradient-informed rules, we obtain explanations with high fidelity scores also on the test data. On the other hand, with the baseline approach, we obtain nearly 15% lower fidelity scores. In addition, explanations are more complex with the baseline approach. This confirms that making use of model parameters by means of gradients acts as an additional useful cue for the rule-based explainability module, thus resulting in more faithful explanations.

We present some examples of explanation rules for the most accurate LSTM classifier for the synthetic dataset in Figure 3. Here, we indicate infection keywords that were used to populate the dataset with a single underline, and the inflammatory response keywords with a double underline. The first rule in the figure indicates that if two inflammatory response criteria are highly important for the network, the term infection is highly important, and phrases negating the presence of infection

GOLD:non_septic PRED:non_septic
percocet 325 one to two tabs one p.o. q .4 -6 h . plan : # altered mental status : several possible etiologies at this point . paracentesis negative for infection . hyperglycemia assessment : hx iddm . he has a resting tachypnea but is not using accessory muscles to breathe . her chest was clear to auscultation bilaterally . tachycardia : multifactorial : sepsis , electrolytes abnormalities # . no elevated white blood count but a left shift without bands . her swan-ganz catheter was left in place for hemodynamic monitoring . patient passed spontaneous breathing test on hospital day two and was extubated . sensation was normal bilateral lower and upper extremities . # leukocytosis : patient is afebrile . the remaining paranasal sinuses visualized are clear . external rewarming for hypothermia , check thyroid function . chest x-ray : mild to moderate cardiac enlargement with prominent left ventricle contour . discharge examination : non-focal with normal speech on arrival to the floor , the patient was comfortable and assymptomatic .

(a) L2 norm

GOLD:non_septic PRED:non_septic
percocet 325 one to two tabs one p.o. q .4 -6 h . plan : # altered mental status : several possible etiologies at this point . paracentesis negative for infection . hyperglycemia assessment : hx iddm . he has a resting tachypnea but is not using accessory muscles to breathe . her chest was clear to auscultation bilaterally . tachycardia : multifactorial : sepsis , electrolytes abnormalities # . no elevated white blood count but a left shift without bands . her swan-ganz catheter was left in place for hemodynamic monitoring . patient passed spontaneous breathing test on hospital day two and was extubated . sensation was normal bilateral lower and upper extremities . # leukocytosis : patient is afebrile . the remaining paranasal sinuses visualized are clear . external rewarming for hypothermia , check thyroid function . chest x-ray : mild to moderate cardiac enlargement with prominent left ventricle contour . discharge examination : non-focal with normal speech on arrival to the floor , the patient was comfortable and assymptomatic .

(b) Sum

GOLD:non_septic PRED:non_septic
percocet 325 one to two tabs one p.o. q .4 -6 h . plan : # altered mental status : several possible etiologies at this point . paracentesis negative for infection . hyperglycemia assessment : hx iddm . he has a resting tachypnea but is not using accessory muscles to breathe . her chest was clear to auscultation bilaterally . tachycardia : multifactorial : sepsis , electrolytes abnormalities # . no elevated white blood count but a left shift without bands . her swan-ganz catheter was left in place for hemodynamic monitoring . patient passed spontaneous breathing test on hospital day two and was extubated . sensation was normal bilateral lower and upper extremities . # leukocytosis : patient is afebrile . the remaining paranasal sinuses visualized are clear . external rewarming for hypothermia , check thyroid function . chest x-ray : mild to moderate cardiac enlargement with prominent left ventricle contour . discharge examination : non-focal with normal speech on arrival to the floor , the patient was comfortable and assymptomatic .

(c) Dot

Figure 2: Heatmap visualization of word importance distribution for a single validation set instance in LSTM classifier with 50 hidden nodes and 100 dimensional word embeddings when L2, sum, and dot pooling techniques are used. Blue reflects positive importance and red indicates negative importance.

Explanation	Eval type	LSTM100,E100	LSTM100,E50	LSTM50,E100	LSTM50,E50
Baseline(sg)	Fidelity	75.65	77.67	83.19	84.30
	Complexity	63	60	26	46
UNRAVEL(sg)	Fidelity	98.90	99.46	99.97	98.24
	Complexity	32	13	2	49
UNRAVEL(1gram)	Fidelity	98.83	99.51	99.97	97.22
	Complexity	23	18	2	51

Table 2: Test set fidelity scores of explanations (in %macro-F1), and number of explanation rules as the measure of explanation complexity for different LSTM classifiers on the **synthetic dataset** using our approach compared to the baseline approach. LSTMx,Ey refers to LSTM with x hidden nodes and y dimensional word embeddings. sg in parenthesis refers to skipgram-based explanations.

(a) **if** *hyperglycemia* = ++ AND *to exclude* = 0 AND *evidence infection .* = 0 AND *infection* = ++ AND *no infection .*= 0 AND *no infection* = 0 AND *negative infection* = 0 AND *or of infection* = 0 AND *fungal infection other* = 0 AND *of infection in the* = 0 AND *altered* = ++ $\Longrightarrow$ septic (17466/17466)

(b) **if** *tachypnea* = 0 AND *meningitis* = 0 AND *urinary tract* = 0 AND *endocarditis* = 0 AND *hyperglycemia* = 0 $\Longrightarrow$ non-septic (16015/16015)

(c) **if** *no* = ++ AND *urinary* = 0 AND *bacterial* = 0 AND *mental* = − $\Longrightarrow$ non-septic (1277/1345)

Figure 3: Example explanation rules for the best LSTM classifier on the **synthetic dataset**. Infection keywords from set I are marked with a single underline, and the corresponding inflammatory response keywords from set $Infl$ are marked with double underline. ++ refers to high positive importance of a term, 0 represents absence of a term, and − means that the term gets a low negative importance, i.e., presence of the term reduces the output probability. The numbers (a/b) mean that b training instances are explained by the rule, of which a are correct. The first two rules are obtained with skipgrams, and the third one is obtained on using only unigrams for explanations.

Dataset	Explanation	Fidelity	N_rules
+discharge	Baseline(sg)	61.7	825
	UNRAVEL(sg)	97.9	16
-discharge	UNRAVEL(sg)	77.3	196
SST2	Anchors	70.3	10
	UNRAVEL(sg)	80.2	87

Table 3: Explanation fidelity (% macro F1) and complexity for **sepsis classification**: 1) With discharge notes 2) Without discharge notes, and on the **SST2** dataset. The baseline method did not converge (in several weeks) for sepsis classification without discharge note and for SST2 classification. Anchors did not scale (in memory usage) to document-level sepsis datasets.

are absent, then the class is recognized as septic. This is similar to the rule we have used to label the synthetic dataset, which requires at least one infection term and at least two inflammatory response criteria to not be negated in the document for being assigned a septic class. In the next rule—applied after all the cases from the previous rule have been excluded from the dataset—if several keyword terms are absent, the document is classified as non-septic. It is useful to remember that *urinary tract* is usually followed by the word *infection* in the dataset, and several instances mentioning *infection* have already been explained by the previous rule and hence have been ignored by this rule. This explanation rule is also in accordance to the synthetic dataset, where 20k documents do not contain any keyword term and are labeled as non-septic.

The third rule is an example rule for the same model when explanations are based on unigrams only as opposed to skipgrams. In this case, we lose the context of the negation marker *no*. When using skipgrams, this context of negation is available, which makes the negation scope clearer. Further, terms like *evidence*, *fungal* and *urinary tract* captured by skipgrams provide additional context for understanding the rules. This illustrates that even though the fidelity scores of explanations are similar, skipgram based explanations are more interpretable than only unigram-based explanations. Hence, we use skipgrams for further analysis.

4.3 Explaining clinical models

We rerun our explainability pipeline on both clinical models for sepsis classification—with and without using discharge notes (Section 3.2). For the first classifier with discharge notes, we again obtain very high fidelity scores of explanations (Table 3).

if *sepsis major surgical* = ++ $\Longrightarrow$ septic (209/209)
if *complaint : sepsis* = 0 AND *chief hypotension major* = ++ $\Longrightarrow$ septic (169/169)

Figure 4: First two explanation rules for the **clinical dataset** that uses discharge notes to classify sepsis. ++ refers to high positive importance, and 0 refers to an absent term in the document. (a/b) in parentheses show that a of b examples explained by this rule are correct.

The baseline explanations have significantly lower fidelity scores while also being extremely complex. On inspecting the corresponding explanation rules given in Figure 4, we find that they refer to the direct mentions of sepsis in the discharge notes. In the first rule, if *sepsis major surgical* is mentioned, the class is directly septic. In the second condition, it first rules out the mention of a complaint of sepsis and then checks for additional conditions. This confirms that not only does the classifier pick up on these direct mentions, but the explanations also recover this information. This illustrates the utility of UNRAVEL in understanding our models, which is the first step towards improving them. For example, if our model is learning direct mentions of sepsis as a discriminating feature, we could remove these direct mentions from the dataset before training new models to ensure that they generalize.

Next, for the more difficult case where we use only the final non-discharge note about patients to classify whether they have sepsis, the fidelity score is 77.33%. Although this score is good as an absolute number, it is much lower than other two cases. Explanations for this model are also much more complex. This highlights that more complex classifiers and explanations have lower explanation fidelity. While manually inspecting these explanations, we find that absence of terms such as *diagnosis : sepsis*, *indication endocarditis . valve*, *indication bacteremia*, *admitting diagnosis fever* and *pyelonephritis* are used to rule out sepsis. These are similar to the explanations of the other two datasets, albeit enriched with information about additional infections and body conditions. This confirms that the synthetic dataset closely models a real clinical use case, and suggests that these explanations rules could result into useful hypothesis generation.

4.4 Explaining sentiment classifier

Results of the SST2 explanations are given in Table 3. Our pipeline provides ~10% more accurate explanations compared to Anchors. Moreover, on

if *?* = 0 AND *bad* . = 0 AND *too* = ++ AND *one* = 0 $\implies$ negative (159/159)

if ? = ++ $\implies$ negative (81/82)

if bad . = 0 AND worst = 0 AND fails = 0 AND feels = ++ $\implies$ negative (54/54)

if bad . = 0 AND worst = 0 AND fails = 0 AND is bad = 0 AND flat = 0 AND mess = 0 AND stupid = 0 AND *suffers* = 0 AND *pointless* = 0 AND *dull* = ++ $\implies$ negative (38/38)

if *bad* . = ++ $\implies$ negative (36/36)

Figure 5: Example explanation rules for the **SST2 dataset**. ++ refers to high positive importance, and 0 refers to an absent term in the document. (a/b) in parentheses show that a of b examples explained by this rule are correct.

if *the* is present $\implies$ negative

if *a* is present $\implies$ positive

if *civility* is present $\implies$ positive

if *of* is present $\implies$ positive

if *this* is present $\implies$ negative

if *just* is present $\implies$ negative

if *good* is present $\implies$ positive

if *with* is present $\implies$ positive

if *no* is present $\implies$ negative

if *little* is present $\implies$ positive

Figure 6: Explanation rules for the LSTM classifier on the SST2 dataset with the Anchors submodular pick algorithm. The rules check the presence of words in the input to map to an output class.

inspecting the explanation rules for our method and Anchors respectively presented in Figures 5 and 6, we find that Anchors rules consist only of single words, as opposed to UNRAVEL, which finds conjunctions of different phrases. Furthermore, explanation rules with UNRAVEL obtain 71% classification accuracy on the original task. This performance drop compared to LSTM is ~7% lower than gradient decomposition-based performance drop reported by Murdoch and Szlam (2017), although the numbers aren't strictly comparable because we explain different classifiers[6].

[6]Their implementation is not openly available for direct comparison.

5 Conclusions and Future Work

We have successfully developed a pipeline to obtain transferable, accurate gradient-informed explanation rules from RNNs. We have constructed a synthetic dataset to qualitatively and quantitatively evaluate the results, and we obtain informative explanations with high fidelity scores. We obtain similar results on clinical datasets and sentiment analysis. Our approach is transferable to all similar neural models. In future, it would be interesting to extend the capabilities of this approach to obtain more accurate, less complex and scalable explanations for classifiers with more complex patterns.

Acknowledgements

This research was carried out within the Accumulate strategic basic research project, funded by the government agency Flanders Innovation & Entrepreneurship (VLAIO) [grant number 150056]. It also received funding from the Flemish Government (AI Research Program). We would like to thank all the anonymous reviewers of this paper whose useful comments have ensured that it is in a better state now compared to the original draft.

References

Julius Adebayo, Justin Gilmer, Michael Muelly, Ian Goodfellow, Moritz Hardt, and Been Kim. 2018. Sanity Checks for Saliency Maps. In *Proceedings of the 32Nd International Conference on Neural Information Processing Systems*, NIPS'18, pages 9525–9536, USA. Curran Associates Inc.

Afra Alishahi, Grzegorz Chrupala, and Tal Linzen. 2019. Analyzing and Interpreting Neural Networks for NLP: A Report on the First BlackboxNLP Workshop. *CoRR*, abs/1904.04063.

Marco Ancona, Enea Ceolini, Cengiz Öztireli, and Markus Gross. 2018. Towards better understanding of gradient-based attribution methods for Deep Neural Networks. In *International Conference on Learning Representations*.

Leila Arras, Franziska Horn, Grégoire Montavon, Klaus-Robert Müller, and Wojciech Samek. 2017. "What is relevant in a text document?": An interpretable machine learning approach. In *PloS one*.

Leila Arras, Ahmed Osman, Klaus-Robert Müller, and Wojciech Samek. 2019. Evaluating Recurrent Neural Network Explanations. In *Proceedings of the 2019 ACL Workshop BlackboxNLP: Analyzing and Interpreting Neural Networks for NLP*, pages 113–126, Florence, Italy. Association for Computational Linguistics.

Trapit Bansal, David Belanger, and Andrew McCallum. 2016. Ask the GRU: Multi-task Learning for Deep Text Recommendations. In *Proceedings of the 10th ACM Conference on Recommender Systems*, RecSys '16, pages 107–114, New York, NY, USA. ACM.

Yonatan Belinkov and James Glass. 2019. Analysis methods in Neural Language Processing: A Survey. *Transactions of the Association for Computational Linguistics*, 7:49–72.

Grzegorz Chrupala and Afra Alishahi. 2019. Correlating Neural and Symbolic Representations of Language. In *Proceedings of the 57th Conference of the Association for Computational Linguistics, ACL 2019, Florence, Italy, July 28- August 2, 2019, Volume 1: Long Papers*, pages 2952–2962.

Misha Denil, Alban Demiraj, and Nando de Freitas. 2014. Extraction of Salient sentences from Labelled Documents. Technical report, University of Oxford.

Benjamin P. Evans, Bing Xue, and Mengjie Zhang. 2019. What's Inside the Black-box?: A Genetic Programming Method for Interpreting Complex Machine Learning Models. In *Proceedings of the Genetic and Evolutionary Computation Conference*, GECCO '19, pages 1012–1020, New York, NY, USA. ACM.

Eibe Frank and Ian H. Witten. 1998. Generating accurate rule sets without global optimization. In *Fifteenth International Conference on Machine Learning*, pages 144–151. Morgan Kaufmann.

Joseph Futoma, Sanjay Hariharan, and Katherine A. Heller. 2017. Learning to detect sepsis with a multitask gaussian process RNN classifier. In *Proceedings of the 34th International Conference on Machine Learning, ICML 2017, Sydney, NSW, Australia, 6-11 August 2017*, pages 1174–1182.

Yotam Hechtlinger. 2016. Interpretation of Prediction Models Using the Input Gradient. *Computing Research Repository*, arXiv:1611.07634.

Sepp Hochreiter and Jürgen Schmidhuber. 1997. Long short-term memory. *Neural computation*, 9(8):1735–1780.

Alon Jacovi, Oren Sar Shalom, and Yoav Goldberg. 2018. Understanding Convolutional Neural Networks for Text Classification. In *Proceedings of the 2018 EMNLP Workshop BlackboxNLP: Analyzing and Interpreting Neural Networks for NLP*, pages 56–65, Brussels, Belgium. Association for Computational Linguistics.

Alistair EW Johnson, Tom J Pollard, Lu Shen, Liwei H Lehman, Mengling Feng, Mohammad Ghassemi, Benjamin Moody, Peter Szolovits, Leo Anthony Celi, and Roger G Mark. 2016. MIMIC-III, a freely accessible critical care database. *Scientific data*, 3:.

Diederik P Kingma and Jimmy Ba. 2014. Adam: A method for stochastic optimization. *Computing Research Repository*, arXiv:1412.6980.

Himabindu Lakkaraju, Ece Kamar, Rich Caruana, and Jure Leskovec. 2017. Interpretable & Explorable Approximations of Black Box Models. *Workshop on Fairness, Accountability, and Transparency in Machine Learning, KDD*, arXiv:1707.01154.

Piyawat Lertvittayakumjorn and Francesca Toni. 2019. Human-grounded Evaluations of explanation methods for text classification. *Computing Research Repository*, arXiv:1908.11355.

Jiwei Li, Xinlei Chen, Eduard Hovy, and Dan Jurafsky. 2016a. Visualizing and understanding neural models in NLP. In *Proceedings of the 2016 Conference of the North American Chapter of the Association for Computational Linguistics: Human Language Technologies*, pages 681–691, San Diego, California. Association for Computational Linguistics.

Jiwei Li, Will Monroe, and Dan Jurafsky. 2016b. Understanding Neural Networks through Representation erasure. *Computing Research Repository*, arXiv:1612.08220.

Scott M Lundberg and Su-In Lee. 2017. A unified approach to interpreting model predictions. In *Advances in Neural Information Processing Systems*, pages 4768–4777.

Yao Ming, Huamin Qu, and Enrico Bertini. 2018. Rulematrix: visualizing and understanding classifiers with rules. *IEEE transactions on visualization and computer graphics*, 25(1):342–352.

Grégoire Montavon, Wojciech Samek, and Klaus-Robert Müller. 2018. Methods for interpreting and understanding deep neural networks. *Digital Signal Processing*, 73:1 – 15.

W. James Murdoch and Arthur Szlam. 2017. Automatic Rule Extraction from Long Short Term Memory Networks. In *5th International Conference on Learning Representations, ICLR 2017, Toulon, France, April 24-26, 2017, Conference Track Proceedings*.

Eliana Pastor and Elena Baralis. 2019. Explaining Black Box Models by Means of Local Rules. In *Proceedings of the 34th ACM/SIGAPP Symposium on Applied Computing*, SAC '19, pages 510–517, New York, NY, USA. ACM.

Jeffrey Pennington, Richard Socher, and Christopher D. Manning. 2014. Glove: Global vectors for word representation. In *Empirical Methods in Natural Language Processing (EMNLP)*, pages 1532–1543.

Nina Poerner, Hinrich Schütze, and Benjamin Roth. 2018. Evaluating neural network explanation methods using hybrid documents and morphosyntactic agreement. In *Proceedings of the 56th Annual Meeting of the Association for Computational Linguistics (Volume 1: Long Papers)*, pages 340–350. Association for Computational Linguistics.

Nikaash Puri, Piyush Gupta, Pratiksha Agarwal, Sukriti Verma, and Balaji Krishnamurthy. 2017. MAGIX: Model Agnostic Globally Interpretable Explanations. *Computing Research Repository*, arXiv:1706.07160.

Marco Tulio Ribeiro, Sameer Singh, and Carlos Guestrin. 2018. Anchors: High-Precision Model-Agnostic Explanations. In *AAAI Conference on Artificial Intelligence (AAAI)*.

Karen Simonyan, Andrea Vedaldi, and Andrew Zisserman. 2013. Deep inside convolutional networks: Visualising image classification models and saliency maps. *Computing Research Repository*, arXiv:1312.6034.

Richard Socher, Alex Perelygin, Jean Wu, Jason Chuang, Christopher D. Manning, Andrew Ng, and Christopher Potts. 2013. Recursive deep models for semantic compositionality over a sentiment treebank. In *Proceedings of the 2013 Conference on Empirical Methods in Natural Language Processing*, pages 1631–1642, Seattle, Washington, USA. Association for Computational Linguistics.

Ergin Soysal, Jingqi Wang, Min Jiang, Yonghui Wu, Serguei Pakhomov, Hongfang Liu, and Hua Xu. 2017. CLAMP — a toolkit for efficiently building customized clinical natural language processing pipelines. *Journal of the American Medical Informatics Association*, page .

Madhumita Sushil, Simon Šuster, and Walter Daelemans. 2018. Rule induction for global explanation of trained models. In *Proceedings of the 2018 EMNLP Workshop BlackboxNLP: Analyzing and Interpreting Neural Networks for NLP*, pages 82–97. Association for Computational Linguistics.

Exploring Word Segmentation and Medical Concept Recognition for Chinese Medical Texts

Yang Liu♠♢, Yuanhe Tian♥, Tsung-Hui Chang♠♢, Song Wu♣△, Xiang Wan♢, Yan Song♠♢†

♠The Chinese University of Hong Kong (Shenzhen), ♥University of Washington

♣PingHu Hospital of Shenzhen University

△Shenzhen Hospital of Shanghai University of Traditional Chinese Medicine

♢Shenzhen Research Institute of Big Data

♠yangliu5@link.cuhk.edu.cn ♥yhtian@uw.edu ♣wusong@szu.edu.cn

♢wanxiang@sribd.cn ♠{changtsunghui, songyan}@cuhk.edu.cn

Abstract

Chinese word segmentation (CWS) and medical concept recognition are two fundamental tasks to process Chinese electronic medical records (EMRs) and play important roles in downstream tasks for understanding Chinese EMRs. One challenge to these tasks is the lack of medical domain datasets with high-quality annotations, especially medical-related tags that reveal the characteristics of Chinese EMRs. In this paper, we collected a Chinese EMR corpus, namely, ACEMR, with human annotations for Chinese word segmentation and EMR-related tags. On the ACEMR corpus, we run well-known models (i.e., BiLSTM, BERT, and ZEN) and existing state-of-the-art systems (e.g., WMSeg and TwASP) for CWS and medical concept recognition. Experimental results demonstrate the necessity of building a dedicated medical dataset and show that models that leverage extra resources achieve the best performance for both tasks, which provides certain guidance for future studies on model selection in the medical domain.[1]

1 Introduction

Medical language processing (MLP), i.e., natural language processing (NLP) for the electronic medical record (EMR), has drawn significant attention over the past few decades (Rector et al., 1991; Friedman et al., 2004; Stevenson et al., 2012; Koleck et al., 2021). EMR normally records the entire process of a patient's examination, diagnosis, and treatment by clinicians in the hospital, and contains a large amount of medical information, which, if extracted properly, can be used to train a machine learning model as an automated tool for auxiliary diagnosis and treatment, forming the foundation of wise information technology of medicine.

Chinese word segmentation (CWS) and medical concept recognition are two important and related tasks for Chinese MLP, which received much attention in previous studies (Xing et al., 2018; Wang et al., 2019). The first task (i.e., CWS) aims to segment Chinese text (i.e., character sequence) into words, which is a necessary step for MLP because the meaning of many medical terms cannot be simply inferred by its component characters. For example, it is hard to infer the meaning of "扁桃体" (*tonsil*) from its components "扁" (*flat*), "桃" (*peach*), and "体" (*body*). The second task (i.e., medical concept recognition) assigns an EMR-related tag (e.g., *Organism* and *Group*) to the segmented words. It is worth noting that the medical concept in this paper includes not only the standard medical named entities but also other categories that are useful for medical text analysis. For example, "*Time*" is a medical concept that can be used to represent the disease history; "*Probability*" is a possible medical concept tag for "考虑" (*consider*), in EMR.

To perform CWS and medical concept recognition in Chinese EMR, researchers face a challenge that existing training data for the tasks is either publicly unavailable or of poor quality. Although one possible solution is to apply models trained in the general domain to the medical text, these models always fail to have good performance because there are many domain-specific medical terms that rarely occur in the general domain. To address these challenges, we collect and annotate a new Chinese EMR corpus, named ACEMR, where texts from 500 EMRs (7K sentences) are annotated with CWS and medical concept recognition labels. In addition, we test several state-of-the-art models for CWS and medical concept recognition on the collected ACEMR corpus. Experimental results show the necessity of constructing an informative Chinese medical corpus and provide certain guidance for the model selection in medical domain.

†Corresponding author.

[1]The resources in this paper are released at https://github.com/cuhksz-nlp/ACEMR.

Proceedings of the BioNLP 2021 workshop, pages 213–220
June 11, 2021. ©2021 Association for Computational Linguistics

基本信息 (Basic Information)	Patient's name, gender, age, reason for admission, time of admission
病历特点 (Case Characteristic)	Detailed symptoms of the patient before admission, past history, physical examination results and auxiliary examination results
初步诊断 (Preliminary Diagnosis)	The type of disease initially judged, usually several disease names
鉴别诊断 (Differential Diagnosis)	According to the main complaint of the patient, distinguish it from other diseases and exclude the possible diagnosis of other diseases
治疗计划 (Treatment Plan)	Medical examinations to be done in the next step, and preliminary treatment plan

Table 1: The major five parts of information contained in one First Course Record in Chinese EMRs.

Class	Sub-class	Count
物体 Thing	Organism (Ogm)	150
	Group (Gr)	3,059
	Health Device (HD)	433
事件 Event	Health Behavior (HB)	3,093
	Events (E)	4,442
身体 Body	Body Parts (BP)	19,004
	Body Substance (BS)	1,103
	Body Function (BF)	5,179
异常 Abnormality	Signs or symptoms (SOS)	21,263
	Disease (Di)	3,543
检查 Examination	Examination Project (EP)	3,201
治疗 Treatment	Treatment Project (TP)	1,579
	Clinical Drug (Drug)	728
概念 Concept	Time (T)	4,514
	Qualitative (Ql)	14,510
	Space (Sp)	7,626
	Presence (Pre)	8,748
	Absence (Ab)	13,642
	Probability (Prob)	388
	Cause and Effect (CE)	1,359
Total	–	107,943

Table 2: The list of all medical concepts and counts.

2 Related Work

NLP for medical text has draw many attentions in the recent years (Xue et al., 2012; Xu et al., 2015; Li et al., 2019; Tian et al., 2019, 2020a; Song et al., 2020; Wang et al., 2020; Chen et al., 2020b), especially for the EMR texts. Among different tasks to process Chinese EMR texts, CWS and medical concept recognition are two fundamental ones that draw much attentions from previous studies. Due to the dramatic performance drop when applying the model trained from open source corpus on the medical field, previous studies (Xu et al., 2014, 2015; Li et al., 2015; Zhang et al., 2016; He et al., 2017) always construct Chinese medical datasets themselves and test their models on the datasets. However, most constructed datasets used for CWS are relatively small, where there are only roughly 100 Chinese EMRs. Besides, the medical concept types in most existing datasets are limited to named entities (e.g., "*Disease*" and "*Symptoms and Signs*"), which fails to consider other medical concept types (e.g., "*Time*") in EMRs that are potentially helpful for Chinese EMR texts analysis.

3 The ACEMR Corpus

3.1 Data Collection

We collected 500 Chinese EMRs from five departments (i.e., Respiratory, Gastroenterology, Urology, Gynecology, and Cardiology) of a local hospital, where one EMR specifically means the *First Course Record* in the inpatient record for one patient. *First Course Record* refers to the first course record written by the treating physician or on-duty physician within eight hours after the patient is admitted to the hospital. It contains seven fields, namely department, ward, basic information, case characteristics, preliminary diagnosis, differential diagnosis, treatment plan, where the last five fields are illustrated in Table 1. We extract the texts in those fields and clean them by anonymizing the text and removing invalid or garbled characters.

3.2 CWS and Medical Concept Annotation

Four specialists participated in the development of the annotation guideline, where two of them are junior doctors, and the other two are PhD students in NLP. For CWS guideline, we refer to the segmentation guidelines of the Chinese Treebank (Xia, 2000) for the general domain as well as the annotation guideline proposed by He et al. (2017) for the medical domain. For medical concept annotation guideline, we refer to the medical taxonomy defined by unified medical language system (UMLS) semantic groups (Lindberg et al., 1993) and define 7 major medical concept classes with 20 sub-classes, which are elaborated in Table 2. Compared to existing medical taxonomies, our proposed medical concept classes are simple and clear with fine-grained medical concept focusing on the characteristics of Chinese EMR texts. Note that,

	Count	Length		
		Avg.	Max.	Min.
Char/Types	326,098/1,595	-	-	-
Word/Types	205,304/4,144	2.54	13	1
Sentences	7,370	43.63	311	4

Table 3: The statistics of the ACEMR corpus.

for segmentation, we do not segment one word if it is a defined medical concept.

According to the annotation guideline, we ask the two junior doctors to annotate the 500 EMRs independently and resolve their disagreements by discussion. The consistency of labeling between two annotators is evaluated by the F value (Hripcsak and Rothschild, 2005). The specific method is to treat the labeling result of one annotator (A1) as the standard answer, and calculate the F value of the labeling result of the other annotator (A2). The annotation agreement evaluated by the F value between two annotators of CWS and medical concept tagging are 0.9409 and 0.9360, respectively. We name the annotated corpus as Annotated Chinese Electronic Medical Record (ACEMR) and report its statistics in Table 3, where the lengths are computed based on Chinese characters. In addition, the number of medical concepts in ACEMR is also reported in the last column of Table 2.

Table 4 shows two annotated example sentences, where Chinese words are split by white spaces[2]. The medical concept tag attached to a specific word is highlighted in red color ("/" is a delimiter between a word and its medical concept tag).

3.3 The Corpus Properties

ACEMR is an informative Chinese medical dataset. It contains 500 Chinese EMR texts that are annotated with CWS labels and medical concepts from 20 sub-classes. Due to space limitations, among 20 sub-classes, we introduce three sub-classes (i.e., *Group*, *Health Behavior*, and *Qualitative*) in the following texts. *Group* includes the patient's gender, age, and name. It generally appears at the beginning of Chinese EMRs as part of the basic information, indicating the group the patient belongs to. In addition, it can also act as a participant in medical and health activities (i.e. patients and doctors). *Health Behavior* means medical-related behaviors. It mainly includes examination behaviors, diagnostic behaviors, and broad non-specific treatment behaviors. E.g., "予" (*given*), "入院治疗" (*admission to hospital for treatment*). *Qualitative* emphasizes a qualitative description of something, rather than a direct measurement and can be used to describe the body, abnormalities, etc. E.g., "胃肠型感冒" (*gastrointestinal cold*) where "胃肠型" (*gastrointestinal*) are *Qualitative* medical concepts.

患者/Gr 老年/Gr 女性/Gr ， 慢性/Ql 病程/Di ， 急性/Ql 加重/SOS 。 患者/Gr 主/Ql 因/CE " 反复/Ql 咳嗽/SOS 、 咳痰/SOS , 加重/SOS 3天/T " 入院/E 。
Patient/Gr elderly/Gr female/Gr , chroic/Ql course/Di , acute/Ql exacerbation/SOS . The main/Ql cause/CE of the patient/Gr was " repeated/Ql cough/SOS and sputum/SOS , which became worse/SOS for *3 days*/T " and was *admitted to the hospital*/E .

Table 4: An example of annotated medical sentence in ACEMR with the corresponding English translations. The abbreviations of tags are used for annotation.

4 Methods

A good text representation is highly important in achieving a promising performance in many NLP tasks (Song et al., 2017; Liu and Lapata, 2018; Song and Shi, 2018). Therefore, we select several well-known models for CWS and medical concept recognition tasks and test them on ACEMR corpus.

4.1 CWS for Chinese EMR

For CWS, we follow the convention in previous CWS studies (Sun and Xu, 2011; Song et al., 2012; Song and Xia, 2013; Chen et al., 2015; Zhang et al., 2016; Qiu et al., 2019) to regard it as a sequence labeling task with the "BIES" scheme. We select four well-know models, namely, BiLSTM, BERT (Devlin et al., 2019), ZEN (Diao et al., 2020), and WMSeg (Tian et al., 2020d) with softmax and CRF decoder. Herein, BERT and ZEN are pre-trained language models that have achieved state-of-the-art performance in many NLP tasks (Liang et al., 2020; Tian et al., 2020c; Yu et al., 2020; Nie et al., 2020; Luoma and Pyysalo, 2020; Chen et al., 2020a; Helwe et al., 2020; Tian et al., 2021a,b). WMSeg is CWS model that leverages key-value memory networks (KVMN) (Miller et al., 2016) to incorporate wordhood information to improve model performance, which achieves state-of-the-art performance on many CWS benchmark datasets.

4.2 Medical Concept Recognition

Similarly, for medical concept recognition, we regard it as a character-based sequence labeling task and perform it in a similar way with named entity

[2] If a Chinese word is translated into multiple English words, we use "*" in the English translation to mark its boundary in Table 4. E.g., "3天" is translated into "*3 days*".

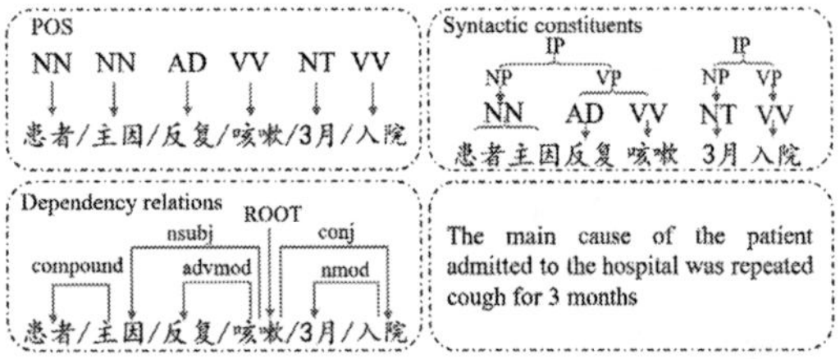

Figure 1: The auto-generated syntactic information (i.e., POS labels, dependency relations, and syntactic constituents) of a sentence with English translation.

Dataset	Word Counts	Word Types
Train	169,047	3,833
Test	36,257	1,529

Table 5: The statistics (i.e., word count and word type) of the training and test sets of ACEMR.

recognition, where the medical concept tags for the input characters follow the "BIOES" scheme. For example, "支气管" ("*virus*") has a medical tag sub-class "BP", and thus the tags for the three characters are "B-BP", "I-BP", and "E-BP", respectively. We try BiLSTM, BERT, ZEN, as well as TwASP (Tian et al., 2020b) with the CRF decoder for medical concept recognition. TwASP is a model that leverages the auto-generated syntactic information (e.g., the POS tags (POS), the dependency relations (Dep.), and the syntactic constituents (Syn.)) through a two-way attention mechanism to improve model performance for sequence labeling tasks. To obtain the syntactic information of the input sentence required by TwASP, we use Stanford CoreNLP Toolkits (Manning et al., 2014) to obtain the POS tags, the dependency tree, and the constituent syntax tree. Figure 1 shows an example sentence (with English translation) and the three types of the auto-generated syntactic information.

5 Experiments

In the experiments, we use two datasets. The first is the in-domain ACMER corpus introduced in Sec. 3; the second is CTB6 (Xue et al., 2005), which is a benchmark CWS dataset of the general domain text. We split the ACMER corpus into training/test sets and report the statistics in Table 5. For all experiments, we use precision (Prec.), recall, and F1 scores to evaluate different models.

5.1 Performance on Medical CWS

For medical CWS, we try BiLSTM, BERT, ZEN, and WMSeg[3]. For BiLSTM, we use pre-trained

Methods	Prec.	Recall	F1
CTB Only			
WMSeg **ZEN is the base model*	77.60	76.85	77.22
CTB+ACEMR			
BiLSTM	98.01	98.09	98.05
+ CRF	98.22	98.30	98.26
+ Tencent Embedding	98.75	98.68	98.72
BERT	98.32	98.65	98.48
+ CRF	98.40	98.66	98.53
+ KVMN	98.55	98.78	98.69
ZEN	98.51	**98.89**	98.70
+ CRF	98.70	98.81	98.76
+ KVMN	**98.86**	98.84	**98.85**
ACEMR Only			
ZEN	99.01	99.00	99.00
+ CRF	98.99	98.91	98.94
+ KVMN	**99.03**	**99.04**	**99.03**

Table 6: CWS performance for different composition of training data where +CRF, +KVMN, +Tencent Embedding represent the use of CRF layer, memory network (WMSeg) and Tencent Embedding respectively.

character embeddings from Tencent Embedding[4] (Song et al., 2018), with the training epoch, batch size, and learning rate set to 50, 32, and 0.001, respectively. For BERT, ZEN, and WMSeg, we use the official settings (e.g., 768 dimensional hidden vectors with 12 multi-head self-attentions for BERT), where the number of training epoch is 50, the batch size is 16, and the learning rate is 1e-5.

The experimental results of CWS are presented in Table 6 with three different settings (i.e., *CTB Only*, *CTB+ACEMR*, and *ACEMR Only*). The *CTB Only* setting displays the results of WMSeg model (with ZEN encoder) when it is trained on CTB6 only and evaluated on the ACEMR test set. The inferior results confirm the big gap between the texts and guidelines in general and medical domains, which indicates the challenge to perform transfer learning from the general domain to the medical domain. The *CTB+ACEMR* setting shows the results of all models trained on the combination of ACEMR and CTB6 datasets, where all models have a high improvement compared with the WMSeg model trained on CTB6 only, emphasizing the necessity of constructing an annotated dataset in medical domain. Compared with BERT and ZEN baseline, adding the KVMN module at the top of the BERT/ZEN encoder to leverage wordhood information (which is exactly the architecture of

[3] https://github.com/SVAIGBA/WMSeg

[4] We use the official release from https://ai.tencent.com/ailab/nlp/zh/embedding.html.

Methods	Prec.	Recall	F1
BiLSTM-CRF	95.65	95.41	95.53
BERT-CRF	97.62	97.84	97.73
ZEN-CRF	97.00	97.87	97.82

Table 7: The results on three different well-known models on medical concept recognition.

Concept	F1	Count	OOV
Top 3 sub-classes			
Probability (Prob)	100.00	372	0.000
Group (Gr)	99.84	2,440	0.177
Absence (Ab)	99.74	10,964	0.053
Bottom 3 sub-classes			
Treatment Project (TP)	90.91	1,380	0.291
Clinical Drug (Drug)	88.20	648	0.462
Body Substance (BS)	74.24	976	0.400
Total	97.82	-	-

Table 8: The top and bottom 3 results of ZEN-CRF on each sub-classes of medical concept recognition, where the number of medical concepts belonging to each sub-class in training set and the out-of-vocabulary (OOV) rate in test set are reported in last two columns.

WMSeg) can improve the performance on CWS. In addition, models with ZEN encoder achieve higher performance than the ones with BERT, which may result from the fact that ZEN leverage n-gram information during pre-training and thus can obtain a better contextual representation. Moreover, if we train the model on ACEMR only (i.e., the *ACEMR only* setting), models with ZEN encoder can be further improved. This observation is not surprising because the texts in CTB6 from the general domain could introduce noise into the model.

5.2 Performance on Concept Recognition

For medical concept recognition (MCR) task where the gold CWS results are given, the results from BiLSTM, BERT, and ZEN encoder with CRF decoder are reported in Table 7, where ZEN-CRF achieves the highest performance. In addition, we rank the F1 scores of all sub-class labels obtained by ZEN-CRF and present the results of the top and bottom 3 ones in Table 8, where the number of medical concepts belonging to each sub-class in the training set as well as the rate of out-of-vocabulary (OOV) medical concepts in the test set is also reported. It is observed that the model does not perform well on sub-classes with fewer training instances and higher OOV rate (e.g., *Body Substance*), which suggests that the OOV issue is a challenge for Chinese medical concept recognition.

Methods	Prec.	Recall	F1
BERT-CRF	97.62	97.84	97.73
TwASP (POS)	97.74	**98.04**	97.89
TwASP (Dep.)	**97.85**	98.02	**97.94**
TwASP (Syn.)	97.65	97.93	97.79
ZEN-CRF	97.00	97.78	97.82
TwASP (POS)	97.77	98.00	97.85
TwASP (Dep.)	97.64	98.01	97.90
TwASP (Syn.)	97.52	97.74	97.63

Table 9: The results of TWASP on medical concept recognition with auto-generated POS labels, dependencies (Dep.), and syntactic constituents (Syn.).

In addition, we run TwASP[5] with three different types of auto-generated syntactic information (i.e., POS labels, dependency relations, and syntactic constituents). The results are reported in Table 9, where we find that MCR can benefit from syntactic information and obtain improvement in most cases, although BERT-CRF and ZEN-CRF baselines have already achieve outstanding performance.

6 Conclusion

In this paper, we collect a new Chinese medical corpus, named ACEMR, which contains 500 EMRs from a local hospital, and annotate the corpus with CWS and medical concept labels. ACEMR features in the rich types of medical concept, in which 20 sub-classes of medical concepts are annotated. We test several state-of-the-art models for CWS and medical concept recognition on the annotated ACEMR. The results on CWS show that models trained on general domain dataset (i.e., CTB6) cannot perform well on medical domain, which confirms the necessity of constructing the ACEMR corpus. Furthermore, WMSeg with wordhood information and TwASP with auto-generated syntactic information outperforms strong baselines on word segmentation and medical concept recognition, respectively, which demonstrates the benefit of leveraging extra resources (i.e., wordhood information and syntactic information) for CWS and medical concept recognition.

Acknowledgements

This work is supported by Chinese Key-Area Research and Development Program of Guangdong Province (2020B0101350001) and also partially supported by NSFC under the project "The Essential Algorithms and Technologies for Standardized Analytics of Clinical Texts" (12026610).

[5]`https://github.com/SVAIGBA/TwASP`

References

Guimin Chen, Yuanhe Tian, and Yan Song. 2020a. Joint Aspect Extraction and Sentiment Analysis with Directional Graph Convolutional Networks. In *Proceedings of the 28th International Conference on Computational Linguistics*, pages 272–279.

Xinchi Chen, Xipeng Qiu, Chenxi Zhu, Pengfei Liu, and Xuanjing Huang. 2015. Long Short-Term Memory Neural Networks for Chinese Word Segmentation. In *Proceedings of the 2015 Conference on Empirical Methods in Natural Language Processing*, pages 1197–1206.

Zhihong Chen, Yan Song, Tsung-Hui Chang, and Xiang Wan. 2020b. Generating Radiology Reports via Memory-driven Transformer. In *Proceedings of the 2020 Conference on Empirical Methods in Natural Language Processing (EMNLP)*, pages 1439–1449, Online.

Jacob Devlin, Ming Wei Chang, Kenton Lee, and Kristina Toutanova. 2019. Bert: Pre-training of Deep Bidirectional Transformers for Language Understanding. In *In Proceedings of the 2019 Conference of the North American Chapter of the Association for Computational Linguistics: Human Language Technologies, Volume 1 (Long and Short Papers)*, pages 4171–4186.

Shizhe Diao, Jiaxin Bai, Yan Song, Tong Zhang, and Yonggang Wang. 2020. ZEN: Pre-training Chinese Text Encoder Enhanced by N-gram Representations. In *Proceedings of the 2020 Conference on Empirical Methods in Natural Language Processing: Findings*, pages 4729–4740.

Carol Friedman, Lyudmila Shagina, Yves Lussier, and George Hripcsak. 2004. Automated Encoding of Clinical Documents based on Natural Language Processing. *Journal of the American Medical Informatics Association*, 11(5):392–402.

Bin He, Bin Dong, Yi Guan, Jinfeng Yang, Zhipeng Jiang, Qiubin Yu, Jianyi Cheng, and Chunyan Qu. 2017. Building a Comprehensive Syntactic and Semantic Corpus of Chinese Clinical Texts. *Journal of biomedical informatics*, 69:203–217.

Chadi Helwe, Ghassan Dib, Mohsen Shamas, and Shady Elbassuoni. 2020. A Semi-Supervised BERT Approach for Arabic Named Entity Recognition. In *Proceedings of the Fifth Arabic Natural Language Processing Workshop*, pages 49–57.

George Hripcsak and Adam S Rothschild. 2005. Agreement, the F-measure, and Reliability in Information Retrieval. *Journal of the American medical informatics association*, 12(3):296–298.

Theresa A Koleck, Nicholas P Tatonetti, Suzanne Bakken, Shazia Mitha, Morgan M Henderson, Maureen George, Christine Miaskowski, Arlene Smaldone, and Maxim Topaz. 2021. Identifying Symptom Information in Clinical Notes Using Natural Language Processing. *Nursing research*.

Xiaozheng Li, Huazhen Wang, Huixin He, Jixiang Du, Jian Chen, and Jinzhun Wu. 2019. Intelligent Diagnosis with Chinese Electronic Medical Records based on Convolutional Neural Networks. *BMC bioinformatics*, 20(1):1–12.

Yu-Bing Li, Xue-Zhong Zhou, Run-Shun Zhang, Ying-Hui Wang, Yonghong Peng, Jing-Qing Hu, Qi Xie, Yan-Xing Xue, Li-Li Xu, Xiao-Fang Liu, et al. 2015. Detection of Herb-Symptom Associations from Traditional Chinese Medicine Clinical Data. *Evidence-Based Complementary and Alternative Medicine*, 2015.

Chen Liang, Yue Yu, Haoming Jiang, Siawpeng Er, Ruijia Wang, Tuo Zhao, and Chao Zhang. 2020. Bond: Bert-Assisted Open-Domain Named Entity Recognition with Distant Supervision. In *ACM SIGKDD International Conference on Knowledge Discovery and Data Mining*.

Donald Lindberg, Betsy Humphreys, and Alexa McCray. 1993. The Unified Medical Language System. *Methods of information in medicine*, 32(4):281.

Yang Liu and Mirella Lapata. 2018. Learning Structured Text Representations. *Transactions of the Association for Computational Linguistics*, 6:63–75.

Jouni Luoma and Sampo Pyysalo. 2020. Exploring Cross-sentence Contexts for Named Entity Recognition with BERT. In *Proceedings of the 28th International Conference on Computational Linguistics*, pages 904–914.

Christopher Manning, Mihai Surdeanu, John Bauer, Jenny Finkel, Steven Bethard, and David McClosky. 2014. The Stanford CoreNLP Natural Language Processing Toolkit. In *Proceedings of 52nd Annual Meeting of the Association for Computational Linguistics: System Demonstrations*, pages 55–60, Baltimore, Maryland.

Alexander Miller, Adam Fisch, Jesse Dodge, Amir-Hossein Karimi, Antoine Bordes, and Jason Weston. 2016. Key-Value Memory Networks for Directly Reading Documents. In *Proceedings of the 2016 Conference on Empirical Methods in Natural Language Processing*, pages 1400–1409.

Yuyang Nie, Yuanhe Tian, Xiang Wan, Yan Song, and Bo Dai. 2020. Named Entity Recognition for Social Media Texts with Semantic Augmentation. In *Proceedings of the 2020 Conference on Empirical Methods in Natural Language Processing (EMNLP)*, pages 1383–1391, Online.

Xipeng Qiu, Hengzhi Pei, Hang Yan, and Xuanjing Huang. 2019. Multi-Criteria Chinese Word Segmentation with Transformer. *arXiv preprint arXiv:1906.12035*.

Alex Rector, Wanda Nowlan, and Shazia Kay. 1991. Foundations for an Electronic Medical Record. *Methods of information in medicine*, 30(03):179–186.

Yan Song, Prescott Klassen, Fei Xia, and Chunyu Kit. 2012. Entropy-based Training Data Selection for Domain Adaptation. In *Proceedings of COLING 2012: Posters*, pages 1191–1200.

Yan Song, Chia-Jung Lee, and Fei Xia. 2017. Learning Word Representations with Regularization from Prior Knowledge. In *Proceedings of the 21st Conference on Computational Natural Language Learning (CoNLL 2017)*, pages 143–152.

Yan Song and Shuming Shi. 2018. Complementary Learning of Word Embeddings. In *Proceedings of the Twenty-Seventh International Joint Conference on Artificial Intelligence, IJCAI-18*, pages 4368–4374.

Yan Song, Shuming Shi, Jing Li, and Haisong Zhang. 2018. Directional Skip-Gram: Explicitly Distinguishing Left and Right Context for Word Embeddings. In *Proceedings of the 2018 Conference of the North American Chapter of the Association for Computational Linguistics: Human Language Technologies, Volume 2 (Short Papers)*, pages 175–180.

Yan Song, Yuanhe Tian, Nan Wang, and Fei Xia. 2020. Summarizing Medical Conversations via Identifying Important Utterances. In *Proceedings of the 28th International Conference on Computational Linguistics*, pages 717–729.

Yan Song and Fei Xia. 2013. A Common Case of Jekyll and Hyde: The Synergistic Effect of Using Divided Source Training Data for Feature Augmentation. In *Proceedings of the Sixth International Joint Conference on Natural Language Processing*, pages 623–631.

Mark Stevenson, Eneko Agirre, and Aitor Soroa. 2012. Exploiting Domain Information for Word Sense Disambiguation of Medical Documents. *Journal of the American Medical Informatics Association: JAMIA*, 19(2):235–240.

Weiwei Sun and Jia Xu. 2011. Enhancing Chinese Word Segmentation Using Unlabeled Data. In *Proceedings of the Conference on Empirical Methods in Natural Language Processing*, pages 970–979.

Yuanhe Tian, Guimin Chen, and Yan Song. 2021a. Aspect-based Sentiment Analysis with Type-aware Graph Convolutional Networks and Layer Ensemble. In *Proceedings of the 2021 Conference of the North American Chapter of the Association for Computational Linguistics: Human Language Technologies, Volume 1 (Long and Short Papers)*.

Yuanhe Tian, Guimin Chen, and Yan Song. 2021b. Enhancing Aspect-level Sentiment Analysis with Word Dependencies. In *Proceedings of the 16th Conference of the European Chapter of the Association for Computational Linguistics: Main Volume*, pages 3726–3739, Online.

Yuanhe Tian, Weicheng Ma, Fei Xia, and Yan Song. 2019. ChiMed: A Chinese Medical Corpus for Question Answering. In *Proceedings of the 18th BioNLP Workshop and Shared Task*, pages 250–260, Florence, Italy.

Yuanhe Tian, Wang Shen, Yan Song, Fei Xia, Min He, and Kenli Li. 2020a. Improving Biomedical Named Entity Recognition with Syntactic Information. *BMC Bioinformatics*, 21:1471–2105.

Yuanhe Tian, Yan Song, Xiang Ao, Fei Xia, Xiaojun Quan, Tong Zhang, and Yonggang Wang. 2020b. Joint Chinese Word Segmentation and Part-of-speech Tagging via Two-way Attentions of Auto-analyzed Knowledge. In *Proceedings of the 58th Annual Meeting of the Association for Computational Linguistics*, pages 8286–8296.

Yuanhe Tian, Yan Song, and Fei Xia. 2020c. Joint Chinese Word Segmentation and Part-of-speech Tagging via Multi-channel Attention of Character N-grams. In *Proceedings of the 28th International Conference on Computational Linguistics*, pages 2073–2084.

Yuanhe Tian, Yan Song, Fei Xia, Tong Zhang, and Yonggang Wang. 2020d. Improving Chinese Word Segmentation with Wordhood Memory Networks. In *Proceedings of the 58th Annual Meeting of the Association for Computational Linguistics*, pages 8274–8285.

Nan Wang, Yan Song, and Fei Xia. 2020. Studying Challenges in Medical Conversation with Structured Annotation. In *Proceedings of the First Workshop on Natural Language Processing for Medical Conversations*, pages 12–21, Online.

Qi Wang, Yangming Zhou, Tong Ruan, Daqi Gao, Yuhang Xia, and Ping He. 2019. Incorporating Dictionaries into Deep Neural Networks for the Chinese Clinical Named Entity Recognition. *Journal of biomedical informatics*, 92:103133.

Fei Xia. 2000. The Segmentation Guidelines for the Penn Chinese Treebank (3.0).

Junjie Xing, Kenny Zhu, and Shaodian Zhang. 2018. Adaptive Multi-task Transfer Learning for Chinese Word Segmentation in Medical Text. In *Proceedings of the 27th International Conference on Computational Linguistics*, pages 3619–3630.

Dong Xu, Meizhuo Zhang, Tianwan Zhao, Chen Ge, Weiguo Gao, Jia Wei, and Kenny Q Zhu. 2015. Data-driven Information Extraction from Chinese Electronic Medical Records. *PloS one*, 10(8):e0136270.

Yan Xu, Yining Wang, Tianren Liu, Jiahua Liu, Yubo Fan, Yi Qian, Junichi Tsujii, and Eric I Chang. 2014. Joint Segmentation and Named Entity Recognition using Dual Decomposition in Chinese discharge summaries. *Journal of the American Medical Informatics Association*, 21(e1):e84–e92.

Naiwen Xue, Fei Xia, Fu-Dong Chiou, and Marta Palmer. 2005. The Penn Chinese Treebank: Phrase Structure Annotation of a Large Corpus. *Natural language engineering*, 11(2):207.

Yajiong Xue, Huigang Liang, Xiaocheng Wu, Hai Gong, Bin Li, and Yuxia Zhang. 2012. Effects of Electronic Medical Record in a Chinese hospital: A Time Series Study. *International journal of medical informatics*, 81(10):683–689.

Juntao Yu, Bernd Bohnet, and Massimo Poesio. 2020. Named Entity Recognition as Dependency Parsing. In *Proceedings of the 58th Annual Meeting of the Association for Computational Linguistics*, pages 6470–6476.

Shaodian Zhang, Tian Kang, Xingting Zhang, Dong Wen, Noémie Elhadad, and Jianbo Lei. 2016. Speculation Detection for Chinese Clinical Notes: Impacts of Word Segmentation and Embedding Models. *Journal of biomedical informatics*, 60:334–341.

BioM-Transformers: Building Large Biomedical Language Models with BERT, ALBERT and ELECTRA

Sultan Alrowili
University of Delaware
Newark, Delaware, USA
alrowili@udel.edu

K. Vijay-Shanker
University of Delaware
Newark, Delaware, USA
vijay@udel.edu

Abstract

The impact of design choices on the performance of biomedical language models recently has been a subject for investigation. In this paper, we empirically study biomedical domain adaptation with large transformer models using different design choices. We evaluate the performance of our pretrained models against other existing biomedical language models in the literature. Our results show that we achieve state-of-the-art results on several biomedical domain tasks despite using similar or less computational cost compared to other models in the literature. Our findings highlight the significant effect of design choices on improving the performance of biomedical language models.

1 Introduction

The amount of biomedical literature has grown substantially in recent years. This growth created a demand for powerful biomedical language models. Transformer-based language models, such as BERT (Devlin et al., 2019), have shown effectiveness in capturing the contextual representation of corpora at large volume. To address the lack of biomedical contextual representation, both BioBERT (Lee et al., 2019), and SciBERT (Beltagy et al., 2019) have adapted BERT to the biomedical domain.

Recently, several Transformer-based models have been introduced, including Megatron (Shoeybi et al., 2020), RoBERTa (Liu et al., 2019), ALBERT (Lan et al., 2020) and ELECTRA (Clark et al., 2020). These models show impressive performance gains over BERT in the general domain leading most NLP leader boards. However, these models have been evaluated with environmental design factors varying in several dimensions (e.g., vocabulary and corpora domain, loss function, training steps, batch size, and model's scale). Understanding the contribution of these factors to the performance of the language models is challenging, especially when our goal is to shift the contextual representations to the biomedical domain.

This challenge motivates us to investigate the impact of design choices on the performance of biomedical language models. Moreover, highlighting this impact is critical when evaluating new applications in BioNLP, where each application may evaluate its performance against other models that use different design setups. In this work, we pretrain and evaluate different variants of large biomedical Transformer-based models across different design factors.

Thus, our contributions in this paper includes :

(i) We pretrain four different variations of Transformer-based models including: $\text{ELECTRA}_{\text{Base}}$, $\text{ELECTRA}_{\text{Large}}$, $\text{BERT}_{\text{Large}}$ and $\text{ALBERT}_{\text{xxlarge}}$ on biomedical domain corpora using Tensor Processing Units TPUs.

(ii) We fine-tune and evaluate our pretrained models on several downstream biomedical tasks. We present a comprehensive evaluation that highlights the impact of design choices on the performance of biomedical language models.

(iii) We released our pretrained models along with our Github repository.[1]

2 Related Work

2.1 Transformer-based Language Models

The introduction of the BERT model (Devlin et al., 2019) has initiated the advancement of Transformer-based models. Consequently, the investigation of the architecture and design choices of BERT introduced new state-of-the-art models. By exploiting the advantage of using the large batch size and increasing the size of the corpus,

[1]Our pre-trained models and our Github repository are accessible at https://github.com/salrowili/BioM-Transformers.

Proceedings of the BioNLP 2021 workshop, pages 221–227
June 11, 2021. ©2021 Association for Computational Linguistics

RoBERTa (Liu et al., 2019) has achieved significant performance gains on all downstream tasks.

The loss function and scalability of BERT were also a subject for investigation by ELECTRA (Clark et al., 2020) and ALBERT (Lan et al., 2020). ELECTRA reaches state-of-the-art results by introducing a binary loss function. This loss function uses generative and discriminative models to accelerate the learning curve. Furthermore, the ALBERT model introduces multiple ideas to the BERT model to improve performance and scalability, including parameter-sharing technique, LAMB optimizer, and factorization of embedding layers. Both ELECTRA and ALBERT are now leading most of NLP benchmarks, including SQuAD (Rajpurkar et al., 2016) and GLUE (Wang et al., 2018).

2.2 Biomedical Language Models

In this section, we will briefly summarize the current state-of-the-art biomedical language models. We should also note that there are other insightful models in literature such as ClinicalBERT (Alsentzer et al., 2019), BlueBERT (Peng et al., 2019), BioELECTRA (Ozyurt, 2020) and BioMedBERT (Chakraborty et al., 2020) .

BioBERT (Lee et al., 2019) is a $\text{BERT}_{\text{Base}}$ model that has been pretrained on biomedical corpora, including PubMed and PMC articles for 23 days on eight V100 GPUs. In our evaluation, we use $\text{BioBERT}_{\text{Base}}$v1.1, which extends the pre-training steps of $\text{BioBERT}_{\text{B}}$ to 1M steps and was trained on PubMed abstracts only.

SciBERT (Beltagy et al., 2019) is a $\text{BERT}_{\text{Base}}$ model that has been pretrained on 1.14M biomedical and computer science papers from Semantic Scholar Corpus .

PubMedBERT (Gu et al., 2021) follows a similar approach of BioBERT by pretraining the BERT model on large biomedical corpora, including PubMed abstracts and PMC articles. PubMedBERT, in contrast to BioBERT, is pretrained using a large batch size (8192) and studies various effects on domain adaptation. The paper also introduces the BLURB benchmark, which is a collection of downstream biomedical tasks.

$\text{BioMegaTron}_{\text{345m}}$ (Shin et al., 2020) is a large-scale model (345m parameters) by NVIDIA based on MegaTron architecture. (Shoeybi et al., 2020). BioMegaTron introduces a variety of large biomedical language models examining the choice of corpora and vocabulary domain.

BioRoBERTa (Lewis et al., 2020) extends the state-of-the-art results by testing different design choices. Similar to BioMegaTron's approach, BioRoBERTa models investigate the effect of vocabulary and corpora domain on the performance of biomedical language model.

3 Pretraining our Language Models

We pretrain all our models using the original implementation of BERT, ALBERT, and ELECTRA. We use TensorFlow 1.15 and TPUv3-512 units to pretrain our large models and TPUv3-32 to pretrain our BioM-$\text{ELECTRA}_{\text{B}}$ model.

3.1 BioM-ALBERT

Initially, we pretrain our model BioM-$\text{ALBERT}_{\text{xxlarge}}$ on PubMed abstracts only. BioM-$\text{ALBERT}_{\text{xxlarge}}$ is based on $\text{ALBERT}_{\text{xxlarge}}$ architecture which has larger hidden layer size (4096) than both BERT_{L} and $\text{ELECTRA}_{\text{L}}$ (1024). We build our specific domain vocabulary, which has a size of 30K words, using the sentence piece model (Kudo and Richardson, 2018). We maintain the same hyperparameters that (Lan et al., 2020) use, except that we increase the batch size to 8192, decrease the initializer range to 0.01. We pretrain BioM-$\text{ALBERT}_{\text{xxlarge}}$ with a learning rate of 1.76e-3 for 264K steps.

Table 1 show the details of our pretrained models compared to the existing model in the literature. The goal to pretrain BioM-$\text{ALBERT}_{\text{xxlarge}}$ is to understand the impact of using ALBERT's techniques on domain adaptation. Moreover, we introduce PMC articles at 264k step, to study the influence of adding PMC articles on the language model. BioM-$\text{ALBERT}_{\text{xxlarge}}$ is the first model that we pretrain and fine-tune among our large models.

3.2 BioM-ELECTRA

We build our BioM-$\text{ELECTRA}_{\text{Base}}$ and BioM-$\text{ELECTRA}_{\text{Large}}$ based on ELECTRA architecture (Clark et al., 2020). We pre-train BioM-$\text{ELECTRA}_{\text{L}}$ on PubMed abstracts only using specific domain vocabulary generated by PubMedBERT, which has a size of 28,895 words. Our evaluation of BioM-$\text{ALBERT}_{\text{xxlarge}}$ on downstream tasks, influences our decision to pretrain BioM-ELECTRA on PubMed abstracts only. We use

Model	Steps	Batch	C	Corpus	Vocabulary
RoBERTa$_{Base}$	500k	8192	4.00x	Web crawl	50K Web crawl
ELECTRA$_{Base++}$	4M	256	1.00x	XLNET Data	30K Wikipedia + Books
SciBERT$_{Base}$	-	-	-	Semantic Scholar	30K PMC+CS
BioBERT$_{Base}$	1M	256	0.25x	PubMed Abstracts	30K Wikipedia + Books
PubMedBERT$_{Base}$	64K	8192	0.50x	PubMed Abstracts	29K PubMed Abstracts
PubMedBERT$_{Base+}$	64K	8192	0.50x	PubMed+PMC	30K PubMed+PMC
BioM-ELECTRA$_{Base}$	500K	1024	0.50x	PubMed Abstracts	29K PubMedBERT
ELECTRA$_{Large}$	1.7M	2048	3.40x	XLNET Data	30K Wikipedia + Books
ALBERT$_{xxlarge}$	1.5M	4096	6.00x	Wikipedia + Books	30k Wikpedia + Books
BioRoBERTa$_{Large}$	500K	8192	4.00x	PubMed+PMC+M	50K PubMed+PMC+M
BioM-BERT$_{Large}$	690K	4096	2.76x	PubMed+PMC	30k Wikipedia + Books
BioM-ELECTRA$_{Large}$	434K	4096	1.73x	PubMed Abstracts	29K PubMedBERT
BioMegaTron$_{345m}$	800K	512	0.40x	PubMed+PMC-CC	50K PubMed Abstracts
BioM-ALBERT$_{xxlarge}$	264K	8192	2.11x	PubMed Abstracts	30k PubMed (ours)

Table 1: Design choices for our pretrained models and state-of-the-art models. The computational ratio (C) represents the ratio between the number of steps multiplied by the batch size where ELECTRA$_{base++}$ is the baseline. XLNet (Yang et al., 2020) data set consist of 33B tokens (130GB) of English corpora. We split the table based on the scale and the domain of language models. CC: Commercial use Collection.

similar pre-training hyperparameters setting described by (Clark et al., 2020) except that we use a larger batch size for BioM-ELECTRA$_{base}$ (1024) and BioM-ELECTRA$_{large}$ (4096). We pretrain our BioM-ELECTRA$_{base}$ for 500K steps and BioM-ELECTRA$_{large}$ model for 434K steps .

The main objective to pretrain BioM-ELECTRA$_{Base}$ is to study the effect of ELECTRA function by comparing its performance with PubMedBERT$_{Base}$ and RoBERTa$_{Base}$. Furthermore, we build our BioM-ELECTRA$_{Large}$ model to study the effect of model scale by comparing it with BioM-ELECTRA$_{Base}$ and PubMedBERT$_{Base}$ where other factors are similar. We should also note that we choose general domain model ELECTRA$_{B++}$ as a baseline model instead of ELECTRA$_{B}$ model. The difference between ELECTRA$_{B}$ and ELECTRA$_{B++}$ is that ELECTRA$_{B}$ is pretrained with less steps (1M) and on smaller corpora (Wikipedia+ Books) (Clark et al., 2020).

3.3 BioM-BERT

We pretrain BioM-BERT$_{Large}$ model on PubMed abstracts and PMC articles using the same vocabulary of BioBERT$_{Base}$. BioBERT$_{Base}$ uses a general domain vocabulary pretrained on English Wikipedia and Books Corpus. Our BioM-BERT$_{Large}$ model aims to study the effect of using general domain vocabulary and PubMed + PMC corpora on downstream biomedical tasks. We use a batch size of 4096, a learning rate of 2e-4, and we set the pretraining steps to 700K. However, since we use preemptible TPUs, our TPUs preempted at 690K. We use the ELECTRA implementation of BERT to pretrain our BERT$_{Large}$ model. This implementation uses a dynamic masking feature without using next-sentence prediction objective.

4 Fine-Tuning

4.1 Downstream Tasks

Our choices of downstream biomedical tasks are similar to (Shin et al., 2020). For Named Entity Recognition (NER) and Relation Extraction (RE), we generate our training, development, and test data using the same script that PubMedBERT uses (Gu et al., 2021).

Named Entity Recognition Our choices for NER tasks including: BC5CDR-Chemical, BC5CDR-Disease (Li et al., 2016) and NCBI-Disease task. (Doğan et al., 2014). These tasks aim to identify chemical and disease entities using IOB tagging format (Ramshaw and Marcus, 1995). For NER tasks, we use entity-Level F1 score, which is a common standard in the literature.

Relation Extraction is a text classification task where we classify each sequence from a list of labels (classes). For RE task, we choose the ChemProt task (Krallinger et al., 2015) , which is a task that classifies chemical-protein interactions. We use micro-level F1 score on the five most common classes. We reproduce the results of BioRoBERTa$_L$ [2] on ChemProt task since BioRoBERTa uses a different pre-processing script than (Gu et al., 2021).

Question Answering We use the same BioASQ7B-factoid dataset that (Lee et al., 2019) use, which is in the format of SQuADv1.1. We use Mean Reciprocal Rank (MMR) as an evaluation metric for this task. Moreover, as it is a common practice, we fine-tune our models on BioASQ task using a checkpoint fine-tuned on SQuAD2.0 task (Rajpurkar et al., 2016).

4.2 Fine-Tuning Hyperparameters

We conduct a hyperparameters grid search using the development data set on TPUv3-8. We use TensorFlow 1.15 to fine-tune our model for all tasks, except that we use Transformers library (Wolf et al., 2020) to fine-tune our BioM-ALBERT on NER tasks. Since we are fine-tuning different architectures, we extend our grid search range to : learning rate (1e-4, 2e-4, 1e-5 - 7e-5), batch size (24, 32, 48, 64, 128) and (2-5) epochs . We fixed our choices of hyperparameters for each set of tasks, model's scale, and architecture. The details of our fine-tuning hyperparameters can be found in Appendix A.1.

5 Results and Discussion

Table 2 shows our evaluation results. We categorize models into four categories based on the domain and the scale of each model. We show the results of BioM-BERT$_L$ and BioM-ALBERT$_{xxlarge}$ at different steps. We report entity-level F1 for NER tasks, micro-level F1 for ChemProt, F1 for SQuAD2.0, and Mean Reciprocal Rank (MMR) for BioASQ. We add SQuAD results to track the direction of contextual representation between the general and biomedical domain.

5.1 ELECTRA Objective

The effect of the ELECTRA objective can be seen from comparing both PubMedBERT$_B$ and BioM-ELECTRA$_B$, where they both use similar design choices, vocabulary set, and C ratio. Our evaluation shows that the ELECTRA function improves the performance on ChemProt, SQuAD, and BioASQ tasks. On the SQuAD task, our BioM-ELECTRA$_B$ exceeds RoBERTa$_B$ despite using biomedical corpora and less C ratio. On NER tasks, BioM-ELECTRA$_B$ performs better on the NCBI-disease and worse on the BC5-CDR task. In contrast, BioM-ELECTRA$_{large}$ performs better than other large models on the BC5-CDR dataset, which excludes the assumption that ELECTRA function negatively affects BioM-ELECTRA$_B$ performance on BC5-CDR tasks

5.2 Named Entity Recognition

Specific domain vocabulary significantly improves the results on NER tasks. Results of BioM-ELECTRA$_L$ and BioRoBERTa$_L$ show that biomedical corpora choices have a marginal effect on NER tasks. Our results also show that the gap between base-scale and large-scale biomedical models on NER tasks is relatively smaller than RE and QA tasks, especially for NCBI-Disease task.

5.3 Relation Extraction

On ChemProt task, BioM-BERT$_{Large}$ achieve 78.8 F1 score at 100K step with a C ratio of 0.4x matching the performance of BioRoBERTa$_L$ which has a C ratio of 4.0x. At 1.6x C ratio (400K), it exceeds by a significant margin all large-scale biomedical models. BioM-BERT$_L$ is the only large model in Table 2 that has PP design choice, which highlights the critical impact of general domain vocabulary on some RE tasks such as ChemProt.

5.4 Question Answering

Our results highlight that question answering tasks are sensitive to out-of-domain corpora. This sensitivity can be clearly seen when we introduce (PP) design to BioM-ALBERT$_{xxlarge}$. The performance decreases significantly on the BioASQ challenge. In contrast, the performance on the SQuAD dataset increase to 88.0%. This increase is not caused by extending the training steps since SQuAD score remains stable at 215K and 264K steps.

Moreover, we can observe a gap of 3.9% in the SQuAD benchmark between BioM-ELECTRA$_{Large}$ and BioM-ELECTRA$_{Base}$. However, this gap is not reflected in the BioASQ benchmark since it is in the format of SQuADv1.1, highlighting the need to have a biomedical questing answering task in the format of SQuADv2.0.

Furthermore, our evaluation shows that ELECTRA$_{B++}$ model achieve state-of-the-art result on BioASQ for base-scale models. We attribute this performance to the fact that we use

[2] BioRoBERTA released their models at `https://github.com/facebookresearch/bio-lm`. We use following hyperparameters to reproduce results (lr: 2e-5 , batch size: 16, epochs : 10, seeds: 10, 42, 1234, 12345, 666).

Model	Design		BC5CDR-		NCBI-	Chem-	QA	
	C	Design	Chem.	Dise.	Dise.	Prot	SQuAD	BioASQ
RoBERTa$_{B}$	4.00x	G	89.4	80.7	86.6	73.0	83.7	-
ELECTRA$_{B++}$	1.00x	G	90.7	83.0	86.3	73.7	86.2	52.5
SciBERT$_{B}$	-	S V	92.5	84.7	88.3	75.0	-	-
BioBERT$_{B}$	0.25x	P	92.6	84.7	**89.1**	76.1	-	41.1
PubMedBERT$_{B}$	0.50x	P V	93.3	85.6	87.9	77.2	79.1	51.6
PubMedBERT$_{B+}$	0.50x	PP V	93.4	85.6	88.3	77.0	80.9	51.9
BioM-ELECTRA$_{B}$	0.50x	P V	93.1	85.2	88.4	77.6	84.4	52.3
ELECTRA$_{L}$	3.40x	G	91.6	84.4	87.6	75.3	90.7	53.0
ALBERT$_{xxlarge}$	6.00x	G	89.7	81.7	85.5	75.8	**90.2**	53.1
BioRoBERTa$_{L}$	4.00x	PPM V	93.7	85.2	89.0	78.8	-	-
BioM-BERT$_{L}$								
100K	0.40x	PP	-	-	87.8	78.8	84.0	-
400K	1.60x	PP	-	-	88.5	79.8	86.5	-
690K	2.76x	PP	92.4	84.5	88.6	**80.0**	87.3	53.4
BioM-ELECTRA$_{L}$	1.73x	P V	**93.8**	85.9	89.0	78.6	88.3	54.1
BioMegaTron$_{345m}$	0.40x	PP V	92.5	**88.5**	87.0	77.0	84.2	52.5
BioM-ALBERT$_{xxlarge}$								
215K	1.70x	P V	-	-	-	79.0	87.0	55.1
264K	2.11x	P V	93.5	85.2	88.7	79.3	87.0	**56.9**
+64K	2.60x	PP V	-	-	-	79.2	88.0	54.5

Table 2: Evaluation results of our pretrained models. For NER and ChemProt, we use reported results of SciBERT$_{B}$, RoBERTa$_{B}$, BioBERT$_{B}$, PubMedBERT$_{B}$, PubMedBERT$_{B++}$ (Gu et al., 2021), BioMegaTron (Shin et al., 2020), BioRoBERTa$_{L}$ (Lewis et al., 2020). We generate QA results for all models, except that we use reported results for BioMegaTron, BioBERT (Shin et al., 2020), RoBERTa$_{B}$ (Dai et al., 2020). BioMegaTron uses sub-tokens evaluation for NER tasks rather than whole-entity evaluation and uses different pre-processed data set for ChemProt task. Our results are the average scores of five different runs. **B**: Base, **L**: Large, **P**: PubMed, **PP**: PubMed+PMC, **PPM**: PubMed+PMC+MMIC, **V**: Specific domain vocabulary, **S**: Semantic Scholar, **G**: General domain model.

a SQuAD fine-tuned checkpoint to fine-tune our models on BioASQ task. In contrast, the gap between the general and biomedical domain is worse on NER and RE tasks since we are not using any general domain fine-tune checkpoints.

5.5 Fine-Tuning Time

Table 3 shows the fine-tuning efficiency. All base-scale models in Table 2 have similar fine-tuning time to BioM-ELECTRA$_{B}$ since they are built on BERT$_{B}$ architecture. Also all models that are based on BERT$_{L}$, such as BioRoBERTa$_{L}$ have similar fine-tuning time to BioM-ELECTRA$_{L}$. Our evaluation shows that hidden layer size (H) significantly influences the fine-tuning time.

Model	H	Time	Ratio
BioM-ELECTRA$_{B}$	768	03:01	0.35x
BioM-ELECTRA$_{L}$	1024	08:27	1.00x
BioM-ALBERT$_{xxlarge}$	4096	31:15	3.67x

Table 3: Fine-Tuning time of our pre-trained models. We fine-tune all models on ChemProt data set for 3 epochs with a batch size of 32 and max seq. length of 128 on 3090RTX GPU with PyTorch (FP16).

6 Conclusion

We introduce four biomedical Transformer-based language models. Our results show that language models with general domain vocabulary and PubMed+PMC corpora perform better on the ChemProt task. Language models with specific domain vocabulary and PubMed abstracts perform better on NER and QA tasks. In the future, we are planning to extend our evaluation to additional biomedical tasks and investigate implementing early existing (Zhou et al., 2020) to reduce the fine-tuning time. Also, we are planning to build an End-to-End ensemble QA system with our large models and Sentence-BERT (Reimers and Gurevych, 2019) to address pandemic issues such as COVID-19.

Acknowledgment

We would like to acknowledge the support we have from Tensorflow Research Cloud (TFRC) team to grant us access to TPUv3 units. The authors also would like to thank Professor Dr. Li Liao for his insightful discussion and comments on BioM-ALBERT model, anonymous reviewers from the BioNLP2021 workshop for their constructive feedback on our initial manuscript, Hoo Chang Shin for clarifying the experimental design of BioMegaTron.

References

Emily Alsentzer, John Murphy, William Boag, Wei-Hung Weng, Di Jindi, Tristan Naumann, and Matthew McDermott. 2019. Publicly available clinical BERT embeddings. In *Proceedings of the 2nd Clinical Natural Language Processing Workshop*, pages 72–78, Minneapolis, Minnesota, USA. Association for Computational Linguistics.

Iz Beltagy, Kyle Lo, and Arman Cohan. 2019. SciBERT: A pretrained language model for scientific text. In *Proceedings of the 2019 Conference on Empirical Methods in Natural Language Processing and the 9th International Joint Conference on Natural Language Processing (EMNLP-IJCNLP)*, pages 3615–3620, Hong Kong, China. Association for Computational Linguistics.

Souradip Chakraborty, Ekaba Bisong, Shweta Bhatt, Thomas Wagner, Riley Elliott, and Francesco Mosconi. 2020. BioMedBERT: A pre-trained biomedical language model for QA and IR. In *Proceedings of the 28th International Conference on Computational Linguistics*, pages 669–679, Barcelona, Spain (Online). International Committee on Computational Linguistics.

Kevin Clark, Minh-Thang Luong, Quoc V. Le, and Christopher D. Manning. 2020. Electra: Pre-training text encoders as discriminators rather than generators.

Zihang Dai, Guokun Lai, Yiming Yang, and Quoc V. Le. 2020. Funnel-transformer: Filtering out sequential redundancy for efficient language processing.

Jacob Devlin, Ming-Wei Chang, Kenton Lee, and Kristina Toutanova. 2019. BERT: Pre-training of deep bidirectional transformers for language understanding. In *Proceedings of the 2019 Conference of the North American Chapter of the Association for Computational Linguistics: Human Language Technologies, Volume 1 (Long and Short Papers)*, pages 4171–4186, Minneapolis, Minnesota. Association for Computational Linguistics.

Rezarta Islamaj Doğan, Robert Leaman, and Zhiyong Lu. 2014. Ncbi disease corpus: a resource for disease name recognition and concept normalization. *Journal of biomedical informatics*, 47:1–10. 24393765[pmid].

Yu Gu, Robert Tinn, Hao Cheng, Michael Lucas, Naoto Usuyama, Xiaodong Liu, Tristan Naumann, Jianfeng Gao, and Hoifung Poon. 2021. Domain-specific language model pretraining for biomedical natural language processing.

Martin Krallinger, Obdulia Rabal, Florian Leitner, Miguel Vazquez, David Salgado, Zhiyong Lu, Robert Leaman, Yanan Lu, Donghong Ji, Daniel M. Lowe, Roger A. Sayle, Riza Theresa Batista-Navarro, Rafal Rak, Torsten Huber, Tim Rocktäschel, Sérgio Matos, David Campos, Buzhou Tang, Hua Xu, Tsendsuren Munkhdalai, Keun Ho Ryu, S. V. Ramanan, Senthil Nathan, Slavko Žitnik, Marko Bajec, Lutz Weber, Matthias Irmer, Saber A. Akhondi, Jan A. Kors, Shuo Xu, Xin An, Utpal Kumar Sikdar, Asif Ekbal, Masaharu Yoshioka, Thaer M. Dieb, Miji Choi, Karin Verspoor, Madian Khabsa, C. Lee Giles, Hongfang Liu, Komandur Elayavilli Ravikumar, Andre Lamurias, Francisco M. Couto, Hong-Jie Dai, Richard Tzong-Han Tsai, Caglar Ata, Tolga Can, Anabel Usié, Rui Alves, Isabel Segura-Bedmar, Paloma Martínez, Julen Oyarzabal, and Alfonso Valencia. 2015. The chemdner corpus of chemicals and drugs and its annotation principles. *Journal of Cheminformatics*, 7(1):S2.

Taku Kudo and John Richardson. 2018. SentencePiece: A simple and language independent subword tokenizer and detokenizer for neural text processing. In *Proceedings of the 2018 Conference on Empirical Methods in Natural Language Processing: System Demonstrations*, pages 66–71, Brussels, Belgium. Association for Computational Linguistics.

Zhenzhong Lan, Mingda Chen, Sebastian Goodman, Kevin Gimpel, Piyush Sharma, and Radu Soricut. 2020. Albert: A lite bert for self-supervised learning of language representations.

Jinhyuk Lee, Wonjin Yoon, Sungdong Kim, Donghyeon Kim, Sunkyu Kim, Chan Ho So, and Jaewoo Kang. 2019. BioBERT: a pre-trained biomedical language representation model for biomedical text mining. *Bioinformatics*, 36(4):1234–1240.

Patrick Lewis, Myle Ott, Jingfei Du, and Veselin Stoyanov. 2020. Pretrained language models for biomedical and clinical tasks: Understanding and extending the state-of-the-art. In *Proceedings of the 3rd Clinical Natural Language Processing Workshop*, pages 146–157, Online. Association for Computational Linguistics.

Jiao Li, Yueping Sun, Robin J. Johnson, Daniela Sciaky, Chih-Hsuan Wei, Robert Leaman, Allan Peter Davis, Carolyn J. Mattingly, Thomas C. Wiegers, and Zhiyong Lu. 2016. BioCreative V CDR task corpus: a resource for chemical disease relation extraction. *Database*, 2016. Baw068.

Yinhan Liu, Myle Ott, Naman Goyal, Jingfei Du, Mandar Joshi, Danqi Chen, Omer Levy, Mike Lewis, Luke Zettlemoyer, and Veselin Stoyanov. 2019. Roberta: A robustly optimized bert pretraining approach.

Ibrahim Burak Ozyurt. 2020. On the effectiveness of small, discriminatively pre-trained language representation models for biomedical text mining. In *Proceedings of the First Workshop on Scholarly Document Processing*, pages 104–112, Online. Association for Computational Linguistics.

Yifan Peng, Shankai Yan, and Zhiyong Lu. 2019. Transfer learning in biomedical natural language processing: An evaluation of bert and elmo on ten benchmarking datasets. In *Proceedings of the 2019 Workshop on Biomedical Natural Language Processing (BioNLP 2019)*, pages 58–65.

Pranav Rajpurkar, Jian Zhang, Konstantin Lopyrev, and Percy Liang. 2016. SQuAD: 100,000+ questions for machine comprehension of text. In *Proceedings of the 2016 Conference on Empirical Methods in Natural Language Processing*, pages 2383–2392, Austin, Texas. Association for Computational Linguistics.

Lance Ramshaw and Mitch Marcus. 1995. Text chunking using transformation-based learning. In *Third Workshop on Very Large Corpora*.

Nils Reimers and Iryna Gurevych. 2019. Sentence-BERT: Sentence embeddings using Siamese BERT-networks. In *Proceedings of the 2019 Conference on Empirical Methods in Natural Language Processing and the 9th International Joint Conference on Natural Language Processing (EMNLP-IJCNLP)*, pages 3982–3992, Hong Kong, China. Association for Computational Linguistics.

Hoo-Chang Shin, Yang Zhang, Evelina Bakhturina, Raul Puri, Mostofa Patwary, Mohammad Shoeybi, and Raghav Mani. 2020. BioMegatron: Larger biomedical domain language model. In *Proceedings of the 2020 Conference on Empirical Methods in Natural Language Processing (EMNLP)*, pages 4700–4706, Online. Association for Computational Linguistics.

Mohammad Shoeybi, Mostofa Patwary, Raul Puri, Patrick LeGresley, Jared Casper, and Bryan Catanzaro. 2020. Megatron-lm: Training multi-billion parameter language models using model parallelism.

Alex Wang, Amanpreet Singh, Julian Michael, Felix Hill, Omer Levy, and Samuel Bowman. 2018. GLUE: A multi-task benchmark and analysis platform for natural language understanding. In *Proceedings of the 2018 EMNLP Workshop BlackboxNLP: Analyzing and Interpreting Neural Networks for NLP*, pages 353–355, Brussels, Belgium. Association for Computational Linguistics.

Thomas Wolf, Lysandre Debut, Victor Sanh, Julien Chaumond, Clement Delangue, Anthony Moi, Pierric Cistac, Tim Rault, Remi Louf, Morgan Funtowicz, Joe Davison, Sam Shleifer, Patrick von Platen, Clara Ma, Yacine Jernite, Julien Plu, Canwen Xu, Teven Le Scao, Sylvain Gugger, Mariama Drame, Quentin Lhoest, and Alexander Rush. 2020. Transformers: State-of-the-art natural language processing. In *Proceedings of the 2020 Conference on Empirical Methods in Natural Language Processing: System Demonstrations*, pages 38–45, Online. Association for Computational Linguistics.

Zhilin Yang, Zihang Dai, Yiming Yang, Jaime Carbonell, Ruslan Salakhutdinov, and Quoc V. Le. 2020. Xlnet: Generalized autoregressive pretraining for language understanding.

Wangchunshu Zhou, Canwen Xu, Tao Ge, Julian McAuley, Ke Xu, and Furu Wei. 2020. Bert loses patience: Fast and robust inference with early exit.

A Appendix

A.1 Fine-Tuning Hyperparameters

Task	Model	E	LR	B
NER	ELECTRA$_B$	5	2e-4	48
NER	BioM-ELECTRA$_B$	5	2e-4	48
NER	BioM-ELECTRA$_L$	5	7e-5	32
NER	ELECTRA$_L$	5	7e-5	32
NER	BioM-BERT$_L$	5	7e-5	32
NER	BioM-ALBERT$_{xxl}$	4	3e-5	16
NER	ALBERT$_{xxl}$	4	3e-5	16
RE	ELECTRA$_B$	4	1e-4	32
RE	BioM-ELECTRA$_B$	4	1e-4	32
RE	BioM-ELECTRA$_L$	4	7e-5	32
RE	ELECTRA$_L$	4	7e-5	32
RE	BioM-BERT$_L$	4	7e-5	32
RE	BioM-ALBERT$_{xxl}$	5	3e-5	128
RE	ALBERT$_{xxl}$	5	3e-5	128
SQ.	PubMedBERT	2	5e-5	32
SQ.	BioM-ELECTRA$_B$	3	1e-4	32
SQ.	BioM-ELECTRA$_L$	3	5e-5	32
SQ.	BioM-BERT$_L$	5	5e-5	48
SQ.	BioM-ALBERT$_{xxl}$	2	3e-5	128
Bio.	BioM-ELECTRA$_B$	4	2e-5	24
Bio.	ELECTRA$_B$	4	2e-5	24
Bio.	BioM-ELECTRA$_L$	4	2e-5	24
Bio.	ELECTRA$_L$	4	2e-5	24
Bio.	PubMedBERT	3	1e-5	128
Bio.	BioM-ALBERT$_{xxl}$	3	1e-5	128
Bio.	ALBERT$_{xxl}$	3	1e-5	128

Table 4: Fine-Tuning hyperparameters of our pre-trained models and base-line general models. We fine-tune all listed models with TensorFlow 1.15 on TPUv3-8 unit. (SQ.: SQuAD2.0, Bio.: BioASQ7B-Factoid, E: Epochs, LR: learning rate, B: Batch size).

Semi-Supervised Language Models for Identification of Personal Health Experiential from Twitter Data: A Case for Medication Effects

Minghao Zhu[1], Keyuan Jiang[2]
[1]School of Computer Science and Technology, Donghua University, Shanghai 201620, China
[2]Department of Computer Information Technology and Graphics,
Purdue University Northwest, Hammond, Indiana 46323, U.S.A.
minghao.zhu0@gmail.com, kjiang@pnw.edu

Abstract

First-hand experience related to any changes of one's health condition and understanding such experience can play an important role in advancing medical science and healthcare. Monitoring the safe use of medication drugs is an important task of pharmacovigilance, and first-hand experience of effects about consumers' medication intake can be valuable to gain insight into how our human body reacts to medications. Social media have been considered as a possible alternative data source for gathering personal experience with medications posted by users. Identifying personal experience tweets is a challenging classification task, and efforts have been made to tackle the challenges using supervised approaches requiring annotated data. There exists an abundance of unlabeled Twitter data, and being able to use such data for training without suffering in classification performance is of great value, which can reduce the cost of laborious annotation process. We investigated two semi-supervised learning methods, with different mixes of labeled and unlabeled data in the training set, to understand the impact on classification performance. Our results from both pseudo-label and consistency regularization methods show that both methods generated a noticeable improvement in F1 score when the labeled set was small, and consistency regularization could still provide a small gain even a larger labeled set was used.

1 Introduction

First-hand experience related to any changes of one's health condition and understanding such experience can play an important role in advancing medical science and healthcare. What has happened since the COVID-19 pandemic started demonstrates potential values and applications of such experiential knowledge, ranging from understanding the symptoms of the viral infection, to learning the effects after vaccination – personal experience shared on social media pertaining to symptoms of infection and side effects of vaccine may help us gain insight into the virus and vaccine, and ultimately advance medical science and clinical practice. Post-market surveillance is an important activity of pharmacovigilance, and experiential information from the users of the therapeutic products can help supplement the knowledge of medication effects gathered with other data sources. Many nations recognized importance of patient reporting of drug effects and its scientific value (van Hunsel et al., 2012), and potential benefits of patient reported drug events were studied (de Langen et al., 2008; Blenkinsopp et al., 2007; Avery et al., 2011; Anderson et al., 2011). Patient reporting could help identify new adverse drug effects sooner than that by healthcare professionals alone (Egberts et al., 1996). A study by McLernon and colleagues found that patient reports contained a higher median number of suspected adverse drug reactions (ADRs) per report, and described reactions in more detail, and they were richer in descriptions of reactions than those from healthcare providers (McLernon et al., 2010). One study showed that consumers reported seven categories of ADRs unreported by the other sources, and the investigators recommended that consumers should be included in systematic drug surveillance systems (Aagaard et al., 2009).

It is a primary concern of monitoring the health conditions as well as the safe use of pharmaceutical products to find a rich and accessible data source and build an efficient system to process and analyze the data.

Proceedings of the BioNLP 2021 workshop, pages 228–237
June 11, 2021.

People share their personal health experience on social media thanks to their prevalence. As such, social media have been considered an alternative and active data source for studying health surveillance. Platforms like Twitter allow users to express their health condition freely online. There exist many studies of using social media for health surveillance, such as influenza outbreak detection (Culotta et al., 2010), public health analyzing (Paul et al., 2011), dental pain surveillance (Heaivilin et al., 2011).

Personal experience tweets (PETs) related to medication use are defined as Twitter posts expressing one's first-hand personal encounters or observations about their health conditions after the administration of pharmaceutical drugs. Medication effects can be undesirable feelings caused by medication's side-effects which exacerbate one's health condition, or beneficial effects which help alleviate one's health condition after medication intake. Below are examples of personal experience tweets (PETs) pertaining to medication use (the medication names are in boldface and experiences are underscored):

> "**codeine** got me feeling sloooow **xanax** got me sleeping"

> "this **vicodin** is putting me to sleep"

> "**morphine** actual makes ur face so itchy think ive scratched ma whole face off"

As a general purpose social media platform, Twitter contains posts on almost all thinkable topics and many of them are unrelated to health, let alone misspellings, incorrect grammas, and creative short texts found in the posts. Therefore, differentiating personal experience tweets (PETs) from other irrelevant or noisy tweets is challenging. Efforts have been made in previous endeavors. Personal pronouns were chosen as the feature to distinguish PETs from irrelevant tweets such advertisements, news, even spams (Jiang and Zheng, 2013). An effort was made to engineer features including Twitter specific features, n-grams, punctuation elements, and topics, but the topic feature was discarded because of its significant efforts required to achieve minimum merit of classification performance improvement (Alvaro et al., 2015). Later, a set of 22 Twitter features including textual data and metadata was engineered by Jiang and his colleagues (2016) and conventional machine learning methods such as decision tree were applied to predict PETs. The concept of deep grammulator was proposed to include a textual feature with expressions in one class but not in the opposite class, to enhance the discriminatory power of the classification (Calix et al., 2017). In recent studies, application of neural embedding and recurrent neural network (LSTM) was investigated to improve the classification performance (Jiang et al., 2018). In the latest development, pre-trained attention-based language model approaches based on BERT and RoBERTa language models were explored to have achieved even better classification performance (Jiang et al., 2019, Zhu et al., 2020).

However, all previous attempts are based on fully supervised learning mechanisms, requiring the laborious effort of annotation which can be cost-prohibitive if a large amount of accurately labeled data is needed with a limited budget.

Unlike labeling text data in formal writing, annotating Twitter posts can be especially challenging, because of various complexities associated with the data such as misspellings, use of nonstandard language, and lack of sufficient context within the limited space. In addition, supervised methods are widely used on social media data in health-related tasks due to their higher accurate than unsupervised approaches, requiring manual annotation of large corpora of data. Furthermore, the subjectivity of labeling social media data is of concern. Inter-annotator agreements tend to be relatively low for social media–based annotation tasks even with domain experts as annotators (O'Connor, 2020).

On social media, there exists an abundance of unannotated data, and being able to use such large amount of unlabeled data for training may improve the classification performance, without spending a significant amount of resources in annotation. In this study, we investigate how the classification performance in predicting personal experience tweets related to medication use will be affected using a relatively small amount of annotated data instances in training an attention-based language model.

2 Background

Data and features determine the upper limit of machine learning, while models and algorithms can only approach this upper limit. For most machine learning tasks, the amount of labeled data directly affects the final learning performance.

In order to obtain a large set of labeled data, researchers usually need to spend a significant amount of time to annotate data, and the cost of annotation process can be drastic and sometimes unaffordable. On the other hand, there is abundance of unlabeled data which are easily accessible. Semi-supervised learning is a promising approach in machine learning which uses the combination of both of labeled and unlabeled data. Studies have shown that it can achieve considerable improvement for various tasks with a small labeled dataset in conjunction with a large set of unlabeled data.

Based on the cluster assumption, semi-supervised learning methods are mainly classified into two different categories: proxy-label and consistency regularization. The proxy-label method uses the supervised model or its variants to generate proxy labels for the unlabeled data, and the proxy labels are mixed with true labels to provide additional features to benefit training process. A typical implementation of such approach is the pseudo-label method as described below (Lee 2013).

Consistency regularization is a relatively new method. In consistency training, models are regularized to be invariant to a small amount of noise applied to inputs or hidden neurons. The invariance can be all or parts of hidden states in the network, or the outputs of the model. Common methods include Temporal Ensembling (Laine et al., 2016), Mean Teachers (Tarvainen et al., 2017), and Unsupervised Data Augmentation (Xie et al., 2019).

3 Method

The pipeline of data processing and analysis of our methods is depicted in Figure 1. Our approach of identifying personal experience tweets was based upon the two semi-supervised learning methods mentioned above: (1) Pseudo-Label which generates pseudo labels for unlabeled tweets and trains the model with labeled tweets together in a supervised behavior, and (2) Consistency Regularization which does not generate any labels for unlabeled data but tries to keep the consistency of the model outputs with the same inputs injected with some stochastic noise.

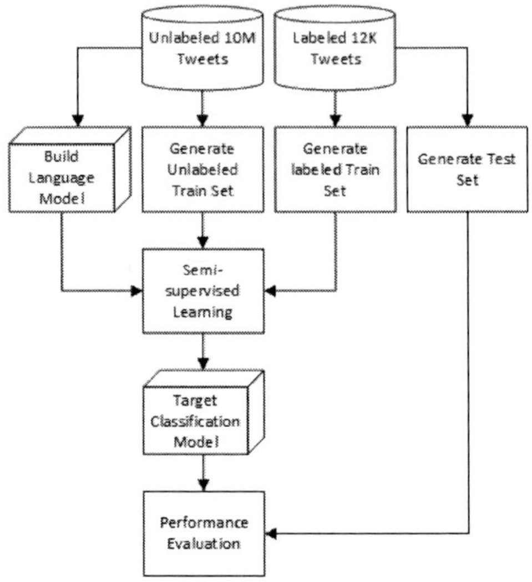

Figure 1. Pipeline of Data Processing and Analysis.

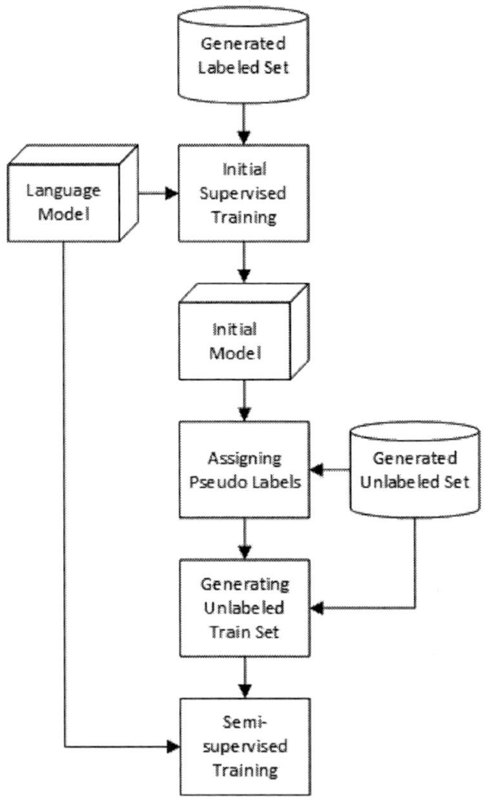

Figure 2. Setup of Pseudo-label Based Semi-supervised Learning.

Our language model is based upon the Google's attention-based Bidirectional Encoder Representations from Transformers (BERT) (Devlin et al., 2018). Although the Google team highly recommends using its pre-trained language model, a Microsoft team demonstrated that the domain-specific BERT language model can perform better (Gu et al., 2020), and we adopted Microsoft's approach in this work.

3.1 Pseudo-Label

Figure 2 shows our pseudo-label based method. A naive but efficient semi-supervised learning structure was implemented in this method by combining both labeled and unlabeled (pseudo-labeled) sets of data. First, the model was trained in a fully supervised manner with the labeled set, and the trained model was used to assign pseudo labels to unlabeled tweets. Later, a subset of unlabeled tweets was chosen for prediction by the initial model. From the prediction results, each unlabeled tweet was assigned a pseudo-label whose class has the maximum predicted probability. Due to the class imbalance of the corpus of labeled tweets (Table 1 below), the composition of the unlabeled train set was made up of the classes which were inversely proportional to the annotated corpus (PET: non-PET = 3:1). Finally, both sets of labeled and pseudo-labeled tweets were combined to train the model.

3.2 Consistency Regularization

For consistency training, the framework of Π-model (Laine et al., 2016) was used for reference. Figure 3 and Algorithm 1 shows our method. In this approach, two parts of the loss function were considered: (1) the classification loss, which is usually the cross entropy, and (2) the consistency loss. The classification loss was only applied to the labeled tweets while the consistency loss was for all. During training, each input was evaluated twice with hidden noise injection, and the difference between the two evaluation results was calculated by the squared error. In combining these two parts of loss, a weight variable was applied to scale the consistency loss. The weight variable was initialed to zero, allowing the training loss to be dominated by classification so that the model could learn from labeled data first, and later the weight variable was recalculated in the training epochs to reflect the consistency loss consistent with the data. The value of this weight would be adjusted to a fixed level according to the number of labeled and unlabeled data used in training. It was an important and tricky step. This is because if the value is too small, the training is more likely to be supervised and prone to overfitting, and on the contrary, the model trained with an overly large weight will noticeably deteriorate, making the predictions less meaningful. Refer to Appendix A for the details of training parameters.

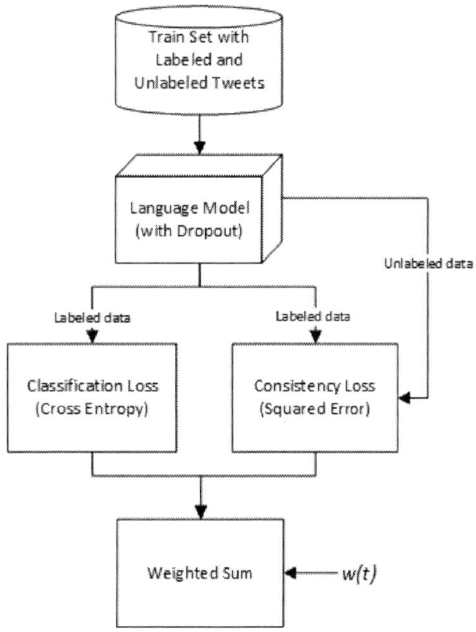

Figure 3. Setup of Consistency Regularization

Require: x_i = training stimuli
Require: L = labeled set
Require: y_i = labels of labeled data
Require: w(t) = weight ramp-up function for consistency loss
Require: $f_\theta(x)$= neural network with dropout and parameter θ

for t **in** [1, num_epochs] **do**
  for each minibatch B **do**
    $z_{i\in B} \leftarrow f_\theta(x_{i\in B})$
    $z'_{i\in B} \leftarrow f_\theta(x_{i\in B})$
    $loss \leftarrow -\frac{1}{|B|}\sum_{i\in(B\cap L)}(y_i \log z_i + (1 - y_i)\log(1 - z_i)) + w(t)\frac{1}{|B|}\sum_{i\in B}|z_i - z'_i|^2$
    Update θ by using Adam
  end for
end for

Algorithm 1. Pseudo code of consistency regularization

Dropout regularization was chosen as the noise injection method in the hidden layer of the model. Because the dropout performed stochastically, the model outputs were different from training even with the same inputs. Therefore, there were two different evaluation results for each input, and the expected goal was to minimize them.

3.3 Network Structure and Baseline

Pre-trained language models such as Bidirectional Encoder Representations from Transformers (BERT) (Devlin et al., 2018), Robustly Optimized BERT Pretraining Approach (RoBERTa) (Liu et al., 2019) and Generative Pre-trained Transformer (GPT) (Radford et al., 2019) have achieved the state-of-the-art performances in many NLP benchmarks and downstream tasks. Efforts have been made to apply the pre-trained BERT and RoBERTa (and its continuous pre-training) language models in identifying PETs (Jiang et al., 2019; Zhu et al., 2020) which demonstrated significant improvement in all classification measures. However, these language models are all pre-trained using general-domain texts such as news, Wikipedia and/or BookCorpus which may not have significant relevance to Twitter posts, especially pertaining to personal health experience. Gu and colleagues (Gu et al., 2020) found that the language model learned with domain specific data can provide substantial gains over the general domain language model, and domain-specific learning from the scratch is better than the one that starts with the general domain model and is updated with the specific domain data. Therefore, we decided to pre-train a domain-specific BERT language model from scratch to investigate its performance. The pre-training process used 10M unlabeled medication-related tweets we collected, and a new set of in-domain sub-word vocabulary was generated. See Appendix A for detail settings. This newly pre-trained language model was utilized as a network backend for both semi-supervised learning approaches as well as the baseline method.

For the baseline, supervised learning was considered. Following the official transfer learning guideline, the domain-specific BERT was transferred for binary classification and trained in a fully supervised way with the same labeled data as semi-supervised methods used.

3.4 Data

A set of 22 million raw tweets was collect with Twitter Streaming APIs from 25 August 2015 to 7 December 2016, and another set of 52 million tweets posted between 2006 as 2017 was collected using a home-made crawler which followed the crawling policy documented in the Twitter.com's robots.txt file. To clean the collected raw data, both sets were filtered by a set of brand and generic medication names, and duplicate as well as non-English tweets were all eliminated. The above pre-processing yielded a total 10 million tweets, among which a collection of 12,331 (12K) tweets was selected and annotated according to the annotation guideline which defines what is a PET and a non-PET. Table 1 lists the composition of labeled tweets.

First, a corpus of 10 million unlabeled tweets was used to build a sub-word vocabulary and pre-training domain-specific BERT language model. To avoid any possible data leakage, the 12K annotated tweets were excluded from the 10 million set. The set of 12K labeled tweets was used for both supervised baseline method and the labeled part of semi-supervised methods. As for the unlabeled part of semi-supervised learning, a stochastic subset of was randomly generated from the 10 million unlabeled corpus.

	PETs	Non-PETs	Total
Count	2,962	9,369	12,331

Table 1. Composition of annotated tweets.

3.5 Implementation

Our two semi-supervised learning methods were evaluated in the task of identifying personal experience tweets related to medication effects. To simulate the situation of the lack of labeled data and investigate how semi-supervised learning would perform for our task, both of our methods were tested with different percentages of the labeled data in our training set, and we evaluated the performance of the semi-supervised methods along with the fully supervised settings as the baseline.

Ten-fold cross-validation was applied to both supervised and semi-supervised approaches, and the mean of each classification measure was collected. For each fold, 10% of labeled tweets was partitioned as the test set which was only used for testing the classification performance. The labeled training set was randomly selected from the remaining 90% tweets by a proportion – we set six different proportions: 10%, 30%, 50%, 70%, 90% and 100% to investigate how the size of labeled data would affect the performance. Note that the initial training of pseudo-labeling method used the

%[1]	Method[2]	Acc. (PET)	Recall (PET)	Prec. (PET)	F1 (PET)	AUC/ROC
10	C.	**0.8597**	0.6239	**0.7500**	0.6786	**0.9079**
	P.	0.8390	**0.7512**	0.6500	**0.6916**	0.8944
	S.	0.8466	0.6607	0.7038	0.6706	0.9048
30	C.	**0.8723**	0.6894	**0.7559**	**0.7206**	**0.9235**
	P.	0.8528	**0.7802**	0.6661	0.7180	0.9155
	S.	0.8638	0.6904	0.7325	0.7081	0.9199
50	C.	**0.8765**	0.6945	**0.7689**	0.7294	**0.9287**
	P.	0.8591	**0.7927**	0.6798	**0.7303**	0.9211
	S.	0.8657	0.7474	0.7198	0.7280	0.9266
70	C.	**0.8797**	0.7134	**0.7678**	**0.7392**	**0.9313**
	P.	0.8646	**0.7765**	0.6963	0.7338	0.9239
	S.	0.8755	0.7171	0.7556	0.7338	0.9289
90	C.	**0.8829**	0.7316	**0.7680**	**0.7488**	**0.9338**
	P.	0.8685	**0.7846**	0.7037	0.7415	0.9266
	S.	0.8758	0.7552	0.7383	0.7452	0.9315
100	C.	**0.8855**	0.7245	**0.7802**	0.7508	**0.9344**
	P.	0.8678	**0.7937**	0.6990	0.7427	0.9278
	S.	0.8792	0.7620	0.7447	**0.7519**	0.9335

1. the percentage of labeled data used.

2. 'C' for consistency regularization, 'P' for pseudo-label, 'S' for supervised baseline.

Table 2. Classification performance

%	Method	F1	AUC/ROC
10	C.	2.749×10^{-1}	5.534×10^{-2}
	P.	$\mathbf{3.676 \times 10^{-2}}$	5.188×10^{-5}
30	C.	$\mathbf{3.653 \times 10^{-2}}$	$\mathbf{6.923 \times 10^{-3}}$
	P.	$\mathbf{3.828 \times 10^{-2}}$	3.372×10^{-4}
50	C.	4.115×10^{-1}	$\mathbf{2.538 \times 10^{-2}}$
	P.	3.002×10^{-1}	4.669×10^{-5}
70	C.	7.326×10^{-2}	$\mathbf{1.331 \times 10^{-2}}$
	P.	4.965×10^{-1}	5.089×10^{-4}
90	C.	1.979×10^{-1}	$\mathbf{4.581 \times 10^{-3}}$
	P.	6.620×10^{-2}	9.336×10^{-4}
100	C.	3.724×10^{-1}	1.018×10^{-1}
	P.	3.222×10^{-2}	3.619×10^{-4}

Table 3. T-test results (p-values) between baseline and semi-supervised learning methods (C. and P.) Boldface figures: < 0.05

same proportion of labeled data as the supervised baseline as well as the consistency regularization method to initial the model for assigning pseudo labels of unlabeled tweets. A fixed random seed was used to ensure that all methods have the same partition of data.

For both semi-supervised learning methods, a collection of 10K of unlabeled tweets was used as the unlabeled training set and mixed up with labeled ones. More specifically, in consistency training, the unlabeled train set was simply generated from the 10M unlabeled corpus by a stochastic sampler. But for pseudo-label, it was time consuming to predict for 10M tweets, and it was observed that the prediction of unlabeled data was imbalanced – the number of non-PETs was about ten times more than that of PETs. In other words, the training set would be more imbalanced if the 10K unlabeled set were to be used before assigning pseudo labels. To address the issue, and keep the training time tolerable, a set of 100K unlabeled tweets was chosen and assigned with pseudo labels, and afterwards, a training set of 10K unlabeled tweets was composed from the 100K tweets with pseudo labels. To balance the labeled set with a 1:3 ratio for PET: non-PET (Table 1), our pseudo-labeled training set was made up of 7,500 tweets with the PET pseudo-label and 2500 the non-PET pseudo label.

4 Results and Discussions

Listed in Table 2 are the measures of classification performance of our semi-supervised methods along with the supervised baseline in different proportions of labeled data (the highest values are in boldface).

As can be seen in Table 2, no single method achieved the best performance in all classification measures. The pseudo-label-based method achieved the best recall but showed the poorest accuracy, precision and AUC/ROC, in all the proportions of the labeled tweets used. On the contrary, the consistency regularization approach demonstrated the completely opposite

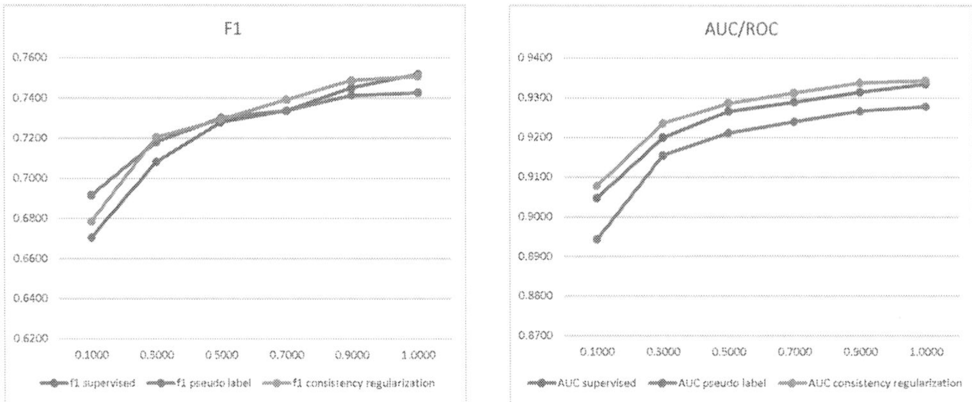

Figure 4. F1 and AUC/ROC Measures for Each Method.

performances, and notably its AUC/ROC, a more comprehensive measure for discriminability, is consistently higher than other two methods (Fawcett, 2006) – this may indicate that the consistency regularization method performed better in correctly predicting each class of data.

It seems to be inconclusive to state which one is the winner on a single classification measure. A higher accuracy does not necessarily indicate that the method is better because the accuracy is calculated based upon the prediction results of both positive and negative classes. The class imbalance in our test set may contribute to a higher accuracy if the majority class dominates. Recall and precision are of equal importance but neither of them alone can measure a network independently because recall focuses on the sensitivity of positive class while precision just measures the percentage of true positive samples in the predicted class. The F1 measure represents the harmonic mean calculated from both recall and precision, along with AUC/ROC, they have been considered as the most comprehensive measure among these five measures. Figure 4 shows the changes of F1 value for each method along with the proportions of the labeled data used. To confirm if the improvement difference does exist, we conducted statistical analysis (paired t-test) on the results between semi-supervised learning methods and baseline. We set the null hypothesis to that the difference between a pair of method does not exist while the labeled data remain the same. Table 3 shows the results of statistical analysis on F1 and AUC/ROC between the baseline and two semi-supervised learning methods. We set the *p*-value threshold to 0.05, meaning that any p-value less than 0.05, and if its corresponding value of performance measure larger than that of baseline, it indicates that the improvement difference does exist with statistical significance and it is *not* due to chance (these values are shown in boldface).

As shown in Figure 4, it is clear that both semi-supervised learning methods performed better than supervised baseline when a small amount of labeled data was used. In other words, semi-supervised learning may help us build a more robust PET prediction network in the situation where only a limited or small amount of the labeled data is available.

More specifically, according to Figure 4 and its corresponding *p*-values, the pseudo-label based method showed an outstanding F1 performance in tiny labeled sets (about 10% and 30% of the labeled data), and its improvement is of statistical significance. However, its performance appeared to deteriorate when the training set contained a large amount of annotated tweets (about more than 70% of labeled data). A possible explanation of this phenomenon might be the mislabeling of unlabeled tweets, which may mislead the training process if it has more labeled data than unlabeled one.

The consistency regularization method appears to be more stable than the pseudo-label method. It demonstrates consistently good performance even with the larger labeled sets (about 90% of labeled data), except that it shows a slight but not significantly lag behind the supervised baseline in F1 when the training set contains all of the labeled data. Although the statistical analysis seems does not show that the improvement in F1 is significant, the constant outperformance in AUC/ROC could be confirmed by the t-test – the larger AUC/ROC values in 30%, 50%, 70% and 90% of labeled data demonstrated their statistical significance. In the case of 10% of labeled data, the t-test results do not confirm the existence of performance differences. This may be attributed to the fact that too many unlabeled instances dominate the training set, which confused models in training. It may be concluded that Consistency Regularization is consistently better if the training data have more than 10% but less than 100% labeled instances and performs equally well with 100% labeled data in this task. However, it seems not performing well as the pseudo-label one in F1 when very few tweets are labeled (about 10%). This may indicate that labels are important in contributing to the performance, and the pseudo-label approach may be more suitable for the training with an extremely small labeled set, whereas consistency regularization seems to perform well in other situations.

5 Conclusion

In this study, we investigated classification performance using semi-supervised learning in identifying personal experience tweets. Two methods of semi-supervision were studied: pseudo-label and consistency regularization. Our results show that either of the methods performs outstandingly well in individual classification measures, in comparison with the supervised baseline method. However, the F1 and AUC/ROC scores show that both could enhance the network performance when a small size of the labeled set was used, and consistency regularization performed consistently well even with the datasets containing high number of labeled instances. In summary, either semi-supervised method performed well in predicting PETs with a small amount of labeled instances in the training set, which could significantly reduce the annotation effort. Although this study focused on the personal experience pertaining to medication effects, it is conceivable that our semi-supervised approach can help other health-related studies where personal experience is needed.

6 Acknowledgement

Authors wish to thank anonymous reviewers for their critiques and constructive comments which helped improved this manuscript, and College of Technology at Purdue University Northwest for providing funding to support this work.

References

Aagaard L, Nielsen LH, Hansen EH. Consumer reporting of adverse drug reactions: a retrospective analysis of the Danish adverse drug reaction database from 2004 to 2006. Drug Saf. 2009;32(11):1067-74.

Alvaro, N., Conway, M., Doan, S., Lofi, C., Overington, J. and Collier, N., 2015. Crowdsourcing Twitter annotations to identify first-hand experiences of prescription drug use. Journal of biomedical informatics, 58, pp.280-287.

Anderson C, Krska J, Murphy E, Avery A, Collaboration YCS. The importance of direct patient reporting of suspected adverse drug reactions: a patient perspective. Br J Clin Pharmacol. 2011;72(5):806-22.

Avery AJ, Anderson C, Bond CM, Fortnum H, Gifford A, Hannaford PC, Hazell L, Krska J, Lee AJ, McLernon DJ, Murphy E, Shakir S and Watson MC. Evaluation of patient reporting of adverse drug reactions to the UK 'Yellow Card Scheme': literature review, descriptive and qualitative analyses, and questionnaire surveys. Health Technol Assess. 2011;15(20):1-234, iii-iv.

Blenkinsopp A, Wilkie P, Wang M, Routledge PA. Patient reporting of suspected adverse drug reactions: a review of published literature and international experience. Br J Clin Pharmacol. 2007;63(2):148-56.

Calix, R.A., Gupta, R., Gupta, M. and Jiang, K., 2017, Novem-ber. Deep gramulator: Improving precision in the classification of personal health-experience tweets with deep learning. In 2017 IEEE International Conference on Bioinformatics and Biomedicine (BIBM) (pp. 1154-1159). IEEE.

Culotta, A. (2010). Detecting influenza outbreaks by analyzing Twitter messages. arXiv preprint arXiv:1007.4748.

de Langen J, van Hunsel F, Passier A, de Jong-van den Berg L, van Grootheest K. Adverse drug reaction reporting by patients in the Netherlands: three years of experience. Drug Saf. 2008;31(6):515-24.

Devlin, J., Chang, M. W., Lee, K., & Toutanova, K. (2018). Bert: Pre-training of deep bidirectional transformers for language understanding. arXiv preprint arXiv:1810.04805.

Egberts TC, Smulders M, de Koning FH, Meyboom RH, Leufkens HG. Can adverse drug reactions be detected earlier? A comparison of reports by patients and professionals. BMJ. 1996;313(7056):530-1.

Fawcett, T. An introduction to ROC analysis." Pattern recognition letters 27.8 (2006): 861-874.

Gu, Y., Tinn, R., Cheng, H., Lucas, M., Usuyama, N., Liu, X., Naumann, T., Gao, J. & Poon, H. (2020). Domain-specific language model pretraining for biomedical natural language processing. arXiv preprint arXiv:2007.15779.

Heaivilin, N., Gerbert, B., Page, J. E., & Gibbs, J. L. (2011). Public health surveillance of dental pain via Twitter. Journal of dental research, 90(9), 1047-1051.

Jiang, K., Calix, R., & Gupta, M. (2016, August). Construction of a personal experience tweet corpus for health surveillance. In Proceedings of the 15th workshop on biomedical natural language processing (pp. 128-135).

Jiang, K., Chen, T., Calix, R. A., & Bernard, G. R. (2019, July). Prediction of personal experience tweets of medication use via contextual word representations. In 2019 41st Annual International Conference of the IEEE Engineering in Medicine and Biology Society (EMBC) (pp. 6093-6096). IEEE.

Jiang, K., Feng, S., Song, Q., Calix, R. A., Gupta, M., & Bernard, G. R. (2018). Identifying tweets of personal health experience through word embedding and LSTM neural network. BMC bioinformatics, 19(8), 67-74.

Jiang, K., & Zheng, Y. (2013, December). Mining twitter data for potential drug effects. In International conference on advanced data mining and applications (pp. 434-443). Springer, Berlin, Heidelberg.

Laine, S., & Aila, T. (2016). Temporal ensembling for semi-supervised learning. arXiv preprint arXiv:1610.02242.

Lee, D. H. (2013, June). Pseudo-label: The simple and efficient semi-supervised learning method for deep neural networks. In Workshop on challenges in representation learning, ICML (Vol. 3, No. 2).

Liu, Y., Ott, M., Goyal, N., Du, J., Joshi, M., Chen, D., Levy, O., Lewis, M., Zettlemoyer, L. & Stoyanov, V. (2019). Roberta: A robustly optimized bert pretraining approach. arXiv preprint arXiv:1907.11692.

McLernon DJ, Bond CM, Hannaford PC, Watson MC, Lee AJ, Hazell L and Avery A. Adverse drug reaction reporting in the UK: a retrospective observational comparison of yellow card reports submitted by patients and healthcare professionals. Drug Saf. 2010;33(9):775-88.

O'Connor, Karen, Abeed Sarker, Jeanmarie Perrone, and G. Gonzalez Hernandez. "Promoting Reproducible Research for Characterizing Nonmedical Use of Medications Through Data Annotation: Description of a Twitter Corpus and Guidelines." J Med Internet Res 22, no. 2 (2020): e15861.

Paul, M., & Dredze, M. (2011, July). You are what you tweet: Analyzing twitter for public health. In Proceedings of the International AAAI Conference on Web and Social Media (Vol. 5, No. 1).

Radford, A., Wu, J., Child, R., Luan, D., Amodei, D., & Sutskever, I. (2019). Language models are unsupervised multitask learners. OpenAI blog, 1(8), 9.

Schuster, M., & Nakajima, K. (2012, March). Japanese and korean voice search. In 2012 IEEE International Conference on Acoustics, Speech and Signal Processing (ICASSP) (pp. 5149-5152). IEEE.

Tarvainen, A., & Valpola, H. (2017). Mean teachers are better role models: Weight-averaged consistency targets improve semi-supervised deep learning results. arXiv preprint arXiv:1703.01780.

van Hunsel F, Härmark L, Pal S, Olsson S, van Grootheest K. Experiences with adverse drug reaction reporting by patients: an 11-country survey. Drug Saf. 2012;35(1):45-60.

Xie, Q., Dai, Z., Hovy, E., Luong, M. T., & Le, Q. V. (2019). Unsupervised data augmentation for consistency training. arXiv preprint arXiv:1904.12848.

Zhu, M., Song, Y., Jin, G., & Jiang, K. (2020, November). Identifying Personal Experience Tweets of Medication Effects Using Pre-trained RoBERTa Language Model and Its Updating. In Proceedings of the 11th International Workshop on Health Text Mining and Information Analysis (pp. 127-137).

A Setup and Training Parameters

All networks were trained by Adam optimizer. For supervised baseline and the supervised initial training of pseudo-label method, we used the parameters suggested by the BERT's official fine-tuning guideline with learning rate being 1e-5, and training with a batch size of 32 for two epochs. For the semi-supervised learning, both methods were trained with a batch size of 128, started with a learning rate of 1e-5, then progressed with linear decay in four (for pseudo-label) and five (for consistency regularization) epochs.

The unsupervised weight variable used in consistency training was set by following the setup of Π-model (Laine et al., 2016). A Gaussian curve $\exp[-5(1 - T)^2]$ was used to ramp-up the weight, where T was increased linearly from zero to one during the ramp-up period. We set the first three epochs as the period to ramp-up the weight, and the maximum value of this weight variable was set to $w_{max} * M / N$ where M is the number of labeled tweets and N is the total number of train set. w_{max} was manually set to 1 in this task.

The structure of domain-specific language model was based on BERT which has 12 layers, 768 hidden neurons and 12 self-attention heads. The pre-training process used masked language model task only and the next sentence prediction was discarded. Following the official guideline of pre-training settings, fifteen percent (15%) of words in each tweet were masked by special [MASK] tokens and the model is trained to predict the masked token correctly. We trained the language model with 10M unlabeled tweets for 400K steps with a batch size of 512. The learning rate started with zero, and warmed up to 1e-4 in first one thousand steps and then linearly decayed to zero in the rest of training steps.

A sub-word vocabulary with about 50K tokens was generated by applying WordPiece algorithm (Schuster et al., 2012) in our 10M unlabeled tweets.

Our implementation was based on TensorFlow (`www.tensorflow.org`) and Transformers (`huggingface.co/transformers`).

Context-aware query design combines knowledge and data for efficient reading and reasoning

Emilee Holtzapple[1†], Brent Cochran[2‡], Natasa Miskov-Zivanov[1,3,4†]
[1]Dept. of Computational and Systems Biology, [2]Dept. of Developmental, Molecular, and Chemical Biology, [3]Dept. of Electrical and Computer Engineering, [4]Dept. of Bioengineering
[†]University of Pittsburgh, Pittsburgh, PA USA.
[‡]Tufts University School of Medicine, Boston, MA USA.
erh87@pitt.edu, Brent.Cochran@tufts.edu, nmzivanov@pitt.edu

Abstract

The amount of biomedical literature has vastly increased over the past few decades. As a result, the sheer quantity of accessible information is overwhelming, and complicates manual information retrieval. Automated methods seek to speed up information retrieval from biomedical literature. However, such automated methods are still too time-intensive to survey all existing biomedical literature. We present a methodology for automatically generating literature queries that select relevant papers based on biological data. By using differentially expressed genes to inform our literature searches, we focus information extraction on mechanistic signaling details that are crucial for the disease or context of interest.

1 Introduction

The number of peer-reviewed publications in molecular biology, biotechnology, and biomedical research increases exponentially every year. There is a considerable number of published papers on any one mainstream biomedical research topic, potentially hundreds of thousands of relevant articles. For many areas of study, simply reading every paper is unrealistic, or even physically impossible. When studying biological systems, such as intracellular signaling networks, this problem is apparent – accurate representation of all relevant signaling events requires extensive, expert knowledge acquired over many years of study. By using natural language processing, machine readers are capable of extracting interactions from hundreds or thousands of papers in a matter of hours, achieving a substantial speedup over manual information extraction (Björne & Salakoski, 2011). For this reason, automated methods for information extraction, such as machine reading, are used to retrieve information about intracellular signaling networks, and this information can then be used for model assembly or extension. While automated methods accelerate model assembly, the time required for processing all selected papers still depends on the number and the type of papers chosen for machine reading (Holtzapple, Telmer, & Miskov-Zivanov, 2020).

To retrieve relevant papers for machine reading, a common method is to query databases that contain biomedical literature. One repository for biomedical literature, MEDLINE, contains over 27 million papers (Fiorini, Lipman, & Lu, 2017), and a common method for retrieving papers from MEDLINE is through its associated search engine, PubMed. Querying MEDLINE through PubMed is particularly useful for identifying papers on a specific context such as disease or cell type. It is also used for identification of individual proteins, signaling pathways, and general cell processes in one specific context. One example of a PubMed query that targets a single pathway in a specific context is '"Hippo pathway" AND "stem cells"'. This query returns 272 papers, many of which describe Hippo pathway signaling trends in cancerous stem cells, as well as non-cancerous stem cells. These papers contain a wealth of information about the mechanistic causes of stemness. However, retrieval of these papers requires a priori knowledge that the Hippo pathway is important in stem cell maintenance and renewal. Additionally, these papers describe one small facet of stem cell signaling, and do not contain all the information needed to understand the system. To widen our perspective, we could retrieve all papers in MEDLINE that concern stem cells by querying PubMed with "stem cells". Here, we encounter two obstacles – this query returns over 271,000 papers, many of which describe morphological or anatomical details, and not signaling pathways.

Proceedings of the BioNLP 2021 workshop, pages 238–246
June 11, 2021.

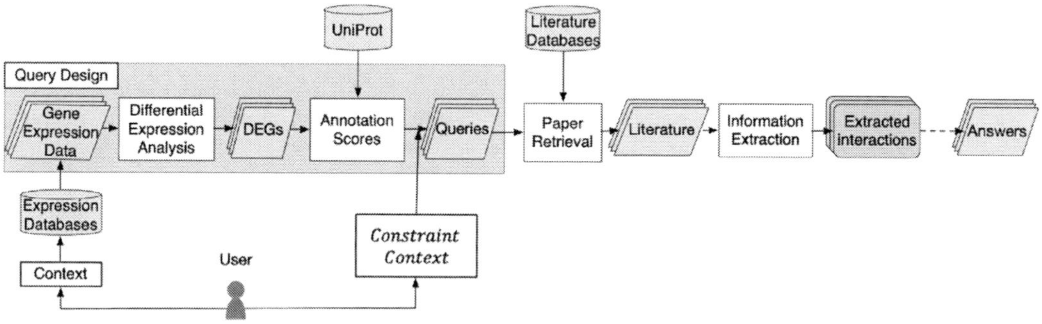

Figure 1. The automated query design methodology for information retrieval in biomedical research.

Our dilemma is that retrieval of relevant, crucial papers requires prior knowledge of which pathways are important.

There is a demand for improvement in methods for patient-specific paper retrieval, as evidenced by the TREC precision medicine track (Roberts et al., 2017). State-of-the-art methods for paper retrieval rely on term lists generated by experts or users, or automated information retrieval of similar papers (Sesagiri Raamkumar, Foo, & Pang, 2017; Wesley-Smith & West, 2016). These methods have several disadvantages. First, paper retrieval may depend on the cooperation of one or more experts in the field. Even for automated techniques that locate papers through related citations, or sematic analysis, some level of prior knowledge is needed. Also, even for experts that are up to date on canonical signaling pathways in a context of interest, novel pathways or signaling events cannot be easily targeted in a literature query. For efficient, thorough, context-aware exploration of cellular signaling, improved methods for literature retrieval are needed.

We present here a methodology for automated query design that does not rely on manual steps of the domain expert. To address the potential role of differentially expressed genes (DEGs) in disease mechanisms, we infer queries from biological data. Under- or over-expressed genes in the disease state are often the ones that play a role in disease progression (Armstrong et al., 2002). Identifying these contextual DEGs and using them as query terms focuses literature reading on genes and proteins that have altered signaling trends, and therefore, it facilitates further exploration of intracellular signaling networks that are potentially affected in disease. Our method utilizes gene expression data to find possible genes of interest based on their relative expression changes in response to disease, infection, etc. These genes of interest are used in the query to narrow down all possible PubMed hits to relevant signaling papers only. Furthermore, we also take into consideration how well-known each gene is, to choose the optimal number of gene terms in a query. Our results show that automated query design using these methods returns relevant signaling papers, and interactions extracted from these papers are informative and useful when reasoning about the queried context. This addresses a well-established problem in precision medicine – altered signaling pathways are often unique to one patient or environment and are difficult to study manually. Our methodology can be used in conjunction with any state-of-the-art model assembly techniques to aid in understanding affected signaling mechanisms in patient or cell line-specific systems. This methodology will provide an automated framework to retrieve research papers and streamline the process of model assembly. Our proposed automated query design methodology is outlined in Figure 1.

2 Query Design Method

In the following sub-sections, we describe our method to identify DEGs in the context of a disease, cell line, tissue type, or other condition (e.g., drug treatments), and for using them to form query terms when searching literature.

2.1 Identification of differentially expressed genes

As shown in Figure 1, the first step in our query design method is to define a context for literature search. Our approach allows a user to automatically design queries for many different contexts, including any biological condition that can be observed long enough to generate gene expression data. The user selects a data source and a relevant dataset from that source. While any kind of gene expression data can be used (microarray, RNA-seq, or single cell RNA-seq), public databases for expression data most frequently include RNA-seq data. Public databases for RNA-seq data include the Cancer Genome Atlas (TCGA) (Weinstein et al., 2013), Gene Expression Omnibus (Clough & Barrett, 2016), and the Expression Atlas (Papatheodorou et al., 2018), all of which contain sufficient expression data to be used in our proposed query generation method.

Once the dataset file is selected and input by the user, our proposed query design method identifies genes that are differentially expressed in the context of interest (e.g., disease state, cell line, etc.), compared to the control. The RNA-seq technique provides insight into the transcriptional activity of a cell population and reveals the number of gene transcripts present at a single point in time. For any gene *X*, we compute its differential expression as the fold change between the amount of its transcript ($X_{transcript}$) in two scenarios, control ($X^{control}_{transcript}$) and disease state ($X^{disease}_{transcript}$), a common method for measuring changes in gene expression (Huang, Zhang, Shen, Wong, & Xie, 2015). Since in this work we are interested in the magnitude of the change from the control, and not the direction of the change (i.e., increase or decrease), we use the absolute value of the change:

$$d_X = \left| \frac{X^{disease}_{transcript}}{X^{control}_{transcript}} \right| \quad (1)$$

We determine the d_X value for all transcripts in the selected RNA-seq dataset. Next, we sort the transcripts in a descending order of their d_X values (i.e., descending magnitude of change), and select a threshold for the d_X value, to ensure that all genes used as query terms are relevant to the dataset context. Specifically, we use 2.0 as a threshold, that is, we remove from the sorted list those transcripts that have $d_X < 2.0$. The standard threshold for d_X is usually 2.0 or 1.5 (Huang et al., 2015), based on what a cell biologist would consider notable or likely due to the effect of the disease or altered state, and not just noise in gene expression. While we use $d_X \geq 2.0$, the user can adjust this threshold to suit the research context (i.e., diseases or cell types with more or less DEGs than expected). We will refer to the transcripts remaining in the sorted list as DEGs. As probable indicators of a disease state, these DEGs become candidates for query terms. To give an estimate of an expected size of the sorted DEG list, previous work on analyzing many RNA-seq datasets over a wide range of conditions, including disease, tissues, cell types, drug treatments, etc., has shown that the median number of DEGs (with $d_X \geq 2.0$) per dataset is 92 (Crow, Lim, Ballouz, Pavlidis, & Gillis, 2019). However, as many as 10,000 DEGs per dataset were also observed, although rarely. We expect to see dozens to hundreds of DEGs (gene transcripts with $d_X \geq 2.0$), out of the 20,000+ genes in an RNA-seq dataset.

2.2 Selection of query terms

The sorted list of selected context-dependent DEGs that is automatically generated as described in Section 2.1, and the list of context terms, $Context$, provided by a user, are inputs to the next step of our proposed query generation method.

Using all DEGs with $d_X > 2.0$ to formulate a query is still not practical, as there can be tens or hundreds of such DEGs (see Section 2.1). Instead, we propose a method to further reduce the size of the sorted DEG list. We determine the number of DEGs to be used as query terms by estimating the number of papers that would be retrieved from a literature database when using the query formed from these terms. For example, in PubMed, the "popularity" of genes varies widely: *TP53* is a well-known oncogene with over 100,000 papers found in PubMed, and therefore, any query containing "p53" will return more papers than a query using a novel gene.

Thus, to estimate the impact of each DEG, as a possible query term, on the number of papers retrieved, we propose to utilize the annotation information provided by the UniProt database (The UniProt Consortium, 2017). This database contains information on the gene itself, known transcripts, as well as information on the gene product, if available. Each gene in the UniProt database has an assigned *annotation score*, which is an amalgamation of evidence of the gene and gene product's existence, including cross-references in other databases, known aliases, experimental evidence, and more. We use this annotation score as a measure of how established a gene is in the literature. For manually annotated genes, where the evidence has been reviewed by an expert, the score is higher. The annotation score has an integer value in the interval between 1 and 5, where score of 5 indicates ample evidence of the protein in existing literature and databases, and score of 1 indicates little to no available information about the protein. For example, the *TP53* gene in humans (UniProt ID P04637), a well-known tumor suppressor, has an annotation score of 5, while the *OATL1* transcript in humans (UniProt ID B4DF03), which has not been observed at the protein level, has an annotation score of 1.

We propose here to use the annotation score together with the d_X value when deciding which DEGs to include as query terms. The combination of these two measures allows the design of queries for different objectives or tasks, for example, to search for literature that contains a few well-known (high annotation score) proteins, or many novel or unstudied (low annotation score) proteins. Furthermore, by incorporating the UniProt annotation score to choose terms, we can automatically design queries that will lead to a selection of a manageable number of papers. In other words, the optimal number of papers would be the one large enough to provide adequate information on the system and small enough to still be processed in a feasible amount of time. What would be considered the "optimal" number of papers depends on both the complexity of the context of interest, as well as the allocated resources for information extraction. For example, a researcher using a machine reader to process literature on diabetes will require many more papers than someone who wants to read papers manually. Additionally, the number of papers found in a literature database as a result of the query will be different for each user depending on the input dataset, annotation score, and the addition of new publications in the literature database, and so this method allows to tailor the query design process to the user's research goals. We will refer to the DEGs that are selected to be used in a query as *query term DEGs*.

Different research tasks, paper contexts, and datasets will require a different number of papers to be read. Therefore, our method allows the user to provide an additional input, $Constraint$, which will influence the number of papers selected for reading. The $Constraint$ input can be either categorical, or a discrete number greater than 0, and is used in our method to determine the cut-off parameter, C. The cut-off C value is in turn used to select those DEGs that will be included in the query. Specifically, we traverse the sorted DEG list, starting with the DEG that has the largest d_X value, and we keep adding DEGs to the query term list, as long as the sum of their annotation scores is smaller than or equal to the cut-off value C.

Table 1. User-input categories, the corresponding cut-off parameter C for the annotation score sum, as well as the expected maximum and minimum number of query term DEGs. (These values do not account for DEGs with no entry in the UniProt database.)

user-input category	C	expected min # of DEGs	expected max # of DEGs
human-readable	15	3	15
automation suggested	35	12	35
automation required	60	20	60

We use three categories to indicate the level of automated reading needed to comprehend all information in the paper set. The first category, "human-readable", results in a selection of a small number of papers, suitable for a human to read in a short time (e.g., hours). The second category, "automation suggested", leads to a medium number of selected papers that is possible for a human to read (e.g., days), but more practical if processed by machine reading. The third category, "automation required", results in a large number of selected papers, only practical for machine reading.

Allowing for two different ways to enter the $Constraint$ input, provides additional flexibility. If the user knows exactly which value they want to use for the cut-off parameter, they can directly enter it. However, in the research process, the users may sometimes be interested in exploring a smaller subset of relevant papers, or doing a more comprehensive exploration of the topic, and the three categories listed above are useful in such cases. The values of the parameter C that correspond to the three categories, and that we used to obtain results and demonstrate our approach, are listed in Table 1.

We note here that, while these values are set internally in the code, they could be easily changed to better suit different domains or research goals. For example, for a "human-readable" reading output, we set C=15, and following our method for selecting query term DEGs given the cut-off value C, this could result in as few as 3 query term DEGs (all with annotation score 5) or as many as 15 query term DEGs (all with annotation score 1).

To this end, it is worth noting that not all DEGs are always found in UniProt, and therefore, the DEGs without a corresponding UniProt entry are assumed to have annotation score value of 0. As this is possible even for DEGs with large d_X value, this could lead, in rare cases, to the actual number of query term DEGs exceeding the cut-off value C (e.g., this would be 15, for our example above). While, in theory, the number of DEGs with $d_X \geq 2.0$ and annotations score of 0 could potentially be very large, we have not encountered such cases. Moreover, our experiments have shown that allowing for DEGs with annotation score 0 to be added to the query term list does not significantly increase the number of selected papers, while at the same time can lead to the retrieval of papers with very novel disease mechanisms. In Table 1, we provide the C values that we use for the three user-input categories, and the corresponding typical minimum and maximum number of query term DEGs. As a guidance, we list in Table 1 the *typical* minimum and maximum numbers that are easily determined from C values, and which consider only those genes with an annotation score greater than 0.

Once the list of the query term DEGs is determined, their official gene names (e.g., *TP53, BRCA1, EGFR*) are combined with a logical **OR**, thus allowing any paper that includes at least one of the query term DEGs to be selected. It is important to note that the official gene name (or another standardized identifier) is already supplied by gene expression datasets, and so we avoid the challenge of using named-entity normalization to automatically standardize the names of DEG query terms. We chose to use a logical **OR** to retrieve the maximum number of relevant papers for each query, since a logical **AND** would make the query more specific, and so restrict the number of papers. Other combinations of logical **AND** and **OR** between the terms in the query are possible and could be informed by the user or inferred if relevant information is available. This is beyond the scope of the work presented here and is one of the next steps that we plan to explore in the future.

Furthermore, since we are interested in creating queries that focus on a particular context, our automated tool adds *Context* to this logical expression as a necessary condition, that is, it combines it with the other terms using a logical **AND**:

$$(gene_1 \textbf{ OR } gene_2 \textbf{ OR } \dots\ gene_N) \textbf{ AND } Context \quad (2)$$

where each $gene_i$ (i=1,..,N) is the official gene name of one of the N query term DEGs. By including only papers that mention the context of interest, we can extract relevant interactions. It is important to note that one context may have multiple aliases (e.g., "coronavirus", "COVID-19", and "SARS-CoV-2" are all referring to the same disease). The user can increase the scope of the retrieved papers by combining all possible context aliases with a logical **OR**.

2.3 Using queries in disease explanation

We discuss in this section the use of automatically generated targeted queries in information extraction conducted by machine readers, followed by automated reasoning about affected signaling networks and biological processes. For each query, we retrieve all machine reading statements in the INDRA database (Gyori et al., 2017) that are associated with at least one paper in our reading set. The INDRA database is a system that incorporates natural language processing tools and standardized databases to collect biomedical signaling events. INDRA relies on several different machine readers to process papers and supply information on signaling events. The interactions output by readers are directed, and therefore, they can be used in the process of assembly or extension of dynamic models, in order to explain mechanisms and timing of the disease. Although the query term DEGs that were selected following our method described in Sections 2.1 and 2.2 are likely to participate in these interactions, it is important to note that the interactions output by readers will include many other relevant genes and proteins. Thus, these extracted interactions are expected to provide the information on intracellular signaling networks that is potentially critical for the context originally selected by the user and included as a term in the generated query (equation 2).

To evaluate the relevance of extracted interactions, we assess what types of biological processes and signaling pathways these interactions are involved in. We use PANTHER (Mi, Muruganujan, Ebert, Huang, & Thomas, 2018) to calculate enriched Gene Ontology (GO) terms (Ashburner et al., 2000) in the protein-protein interactions within our interaction sets for each query. In the GO database, genes and proteins are annotated with known cellular functions. Each

Table 2. Eight automatically formulated queries for four diseases. Each disease has two associated queries, which are expected to retrieve different sized reading sets.

Context	# of DEGs with $d_X \geq 2.0$	C	Query
Thyroid carcinoma	5026	15	**Q1**: (GABRB2 or LIPH or KLHDC8A or LIPI) and "thyroid carcinoma"
		60	**Q2**: (GABRB2 or LIPH or KLHDC8A or LIPI or LINC02471 or PRR15 or MTRNR2L12 or CIDEC or RTL4 or SLIT1 or ZCCHC12 or TRPC5 or LRP4 or RXRG or METTL7B or CDH3) and "thyroid carcinoma"
Ulcerative colitis	1476	15	**Q3**: (AL035661.1 or HP or NECAB1 or MCEMP1) and "ulcerative colitis"
		60	**Q4**: (AL035661.1 or HP or NECAB1 or MCEMP1 or ANKRD22 or ARG1 or BMX or MMP9 or S100A12 or SCART1 or SLC2A14 or SLC1A3 or SLC12A5-AS1 or OLAH or ACHE) and "ulcerative colitis"
COVID-19	32	15	**Q5**: (MX1 or DDX60 or PARP9) and (SARS-CoV2 or COVID-19 or coronavirus)
		60	**Q6**: (MX1 or DDX60 or PARP9 or DDX58 or HELZ2 or CMPK2 or OAS3 or STAT1 or HERC6 or DTX3L or IFIT1 or SAMD9 or AL445490.1 or TAS2R4 or AC147651.1 or AC004253.1) and (SARS-CoV2 or COVID-19 or coronavirus)
Glioblastoma	3300	15	**Q7**: (HOXD9 or PLA2G2A or HOXD10) and (Glioblastoma or GBM)
		60	**Q8**: (HOXD9 or PLA2G2A or HOXD10 or HOXD13 or HOXA5 or HOXD8 or DLGAP5 or HOXA10 or SAA1 or HOXC10 or AC092017.1 or AC011742.1 or AL160286.1 or MIR663AHG or ELL2P1 or TOP2A or IGHA1) and (Glioblastoma or GBM)

GO term has a list of proteins involved in the biological process, and PANTHER calculates representation of all known GO terms for each interaction set. For GO terms that have a greater number of genes found in the interaction set than would be expected by chance, we consider this GO term statistically enriched. To assess whether enriched GO terms are similar, we use NaviGo to calculate the Resnik similarity score between all GO terms (described in (Wei, Khan, Ding, Yerneni, & Kihara, 2017)). By determining highly enriched GO terms, we can draw conclusions about what signaling pathways and biological processes are represented in our paper sets for each query.

3 Results

To demonstrate the usefulness of our automated query design methodology, we show results for four different contexts. For each context, we automatically design two queries, one with an expected large number of output papers, and one with an expected small number of output papers. These results illustrate how DEGs can be used to formulate queries that output relevant papers, and how the annotation score affects the volume of papers. We also show that the papers contain interactions that are closely related and are involved in the same GO biological processes.

3.1 Case studies

Using the Expression Atlas (Papatheodorou et al., 2018), we selected four publicly available RNA-seq datasets. These four datasets provide gene expression data for both control and disease state in SARS-CoV-2 (Blanco-Melo et al., 2020), ulcerative colitis (Mo et al., 2018), glioblastoma multiforme (Gill et al., 2014), and thyroid carcinoma (Costa et al., 2015). All four datasets express transcription in transcripts per million (TPM) and include the d_X values computed for the disease state with respect to the control state. In the following studies, we use the d_X values that are provided with selected datasets. With these case studies, we cover three substantial topics in biomedical research – autoimmune disorders, cancer, and viral infections. These diseases differ in the number of expected publications, largely due to the awareness of the disease itself. We chose several well-studied diseases, as well as several relatively unknown diseases as case studies, to show the utility of our methodology, regardless of the recognition of the system at hand. Using differential gene expression data from these diseases, we illustrate how biological data can provide valuable information for automatically designed targeted queries.

3.2 Selection of queries

To design a query that retrieves a small reading set, as discussed in Section 2.2, we explored the effect of the cut-off value C=15 for the annotation score sum, and to design a query that retrieves a large reading set, we use the cut-off value C=60. The queries generated for all four contexts for these two cut-off values are listed in Table 2. Notably, the same cut-off value C for different datasets may result in queries with a different number of terms. This can be explained by the UniProt annotation score of the top (with large d_X) DEGs in the datasets. Due to differences in experiment techniques, environmental conditions, or other

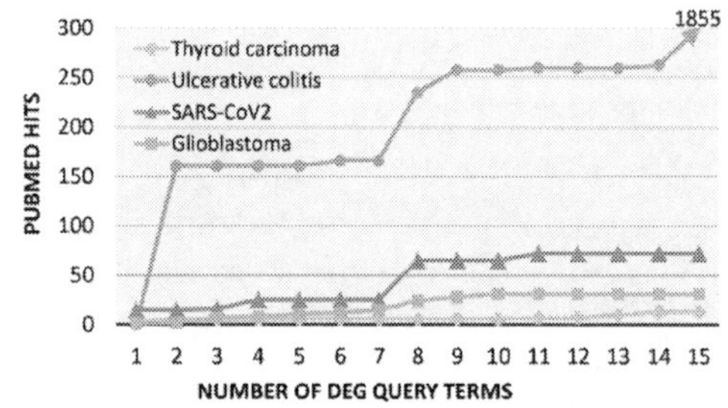

(a)

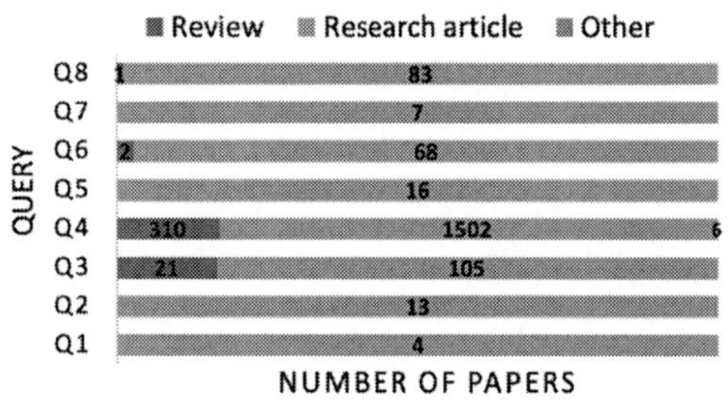

(b)

Figure 2. Number of papers found in PubMed, based on how many of the top DEGs were used as query terms. (b) Distribution of paper types by query.

factors, gene expression datasets from different samples and labs will likely show differences in the top DEGs. Consider a hypothetical example where we formulate queries based on two pancreatic cancer datasets (another example, not listed in Table 2), A and B, and choose the cut-off C=10. For dataset A, this value is achieved after adding two DEG query terms, since the DEGs with highest d_X values are P53 and MDM2, which are both very well-known proteins with an annotation score of 5. For dataset B, the threshold is not passed until five DEG query terms are added. The top five most differentially expressed genes are small non-coding RNAs, which are generally poorly studied, and each has an annotation score of 2.

3.3 Paper retrieval

In our studies, we used PubMed (Fiorini et al., 2017) as the most up-to-date and comprehensive source for biomedical literature. We do not apply any filters for article type, year, or journal. However, we restrict our results to only those papers with valid PMCIDs, to ensure that all papers can be processed with state-of-the-art machine readers.

Once we have formulated queries for each disease, we can use them to search PubMed. In Figure 2a, we show the number of papers retrieved as a function of how many of the top DEGs are used as query terms. As expected, as the number of terms increase, so does the number of retrieved papers. However, many query terms, in conjunction with the context term, add no additional papers to the reading set. This indicates that some of these DEGs have not been explored much or mentioned in papers in the context of the relevant disease, and therefore, they may be a fruitful avenue for exploration.

We also show in Figure 2b that, as the number of extracted papers in the reading output increases, the distribution of article types also changes. We examine the composition of the reading set by classifying each paper as either a research article, review, or other (books, documents, etc.). In large reading sets, reviews are slightly more common than in small reading sets, which is due to one or more query term DEGs having better representation in PubMed. Well-studied genes and proteins are more likely to be included in reviews than novel, relatively unknown genes. Since the scope of reviews and research articles differ drastically, we expect them to contribute differently to the number of extracted interactions.

3.4 Validation of extracted interactions

To validate the paper sets retrieved from each query, we analyzed the statements from the INDRA database (described in Section 2.3). In Figure 3a, we show the number of extracted interactions for each query. The number of interactions is dependent upon the number of papers, as well as the representation of the context and DEG query terms in PubMed. For each query, we also determined the top 10 enriched GO terms, sorted using the false discovery rate (FDR) (Benjamini & Hochberg, 1995). We show the average Resnik similarity score between the top 10 GO terms for each of our eight queries, where a higher score indicates more similarity between GO terms. Finally, in Figure 3b, we show the percent of DEG query terms that are present in the list of extracted interactions. These results, taken together, show that these queries retrieve papers that contain relevant signaling events that can be interpreted by machine readers, and describe highly related biological processes. In general, our method of increasing the cut-off value C not only retrieves

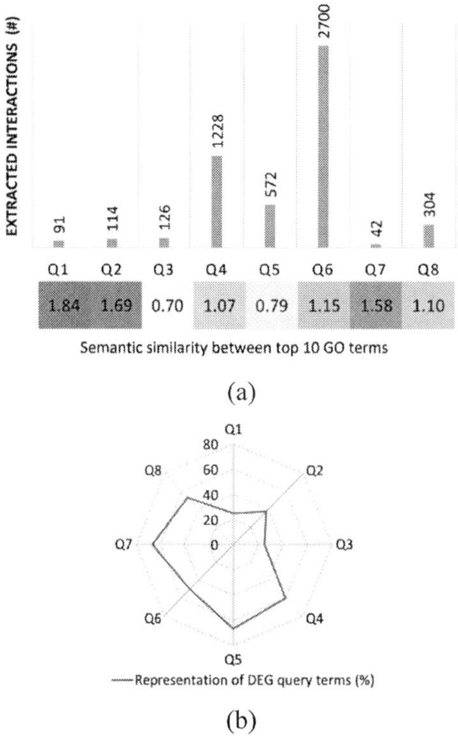

Figure 3 (a) Number of interactions extracted from INDRA for each query, as well as the average pairwise Resnik similarity score for the top 10 enriched GO terms. (b) The percent of DEGs used as query terms in each case study that are present in the set of extracted interactions.

more papers, but it also increases the number of signaling events extracted by readers, without a sizeable cost to relevance, as assessed by GO term semantic similarity.

4 Conclusions

While automated methods for extracting interactions from literature have improved the speed of information extraction, this process still has its pitfalls. Specifically, finding all relevant literature for the context at hand can be difficult, and brute force methods for selecting papers are too slow. By incorporating biological data in our queries, we can select relevant literature, and control the size of the reading sets.

Our results show that using DEGs to formulate queries allows for targeting literature that could help explain differentially regulated pathways in disease. One side effect of this method is identification of DEGs in disease where there is little to no literature presence. In such cases, our proposed method could become critical, as it automatically identifies the gaps in our collective knowledge of certain diseases, and thus, suggests important research directions. For DEGs that return no additional results when used as a query term, this indicates the gene has an undiscovered role in the context of interest.

Future directions include refining the query formulation methodology, as well as expanding our results. The relative presence of different diseases in PubMed affects the size of the reading set, independent of the number of gene query terms. By incorporating preliminary data on the presence of a disease or context in PubMed, we can adjust the annotation score. Additionally, since this method hinges on a list of affected genes or proteins with quantifiable differences from a control state, other measures of relative changes in cell function could also be used. Data on changes in post-translational modification of proteins, changes in epigenetic markers such as methylation, open chromatin, or histone modifications, or even somatic mutations could also be used, especially as such entities and events can be output by the state-of-the-art machine reading. Testing our methods on different datasets would help showcase the usefulness of our approach. In the future, we would also like to compare our method to a literature corpus assembled by an expert. However, since our queries are based on cell line-specific gene expression datasets, there are no existing corpuses for comparison. Future work includes assembling said corpuses and comparing to our method presented in this work.

Acknowledgements

Thanks to Yasmine Ahmed and Casey Hansen for helpful discussions. Funded by Defense Advanced Research Projects Agency (W911NF-17-1-0135).

References

Armstrong, Peter J, Johanning, Jason M, Calton Jr, William C, Delatore, Jason R, Franklin, David P, Han, David C, . . . Elmore, James R. (2002). Differential gene expression in human abdominal aorta: aneurysmal versus occlusive disease. *Journal of vascular surgery, 35*(2), 346-314.

Ashburner, Michael, Ball, Catherine A., Blake, Judith A., Botstein, David, Butler, Heather, Cherry, J. Michael, . . . Sherlock, Gavin. (2000). Gene Ontology: tool for the unification of biology. *Nature Genetics, 25*, 25. doi:10.1038/75556

Benjamini, Yoav, & Hochberg, Yosef. (1995). Controlling the false discovery rate: a practical and powerful approach to multiple testing. *Journal of the Royal statistical society: series B (Methodological), 57*(1), 289-300.

Björne, Jari, & Salakoski, Tapio. (2011). *Generalizing biomedical event extraction.* Paper presented at the Proceedings of the BioNLP Shared Task 2011 Workshop, Portland, Oregon.

Blanco-Melo, Daniel, Nilsson-Payant, Benjamin E., Liu, Wen-Chun, Møller, Rasmus, Panis, Maryline, Sachs, David, . . . tenOever, Benjamin R. (2020). SARS-CoV-2 launches a unique transcriptional signature from in vitro, ex vivo, and in vivo systems. *bioRxiv*, 2020.2003.2024.004655. doi:10.1101/2020.03.24.004655

Clough, Emily, & Barrett, Tanya. (2016). The Gene Expression Omnibus Database. *Methods in molecular biology (Clifton, N.J.), 1418*, 93-110. doi:10.1007/978-1-4939-3578-9_5

Costa, Valerio, Esposito, Roberta, Ziviello, Carmela, Sepe, Romina, Bim, Larissa Valdemarin, Cacciola, Nunzio Antonio, . . . Ciccodicola, Alfredo. (2015). New somatic mutations and WNK1-B4GALNT3 gene fusion in papillary thyroid carcinoma. *Oncotarget, 6*(13), 11242-11251. doi:10.18632/oncotarget.3593

Crow, Megan, Lim, Nathaniel, Ballouz, Sara, Pavlidis, Paul, & Gillis, Jesse. (2019). Predictability of human differential gene expression. *Proceedings of the National Academy of Sciences, 116*(13), 6491. doi:10.1073/pnas.1802973116

Fiorini, Nicolas, Lipman, David J., & Lu, Zhiyong. (2017). Towards PubMed 2.0. *eLife, 6*, e28801. doi:10.7554/eLife.28801

Gill, Brian J., Pisapia, David J., Malone, Hani R., Goldstein, Hannah, Lei, Liang, Sonabend, Adam, . . . Canoll, Peter. (2014). MRI-localized biopsies reveal subtype-specific differences in molecular and cellular composition at the margins of glioblastoma. *Proceedings of the National Academy of Sciences, 111*(34), 12550-12555. doi:10.1073/pnas.1405839111

Gyori, Benjamin M., Bachman, John A., Subramanian, Kartik, Muhlich, Jeremy L., Galescu, Lucian, & Sorger, Peter K. (2017). From word models to executable models of signaling networks using automated assembly. *Molecular systems biology, 13*(11), 954-954. doi:10.15252/msb.20177651

Holtzapple, Emilee, Telmer, Cheryl A, & Miskov-Zivanov, Natasa. (2020). FLUTE: Fast and reliable knowledge retrieval from biomedical literature. *Database, 2020*. doi:10.1093/database/baaa056

Huang, H., Zhang, S., Shen, W. J., Wong, H. S., & Xie, D. (2015). Gene set enrichment ensemble using fold change data only. *J Biomed Inform, 57*, 189-203. doi:10.1016/j.jbi.2015.07.019

Maugeri-Saccà, M., & De Maria, R. (2018). The Hippo pathway in normal development and cancer. *Pharmacol Ther, 186*, 60-72. doi:10.1016/j.pharmthera.2017.12.011

Mi, Huaiyu, Muruganujan, Anushya, Ebert, Dustin, Huang, Xiaosong, & Thomas, Paul D. (2018). PANTHER version 14: more genomes, a new PANTHER GO-slim and improvements in enrichment analysis tools. *Nucleic Acids Research, 47*(D1), D419-D426. doi:10.1093/nar/gky1038

Mo, Angela, Marigorta, Urko, Arafat, Dalia, Chan, Lai, Ponder, Lori, Jang, Se Ryeong, . . . Gibson, Greg. (2018). Disease-specific regulation of gene expression in a comparative analysis of juvenile idiopathic arthritis and inflammatory bowel disease. *Genome Medicine, 10*. doi:10.1186/s13073-018-0558-x

Papatheodorou, Irene, Fonseca, Nuno A., Keays, Maria, Tang, Y. Amy, Barrera, Elisabet, Bazant, Wojciech, . . . Petryszak, Robert. (2018). Expression Atlas: gene and protein expression across multiple studies and organisms. *Nucleic Acids Research, 46*(D1), D246-D251. doi:10.1093/nar/gkx1158

Park, J. H., Shin, J. E., & Park, H. W. (2018). The Role of Hippo Pathway in Cancer Stem Cell Biology. *Mol Cells, 41*(2), 83-92. doi:10.14348/molcells.2018.2242

Roberts, Kirk, Demner-Fushman, Dina, Voorhees, Ellen M., Hersh, William R., Bedrick, Steven, Lazar, Alexander J., & Pant, Shubham. (2017). Overview of the TREC 2017 Precision Medicine Track. *The ... text REtrieval conference : TREC. Text REtrieval Conference, 26.*

Sesagiri Raamkumar, Aravind, Foo, Schubert, & Pang, Natalie. (2017). Using author-specified keywords in building an initial reading list of research papers in scientific paper retrieval and recommender systems. *Information Processing & Management, 53*(3), 577-594. doi:https://doi.org/10.1016/j.ipm.2016.12.006

The UniProt Consortium. (2017). UniProt: the universal protein knowledgebase. *Nucleic Acids Research, 45*(D1), D158-D169. doi:10.1093/nar/gkw1099

Wei, Qing, Khan, Ishita K., Ding, Ziyun, Yerneni, Satwica, & Kihara, Daisuke. (2017). NaviGO: interactive tool for visualization and functional similarity and coherence analysis with gene ontology. *BMC Bioinformatics, 18*(1), 177. doi:10.1186/s12859-017-1600-5

Weinstein, John N, Collisson, Eric A, Mills, Gordon B, Shaw, Kenna R Mills, Ozenberger, Brad A, Ellrott, Kyle, . . . Network, Cancer Genome Atlas Research. (2013). The cancer genome atlas pan-cancer analysis project. *Nature Genetics, 45*(10), 1113.

Wesley-Smith, Ian, & West, Jevin D. (2016). *Babel: A Platform for Facilitating Research in Scholarly Article Discovery.* Paper presented at the Proceedings of the 25th International Conference Companion on World Wide Web, Montréal, Québec, Canada. https://doi.org/10.1145/2872518.2890517

Measuring the relative importance of full text sections for information retrieval from scientific literature.

Lana Yeganova, Won Kim, Donald C. Comeau, W. John Wilbur, and Zhiyong Lu
National Center for Biotechnology Information (NCBI),
National Library of Medicine (NLM), National Institutes of Health (NIH), USA

Abstract

With the growing availability of full-text articles, integrating abstracts and full texts of documents into a unified representation is essential for comprehensive search of scientific literature. However, previous studies have shown that naïvely merging abstracts with full texts of articles does not consistently yield better performance. Balancing the contribution of query terms appearing in the abstract and in sections of different importance in full text articles remains a challenge both with traditional bag-of-words IR approaches and for neural retrieval methods.

In this work we establish the connection between the BM25 score of a query term appearing in a section of a full text document and the probability of that document being clicked or identified as relevant. Probability is computed using Pool Adjacent Violators (PAV), an isotonic regression algorithm, providing a maximum likelihood estimate based on the observed data. Using this probabilistic transformation of BM25 scores we show an improved performance on the PubMed Click dataset developed and presented in this study, as well as on the 2007 TREC Genomics collection.

1 Introduction

PubMed (https://pubmed.gov) is a search engine providing access to a collection of more than 30 million biomedical abstracts. Of these, about 5 million have full text available in PubMed Central (PMC; https://www.ncbi.nlm.nih.gov/pmc). Millions of users search PubMed and PMC daily (Fiorini, Canese, et al., 2018). However, it is not currently possible for a user to simultaneously query the contents of both databases with a single integrated search.

With the growing availability of full-text articles, integrating these two rich resources to allow a unified retrieval becomes an essential goal, which has potential for improving information retrieval and the user search experience (Fiorini, Leaman, Lipman, & Lu, 2018). An obvious benefit is improving the handling of queries that produce limited or no retrieval in PubMed. In many instances, incorporating full text information can yield useful retrieval results. For example, the query *cd40 fmf* retrieves no articles in PubMed, but finds 60 articles in PMC discussing protein *cd40* and a computational technique of flow microfluorometry (FMF).

A number of studies have pointed out the benefits of full text for a range of text mining tasks (Cejuela et al., 2014; Cohen, Johnson, Verspoor, Roeder, & Hunter, 2010; J. Kim, Kim, Han, & Rebholz-Schuhmann, 2015; Westergaard, Stærfeldt, Tønsberg, Jensen , & Brunak, 2018) and demonstrated improved performance on named entity recognition, relation extraction, and other natural language processing tasks (Wei, Allot, Leaman, & Lu, 2019). For information retrieval, however, combining the full text of some papers with only the abstracts of others is not a trivial endeavor. Naïvely merging the body text of articles with abstract data, naturally increases the recall, but at a cost in precision, generally degrading the overall quality of the combined search (W. Kim, Yeganova, Comeau, Wilbur, & Lu, 2018; Jimmy Lin, 2009). This can be explained by several complexities associated with full texts, such as multiple subtopics often being discussed in a full-length article or information being mentioned in the form of conjecture or a proposal for future work. In addition, not every record matching the query is focused on the query subject, as query words may be mentioned in passing, which is more common in full text. Another challenge in incorporating full text in retrieval is merging sources of information with different characteristics: the abstract, generally a concise summary on the topic of the study, versus a lengthy detailed description provided in full text. To address that, recent studies have attempted to use full text in a more targeted way — by performing paragraph-level retrieval (Hersh, Cohen, Ruslen, & Roberts, 2007; Jimmy Lin, 2009), passage-level retrieval (Sarrouti & El Alaoui, 2017) or sentence-level retrieval (Allot et al., 2019; Blanco &

Proceedings of the BioNLP 2021 workshop, pages 247–256
June 11, 2021. ©2021 Association for Computational Linguistics

Zaragoza, 2010). LitSense (Allot et al., 2019), for example, searches over a half-billion sentences from the combined text of 30+ million PubMed records and ~3 million open access full-text articles in PMC.

Towards the overarching goal of improving PubMed document retrieval by incorporating the full texts of articles in PMC, in this work we lay the groundwork by studying strategies for integrating full text information with abstract for one query token at a time. We choose to use BM25, a classical term weighting approach, as a base token score. We, however, observe that token BM25 scores are not directly comparable between the sections of a full text article – the same BM25 score may have a different significance depending on the section. To address variable significance of sections, we propose converting BM25 section scores into probabilities of a document being clicked and using these probabilities to compute the overall token score. To summarize, given a single token in a query, we 1) define how to compute section scores, 2) examine the relative importance of different sections in the full text, and 3) study how to combine section scores from a document.

To examine these questions, we use two evaluation datasets. One is a standard TREC dataset frequently used for evaluating ad-hoc information retrieval. The second is a dataset we created based on PubMed user click information. The dataset is constructed from PubMed queries and clicks under the assumption that a clicked document is relevant to a user issuing a query. The dataset is used for both training and evaluation.

Neural retrieval models have been extensively studies in recent years in the context of Information Retrieval (Guo et al., 2020; Jimmy Lin et al., 2021). However, despite significant advances, they show no consistent improvement over traditional *bag of words* IR methods (Chen & Hersh, 2020; Zhang et al., 2020). BM25 remains in the core of most production search systems, including Lucene's search engine and PubMed. In addition, many relevance ranking algorithms rely on BM25 as a preliminary retrieval step, followed by re-ranking of the top scoring documents (Fiorini, Canese, et al., 2018).

In the next section, we describe the evaluation datasets, and lay out a retrieval framework for studying the problem at hand. Then, we describe our approach of converting the raw BM25 section score into the probability of document relevance. Such probabilities are comparable across the sections of full text documents, including the abstract. In section 4 we learn how to combine them in a way which accounts for the relative importance of sections. Results are presented in section 5, followed by the Discussion and Conclusions section.

2 Evaluation Datasets

Retrieval methods are generally evaluated based on how the retrieval output compares to a gold standard. A gold standard is a set of records judged for relevance to a query that provides a benchmark against which to measure the quality of search results. This approach is used at the annual Text Retrieval Conference (TREC), run by the National Institute of Standards and Technology (NIST) (Voorhees, 2001). NIST develops a list of queries, called topics, and provides large test collections and uniform scoring procedures. The difficulty with this approach is that a gold standard is created by human experts which makes the evaluation expensive, time consuming, and therefore not available for large scale experiments involving thousands of queries. To compare different retrieval approaches without a manually created gold standard we describe semi-automatically created test data based on indirect human judgements that can be utilized in our setting. The PubMed User Click dataset is created based on retrospective analysis of PubMed queries under the assumption that a clicked document is relevant to a user issuing a query. In our study we use both, the TREC 2007 Genomics and PubMed user click datasets.

TREC 2007 Genomics dataset. The Genomics dataset (Hersh et al., 2007) consists of 36 queries, called *topics*, and 162,259 full-text articles from Highwire Press (http:// highwire.stanford.edu/). 160K of these documents were successfully mapped to their corresponding PubMed Identifiers (PMIDs) and are the basis of our experiments. Each document is split into *legal spans* corresponding to paragraphs in the articles, amounting to over 12 million legal spans. For each of the 36 *topics* human relevance judgements are provided on the paragraph level. Following previous studies, a document is labeled positive, if it contains at least 1 paragraph judged to be relevant to the query.

The query topics are presented in the form of biological questions, such as:

What toxicities are associated with etidronate?
What signs or symptoms are caused by human parvovirus infection?

These question-like topic formulations contain generic words, that are not representative of the specific information need of a user, such as "what", "associated", etc. We applied a combination of frequency-based techniques and manual validation to filter these stop words out and used the remaining 165 content terms for our analysis.

PubMed Click Dataset. The dataset is constructed from PubMed queries and clicks, under the general assumption that a clicked document is relevant to a user issuing a query.

The presence of a query token in the title is known to present a strong signal associated with a document being clicked (W. Kim et al., 2018; Resnick, 1961). Users searching PubMed only see the title of the document on the DocSum page and not the abstract or the full text. If query tokens do not appear in the title, then predictions on the abstract or the full text can only be effective to the extent they predict something about the title that makes the user choose to click. This is a weaker signal and would be obscured by query words appearing in a title. To remove this bias, we only consider documents for which none of the query tokens appear in the title. Note that since the document is retrieved via PubMed, all query tokens must be found in the title, abstract or article citation information. We collect only retrieved documents for which none of the query tokens appear in the title and all of them appear in the abstract.

Clicked documents are assumed to be relevant to the user issuing the query, and we label a clicked document as a positive instance. We further assumed that documents displayed above the clicked document were seen by the user and rejected. These documents are labeled negative. Clicks on the top rank are ignored as a precaution, as those clicks might simply represent a user's urge to click on something indiscriminately. Documents displayed below the lowest clicked document on the document summary page are ignored as the user may not have considered them.

The same query string may be searched multiple times within a period of time and subsequently may result in different articles displayed and different documents clicked. In addition, a query within a single search may receive multiple clicks on the same page. To account for these user search actions, we merge the data for the evaluation dataset as follows. Given a unique query string, we collect all positive and negative data points associated with each click instance, and remove from the negative set those documents that also appear as positives following the reasoning: if a document is thought to be relevant by at least one user we consider it relevant for that query string.

Using this dataset, for each query token we wish to compare its score coming from a document's abstract versus the body text. First, to directly measure the benefit of full text, for each query in the PubMed Click Dataset, we perform this comparison on a subset of documents in the dataset that have full text available in PMC. Second, for each query in the dataset, we perform the comparison on all documents available in the PubMed Click dataset. This includes documents that do and do not have the full text available, as in production PubMed.

We randomly sampled 2 million unique queries from the PubMed query log in 2017, which retrieved at least one positive document. On average there are 6.60 documents collected for each query, an average of ~30% of which are labeled positive. Of 6.60 documents available for each query, only 2.65 documents have full text available in PubMed Central (~40%). We separated two thirds of queries for training PAV functions described in the next section, and one third for testing. 634,364 queries along with collected labeled documents comprise the test portion of the PubMed click dataset. A subset of that dataset that includes queries for which all retrieved documents have full text available constitutes 232,636 queries, and will be referred to as Set_FT.

3 Methods – Using Full Text to score a query token

Here we examine how to optimally use BM25 scores coming from the abstracts and full text paragraphs to improve retrieval performance. We first define the score of a token within a full-text section, which then we transform into a probability of that document being relevant given the score and the section. We then learn how to combine these section-based token scores into an overall score predicting the probability of a document being relevant.

3.1 Obtaining Full Text

We obtain full text documents from the PubMed Central full text collection in BioC (https://www.ncbi.nlm.nih.gov/research/bionlp/APIs) (Comeau, Wei, Doğan, & Z., 2019). This collection contains about 5 million full text manuscripts. BioC allows one to obtain full text information by paragraphs.

Full-text articles are typically comprised of sections presented in a logical sequence. Sections such as Introduction, Materials and Methods, Results, and Discussion predominantly appear as they represent the logical sequence in scientific writing. Frequently, however, sections carrying similar types of information are referred to differently depending on the journal, the requirements of the publishing entity, and author writing style. For example, Introduction and Background section titles are used interchangeably. Results sections can be also referred to as Results and Experiments, etc. Using BioC provided section type identifiers that are based on the labels and regular expressions found in (Kafkas et al., 2015). To normalize section titles, we concentrate on the following section types: Abstract, Abbreviation, Caption, Discussion, Case, Keyword, Conclusion, Result, Methods, Introduction, Generic Section Title, Supplement, and Appendix. In what follows, all the sections other than the Abstract text will be referred to as body sections or full text sections.

3.2 Defining the score of a token in a section

Given a token t we can compute a BM25 score s_t representing relevance of the token to a paragraph of text. The score is a product of the IDF weight and a local weighting factor that is zero if t does not occur in the paragraph. Using BM25 scoring of tokens in paragraphs, our goal is to devise a number representing the full text and its contribution to an overall document score that predicts user clicks based on each token in a query.

Since there are generally multiple paragraphs within each section of a paper, we keep the largest BM25 score for a token in a section paragraph and call it the BM25 score of the section type (*stype*) in a full text document and denote it s_t^{stype}. Keeping the maximum score is plausible because it is not affected by the size of the section (Jimmy Lin, 2009). Thus, given a token, for any document we have potentially thirteen different BM25 scores for that token, one from each section type.

Because of the structure of full text documents, the appearance of a token in different sections makes different contributions to the relevance of the document. The same BM25 score may have a different significance depending on the section. For example, a high score in the Results section would likely be more indicative of importance than if it occurred in the Methods section of a paper. To address the issue of variable significance of sections, we convert these BM25 section scores into probabilities of a document being clicked. The Pool Adjacent Violators (PAV) Algorithm (Ayer, Brunk, Ewing, Reid, & Silverman, 1954; Hardle, 1991; Wilbur, Yeganova, & Kim, 2005) is ideal for this purpose.

3.3 Training a PAV Function

Given a set of labeled data points along with their scores with the property that the higher the score the more likely a point is in the positive class, PAV is a simple and efficient algorithm that derives from such data a monotonically non-decreasing estimate of the probability that a point is in the positive class. Among non-decreasing functions that estimate the probability of a point being positive as a function of score, the PAV function assigns the highest likelihood to the actual observed class of the data points. Using training data, we apply PAV to the BM25 scores coming from each section type and obtain a function, p_{stype}, that predicts the probability of relevance. By nature of the monotonically non-decreasing estimate, the probabilities satisfy:

$$s_{t_i}^{stype} \le s_{t_j}^{stype} \Rightarrow p_{stype}\left(s_{t_i}^{stype}\right) \le p_{stype}\left(s_{t_j}^{stype}\right).$$

All scores from single tokens from queries appearing in training documents are distinct data points included for learning these PAV-derived probabilities. The stepwise linear PAV function for each of the thirteen document sections are presented in Figure 1. Results are presented in four blocks, each block comparing three body section PAV probability functions to the abstract probability function. The figures show that there is a difference between the sections in their relative importance. Given two sections, a higher BM25 token score from one section does not necessarily translate to a higher probability of relevance compared to the other section. If one section is more important for retrieval than the other, the same BM25 score in each section will lead to a higher probability in a more important section. Abrupt jumps may be due to sparseness of data This will have implications for retrieval.

The PAV-based probabilistic transformation allows one to directly compare the value of section scores to each other. A clear conclusion here is that the raw BM25 scores do not well reflect the relative importance of different body sections, as expected.

3.4 Combining Scores from Different Sections of the Body Text

Now we examine how to combine these probability scores coming from different sections into a single document score that predicts the document being relevant. Let us denote the probability of relevance given BM25 section scores as $p(rel|BM25 \text{ section scores})$. Then, the log odds ratio, defined as

$$\log\left[\frac{p(rel|BM25 \text{ section scores})}{p(\neg rel|BM25 \text{ section scores})}\right] \qquad (1)$$

is monotonically related to the probability of relevance. We apply Bayes' Theorem.

$$\log\left[\frac{p(rel|BM25 \text{ section scores})}{p(\neg rel|BM25 \text{ section scores})}\right] \qquad (2)$$
$$= \log\left[\frac{p(BM25 \text{ section scores}|rel)p(rel)}{p(BM25 \text{ section scores}|\neg rel)p(\neg rel)}\right]$$

The naïve Bayes' assumption will allow us to factor the right side of (2) as

$$\log\left[\frac{p(BM25 \text{ section scores}|rel)p(rel)}{p(BM25 \text{ secttion scores}|\neg rel)p(\neg \text{rel})}\right] \qquad (3)$$
$$= \log\left[\frac{\prod_{stype} p(s^{stype}|rel)}{\prod_{stype} p(s^{stype}|\neg rel)}\right] + \log\left[\frac{p(rel)}{p(\neg rel)}\right].$$

The second term on the right in equation 3 is a constant and can be disregarded, as it will not affect the ranking. The first term on the right side of equation 3 can be rewritten as:

$$\log\left[\frac{\prod_{stype} p(s^{stype}|rel)}{\prod_{stype} p(s^{stype}|\neg rel)}\right] \qquad (4)$$
$$= \sum_{stype} \log\left[\frac{p(rel|s^{stype})}{1 - p(rel|s^{stype})} \Big/ \frac{p(rel)}{1 - p(rel)}\right].$$

The right side of equation 4 is monotonically related to the left side of equation 2, and consequently should rank documents in the order of their probability of being relevant. This is the ideal ranking according to the probability ranking principle (Robertson, Walker, Jones, Hancock-Beaulieu, & Gatford, 1994). Here $p(rel|s^{stype}) = p_{stype}(s^{stype})$ is the PAV determined probability estimate for the section type, while $p(rel)$ is the fraction of positive documents in the training set. Based on these results we define the log odds score of a token in a section as

$$\log_{odds^{stype}(t)} = \log\left[\frac{p_{stype}(s_t^{stype})}{1-p_{stype}(s_t^{stype})} \Big/ \frac{p_{random}}{1-p_{random}}\right]. \qquad (5)$$

where $p_{random} = p(rel)$. Such scores for tokens can be added if the naive assumption of independence of the BM25 scores on which they are based is reasonably accurate.

Now we test different ways of combining scores of a token from different sections to derive a full-text score for the token. In (Jimmy Lin, 2009), the author found that computing the article score as the maximum score over all spans is superior to computing the score for an article as sum of scores over all spans. Spans in that work were paragraphs of full text documents from the TREC genomics collection, which consists of 36 topics (query questions) and manually annotated spans representing 2,477 full-text articles. In contrast, (Hearst & Plaunt, 1993) found that using the sum of scores over all spans in scoring a document produces a superior ranking when evaluated on a data set of 43 queries and 274 full text documents. Spans in (Hearst & Plaunt, 1993) are computed segments correlating with subtopics of a full text paper and are different from paragraphs.

Taking these references into consideration, we study and compare the *Sum* and *Max* scoring strategies using BM25 raw scores and log odds of BM25 scores. BM25 on Abstracts is also computed as it is used in the PubMed search system.

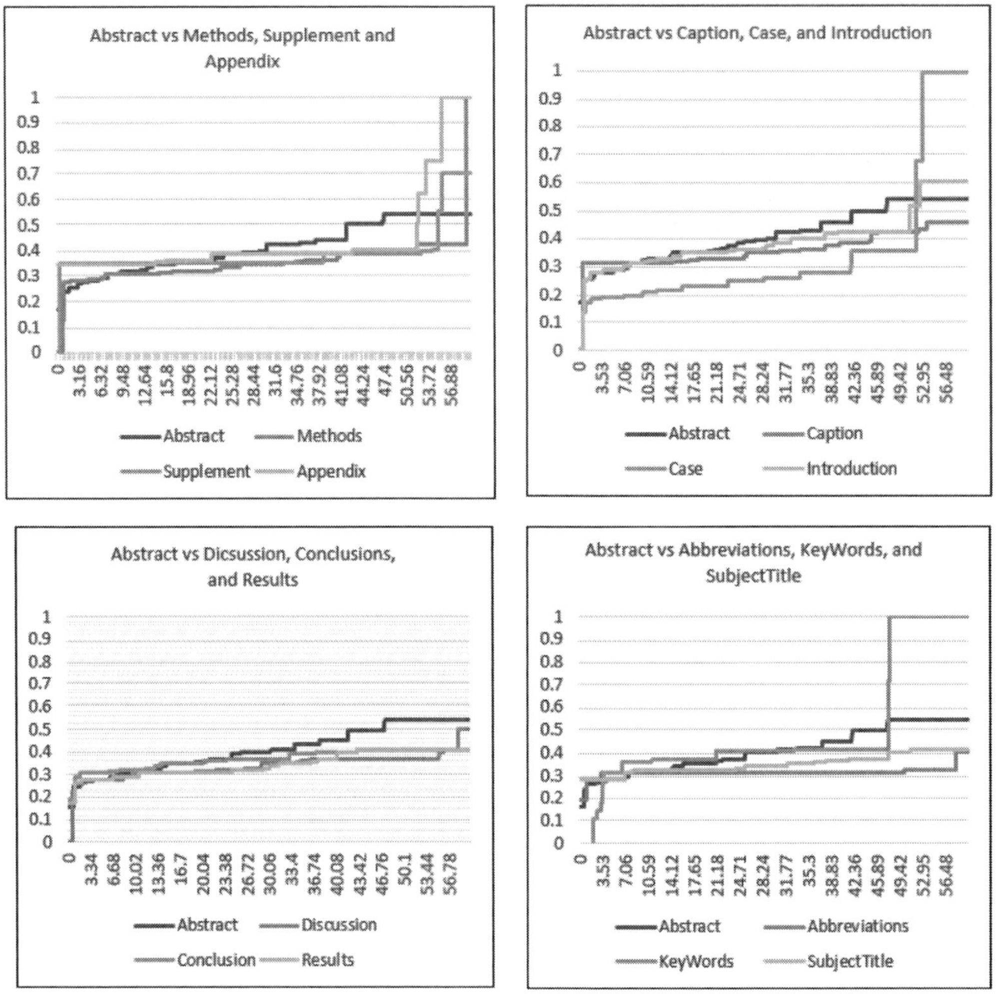

Fig 1. In these four graphs 12 PAV functions for 12 different body sections are compared to the abstract PAV function. X-axis represents the BM25 score across all four graphs, Y axis represents the Probability of a click based on the section term score.

Sum LogOdds: The score of token t in document d is computed as the sum of log odds scores, as defined in (5), coming from sections within the full text document:

$$sco_{Sum\_LogOdds}(d,t) = \sum_{stype \in d} \log\_odds^{stype}(t) \quad (6)$$

Max LogOdds: The score of token t in document is computed as the maximum log odds score coming from sections within the full text document:

$$sco_{Max\_LogOdds}(d,t) = \max\{\log\_odds^{stype}(t) \mid stype \in d\} \quad (7)$$

Abstract BM25: The score of token t in document d is computed as the raw BM25 token score of the abstract

$$sco_{BM25\_Abs}(d,t) = s_t^{abs} \quad (8)$$

Sum BM25: The score of token t in document d is computed as the sum of BM25 section token scores within the full text document

$$sco_{Sum\_BM25}(d,t) = \sum_{stype \in d} s_t^{stype} \quad (9)$$

Max BM25: The score of token t in document d is computed as the highest BM25 section token score within the full text document

$$sco_{Max\_BM25}(d,t) = max_{stype \in d}\{s_t^{stype}\} \quad (10)$$

After trying scoring based directly on log odds using formulas (6) and (7), it was evident that we are dealing with two kinds of documents, which behave differently. Those documents that contain the search token only in the abstract receive a single score from the abstract, and *Sum* and *Max* really don't play a role. But for those documents having the term in multiple sections, *Sum* and *Max* do play a role, and the log odds scores are higher. In order to balance the scores for best results, we found it necessary to create PAV curves for *Sum* and *Max* scores just on documents with multiple sections providing scores. We simply use the probabilities based on a PAV curve for each type of document to rank the different types in the same ranking for retrieval. In what follows, we will continue to use the term LogOdds to refer to this scoring.

4 Results

Proposed methods are tested on the PubMed Click Dataset and on the TREC Genomics collection (Hersh et al., 2007).

4.1 The PubMed Click Dataset

To directly measure the benefit of full text, for each query in the PubMed Click Dataset we first compare the proposed scoring techniques on Set_FT. Set_FT is a subset of the PubMed Click dataset that includes queries for which all labeled documents in the evaluation dataset have full text available. Second, we extend this analysis to the whole test portion of the PubMed Click dataset. It contains queries and labeled documents, which may or may not have full text available. For each query token, we score its corresponding retrieved documents in the evaluation dataset and compute the average Precision using labels in the evaluation dataset. These are averaged over all tokens in a query, and then average over all queries producing the MAP results presented in Figure 2.

Figure 2 demonstrates our findings computed on the complete set of tokens available in the two test sets. We observe that the LogOdds probabilistic scoring approach significantly outperforms the BM25 scoring for both *Sum* and *Max* variants for the PubMed click data and Set_FT. A bigger difference is observed on Set_FT, where full text is available for every participating document. Additionally, we observe that LogOdds Sum computed on article full text outperforms the abstract score and the difference although small is statistically significant.

We conducted pairwise statistical tests for all methods to verify if the differences in performance for each pair of tests is significant. We used the "Percentile bootstrap" test at the 5% significance level which works well for our study because the distribution is symmetric around the MAP value (https://en.wikipedia.org/wiki/Bootstrapping). Differences between all pairs of methods are statistically significant, except for the Max LogOdds and the Abs BM25 for the Set_FT subset of PubMed Click Dataset.

Based on these results we believe that log odds scoring is a useful approach for retrieval incorporating body text. The intuition behind it is that BM25 scores have a different meaning depending on the sections from which they are derived as illustrated in Fig 1. For a single query token, results in Figure 2 also suggest that the *Sum* scoring approach provides a better estimate of token importance than the *Max* scoring approach when using the log odds scoring for the Click dataset. If sections within a full text document were truly independent from each other, Sum LogOdds would be the ideal method to score a single query token over the multiple sections in a document.

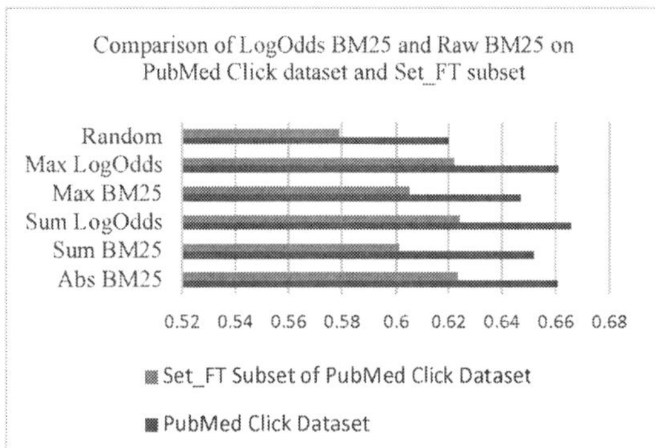

Figure 2. Average Precision for all query tokens is computed, averaged for each query and then over all queries for the PubMed Click dataset and its subset Set_FT. For both datasets, LogOdds Sum and LogOdds Max scoring methods demonstrate a significantly improved performance compared to Sum and Max on raw BM25 scores.

4.2 The TREC Genomics Dataset

We apply the proposed methods to each query token in the TREC dataset. We score the retrieved documents in the evaluation dataset and compute the Average Precision using gold standard labels. These are then averaged over all query tokens, and the MAP results are presented in Figure 3. Leave-one-out training strategy was used for each topic.

Figure 3 demonstrates our findings computed on non-stop word query tokens in the TREC Genomics Dataset. We observe that the Sum LogOdds probabilistic scoring significantly outperforms Sum BM25 scoring. Similarly, the Max LogOdds probabilistic scoring significantly outperforms Max BM25 scoring. Similar to the PubMed Click Dataset, here we observe that Sum LogOdds has a slight advantage over Max LogOdds, and both are competitive with the abstract BM25 score.

We conducted Wilcoxon signed rank test (https://en.wikipedia.org/wiki/Wilcoxon_signed-rank_test) at 5% significance level to verify if the differences in performance for each pair of tests is significant. The differences between Max LogOdds and Abs BM25 as well as Sum LogOdds and Abs BM25 are not statistically significant. The differences between all other pairs of methods are statistically significant.

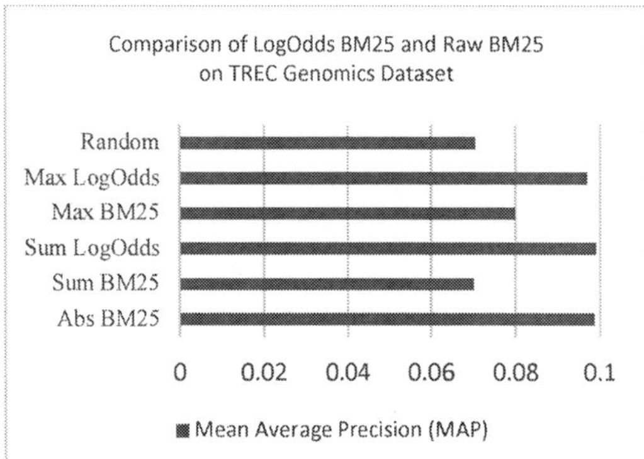

Figure 3. Mean Average Precision on TREC Genomics Dataset is computed on single tokens and averaged for all tokens in the experiment. Sum LogOdds and Max LogOdds demonstrate a significantly improved performance compared to those on raw BM25 scores.

5 Conclusions and Discussion

Based on the PubMed Click dataset and the TREC genomics dataset, we studied how to integrate full text and abstract information for scoring a query token. The main contribution of this work is to study the benefits of log odds of BM25 compared to raw BM25 scores. Our experimental results on both datasets support these important conclusions:

1. PAV based log odds scoring is a useful way to compare the contribution of a token in different sections of a document for predicting clicks. BM25 scores are not directly comparable with each other for making such predictions. The same BM25 score is of different value depending on the section type in which it is found.

2. We proposed two methods to compute the log odds body score by taking the sum or max of scores. In both cases, PAV based LogOdds scoring is significantly better than ranking based on raw BM25 scores. The difference between *Sum* and *Max* scoring is small.

For the PubMed Click dataset, using the Sum LogOdds score from the whole document for a query token produces better results than using only the abstract score. In the TREC genomics dataset, the performance of full text LogOdds is comparable to abstract only score. This is an important contribution and meaningful building block towards improving full text retrieval in PubMed. Our immediate plan is to extend this single token analysis to full queries.

6 Conclusions and Discussion

Based on the PubMed Click dataset and the TREC genomics dataset, we studied how to integrate full text and abstract information for scoring a query token. The main contribution of this work is to study the benefits of log odds of BM25 compared to raw BM25 scores. Our experimental results on both datasets support these important conclusions:

1. PAV based log odds scoring is a useful way to compare the contribution of a token in different sections of a document for predicting clicks. BM25 scores are not directly comparable with each other for making such predictions. The same BM25 score is of different value depending on the section type in which it is found.

2. We proposed two methods to compute the log odds body score by taking the sum or max of scores. In both cases, PAV based LogOdds scoring is significantly better than ranking based on raw BM25 scores. The difference between *Sum* and *Max* scoring is small.

For the PubMed Click dataset, using the Sum LogOdds score from the whole document for a query token produces better results than using only the abstract score. In the TREC genomics dataset, the performance of full text LogOdds is comparable to abstract only score. This is an important contribution and meaningful building block towards improving full text retrieval in PubMed. Our immediate plan is to extend this single token analysis to full queries.

FUNDING

This work was supported by the Intramural Research Program of the National Library of Medicine, National Institutes of Health.

References

Allot, A., Chen, Q., Kim, S., Vera Alvarez, R., Comeau, D. C., Wilbur, W. J., & Lu, Z. (2019). LitSense: making sense of biomedical literature at sentence level. *Nucleic Acids Research, 47*(Web Server issue).

Ayer, M., Brunk, H., Ewing, G., Reid, W., & Silverman, E. (1954). An empirical distribution function for sampling with incomplete information. *Ann Math Stat, 26*, 641-647.

Blanco, R., & Zaragoza, H. (2010). Finding Support Sentences for Entities. *SIGIR '10 Proceedings of the 33rd international ACM SIGIR conference on Research and development in information retrieval*

Cejuela, J., McQuilton, P., Ponting, L., Marygold, S., Stefancsik, R., Millburn, G., . . . Consortium, F. (2014). tagtog: interactive and text-mining-assisted annotation of gene mentions in PLOS full-text articles. *Database (Oxford)*. doi:10.1093/database/bau033. Print 2014.

Chen, J., & Hersh, W. R. (2020). A Comparative Analysis of System Features Used in the TREC-COVID Information Retrieval Challenge (Publication no. https://doi.org/10.1101/2020.10.15.20213645).

Cohen, K. B., Johnson, H., Verspoor, K., Roeder, C., & Hunter, L. (2010). The structural and content aspects of abstracts versus bodies of full text journal articles are different. *BMC Bioinformatics, 11*(492).

Comeau, D. C., Wei, C.-H., Doğan, R. I., & Z., L. (2019). PMC text mining subset in BioC: about 3 million full text articles and growing. *Bioinformatics, doi:10.1093/bioinformatics/btz070.*

Fiorini, N., Canese, K., Starchenko, G., Kireev, E., Kim, W., Miller, V., . . . Lu, Z. (2018). Best Match: New relevance search for PubMed. *PLOS Biology, 16*(8). doi:doi: 10.1371/journal.pbio.2005343

Fiorini, N., Leaman, R., Lipman, D. J., & Lu, Z. (2018). How user intelligence is improving PubMed. *Nature Biotechnology, 36*, 937–945.

Guo, J., Fan, Y., Pang, L., Yang, L., Ai, Q., Zamani, H., . . . Cheng, X. (2020). A Deep Look into Neural Ranking Models for Information Retrieval. *Journal of Information Processing and Management, 57*(6).

Hardle, W. (1991). *Smoothing techniques: with implementation in S*. New York: Springer-Verlag.

Hearst, M. A., & Plaunt, C. (1993). *Subtopic structuring for full-length document access*. Paper presented at the SIGIR93: 16th International ACM/SIGIR '93 Conference on Research and Development in Information Retrieval, Pittsburgh PA USA.

Hersh, W., Cohen, A., Ruslen, L., & Roberts, P. (2007). TREC 2007 Genomics Track Overview *Proceedings of the Sixteenth Text REtrieval Conference (TREC 2007).*

Kafkas, Ş., Pi, X., Marinos, N., Talo', F., Morrison, A., & McEntyre, J. R. (2015). Section level search functionality in Europe PMC. *Journal of Biomed Semantics*. doi:doi: 10.1186/s13326-015-0003-7

Kim, J., Kim, J., Han, X., & Rebholz-Schuhmann, D. (2015). Extending the evaluation of Genia Event task toward knowledge base construction and comparison to Gene Regulation Ontology task. *BMC Bioinformatics*. doi:10.1186/1471-2105-16-S10-S3. Epub 2015 Jul 13.

Kim, W., Yeganova, L., Comeau, D. C., Wilbur, W. J., & Lu, Z. (2018). MeSH-based dataset for measuring the relevance of text retrieval. *Proceedings of the BioNLP 2018 workshop.*

Lin, J. (2009). Is searching full text more effective than searching abstracts? *BMC Bioinformatics, 10*(46).

Lin, J., Ma, X., Lin, S.-C., Yang, J.-H., Pradeep, R., & Nogueira, R. (2021). Pyserini: An Easy-to-Use Python Toolkit to Support Replicable IR Research with Sparse and Dense Representations.

Resnick, A. (1961). Relative effectiveness of document titles and abstracts for determining relevance of documents. *Science, 134*(3484), pp. 1004–1006.

Robertson, S., Walker, S., Jones, S., Hancock-Beaulieu, M., & Gatford, M. (1994). Okapi at TREC-3. *Proceedings of the 3rd Text REtrieval Conference (TREC-3).*

Sarrouti, M., & El Alaoui, S. O. (2017). A passage retrieval method based on probabilistic information retrieval model and UMLS concepts in biomedical question answering. . *Journal of Biomedical Informatics, 68.*

Voorhees, E. (2001). The philosophy of information retrieval evaluation. *CLEF 2001: Evaluation of Cross-Language Information Retrieval Systems, Volume 2406*, pp. 355–370.

Wei, C.-H., Allot, A., Leaman, R., & Lu, Z. (2019). PubTator central: automated concept annotation for biomedical full text articles. *Nucleic Acids Research, 47*(W1).

Westergaard, D., Stærfeldt, H.-H., Tønsberg, C., Jensen , L. J., & Brunak, S. (2018). A comprehensive and quantitative comparison of text-mining in 15 million full-text articles versus their corresponding abstracts. *Plos Computational Biology, 14*(2).

Wilbur, W. J., Yeganova, L., & Kim, W. (2005). The Synergy Between PAV and AdaBoost. *Machine Learning, 61*, 71-103.

Zhang, E., Gupta, N., Tang, R., Han, X., Pradeep, R., Lu, K., . . . Lin, J. (2020). Covidex: Neural Ranking Models and Keyword Search Infrastructure for the COVID-19 Open Research Dataset. *Proceedings of the First Workshop on Scholarly Document Processing*.

UCSD-Adobe at MEDIQA 2021: Transfer Learning and Answer Sentence Selection for Medical Summarization

Khalil Mrini[1], Franck Dernoncourt[2], Seunghyun Yoon[2], Trung Bui[2], Walter Chang[2], Emilia Farcas[1], and Ndapa Nakashole[1]

[1]University of California, San Diego, La Jolla, CA 92093
{khalil, efarcas, nnakashole}@ucsd.edu
[2]Adobe Research, San Jose, CA 95110
{franck.dernoncourt, syoon, bui, wachang}@adobe.com

Abstract

In this paper, we describe our approach to question summarization and multi-answer summarization in the context of the 2021 MEDIQA shared task (Ben Abacha et al., 2021). We propose two kinds of transfer learning for the abstractive summarization of medical questions. First, we train on HealthCareMagic, a large question summarization dataset collected from an online healthcare service platform. Second, we leverage the ability of the BART encoder-decoder architecture to model both generation and classification tasks to train on the task of Recognizing Question Entailment (RQE) in the medical domain. We show that both transfer learning methods combined achieve the highest ROUGE scores. Finally, we cast the question-driven extractive summarization of multiple relevant answer documents as an Answer Sentence Selection (AS2) problem. We show how we can preprocess the MEDIQA-AnS dataset such that it can be trained in an AS2 setting. Our AS2 model is able to generate extractive summaries achieving high ROUGE scores.

1 Introduction

The 2021 **Medi**cal NLP and **Q**uestion **A**nswering (MEDIQA) shared task (Ben Abacha et al., 2021) is comprised of three tasks, centered around summarization in the medical domain: Question Summarization, Multi-Answer Summarization, and Radiology Report Summarization. In this paper, we focus on the first two tasks. In Question Summarization, the goal is to generate a one-sentence formal question summary from a consumer health question – a relatively long question asked by a user. In Multi-Answer Summarization, we are given a one-sentence question and multiple relevant answer documents, and the aim is to compose a question-driven summary from the answer text.

In this paper, we first show that transfer learning from pre-trained language models can achieve very high results for question summarization. Sequence-to-sequence language model BART (Lewis et al., 2020) has achieved state-of-the-art results in various NLP benchmarks, including in the CNN-Dailymail news article summarization dataset (Hermann et al., 2015). We leverage this success and train BART on summarization datasets from the medical domain (Ben Abacha and Demner-Fushman, 2019; Zeng et al., 2020; Mrini et al., 2021). Moreover, we find that training on a different task in the medical domain – Recognizing Question Entailment (RQE) (Ben Abacha and Demner-Fushman, 2016) – can yield better improvements, especially in terms of ROUGE precision scores.

Second, we tackle the extractive track of the multi-answer summarization task, and we cast multi-answer extractive summarization as an Answer Sentence Selection (AS2) problem. A limitation of BART is that the input to its abstractive summarization cannot be as long as the multiple documents in this task. We therefore propose to mitigate this weakness by proposing to cut up the input into pairs of sentences, where the first sentence is the input question, and the second one is a candidate answer. We then train our BART model to score the relevance of each candidate answer with regards to its corresponding question. We also describe in this paper the algorithm used to extract an AS2 dataset from an multi-document extractive summarization dataset.

2 Question Summarization

Our approach to question summarization involves two kinds of transfer learning. First, we train our model to learn from medical summarization datasets. Second, we show that transfer learning from other tasks in the medical domain increases ROUGE scores.

Proceedings of the BioNLP 2021 workshop, pages 257–262
June 11, 2021. ©2021 Association for Computational Linguistics

2.1 Training Details

We adopt the BART Large architecture (Lewis et al., 2020), as it set a state of the art in abstractive summarization benchmarks, and allows us to train a single model on generation and classification tasks.

We use a base model, which is trained on BART's language modeling tasks and the XSum abstractive summarization dataset (Narayan et al., 2018). We use a learning rate of $3 * 10^{-5}$ for summarization tasks and $1 * 10^{-5}$ for the recognizing question entailment task. We use 512 as the maximum number of token positions.

Following the MEDIQA instructions and leaderboard, we use precision, recall and F1 scores for the ROUGE-1, ROUGE-2 and ROUGE-L metrics (Lin, 2004).

2.2 Transfer Learning from Medical Summarization

2.2.1 Summarization Datasets

In addition to the XSum base model, we train on two additional datasets. The first dataset is MeQSum (Ben Abacha and Demner-Fushman, 2019). It is an abstractive medical question summarization dataset, which consists of 1,000 consumer health questions (CHQs) and their corresponding one-sentence-long frequently asked questions (FAQs). It was released by the U.S. National Institutes of Health (NIH), and the FAQs are written by medical experts. Whereas Ben Abacha and Demner-Fushman (2019) use the first 500 datapoints for training and the last 500 for testing, participants in this shared task are encouraged to use the entire MeQSum dataset for training.

We also use the HealthCareMagic (HCM) dataset. It is also a medical question summarization dataset, but it is a large-scale dataset consisting of 181, 122 training instances. In contemporaneous work of ours (Mrini et al., 2021), we extract this dataset from the MedDialog dataset (Zeng et al., 2020), a medical dialog dataset collected from `HealthCareMagic.com` and `iCliniq.com`, two online platforms of healthcare service.

The dialogues in the MedDialog dataset consist of a question from a user, a response from a doctor or medical professional, and a summary of the question from the user. We form a question summarization dataset by taking the user question and its corresponding summary, and we discard the answers. We choose to work with HealthCareMagic as the questions are abstractive and resemble the formal style in the FAQs of the U.S. National Library of Medicine (NLM), whereas iCliniq question summaries are noisier and more extractive.

Given that MeQSum is 180 times smaller than HealthCareMagic, we train for 100 epochs on MeQSum, and 10 epochs for HealthCareMagic. We use the validation set of the MEDIQA question summarization task to select the best parameters.

2.2.2 Results and Discussion

We show the validation results in Table 1 and the test results in Table 2. In all test results, we follow approaches of 2019 MEDIQA participants (Zhu et al., 2019), and add the validation set to training for the leaderboard submissions only.

We notice that the validation results for the BART + XSum base model are significantly lower than other models. The corresponding test results are also the lowest-ranking, even though the difference is not as large as we trained on the validation set. These results show that training on an out-of-domain abstractive summarization dataset is not efficient for this task.

We consider now the training on the medical question summarization datasets. First, the validation results show that training on MeQSum achieves comparable F1 scores as training on HealthCareMagic. The main contrasting point is that training on HealthCareMagic yields higher precision, whereas training on MeQSum yields higher recall. This means that training on HealthCareMagic generates summaries with more relevant content, whereas training on MeQSum generates summaries with higher coverage of the content of the reference summaries. However, the corresponding test results show similar recall, but higher precision for HealthCareMagic. Accordingly, ROUGE F1 test scores are higher when training with HealthCareMagic compared to training with MeQSum.

Finally, we consider the results of training on HealthCareMagic followed by MeQSum (HCM + MeQSum). On the validation set, we notice this method generally scores lower precision than just training on HealthCareMagic, but significantly higher recall than any previous training method, therefore achieving higher F1 across all three ROUGE metrics. On the test set, scores are generally comparable with training on HealthCareMagic only.

Metric →	ROUGE-1			ROUGE-2			ROUGE-L		
Model ↓	**P**	**R**	**F1**	**P**	**R**	**F1**	**P**	**R**	**F1**
BART + XSum	14.64	27.59	18.48	4.73	9.16	5.97	12.26	23.11	15.46
BART + XSum + MeQSum	27.08	37.05	30.46	10.66	14.43	11.92	25.03	34.37	28.20
BART + XSum + HCM	35.33	27.81	29.64	14.56	10.22	11.40	33.82	26.31	28.16
BART + XSum + HCM + MeQSum	32.14	**40.80**	**35.22**	14.84	18.01	15.92	28.94	**36.66**	31.66
BART + XSum + HCM + RQE	**38.86**	32.97	34.10	**20.31**	15.69	**16.88**	**37.89**	31.98	**33.15**
BART + XSum + HCM + RQE + MeQSum	31.81	40.22	34.52	14.60	**18.22**	15.78	28.82	36.57	31.29

Table 1: Validation results for Question Summarization. HCM is the HealthCareMagic dataset, and RQE is the Recognizing Question Entailment dataset.

Metric →	ROUGE-1			ROUGE-2			ROUGE-L		
Model ↓	**P**	**R**	**F1**	**P**	**R**	**F1**	**P**	**R**	**F1**
BART + XSum	28.89	32.86	29.56	10.78	12.19	10.94	26.16	29.65	26.71
BART + XSum + MeQSum	29.88	34.73	30.70	11.69	13.16	11.87	26.71	30.82	27.38
BART + XSum + HCM	31.83	34.31	31.61	13.21	13.81	12.82	28.58	30.75	28.32
BART + XSum + HCM + MeQSum	31.85	35.58	32.00	12.77	13.59	12.51	28.41	31.68	28.53
BART + XSum + HCM + RQE	33.58	35.43	32.65	**14.23**	14.16	13.46	29.51	31.06	28.73
BART + XSum + HCM + RQE + MeQSum	**33.82**	**39.10**	**34.63**	13.91	**15.80**	**14.14**	**29.91**	**34.62**	**30.65**

Table 2: Test results for Question Summarization. All models are trained on the provided validation set as well.

2.3 Transfer Learning from Medical Question Entailment

We consider transfer learning using another task in the medical domain: Recognizing Question Entailment (RQE). Ben Abacha and Demner-Fushman (2016) introduce the RQE task as a binary classification problem, where the goal is to predict whether – given two questions A and B – A entails B. Ben Abacha and Demner-Fushman (2016) further define question entailment as the following: question A entails question B if every answer to B is a correct answer to A, whether partially or fully.

The BART architecture enables us to train on the RQE task using the checkpoint of the question summarization models. BART is an encoder-decoder model that can train, on top of generation tasks, classification tasks as well, such as RQE. We feed the entire RQE question pair as input to both the encoder and the decoder. We add a classification head to be able to predict the entailment score.

2.3.1 Entailment Dataset

For the RQE task, we use the RQE dataset from the 2019 MEDIQA shared task (Ben Abacha et al., 2019). The training set was introduced in Ben Abacha and Demner-Fushman (2016). Similarly to MeQSum, this dataset is released by the U.S. National Institutes of Health. The MEDIQA-RQE dataset contains 8,588 training question pairs. We train for 10 epochs and choose the best parameters using the validation set of the 2021 MEDIQA question summarization task.

2.3.2 Results and Discussion

Similarly to training on HealthCareMagic, we notice in Table 1 that the validation set for training on MEDIQA-RQE yields very high precision scores. This method produces the highest precision scores across all trialled methods, and achieves the highest F1 scores for ROUGE-2 and ROUGE-L. Adding MeQSum to the training (RQE + MeQSum) seems to decrease precision, increase recall, achieve similar ROUGE-1 F1, but lower ROUGE-2 and ROUGE-L F1 scores.

In Table 2, we notice that the test results that the RQE + MeQSum model is the clear winner, providing the highest scores across the board, with the exception of ROUGE-2 precision. Overall, it seems that pre-training on a similar task in the medical domain is beneficial for this medical question summarization task.

3 Multi-Answer Extractive Summarization

3.1 Dataset

The dataset for this task is the MEDIQA-AnS dataset (Savery et al., 2020). It contains 156 user-written medical questions, and answer articles to these questions, such that one question usually has more than one answer article. There are also manually-written abstractive and extractive summaries for the individual answer articles, as well as for the overall question.

3.2 Casting as Answer Sentence Selection

Given that state-of-the-art summarizer BART can only take relatively short sequences of text as input, we cannot summarize directly from the long answer articles to generate the overall answer summary. We considered summarizing in stages: first training BART to generate summaries for individual answer articles, and then summarize the concatenation of those summaries to generate the answer summary for the user question. However, we only have reference summaries of individual answer articles in the training set of this task, not in the validation or test set. We notice that extractive answer summaries for questions consist of sentences extracted fully from the answer articles. Therefore, we decide to tackle the extractive track of this task, and cast multi-answer extractive summarization as an Answer Sentence Selection (AS2) problem. Similarly to RQE, AS2 is a binary classification task, and as such we are able to train it using BART.

In the AS2 setting, we train BART to predict the relevance score of a candidate answer given a question. To obtain the pairs of questions and candidate answers from the MEDIQA-AnS dataset, we proceed as follows. First, we concatenate for each question the text data of its corresponding answer articles. Then, we use the NLTK sentence tokenizer (Loper and Bird, 2002) to split this text data into individual sentences. Finally, we form question-sentence pairs for AS2 by pairing the user question with each sentence from the corresponding answer article text data.

In this training context, AS2 is a binary classification task, where each pair of question and candidate answer is labeled as relevant (1) or irrelevant (0). We use cross-entropy as the loss function. We label sentences contained in the reference extractive summary as relevant. We notice that some sentences in the reference summary may appear slightly changed in the answer articles, or in exceptional cases may not appear at all. We decide to allow a margin of difference between a reference summary sentence and an answer article sentence, such that if the max-normalized Levenshtein distance between both sentences is 25% or less, we consider the answer article sentence to be relevant. In the rare cases when the reference summary sentence does not appear at all in the answer articles, we add it to our training set and label the sentence as relevant. We show the statistics of the resulting dataset in Table 3.

Set	# sentences	# relevant	% relevant
Train	48,317	3,995	8.27
Dev	2,494	692	27.8

Table 3: Statistics for MEDIQA-AnS cast as an Answer Sentence Selection dataset.

Metric → / Model ↓	Acc.	MAP	MRR
BART + XSum + MEDIQA-AnS	71.52	58.63	68.61
BART + XSum + HCM + RQE + MeQSum + MEDIQA-AnS	72.09	57.08	68.52

Table 4: Validation results for Multi-Answer Extractive Summarization, cast as an Answer Sentence Selection problem. We use accuracy and Information Retrieval metrics like Mean Average Precision (MAP) and Mean Reciprocal Rank (MRR).

3.3 Results and Discussion

In Answer Sentence Selection, we use two Information Retrieval metrics for evaluation: Mean Average Precision (MAP) and Mean Reciprocal Rank (MRR). MAP measures how many of the top-ranked answers are relevant, whereas MRR measures how highly a first relevant answer is ranked. We compute the scores as follows, given a set Q of questions:

$$\text{MAP}(Q) = \frac{\sum_{q \in Q} \text{average\_precision}(q)}{|Q|} \quad (1)$$

$$\text{MRR}(Q) = \frac{\sum_{q \in Q} \frac{1}{\text{rank}(q)}}{|Q|} \quad (2)$$

We take as base models the BART + XSum model, as well as the best-performing model in the test set of the question summarization task, as shown in Table 2. We train for 10 epochs on the AS2 version of the MEDIQA-AnS dataset. We show classification and AS2 validation results in Table 4. We notice that both models perform somewhat similarly. Accuracy, MAP and MRR scores are independent of the extractive summary.

We now evaluate the same two models on Multi-Answer Summarization. To form an extractive summary of k sentences, we concatenate the top k most relevant sentences, in the order in which they appeared in the answer articles. We consider two options. First, we generate extractive summaries of

Metric →		ROUGE-1			ROUGE-2			ROUGE-L		
Model ↓	**# sentences ↓**	**P**	**R**	**F1**	**P**	**R**	**F1**	**P**	**R**	**F1**
BART + XSum + MEDIQA-AnS	Same as ref.	**70.89**	61.48	**65.17**	**53.82**	47.43	**49.99**	**40.28**	34.86	37.00
BART + XSum + MEDIQA-AnS	11	65.13	66.65	61.10	50.45	54.37	48.49	36.57	39.26	35.00
BART + XSum + HCM + RQE + MeQSum + MEDIQA-AnS	Same as ref.	68.53	63.28	65.06	52.09	48.41	49.65	40.10	36.40	**37.77**
BART + XSum + HCM + RQE + MeQSum + MEDIQA-AnS	11	61.84	**67.83**	60.52	46.72	**54.57**	47.08	35.64	**40.53**	35.36

Table 5: Validation results for Multi-Answer Extractive Summarization.

Metric →		ROUGE-1			ROUGE-2			ROUGE-L		
Model ↓	**# sentences ↓**	**P**	**R**	**F1**	**P**	**R**	**F1**	**P**	**R**	**F1**
BART + XSum + MEDIQA-AnS	11	**61.57**	**67.19**	**60.74**	**47.33**	**53.09**	**47.20**	**43.27**	**48.07**	**42.90**
BART + XSum + HCM + RQE + MeQSum + MEDIQA-AnS	11	59.74	66.34	59.22	45.87	52.21	45.95	42.08	46.98	41.70

Table 6: Test results for Multi-Answer Extractive Summarization.

the same number of sentences as the corresponding reference extractive summary. Second, we generate extractive summaries of 11 sentences, as the average number of sentences in the reference extractive summaries is 10.66. We show validation results in Table 5 and test results in Table 6. For the test results, we are not able to match the number of sentences since we do not have access to the reference summaries. In addition, we train on the validation set as well to report test results, following the approach of MEDIQA 2019 participants (Zhu et al., 2019).

The summarization results on the validation set show that extractive summaries with the same number of sentences as the corresponding reference summaries have higher precision, whereas the 11-sentence extractive summaries have higher recall. Overall, the model trained on BART + XSum fares better than the one fine-tuned on top of question summarization. The test results in Table 6 display the same trend, as the model trained on BART + XSum achieves higher scores across the board. It seems that for this task, transfer learning from other medical datasets was not as useful as for medical question summarization.

4 Conclusions

This paper describes the approach taken by our team, UCSD-Adobe, at the 2021 MEDIQA shared task. We tackle the tasks of question summarization and multi-answer summarization.

For question summarization, we propose two kinds of transfer learning. First, we propose to pre-train on a large-scale dataset of abstractive summarization of medical questions, HealthCareMagic. Our results show that training on this dataset enhances performance in both validation and test sets. Then, we propose to transfer from another medical question-based task: recognizing question entailment. This binary classification task increases performance, and precision scores in particular. In the test results, the highest ROUGE scores are achieved by a model trained on both transfer learning methods.

We tackle the extractive track of the multi-answer summarization task. We propose to cast the question-driven extractive summarization of multiple answer documents as an answer sentence selection problem. We show how we can transform the MEDIQA-AnS dataset into an AS2 dataset. We show that we achieve good ROUGE scores with and without transfer learning from question summarization on the validation set. In the test results, the model without question summarization training achieves the highest ROUGE scores.

Acknowledgments

We gratefully acknowledge the award from NIH/NIA grant R56AG067393. Khalil Mrini is additionally supported by unrestricted gifts from Adobe Research. We thank the anonymous reviewers for their feedback.

References

Asma Ben Abacha and Dina Demner-Fushman. 2016. Recognizing question entailment for medical question answering. In *AMIA Annual Symposium Proceedings*, volume 2016, page 310. American Medical Informatics Association.

Asma Ben Abacha and Dina Demner-Fushman. 2019. On the summarization of consumer health questions. In *Proceedings of the 57th Annual Meeting of the Association for Computational Linguistics*, pages 2228–2234.

Asma Ben Abacha, Yassine Mrabet, Yuhao Zhang, Chaitanya Shivade, Curtis Langlotz, and Dina Demner-Fushman. 2021. Overview of the mediqa 2021 shared task on summarization in the medical domain. In *Proceedings of the 20th SIG-BioMed Workshop on Biomedical Language Processing, NAACL-BioNLP 2021*. Association for Computational Linguistics.

Asma Ben Abacha, Chaitanya Shivade, and Dina Demner-Fushman. 2019. Overview of the mediqa 2019 shared task on textual inference, question entailment and question answering. In *Proceedings of the 18th BioNLP Workshop and Shared Task*, pages 370–379.

Karl Moritz Hermann, Tomás Kociský, Edward Grefenstette, Lasse Espeholt, Will Kay, Mustafa Suleyman, and Phil Blunsom. 2015. Teaching machines to read and comprehend. In *NIPS*, pages 1693–1701.

Mike Lewis, Yinhan Liu, Naman Goyal, Marjan Ghazvininejad, Abdelrahman Mohamed, Omer Levy, Veselin Stoyanov, and Luke Zettlemoyer. 2020. Bart: Denoising sequence-to-sequence pre-training for natural language generation, translation, and comprehension. In *Proceedings of the 58th Annual Meeting of the Association for Computational Linguistics*, pages 7871–7880.

Chin-Yew Lin. 2004. Rouge: A package for automatic evaluation of summaries. In *Text summarization branches out*, pages 74–81.

Edward Loper and Steven Bird. 2002. Nltk: The natural language toolkit. In *Proceedings of the ACL-02 Workshop on Effective Tools and Methodologies for Teaching Natural Language Processing and Computational Linguistics*, pages 63–70.

Khalil Mrini, Franck Dernoncourt, Walter Chang, Emilia Farcas, and Ndapa Nakashole. 2021. Joint summarization-entailment optimization for consumer health question understanding. In *Proceedings of the Second Workshop on Natural Language Processing for Medical Conversations, NAACL-NLPMC 2021*, Online. Association for Computational Linguistics.

Shashi Narayan, Shay B Cohen, and Mirella Lapata. 2018. Don't give me the details, just the summary! topic-aware convolutional neural networks for extreme summarization. In *Proceedings of the 2018 Conference on Empirical Methods in Natural Language Processing*, pages 1797–1807.

Max Savery, Asma Ben Abacha, Soumya Gayen, and Dina Demner-Fushman. 2020. Question-driven summarization of answers to consumer health questions. *Scientific Data*, 7(1):1–9.

Guangtao Zeng, Wenmian Yang, Zeqian Ju, Yue Yang, Sicheng Wang, Ruisi Zhang, Meng Zhou, Jiaqi Zeng, Xiangyu Dong, Ruoyu Zhang, et al. 2020. Meddialog: A large-scale medical dialogue dataset. In *Proceedings of the 2020 Conference on Empirical Methods in Natural Language Processing (EMNLP)*, pages 9241–9250.

Wei Zhu, Xiaofeng Zhou, Keqiang Wang, Xun Luo, Xiepeng Li, Yuan Ni, and Guotong Xie. 2019. Panlp at mediqa 2019: Pre-trained language models, transfer learning and knowledge distillation. In *Proceedings of the 18th BioNLP Workshop and Shared Task*, pages 380–388.

ChicHealth @ MEDIQA 2021: Exploring the limits of pre-trained seq2seq models for medical summarization

Liwen Xu, Yan Zhang, Lei Hong, Yi Cai*, Szui Sung
Chic Health, Shanghai, China

Abstract

In this article, we will describe our system for MEDIQA2021 shared tasks. First, we will describe the method of the second task, multiple answer summary (MAS). For extracting abstracts, we follow the rules of Xu and Lapata (2020). First, the candidate sentences are roughly estimated by using the Roberta model. Then the Markov chain model is used to evaluate the sentences in a fine-grained manner. Our team won the first place in overall performance, with the fourth place in MAS task, the seventh place in RRS task and the eleventh place in QS task. For the QS and RRS tasks, we investigate the performanceS of the end-to-end pre-trained seq2seq model. Experiments show that the methods of adversarial training and reverse translation are beneficial to improve the fine tuning performance.

1 Introduction

The mediqa 2021 shared tasks aim to investigate the most advanced summary models, especially their performance in the medical field. There are three tasks. The first is question summary (QS), which classifies long and complex consumer health problems into simple ones, which has been proved to be helpful to answer questions automatically (Abacha and Demner-Fushman, 2019). The second task is multiple answer summary (MAS) (Savery et al., 2020). Different answers can bring complementary views, which may benefit the users of QA system. The goal of this task is to develop a system that can aggregate and summarize answers scattered across multiple documents. The third task is radiology report summary (RRs) (Zhang et al., 2018, 2020b), which generates radiology impression statements by summarizing the text results written by radiologists.

Automatic summarization is an important task in the field of medicine. When users use Google, MEDLINE and other search engines, they need to read a large number of medical documents about a certain topic and get a list of possible answers, which is very time-consuming. First, the content may be too specialized for laymen to understand. Second, one document may not be able to fully answer queries, and users may need to summarize conclusions across multiple documents, which may lead to a waste of time or misunderstanding. In order to improve the user experience when using medical applications, automatic summarization technology is needed.

In the MAS task, we improve upon (Xu and Lapata, 2020) via three methods. First, during the coarse ranking of a sentence in one of the given documents, we also add the surrounding sentences as input and use two special tokens marking the positions of the sentence. This modification improves the coarse ranking with a large margin. Second, due to the low resource settings of this task, we find that applying a RoBERTa (Liu et al., 2019) model which is already fine-tuned on the GLUE benchmark (Wang et al., 2018) can be beneficial.

In the MAS task, we use two methods to improve (?). First, when we rank a sentence coarsely in a given document, we add the surrounding sentences as input. This modification greatly improves the efficiency of coarse ranking. Secondly, due to the low resource setting of this task, we find that it is beneficial to apply the Roberta (Liu2019RoBERTaAR) model, which has been fine tuned on the glue benchmark (Wang2018GLUEAM).

For the other two tasks, we mainly discuss how the pre trained seq2seq model, such as Bart (Lewis et al., 2020), Pegasus (Zhang et al., 2020a), can be implemented in these tasks. You can make two takeout. First, for tasks with smaller datasets, freezing part of the parameters is beneficial. Second, backtranslation is beneficial for generalization.

Our team ChicHealth participated in all three tasks and won the first place for the overall per-

*Contact author: yi.cai@chic-health.com.

Proceedings of the BioNLP 2021 workshop, pages 263–267
June 11, 2021.

formances. Experiments show that our methods are beneficial for pre-trained models' downstream performances.

2 Extractive MDS

Let Q denote a query, and $D = \{d_1, d_2, ..., d_M\}$ a set of documents. We have implemented multi granularity MDS following the implementation of Xu and Lapata (2020). We first break down the document into paragraphs, which are sentences. Then, a trained Roberta model quantifies the semantic similarity between the selected sentence and the query, and estimates the importance of the sentence (evidence estimator) according to the sentence itself or the local context of the sentence. Thirdly, in order to give the global estimation of the importance of each part in the summary, we use the centrality estimator based on the Markov chain.

2.1 Evidence Estimator

Let $\{S_1, S_2, ..., S_N\}$ as the candidate answer set. Our training goal is to find the right answers in this group. We use Roberta as our sequence encoder

We concatenate query Q after candidate sentence S into a sequence $< /s >$, S, $< /s > < s >$, Q, $< /s >$, as the input to the RoBERTa encoder. The starting $< s >$ token's vector representations t serves as input to a single layer feed forward layer to obtain the distribution over positive and negative classes, where the positive class denotes that a sentence contains the answer and 0 otherwise.

We connect the query Q to the sequence $< s >$, S, $< /s >$, Q, $< /s >$ after the candidate statement $s$as the input of the Roberta encoder. The vector of the starting $< s >$ is used as the input of the single feed-forward layer to obtain the distribution on the positive and negative classes, where the positive class indicates that the sentence contains the answer, otherwise it is 0. We can improve the performance of the evidence estimator by adding the surrounding sentences of S into the model during training.

After fine-tuning, we take the probability of positive class as the score of local evidence, and we will use it to sort all sentences of each query.

2.2 Centrality Estimator

In order to obtain a global estimate of the score of each candidate sentence, we apply a global estimator following Xu and Lapata (2020). The centrality estimator is essentially an extension of the famous LexRank algorithm (Erkan and Radev, 2004).

For each document cluster, i.e., the collections of documents for each query in our tasks, LexRank builds a graph $G = (V; E)$ with nodes V corresponding to sentences and undirected edges E whose weights are computed based on a certian similarity metric. The original LEXRANK algorithm uses TF-IDF (Term Frequency Inverse Document Frequency). (Xu and Lapata, 2020) proposes to use TF-ISF (Term Frequency Inverse Sentence Frequency), which is similar to TF-IDF but operates at the sentence level.

Following ((Xu and Lapata, 2020)), the similarity matrix E is combined with the evidence estimator's , that is,

$$\tilde{E} = w * [\tilde{q}; ...; \tilde{q}] + (1 - w) * E, \tag{1}$$

where $w \in (0, 1)$ controls the extent to which the evidence estimator can influence the final summarization, and $\tilde{q}$ is obtained by normalizing the the evidence scores,

$$\tilde{q} = \frac{q}{\sum_v^{|V|} q_v}. \tag{2}$$

We run a Markov Chain on the graph and the final stationary distribution $\tilde{q}^*$ of this Markov chain serves as the final scores of each sentence.

3 Abstractive summarization

Pre-trained models. In this section, we investigate the pretrained Seq2Seq models to obtain abstractive summarizations, after finetuning their on our datasets. We mainly investigate two types of models, BART ((Lewis et al., 2020)) and PEGASUS ((Zhang et al., 2020a)). And experiments show the PEGASUS model is better

Finetuning techniques. In order to fine tune the pre-trained seq2seq model, we test some methods/techniques that can improve the performance of downstream tasks:

- Freezing a proportion of the parameters of the model;
- Advarsarial training method, i.e., Projected Gradient Descent (PGD, (Madry et al., 2018)).
- Backtraslation (from English to Thai, and then Thai to English) is applied for data augmentation.

model	ROUGE-2 F1
BART-base	11.47
BART-large	13.73
PEGASUS-large	16.37

Table 1: Comparison of different pretrained models on valid set in Task 1.

model	# layers to freeze	ROUGE-2
PEGASUS-large	3	16.37
PEGASUS-large	0	15.80
PEGASUS-large	6	14.98
PEGASUS-large	9	15.64
PEGASUS-large	12	9.85

Table 2: Results of PEGASUS-large model, when we freeze different numbers of lower layers of the encoder and decoder.

with Adv training?	ROUGE-2
Yes	16.37
No	15.46

Table 3: Results of PEGASUS-large model, with or without adversarial training.

evidence estimator	centrality estimator	ROUGE-2
dev set		
roberta-base	No	44.32
roberta-large	No	46.48
roberta-large + GLUE finetuning	No	47.13
roberta-large + GLUE finetuning	LexRank	48.24
ensemble models	LexRank	49.18

Table 4: Comparison of different models on dev set of the MAS task.

model	ROUGE-2
BERT-abs	34.95
T5-small	45.46
T5-base	49.41
T5-large	50.68
BART-base	49.65
BART-large	49.81
PEGASUS-pubmed	45.93
PEGASUS-large	**51.95**

Table 5: The results of different summarization models.

4 Experiments

4.1 dataset statistics

For QS tasks (Figure 1 and 2), the source length distribution is consistent on the train/Val/test set, and the target length distribution is also consistent. For RRS tasks (7 and 8), we can observe that the sequence length distribution of train/ val/test set is different, which may lead to skewed model. For task 2, the length of the document varies, which is too long for pre-trained models like Pegasus. Therefore, for task 2, abstractive summaries are generated from extractive summaries.

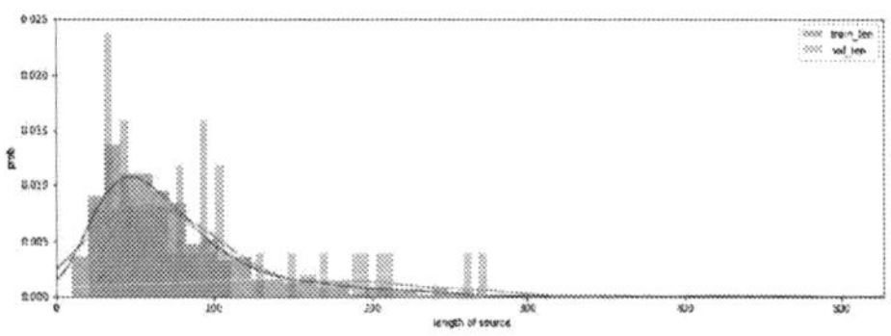

Figure 1: Source sequence length of QS.

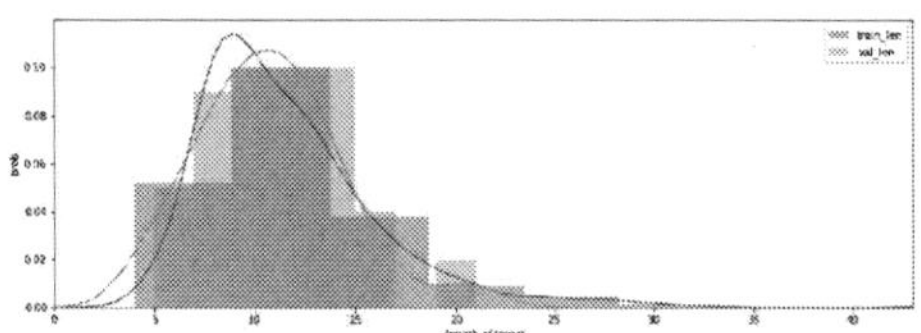

Figure 2: Target sequence length of QS.

4.2 Results on QS

We first report the results on the QS task. First, we compare BART and PEGASUS (Table 1), and find that PEGASUS performs significantly better than BART. Second, we compare PEGASUS with different number of layers freezed (Table 2), and find that freezing three 3 layers obtains the best dev performance. Third, we compare the model with or without adversarial training (Table 3), and show that adversarial training is important for this task.

4.3 Results on MAS

Now we report results on the MAS task (Table 4). RoBERTa large performs better on coarse ranking than RoBERTa base. And using a model finetuned on GLUE also helps to improve the fine-tuning task. After centrality ranking with LexRank, the score improve by more than one percent. And our best score is obtained by using ensemble on the evidence estimators.

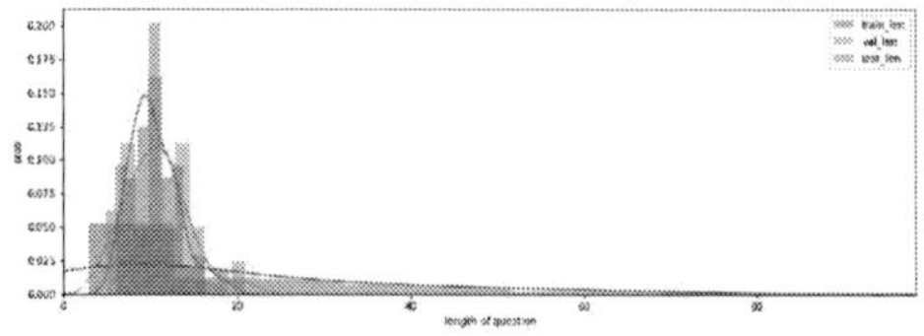

Figure 3: Query length of MAS.

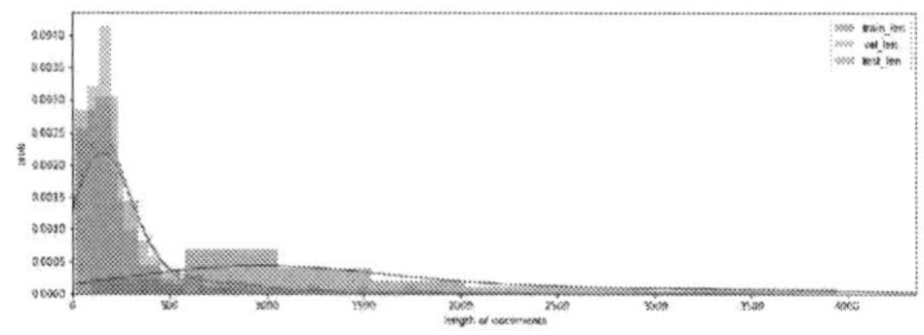

Figure 4: Document length of MAS.

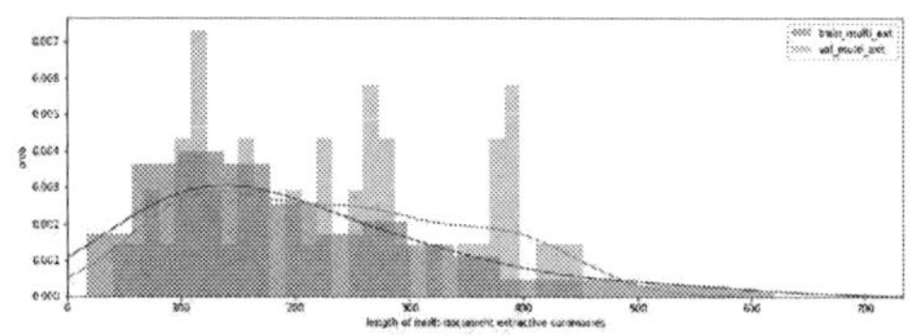

Figure 5: Extractive summary length of MAS.

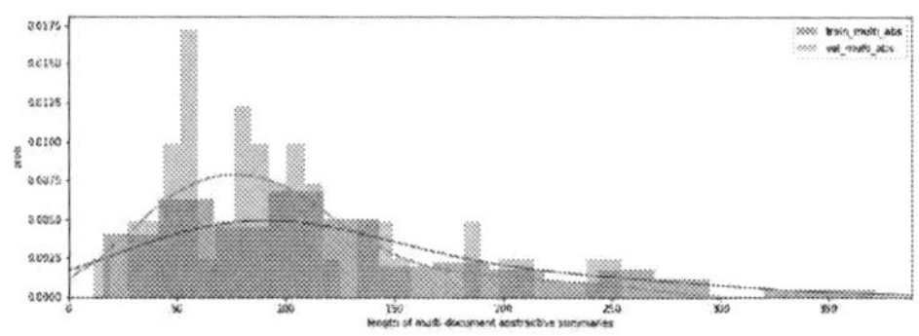

Figure 6: Abstractive summary length of MAS.

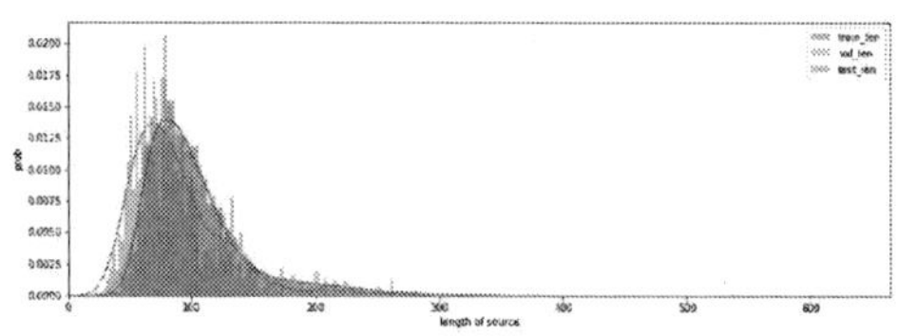

Figure 7: source length of task3 using PEGASUS tokenizer

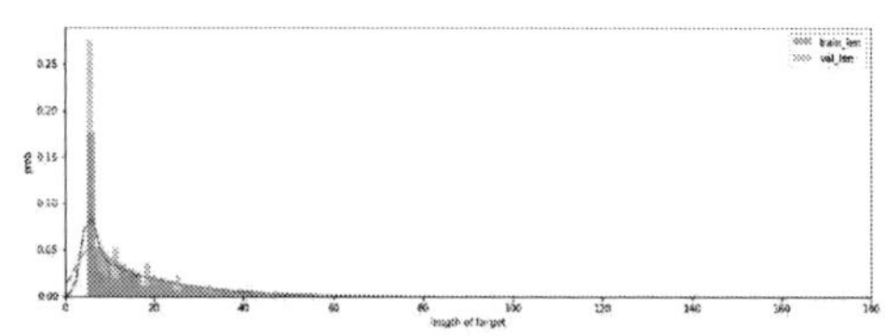

Figure 8: target length of task2 using PEGASUS tokenizer

4.4 Results on RRS

Now we report results on the RRS task. We compare 4 groups of models, BERT-abs, T5 (Raffel et al., 2020), BART and PEGASUS (Table 5). PEGASUS also performs best, like in the QS task. However, we find that the PEGASUS trained on PubMed performs significant worse, which is contradictory to our hypothesis that fine-tuning on related domain corpus is beneficial for downstream tasks.

5 Conclusion

In this work, we elaborate on the methods we employed for the three tasks in the MEDIQA 2021 shared tasks. For the extractive summarization of MAS task, we build upon Xu and Lapata (2020), and achieve improvements by adding contexts and sentence position markers. For generating abstractive summaries, we leverage the pre-trained seq2seq models. To improve the fine-tuning performances on the downstream tasks, we implement a few techniques, like freezing part of the models, adversarial training and back-translation. Our team achieves the 1st place for the overall performances.

In this work, we elaborate the methods used in the three shared tasks of mediqa 2021. For MAS task, we employ the methods that are similar to Xu and Lapata (2020). In order to generate abstract abstracts, we take advantages of the pre-trained seq2seq model. In order to improve the fine-tuning performance of downstream tasks, we use freezing part of the model, adversarial training. Our team ranks first in the overall performances of the three task.

References

Asma Ben Abacha and Dina Demner-Fushman. 2019. On the role of question summarization and information source restriction in consumer health question answering. *AMIA Joint Summits on Translational Science proceedings. AMIA Joint Summits on Translational Science*, 2019:117–126.

Günes Erkan and Dragomir R. Radev. 2004. Lexrank: Graph-based lexical centrality as salience in text summarization. *J. Artif. Intell. Res.*, 22:457–479.

M. Lewis, Yinhan Liu, Naman Goyal, Marjan Ghazvininejad, A. Mohamed, Omer Levy, Ves Stoyanov, and Luke Zettlemoyer. 2020. Bart: Denoising sequence-to-sequence pre-training for natural language generation, translation, and comprehension. *ArXiv*, abs/1910.13461.

Y. Liu, Myle Ott, Naman Goyal, Jingfei Du, Mandar Joshi, Danqi Chen, Omer Levy, M. Lewis, Luke Zettlemoyer, and Veselin Stoyanov. 2019. Roberta: A robustly optimized bert pretraining approach. *ArXiv*, abs/1907.11692.

A. Madry, Aleksandar Makelov, L. Schmidt, D. Tsipras, and Adrian Vladu. 2018. Towards deep learning models resistant to adversarial attacks. *ArXiv*, abs/1706.06083.

Colin Raffel, Noam Shazeer, Adam Roberts, Katherine Lee, Sharan Narang, Michael Matena, Yanqi Zhou, W. Li, and Peter J. Liu. 2020. Exploring the limits of transfer learning with a unified text-to-text transformer. *J. Mach. Learn. Res.*, 21:140:1–140:67.

Max E. Savery, Asma Ben Abacha, Soumya Gayen, and Dina Demner-Fushman. 2020. Question-driven summarization of answers to consumer health questions. *Scientific Data*, 7.

Alex Wang, Amanpreet Singh, Julian Michael, Felix Hill, Omer Levy, and Samuel R. Bowman. 2018. Glue: A multi-task benchmark and analysis platform for natural language understanding. In *BlackboxNLP@EMNLP*.

Yumo Xu and Mirella Lapata. 2020. Coarse-to-fine query focused multi-document summarization. In *EMNLP*.

Jingqing Zhang, Y. Zhao, Mohammad Saleh, and Peter J. Liu. 2020a. Pegasus: Pre-training with extracted gap-sentences for abstractive summarization. In *ICML*.

Yuhao Zhang, D. Ding, Tianpei Qian, Christopher D. Manning, and C. Langlotz. 2018. Learning to summarize radiology findings. In *Louhi@EMNLP*.

Yuhao Zhang, Derek Merck, E. Tsai, Christopher D. Manning, and C. Langlotz. 2020b. Optimizing the factual correctness of a summary: A study of summarizing radiology reports. In *ACL*.

NCUEE-NLP at MEDIQA 2021: Health Question Summarization Using PEGASUS Transformers

Lung-Hao Lee, Po-Han Chen, Yu-Xiang Zeng, Po-Lei Lee, and Kuo-Kai Shyu
Department of Electrical Engineering, National Central University, Taiwan
Pervasive Artificial Intelligence Research (PAIR) Labs, Taiwan

Abstract

This study describes the model design of the NCUEE-NLP system for the MEDIQA challenge at the BioNLP 2021 workshop. We use the PEGASUS transformers and fine-tune the downstream summarization task using our collected and processed datasets. A total of 22 teams participated in the consumer health question summarization task of MEDIQA 2021. Each participating team was allowed to submit a maximum of ten runs. Our best submission, achieving a ROUGE2-F1 score of 0.1597, ranked third among all 128 submissions.

1 Introduction

Consumers increasingly use online resources to meet their health information needs. However, health information needs are usually complex and to be expressed in natural language (Kilicoglu et al., 2018). In general, health questions tend to consist of considerable contextual information that may hinder automatic Question Answering (QA) systems. Paraphrasing and summarizing the questions has been shown to substantially improve QA performance (Ben Abacha and Demner-Fushman, 2019a). Therefore, effective summarization methods for consumer health questions could play an important role in enhancing medical QA performance.

Automatic text summarization is the process of computationally shortening texts to find or generate the most informative sentences that represent the most important or relevant information within the original content. There are two general approaches to summarization: extraction and abstraction. In extractive summarization methods, a summary is extracted from the original texts, but the extracted sentences are not modified in any way. Abstractive summarization methods learn a semantic representation of the original content, and then use this representation to generate a summary that is closer to what a human might express in terms of original content.

MEDIQA 2021 is the second edition of the MEDIQA challenge collocated with the BioNLP 2021 workshop, focusing on summarization in the medical domain with three tasks: consumer health question summarization, multi-answer summarization, and radiology report summarization. We only participated the first Question Summarization (QS) task, in the domain of abstractive summarization. The goal of this task is to promote the development of new summarization methods that specifically address the challenges of long and complex consumer health questions. The recently developed transformer in NLP is a novel neural architecture that aims to solve sequence-to-sequence tasks in handling long dependencies and usually achieves promising results. This achievement motivates us to explore the use of a transformer-based model to tackle the question summarization problem in the medical domain.

This paper describes the **NCUEE-NLP** (**N**ational **C**entral **U**niversity, Dept. of **E**lectrical **E**ngineering, **N**atural **L**anguage **P**rocessing Lab) system for the QS task of the MEDIQA challenge at the BioNLP 2021 workshop. Our solution explores the use of pre-trained PEGASUS Transformers (Zhang et al., 2020a) and fine-tuning on the downstream question summarization task using our collected and processed datasets. A total of 22 teams participated in this task. Each participating team was allowed to submit a maximum of ten runs. Our best submission had a ROUGE2-F1 score of 0.1597, ranking third among all 128 submissions.

Proceedings of the BioNLP 2021 workshop, pages 268–272

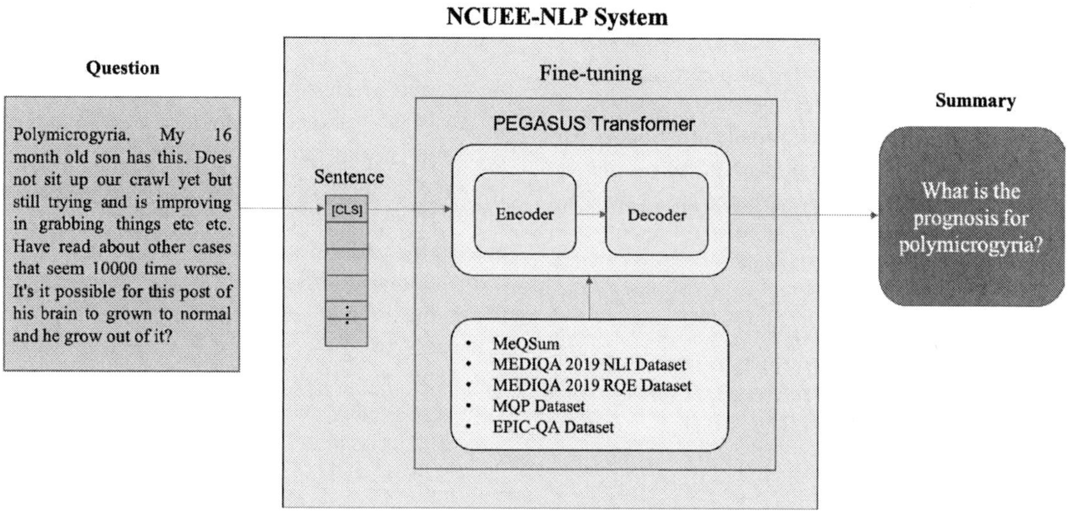

Figure 1: Our NCUEE-NLP system architecture for the QS task.

The rest of this paper is organized as follows. Section 2 describes the NCUEE-NLP system for the question summarization task. Section 3 presents the evaluation results and performance comparisons. Conclusions are finally drawn in Section 4.

2 The NCUEE-NLP System

Figure 1 shows our NCUEE-NLP system architecture for the QS task. Specifically, our system is comprised of two main parts: 1) PEGASUS transformers, and 2) fine-tuning. Details are introduced as follows.

2.1 PEGASUS Transformers

Zhang et al. (2020a) proposed **PEGASUS** (**P**retraining with **E**xtracted **G**ap-sentences for **A**bstractive **SU**mmarization **S**equence-to-sequence) method that pre-trains large transformer-based encoder-decoder models on massive text corpora. New self-supervised objectives called Gap Sentences Generation (GSG) and classical Mask Language Models (MLM) were applied simultaneously as pre-training objectives. The PEGASUS model was evaluated on 12 downstream summarization tasks spanning news, science, stories, instructions, emails, patents, and legislative bills. Experimental results showed that good abstractive summarization performance can be achieved across broad domains by fine-tuning the model, outperforming previous state-of-the-art approaches on many tasks.

These achievements motivate us to explore the use of the PEGASUS transformers and fine-tuning on the downstream QS task in the medical domain.

2.2 Fine-tuning

Many summarization datasets contain original texts and their referenced summarizes written in declarative sentences. Question summaries written in interrogative sentences are relatively rare. Hence, in addition to the training set provided by task organizers, we also collected and processed the following datasets to fine-tune the QS task.

- MEDIQA 2019 – NLI Dataset (Ben Abacha et al., 2019)

The Natural Language Inference (NLI) task of the MEDIQA 2019 challenge identifies three relations between two sentences including entailment, neutral, and contradiction. We only use the entailment relation that was annotated from the training, validation and test datasets. Comparing the lengths of two the sentences in each pair, the longer sentences will be regarded as a question, while the other is used as the corresponding summary. A total of 4,683 pairs were collected from this dataset.

- MEDIQA 2019 – RQE Dataset (Ben Abacha et al., 2019)

The Recognizing Question Entailment (RQE) task of the MEDIQA 2019 challenge focuses on identifying entailments between two questions. We use the positive question-pairs (annotated as "entailment") from the training, validation and test datasets. However, some questions are not written using valid interrogative sentences such as a declarative sentences followed by "Right?". We exclude these cases and only use questions that start with wh-words, be verbs, and auxiliary verbs. Similarly, the shorter question in each question-pairs is regarded as a reference summary. This resulted in a final subset of 4,011 pairs.

- MQP Dataset (McCreery et al., 2020)

The Medical Question Pairs (MQP) dataset contains similar and dissimilar medical question pairs hand-generated and labeled by doctors. A list of 1,524 patient-asked questions were randomly sampled. Doctors as the labelers had rewritten the original question in different ways while maintaining the same intent, and used similar key words to write related but dissimilar questions for which the answer would be wrong or irrelevant. Hence, each question results in one similar and one different pair. We only use the similar question pairs to fine-tune the transformers. In the same way, the longer questions are used as original questions and the shorter ones are their reference summaries.

- EPIC-QA Dataset on COVID-19 (Goodwin et al., 2020)

In response to the COVID-19 pandemic, the Epidemic Question Answering (EPIC-QA) track in TREC 2020 conference focuses on developing systems capable of automatically answering questions about COVID-19. In the question part of EPIC-QA data, two prepared sets of approximately 45 questions were provided: one for expert-level questions and one for consumer-level questions. Without considering the question levels, we regard the longer questions as original questions and the corresponding shorter question are their summaries.

3 Evaluation

3.1 Data

The experimental datasets were mainly provided by task organizers (Ben Abacha et al., 2021). The training, validation and test sets were composed of data from an independent set of consumer health questions. The MeQSum Dataset of consumer health questions and their summaries can be used for training (Ben Abacha and Demner-Fushman, 2019b). The validation and test sets consist of consumer health questions received by the U.S. National Library of Medicine (NLM) in December 2020. Their associated summaries were manually created by medical experts for evaluation.

In summary, during the system development phase, the training and validation sets respectively consisted of 1,000 and 50 consumer health questions and their associated summaries for system designing and implementation. In total, only 100 consumer health questions in the test dataset were used for final performance evaluation.

3.2 Settings

The pre-trained PEGASUS models were downloaded from the HuggingFace (Wolf et al., 2019). A PEGASUS model was trained with sampled gap sentence ratios on both C4 (Raffel et al., 2020) and HugeNews datasets, and important sentences were sampled stochastically. We selected the PEGASUS-Large model and its mixed and stochastic model (denoting PEGASUS-Large-XSum) on the XSum (Narayan et al., 2018) datasets, containing 227k BBC news articles from 2010 to 2017 covering a wide variety of subjects along with professionally written single-sentence summarizes.

To confirm model performance, we compared the previous state-of-the-art BART method (Lewis et al., 2019) that uses a denoising autoencoder to pre-train sequence-to-sequence models. We also downloaded the pre-trained BART-Large and BART-Large-XSum models from the HuggingFace (Wolf et al., 2019).

On an Nvidia DGX-1 server using a V100 GPU with the same settings, the hyper-parameter values for our model implementation were optimized as follows: maximum sequence length 512; learning rate 0.00005; batch size 6 and gradient accumulation steps 128 for both BART models; and batch size 8 and gradient accumulation steps 512 for both PEGASUS models.

3.3 Metrics

ROUGE is used to measure summarization performance (Lin, 2004). ROUGE stands for Recall-Oriented Understudy for Gisting Evaluation,

Models	R1-P	R1-R	R1-F1	R2-P	R2-R	R2-F1	RL-P	RL-R	RL-F1
BART-Large	0.3165	0.3255	0.3209	0.1355	0.1438	0.1395	0.3090	0.3182	0.3135
BART-Large-XSum	**0.3299**	0.3194	0.3246	**0.1435**	0.1488	0.1461	**0.3215**	0.3127	**0.3170**
PEGASUS-Large	0.3153	**0.3368**	**0.3257**	0.1307	**0.1593**	0.1436	0.3029	**0.3285**	0.3152
PEGASUS-Large-XSum	0.3159	0.3269	0.3213	0.1393	0.1553	**0.1469**	0.3017	0.3157	0.3085

Table 1: Results of summarization models on the QS validation dataset.

Models	R1-P	R1-R	R1-F1	R2-P	R2-R	R2-F1	RL-P	RL-R	RL-F1
BART-Large	0.3526	0.3159	0.3132	0.1452	0.1236	0.1268	0.3187	0.2865	0.2842
BART-Large-XSum	0.3308	0.3253	0.3116	0.1212	0.1150	0.1125	0.2976	0.2891	0.2784
PEGASUS-Large	0.3173	**0.3426**	0.2936	0.1377	0.1346	0.1217	0.2821	0.2934	0.2579
PEGASUS-Large-XSum	**0.3869**	0.3316	**0.3352**	**0.1850**	**0.1573**	**0.1597**	**0.3576**	**0.3030**	**0.3090**

Table 2: Results of summarization models on the QS test dataset.

including several automatic evaluation methods that measure the similarity between summaries. ROUGE-N is an n-gram recall between a candidate summary and a set of reference summaries. ROUGE-L accounts for the union Longest Common Sequence (LCS) in matching between a reference summary sentence and every candidate summary sentence.

In the QS task of MEDIQA 2021 challenge, ROUGE-1 (denoted as R1), ROUGE-2 (R2), and ROUGE-L (RL) were adopted as measure metrics. The F1 score, which is a harmonic mean of precision (short in P) and recall (R), of R2 was regarded as the official score to rank the participating teams' performance in the leaderboard.

3.4 Results

Table 1 shows the results on the QS validation set of MEDIQA 2021 challenge. Both PEGASUS models outperformed the BART models in a half of the metrics. The mixed and stochastic models on the XSum datasets usually outperformed than those without the XSum optimization using both BART and PEGASUS transformers. The PEGASUS-Large-XSum model obtained the best overall score of 0.1469 in R2-F1, considered as the ranking metric.

During the final testing phase of the QS task, we used the training set and collected datasets to fine-tune the models and the validation set for parameter optimization. Each participating team was allowed to submit a maximum of ten runs for each task. We submitted the four above-mentioned models. Table 2 shows the results of our testing models. The PEGASUS-Large-XSum model clearly outperformed the others than the others in almost all evaluation metrics.

A total of 22 teams participated in the QS task, each submitting at least one entry. Our best submission achieved an R2-F1 score of 0.1597, significantly outperforming the baseline model with a score of 0.1373 and ranking third place among all 128 submissions.

In addition to ROUGE metrics, task organizers also use several evaluation metrics that may be better adapted to the QS task. Our best submission also achieved a HOLMS score (Mrabet and Demner-Fushman, 2020) of 0.5783, ranking first among all 128 submissions. Our best submission had a BERTScore-F1 (Zhang et al., 2020b) of 0.6960, ranked ninth among all submissions.

4 Conclusions

This study describes the NCUEE-NLP system in the consumer health question summarization task of the MEDIQA 2021 challenge, including system design, implementation and evaluation. We used the PEGASUS transformers and fine-tuned the downstream summarization task using our collected and processed datasets. A total of 22 teams participated in the task, each submitting at least one entry. Our best submission had a ROUGE2-F1 score of 0.1597, ranking third place among all 128 submissions.

Acknowledgments

This study is partially supported by the Ministry of Science and Technology, under the grant MOST 108-2218-E-008-017-MY3 and MOST 108-2634-

F-008-003- through Pervasive Artificial Intelligence Research (PAIR) Labs, Taiwan.

References

Asma Ben Abacha, and Dina Demner-Fushman. 2019a. On the role of question summarization and information source restriction in consumer health question answering. *AMIA Summits on Translational Science Proceedings*, 2019:117-128.

Asma Ben Abacha, and Dina Demner-Fushman. 2019b. On the summarization of consumer health questions. In *Proceedings of the 57th Annual Meeting of the Association for Computational Linguistics*. Association for Computational Linguistics, pages 2228-2234. http://dx.doi.org/10.18653/v1/P19-1215

Asma Ben Abacha, Chaitanya Shivade, and Dina Demner-Fushman. 2019. Overview of the MEDIQA 2019 shared task on textual inference, question entailment and question answering. In *Proceedings of the 2019 Workshop on Biomedical Natural Language Processing*. Association for Computational Linguistics, pages 370-379. http://dx.doi.org/10.18653/v1/W19-5039

Asma Ben Abacha, Yassine Mrabet, Yuhao Zhang, Chaitanya Shivade, Curtis Langlotz, and Dina Demner-Fushman. 2021. Overview of the MEDIQA 2021 shared task on summarization in the medical domain, In *Proceedings of the 20th SIGBioMed Workshop on Biomedical Language Processing*, NAACL-BioNLP 2021, Association for Computational Linguistics.

Clara H. McCreery, Namit Katariya, Anitha Kannan, Manish Chablani, and Xavier Amatrain. 2020. Effective transfer learning for identifying similar questions: matching user questions to COVID-19 FAQs. In *Proceedings of the 26th ACM SIGKDD International Conference on Knowledge Discovery & Data Mining*, pages 3458-3465. https://doi.org/10.1145/3394486.3412861

Chin-Yew Lin. 2004. ROUGE: a package for automatic evaluation of summaries. In *Proceedings of the Workshop on Text Summarization Branches Out*. Association for Computational Linguistics, pages 74-81.

Colin Raffel, Noam Shazeer, Adam Roberts, Katherine Lee, Sharan Narang, Michael Matena, Yanqi Zhou, Wei Li, and Peter J. Liu. 2020. Exploring the limits of transfer learning with a unified text-to-text transformer. *Journal of Machine Learning Research*, 21(140):1-67.

Halil Kilicoglu, Asma Ben Abacha, Yassine Mrabet, Sonya E. Shooshan, Laritza Rodriguez, Kate Masterton, and Dina Demner-Fushman. 2018. Semantic annotation of consumer health questions. BMC Bioinformatics, 19, 34(2018). https://doi.org/10.1186/s12859-018-2045-1

Jingqing Zhang, Yao Zhao, Mohammad Saleh, and Peter Liu. 2020a. PEGASUS: pre-training with extracted gap-sentences for abstractive summarization. In *Proceedings of the 37th International Conference on Machine Learning*, PMLR 119:11328-11339.

Mike Lewis, Yinhan Liu, Naman Goyal, Marjan Ghazvininejad, Abdelrahman Mohamed, Omer Levy, Ves Stoyanov, and Luke Zettlemoyer. 2020. BART: denoising sequence-to-sequence pre-training for natural language generation, translation and comprehension. In *Proceedings of the 58th Annual Meeting of the Association for Computational Linguistics*. Association for Computational Linguistics, pages 7871-7880. http://dx.doi.org/10.18653/v1/2020.acl-main.703

Shashi Narayan, Shay B. Cohen, and Mirella Lapata. 2018. Don't give me the details, just the summary! Topic-aware convolutional neural networks for extreme summarization. In *Proceedings of the 2018 Conference on Empirical Methods in Natural Language Processing*. Association for Computational Linguistics, pages 1797-1807. http://dx.doi.org/10.18653/v1/D18-1206

Thomas Wolf, Lysandre Debut, Victor Sanh, Julien Chaumond, Clement Delangue, Anthony Moi, Pierric Cistac, Tim Rault, Remi Louf, Morgan Funtowicz, Joe Davison, Sam Shleifer, Patrick von Platen, Clara Ma, Yacine Jernite, Julien Plu, Canwen Xu, Teven Le Scao, Sylvain Gugger, Mariama Drame, Quentin Lhoest, and Alexander M. Rush. 2019. HuggingFace's transformers: state-of-the-art natural language processing. *arXiv preprint*. https://arxiv.org/abs/1910.03771

Tianyi Zhang, Varsha Kishore, Felix Wu, Kilian Q. Weinberger, and Yoav Artzi. 2020b. BERTScore: evaluating text generation with BERT. Published as a conference paper at ICLR 2020. *arXiv preprint*. https://arxiv.org/abs/1904.09675

Travis Goodwin, Dina Demner-Fushman, Kyle Lo, Lucy Lu, William R. Hersh, Hoa T. Dang, and Ian M. Soboroff. 2020. EPIC-QA dataset on COVID-19. https://bionlp.nlm.nih.gov/epic_qa/

Yassine Mrabet, and Dina Demner-Fushman. 2020. HOLMS: alternative summary evaluation with large language models. In *Proceedings of the 28th International Conference on Computational Linguistics*. International Committee on Computational Linguistics, pages 5679-5688. http://dx.doi.org/10.18653/v1/2020.coling-main.498

SB_NITK at MEDIQA 2021: Leveraging Transfer Learning for Question Summarization in Medical Domain

Spandana Balumuri, Sony Bachina and Sowmya Kamath S
Healthcare Analytics and Language Engineering (HALE) Lab,
Department of Information Technology,
National Institute of Technology Karnataka, Surathkal, Mangalore 575025, India
{spandanabalumuri99, bachina.sony}@gmail.com
sowmyakamath@nitk.edu.in

Abstract

Recent strides in the healthcare domain, have resulted in vast quantities of streaming data available for use for building intelligent knowledge-based applications. However, the challenges introduced to the huge volume, velocity of generation, variety and variability of this medical data have to be adequately addressed. In this paper, we describe the model and results for our submission at MEDIQA 2021 Question Summarization shared task. In order to improve the performance of summarization of consumer health questions, our method explores the use of transfer learning to utilize the knowledge of NLP transformers like BART, T5 and PEGASUS. The proposed models utilize the knowledge of pre-trained NLP transformers to achieve improved results when compared to conventional deep learning models such as LSTM, RNN etc. Our team SB_NITK ranked 12th among the total 22 submissions in the official final rankings. Our BART based model achieved a ROUGE-2 F1 score of 0.139.

1 Introduction

The Question Summarization (QS) task aims to promote the development of new summarization models that are able to summarize lengthy and complex consumer health questions. The consumer health questions can have a variety of subjects like medications, diseases, effects, medical treatments and procedures. The medical questions can also contain a lot of irrelevant information that makes automated question summarization a difficult and challenging task (Mayya et al., 2021). It is also often cumbersome to go through lengthy questions during the question answering process and then formulate relevant answers (Upadhya et al., 2019). The automated summarization approaches for consumer health questions thus have many medical applications. An effective automated summarization approach for obtaining simplified medical health questions can be crucial to improving medical question answering systems.

The MEDIQA 2021 (Ben Abacha et al., 2021) proposes three different shared tasks to promote the development, performance improvement and evaluation of text summarization models in the medical domain:

- *Consumer Health Question Summarization (QS)* - Development of summarization models to produce the shortened form of consumer health related questions.
- *Multi-Answer Summarization* - Development of summarization models to aggregate and summarize multiple answers to a medical question.
- *Radiology Report Summarization* - Development of summarization models that can produce radiology impression statements by summarising text-based observations.

The role of question summarization or simplification in answering consumer health questions is not explored extensively when compared to the summarization of documents and news articles (George et al., 2021). Ishigaki et al. (2017) explored various extractive and abstractive methods for summarization of questions that are posted on a community question answering site. The results showed that abstractive methods with copying mechanism performed better than extractive methods. Agrawal et al. (2019) proposed a closed-domain Question Answering technique that uses Bi-directional LSTMs trained on the SquAD dataset to determine relevant ranks of answers for a given question. Ben Abacha and Demner-Fushman (2019) proposed sequence-to-sequence attention models with pointer generator network for summarization of consumer health questions collected from MeQSum, Quora question pairs dataset and other sources. The addition of pointer generator and cov-

Proceedings of the BioNLP 2021 workshop, pages 273–279
June 11, 2021.

erage mechanisms on the sequence-to-sequence has improved the ROUGE scores considerably.

In this paper, we describe the different models and experiments that we designed and evaluated for the Consumer Health Question Summarization (QS) task. The proposed models utilize the knowledge of pre-trained NLP transformers to achieve improved results when compared to conventional deep learning models such as LSTM, RNN etc. The proposed models are based on transfer learning and fine tuning the dataset on different versions of NLP transformers like BART (Lewis et al., 2019), T5 (Raffel et al., 2020) and PEGASUS (Zhang et al., 2020). We have also benchmarked all the proposed models against traditional Seq2Seq LSTM encoder-decoder networks with attention.

The rest of this article is organized as follows. In Section 2, we provide information about the data used such as description of datasets, dataset augmentation and pre-processing. Section 3 gives an overview of transformer architecture and transfer learning. In Section 4, we describe and compare results obtained from fine-tuning various transformer models on our augmented dataset. In Section 5, we compare the performance of our proposed models with different transformer models in detail, followed by conclusion and directions for future work.

2 Data

2.1 MeQSum Dataset Description

The main dataset for the task was provided by the organizers of MEDIQA 2021 (Ben Abacha et al., 2021). The training set comprised of consumer health questions (CHQs) and the corresponding summaries. The validation set consisted of National Library of Medicine (NLM) consumer health questions and their respective summaries. In addition to the questions and summaries, the validation set contains question focus and question type for each question. The MeQSum training corpus consists of 1000 question-summary pairs while the validation dataset provided has 50 NLM question-summary pairs. To improve the performance, the question focus in validation pairs has been appended to the beginning of each question.

2.2 Dataset Augmentation

As the provided training and validation datasets for the task add up to only a 1,050 question-summary pairs, we decided to augment the data to achieve better performance and solve over-fitting problems. The following three datasets were added to the training and validation datasets to broaden the coverage.

TREC-2017 LiveQA: Medical Question Answering Task Dataset. The LiveQA dataset is used for training consumer health question answering systems. The question pairs in this dataset are very similar to those given for the task, however, its small size was not conducive to performance improvement. The test dataset (Ben Abacha et al., 2017) comprises of 104 NLM Questions, out of which 102 of them have an associated summary annotation. Additionally, each question has focus, type, and keyword annotations associated with it. To increase the weight of significant parts of the question, we added the question focus and keyword annotations to the beginning of each question.

Recognizing Question Entailment (RQE) Dataset. The RQE dataset (Ben Abacha and Demner-Fushman, 2016) is used for automatic question answering by recognizing similar questions in the medical domain. Out of the 8,588 training pairs and 302 validation pairs available in the RQE corpus, we chose only those pairs which entail each other, which resulted in 4,655 training pairs and 129 validation pairs. Moreover, to ensure that one of the questions in the pair is a summary of the other, we selected those pairs where one question has at least 2 sentences and the other has only one sentence. This resulted in a total of 2,078 question-summary pairs. However, one of the issues faced with this dataset is that the questions in some pairs are almost similar to each other.

Medical Question Pairs (MQP) Dataset. The MQP dataset (McCreery et al., 2020) consists a total of 3,048 pairs of related and unrelated medical questions. Half of the total questions i.e., 1,524 pairs are labeled as similar to each other. Among the similar question pairs, we chose those pairs where at least one of the questions has only one sentence. In case both the questions have only one sentence each, the question with lesser number of words is considered as the summary. Finally, the dataset resulted in 1,057 pairs. The advantage of MQP dataset lies in the fact that it has more generalized medical questions in contrast to the previously mentioned datasets, which have many esoteric terms.

2.3 Dataset Preprocessing

The dataset preprocessing largely depends on the data at hand and the type of output we anticipate. Some of the common techniques that we incorporated include text case-folding to lowercase, removal of special characters, numbers and stop words etc. However, upon analyzing the summaries, we found that they include uppercase letters, certain special characters, numbers and stop words. Therefore we did not proceed with extensive data preprocessing, except for removing special characters which are absent the summaries. The final cleaned corpus comprises of 4,287 question-summary pairs.

3 System Description

3.1 Transformer Architecture

Transformers have now become the state-of-the-art solution for a variety of NLP tasks including language understanding, translation, text generation, text classification, question answering and sentiment analysis. Transformers continue to outperform other neural network architectures (RNN and LSTM) by maintaining the attention while handling sequences in parallel, i.e., they handle all words at once (considered bidirectional) rather than one by one and effectively learning inter-dependencies, especially in the case of long sentences.

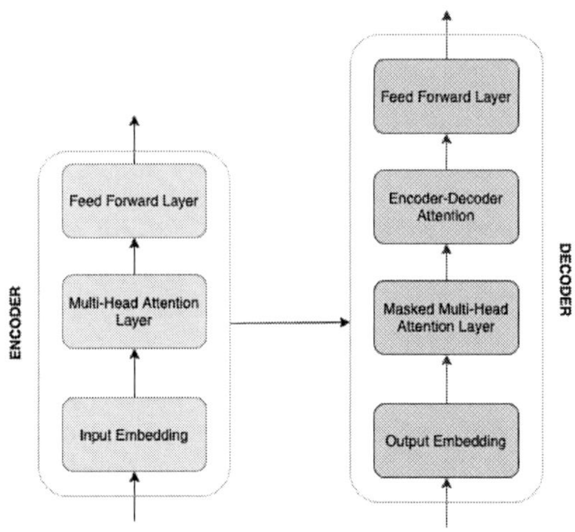

Figure 1: Encoder-Decoder transformer architecture used by PEGASUS, BART and T5.

The transformer architecture as shown in Fig. 1 consists of the encoder and decoder mechanisms, where the segments are connected by a cross-attention layer. An encoder segment consists of a stack of encoders in which each encoder reads the text input and generates embedding vectors. It outputs contextual and positional vectors of the input sequence using attention mechanism. Similarly, the decoder part is a stack of decoders where each decoder takes target sequence and encoder output as input. It generates contextual information from the target sequence and then combines encoder output with it. It models the conditional probability distribution of the target vector sequence based on the previous target vectors to produce an output vector.

The sequence of input tokens is fed into the transformer encoder, which are then embedded into vectors and processed by the neural network. The decoder produces a series of vectors, corresponding to each token in the input sequence. Few examples of existing transformers are BART, T5 etc. As deep neural networks have a large number of parameters, the majority of labelled text datasets are insufficient for training these networks as training them on limited datasets would result in over-fitting. Therefore, for downstream NLP tasks, we can utilize the knowledge of transformers which are pre-trained on large datasets using transfer learning. Transfer learning is a method of using a deep learning model that has been pre-trained on a huge corpus to perform similar NLP tasks by fine-tuning on a different dataset.

For fine-tuning the model with a different dataset, we modify the model parameters like hidden states and weights of the existing model to suit our dataset. Towards this, we have fine-tuned transformer models such as BART, T5 and PEGASUS with our augmented dataset to perform question summarization, for the given task. Fine tuning BART transformer for question summarization with our dataset achieved the best ROUGE-2 scores when compared to other transformer models. The details of experiments and analysis of different models are discussed in Section 4.

4 Models and Results

During the system development phase, we experimented with various models for the task of question summarization. The ranking for the task is based on the ROUGE2-F1 score. ROUGE-2 (Recall-Oriented Understudy for Gisting Evaluation), is a metric which measures the overlap of bigrams between the model-generated and reference summaries in a summarization task. In the following sections, we discuss the various versions of the models that we fine-tuned for the Question Summarization task.

4.1 Seq2Seq models

This model uses a seq2seq bidirectional LSTM based encoder and decoder. The encoder network is combination of an embedding layer followed by a stack of 3 bidirectional LSTM layers each with 128 hidden units and a dropout value of 0.2. The encoder output and encoder states from the LSTM network is given as input to the attention layer (Bahdanau et al., 2016) to generate context vector and attention weights. The generated vectors from attention layer are given as input to decoder. The decoder network is similar to the encoder, having a combination of an embedding layer followed by a stack of bidirectional LSTMs of 128 hidden units and a softmax layer. The output from the decoder network is a vector of tokens' indexes from the vocabulary.

We have experimented with the following variations of seq2seq - attention - coverage model.

1. *Seq2seq + attention + coverage* model with Word2vec ($N \times 300$) embeddings.
2. *Seq2seq + attention + coverage* model with Scibert ($N \times 768$) embeddings.
3. *Seq2seq + attention + coverage* model with Glove ($N \times 300$) embeddings.

However, the above mentioned seq2seq models were not submitted for final evaluation because of the lack of sufficient data to train such models from scratch. Since the size of our training dataset is small (4,287 question-summary pairs), these seq2seq models did not provide acceptable results, hence we omitted them from our submissions for the question summarization task.

4.2 T5

Google's T5 (Text-to-Text Transfer Transformer) is a pre-trained encoder-decoder model that has been trained on C4 (Colossal Clean Crawled Corpus) dataset for unsupervised and supervised tasks. The T5 transformer consists of an encoder, a cross attention layer and an auto-regressive decoder. In T5, every NLP problem is converted to a text-to-text format and the data is augmented with a prefix e.g., for summarization: '*summarize*: ', for translation: "*translate English to French:* ". T5 achieves benchmark performance for various tasks like summarization, question answering, text classification etc, and both supervised and unsupervised methods can be applied for training. Two different versions of T5 were finetuned for our augmented dataset for the summarization task.

1. *t5-base* : T5 model with 12 encoder and decoder layers, trained on C4 dataset, with 220M parameters.
2. *t5-small* : T5 model with 6 encoder and decoder layers, trained on C4 dataset, with 60M parameters.

Table 1 shows the comparison of ROUGE scores obtained for the T5 models we experimented with. The model t5-small obtained a better ROUGE-2-F1 score when compared to t5-base. We submitted a run each for the two models. In addition to these two models, we also experimented with other variations of T5, such as *t5-large* and *t5-base-finetuned-summarize-news*. On comparison of the summaries produced by the various T5 models, t5-small generated the best summaries.

4.3 PEGASUS

Google AI released the PEGASUS model which implements the sequence-to-sequence architecture. The specialty of this model is its self-supervised pre-training objective termed as "gap-sentence generation", where, certain sentences are masked in the input for pre-training. The advantage is gained by keeping the pre-training self-supervised objective closer to the required down-stream task. We mainly focused on the following two versions of the PEGASUS models and fine-tuned them on our augmented dataset.

1. *pegasus-xsum*: pegasus-large model fine-tuned on the XSum dataset having a size of 226k records.
2. *pegasus-wikihow*: pegasus-large model fine-tuned on the WikiHow dataset having a size of 168k records.

Table 1 shows the ROUGE scores obtained for the PEGASUS models finetuned in our work. Among the two, pegasus-wikihow gives better scores than pegasus-xsum. We submitted one run for each of the models. Additionally, we also experimented with other pre-trained PEGASUS models such as, *pegasus-pubmed*, *pegasus-cnn_dailymail* and *pegasus-multi_news*. The summaries produced by these pegasus-cnn_dailymail and pegasus-multi_news were almost similar and acceptable, while those generated by pegasus-pubmed were not up to the mark.

Table 1: Scores and ROUGE values for various models benchmarked for the Question Summarization task

Model	Score	R1-P	R1-R	R1-F1	R2-P	R2-R	R2-F1	RL-R	RL-F1
bart-large-xsum	**0.139**	**0.358**	0.346	**0.333**	**0.152**	**0.144**	**0.139**	**0.318**	**0.308**
bart-large-cnn	0.12	0.339	0.299	0.301	0.137	0.117	0.12	0.274	0.276
pegasus-xsum	0.107	0.329	0.284	0.289	0.128	0.104	0.107	0.261	0.267
pegasus-wikihow	0.129	0.321	**0.349**	0.307	0.143	0.142	0.129	0.304	0.271
t5-base	0.112	0.343	0.297	0.3	0.133	0.107	0.112	0.268	0.273
t5-small	0.114	0.293	0.31	0.281	0.124	0.121	0.114	0.272	0.25

4.4 BART

BART (Bidirectional and Auto-Regressive Transformers) is based on the standard transformer architecture proposed by Facebook, having BERT (Devlin et al., 2019) like encoder and GPT (Radford et al., 2019) like decoder. The denoising objective of the encoder while the decoder that works to reproduce the original sequence, using the previously produced tokens and the encoder output, bring the best of the two models. We experimented with the following different BART pre-trained models by fine-tuning them of our augmented dataset.

1. *bart-large-xsum* : bart-large (BART with 12 encoder & decoder layers) fine-tuned on Xsum dataset with 400M parameters.

2. *bart-large-cnn* : bart-large (BART with 12 encoder & decoder layers) fine-tuned on CNN/Dailymail dataset with 400M parameters.

The ROUGE scores obtained for both the BART based models are tabulated in Table 1. The bart-large-xsum model gives a better performance than the bart-large-cnn model. We have submitted 3 runs for each of the two models, by varying the hyper-parameters such as the summary length, learning rate, length penalty and epochs. The best ROUGE scores were obtained at a learning rate of 3e-5, summary length of 30 and with no length penalty running for 3 epochs. Besides these two models, we have also experimented with other BART models, such as *bart-large-mnli* and *bart-large-gigaword*, however, the summaries generated were not at par with those of the earlier two models.

5 Comparative Evaluation

During the testing phase, we experimented with various models based on the transformer architecture, such as BART, T5 and PEGASUS as mentioned previously. We were allowed to submit a maximum of 10 runs per task. Therefore, we submitted two runs each for T5 and PEGASUS models, and six runs for various approaches of the BART model. The test set provided for the Question Summarization task comprises of 100 NLM questions with their associated question ids. The test set was pre-processed in a similar fashion as the augmented dataset we had used for training. Additionally, certain tokens such as "[NAME]", "[LOCATION]", "[CONTACT]", "[DATE]", "SUBJECT: " and "MESSAGE: " were removed from the test dataset to avoid their appearance in the generated summaries.

Table 2 shows the summaries generated by various transformer based models for a sample question in the test set. From the table it can be observed that, the summaries generated by t5-base and t5-small are almost similar and don't actually capture the main focus of the question. The summary generated by pegasus-xsum is similar but longer than those produced by the T5 models. However, the summary generated by the pegasus-wikihow model is quite apt. The bart-large-cnn model produced a summary which is although grammatically correct, the meaning is incorrect. The bart-large-xsum generated the best summary amongst all the models, because it is both precise and short in length.

The HOLMS (Mrabet and Demner-Fushman, 2020) and BERTScores (Zhang* et al., 2020) for the different models used are referenced in Table 3. Based on the experiments, it was observed that the bart-large-xsum model achieved the best performance in terms of both metrics. Based on this performance, our team ranked 2nd in the BERTScore metric and secured 6th position in HOLMS score, on the leaderboard.

Table 2: Sample summary generated by various models for the test question: *"Gadolinum toxicity and MCS relationship? I have 2 Genovia Labs test results years apart with seriously high Gadolinum toxicity. AND I am very VERY VERY very challenged by MCS - Multiple Chemical Sensitivity. My question is: If I had multiple MARs after an auto accident. And since then the MCS is debilitating. Certainly the symptoms of Gas level in my body cause symptoms as well. But I am debilitated by Synthetic chemicals in the air. How can I find out if the Gas exhaserbated my reaction to exhaust fumes, air fresheners, perfumes, dryer sheets(!!!!), food additives, and much more. Many Thanks"*

Model	Generated Summary
bart-large-xsum	What is the relationship between Gadolinum toxicity and MCS?
bart-large-cnn	What are the causes of and treatments for Multiple Chemical Sensitivity?
pegasus-xsum	How can I find out if synthetic chemicals in the air cause my reaction to exhaust fumes, air fresheners, perfumes, dryer sheets, food additives?
pegasus-wikihow	Where can I find information on Gadolinum toxicity and MCS relationship?
t5-base	How can I find out if gas exhaserbated my reaction to exhaust fumes, air fresheners, perfumes,?
t5-small	How can I find out if the Gas exhaserbated my reaction to exhaust fumes, air fresheners, perfumes?

Table 3: HOLMS and BERTScore F1 performance of the proposed models, for the Question Summarization task

Model	HOLMS	BERTScore-F1
bart-large-xsum	**0.566**	**0.702**
bart-large-cnn	0.556	0.692
pegasus-xsum	0.544	0.674
pegasus-wikihow	0.535	0.665
t5-base	0.550	0.681
t5-small	0.537	0.633

6 Conclusion and Future Work

In this paper, we presented models that explore the use of transfer learning to utilize the knowledge of NLP transformers like BART, T5 and PEGASUS for the task of question summarization. The observed scores and the sample summaries generated by different transformer architecture based models clearly delineated the best performing model among the ones proposed. The summaries produced by the bart-large-xsum achieved the best score, followed by the pegasus-wikihow model. This can be largely attributed to the transfer learning technique that was adapted, by utilizing models which are pre-trained on massive datasets. As part of future work for the question summarization task, we plan to exploit question type feature, in addition to the currently used question focus feature for further enhancing the performance.

References

Anumeha Agrawal, Rosa Anil George, Selvan Suntiha Ravi, Sowmya Kamath, and Anand Kumar. 2019. Ars_nitk at mediqa 2019: Analysing various methods for natural language inference, recognising question entailment and medical question answering system. In *Proceedings of the 18th BioNLP Workshop and Shared Task*, pages 533–540.

Dzmitry Bahdanau, Kyunghyun Cho, and Yoshua Bengio. 2016. Neural machine translation by jointly learning to align and translate.

Asma Ben Abacha, Eugene Agichtein, Yuval Pinter, and Dina Demner-Fushman. 2017. Overview of the medical question answering task at trec 2017 liveqa. In *TREC 2017*.

Asma Ben Abacha and Dina Demner-Fushman. 2016. Recognizing question entailment for medical question answering. In *AMIA 2016, American Medical Informatics Association Annual Symposium, Chicago, IL, USA, November 12-16, 2016*.

Asma Ben Abacha and Dina Demner-Fushman. 2019. On the summarization of consumer health questions. In *Proceedings of the 57th Annual Meeting of the*

Association for Computational Linguistics, pages 2228–2234, Florence, Italy. Association for Computational Linguistics.

Asma Ben Abacha, Yassine Mrabet, Yuhao Zhang, Chaitanya Shivade, Curtis Langlotz, and Dina Demner-Fushman. 2021. Overview of the mediqa 2021 shared task on summarization in the medical domain. In *Proceedings of the 20th SIG-BioMed Workshop on Biomedical Language Processing, NAACL-BioNLP 2021*. Association for Computational Linguistics.

Jacob Devlin, Ming-Wei Chang, Kenton Lee, and Kristina Toutanova. 2019. Bert: Pre-training of deep bidirectional transformers for language understanding.

Rosa George, Selvan Sunitha, and S Sowmya Kamath. 2021. Benchmarking semantic, centroid, and graph-based approaches for multi-document summarization. In *Intelligent Data Engineering and Analytics*, pages 255–263. Springer.

Tatsuya Ishigaki, Hiroya Takamura, and Manabu Okumura. 2017. Summarizing lengthy questions. In *Proceedings of the Eighth International Joint Conference on Natural Language Processing (Volume 1: Long Papers)*, pages 792–800, Taipei, Taiwan. Asian Federation of Natural Language Processing.

Mike Lewis, Yinhan Liu, Naman Goyal, Marjan Ghazvininejad, et al. 2019. Bart: Denoising sequence-to-sequence pre-training for natural language generation, translation, and comprehension.

Veena Mayya, Sowmya Kamath, Gokul S Krishnan, and Tushaar Gangavarapu. 2021. Multi-channel, convolutional attention based neural model for automated diagnostic coding of unstructured patient discharge summaries. *Future Generation Computer Systems*, 118:374–391.

Clara H. McCreery, Namit Katariya, Anitha Kannan, Manish Chablani, and Xavier Amatriain. 2020. Effective transfer learning for identifying similar questions: Matching user questions to covid-19 faqs.

Yassine Mrabet and Dina Demner-Fushman. 2020. HOLMS: Alternative summary evaluation with large language models. In *Proceedings of the 28th International Conference on Computational Linguistics*, pages 5679–5688, Barcelona, Spain (Online). International Committee on Computational Linguistics.

Alec Radford, Jeffrey Wu, Rewon Child, David Luan, Dario Amodei, and Ilya Sutskever. 2019. Language models are unsupervised multitask learners. *OpenAI blog*, 1(8):9.

Colin Raffel, Noam Shazeer, Adam Roberts, Katherine Lee, Sharan Narang, Michael Matena, Yanqi Zhou, Wei Li, and Peter J. Liu. 2020. Exploring the limits of transfer learning with a unified text-to-text transformer.

Akshay Upadhya, Swastik Udupa, and S Sowmya Kamath. 2019. Deep neural network models for question classification in community question-answering forums. In *2019 10th International Conference on Computing, Communication and Networking Technologies (ICCCNT)*, pages 1–6. IEEE.

Jingqing Zhang, Yao Zhao, Mohammad Saleh, and Peter Liu. 2020. Pegasus: Pre-training with extracted gap-sentences for abstractive summarization. In *International Conference on Machine Learning*, pages 11328–11339. PMLR.

Tianyi Zhang*, Varsha Kishore*, Felix Wu*, Kilian Q. Weinberger, and Yoav Artzi. 2020. Bertscore: Evaluating text generation with bert. In *International Conference on Learning Representations*.

Optum at MEDIQA 2021: Abstractive Summarization of Radiology Reports using simple BART Finetuning

Ravi Kondadadi and **Sahil Manchanda** and **Jason Ngo** and **Ronan McCormack**

Optum

Abstract

This paper describes experiments undertaken and their results as part of the BioNLP MEDIQA 2021 challenge. We participated in Task 3: Radiology Report Summarization. Multiple runs were submitted for evaluation, from solutions leveraging transfer learning from pre-trained transformer models, which were then fine tuned on a subset of MIMIC-CXR, for abstractive report summarization. The task was evaluated using ROUGE and our best performing system obtained a ROUGE-2 score of 0.392.

1 Introduction

A BioNLP 2021 shared task, the MEDIQA challenge aims to attract research efforts in NLU across three summarization tasks in the medical domain: multi-answer summarization, and radiology report summarization. We participated in the radiology report summarization and offer experiments and results. A radiology report describes an exam and patient information resulting from trained clinicians(radiologists) interpreting imaging studies during routine clinical care (Zhang et al., 2018). The primary purpose of the report is for radiologists to communicate imaging results to ordering physicians (Gershanik et al., 2011). A standard report will consist of a Background section which will contain details of the patient and describe the examination undertaken, A findings section, in which the radiologist has dictated the initial results into the report, and an Impression section. The Impression section consists of a concise summarization of the most relevant details from the exam based on the dictated findings. Although guidelines for the practice of generating radiology reports are outlined by the American College of Radiology (ACR), there is flexibility in the document in the usage of terms for describing findings and where they are documented. This can lead to referring physicians focusing on just the impressions section of the document (Hall, 2012). Additionally, the process of writing the impressions from the dictation of the findings is time-consuming and repetitive. In this work we propose experiments to automate the generation of the impressions section from the findings of the radiology report, accelerating the radiology workflow and improving the efficiency of clinical communications. Experiments were performed implementing sequence to sequence models with encoder-decoder architecture like BART (Lewis et al., 2019), Pegasus (Zhang et al., 2020a), and T5 (Raffel et al., 2020). These models were then further fine-tuned on a subset of MIMIC-CXR Dataset (Johnson et al., 2019), to generate abstractive summaries from the findings section of the report. MIMIC-CXR is de-identified and Protected health information (PHI) removed, large publicly available dataset of chest radiographs in DICOM format with free-text radiology reports. A subset of MIMIC-CXR and Indiana datasets [1] used for validation carried out using standard ROUGE (Lin, 2004) metrics.

2 Related Work

Initial efforts on summarization were mainly focused on Extractive summarization. Extractive summarization is the process involving extraction of noteworthy words from the text to form a summary. (Luhn, 1958; Kupiec et al., 1995) The advent of Neural network models enabled Abstractive summarization, which involves producing new words to convey the meaning of the text. This involves rephrasing the text in a shorter and more succinct form using similar but not the exact words used in the main text.

Nallapati et al. (2016) proposed an RNN based approach to not only achieve state-of-the-art results in extractive summarization but also enable this model to be trained on abstractive summaries.

[1] https://openi.nlm.nih.gov/faq/collection

Proceedings of the BioNLP 2021 workshop, pages 280–284
June 11, 2021.

Rush et al. (2015) described an attention-based summarization approach where an encoder and a generator model are jointly trained on article pairs. Their work builds on attention-based encoders that are used in neural machine translation (Bahdanau et al. (2016)). Fan et al. (2018) build on the previous work on abstractive summarization to create length constrained summaries and summaries concentrated on particular entities and subjects in the text. Paulus et al. (2017) used intra-temporal attention to produce state-of-the-art results on CNN/Daily Mail dataset.

The work on summarizing radiology reports started with the extraction of information from the text (Friedman et al., 1995; Hassanpour and Langlotz, 2016). For instance, Cornegruta et al. (2016) proposed using clinical language understanding of a radiology report to extract Named entities. A Bidirectional LSTM architecture was used to achieve this. Zhang et al. (2018) describes one of the first attempts at automatic summarization of radiology reports. This work describes an encoder-decoder architecture. Both the encoder and decoder sides are made of Bidirectional LSTMs using the attention framework (Bahdanau et al., 2016).

With the advent of transformers, Pretraining based language generation has been the norm in summarization. Zhang et al. (2019) and Liu (2019) used BERT (Devlin et al., 2019), a pre-trained transformer model on extractive summarization, and achieved state of the art results. Sotudeh et al. (2020) proposed an approach to content selection for abstractive text summarization in clinical notes. Zhang et al. (2020b) presented a general framework and a training strategy to improve the factual correctness of neural abstractive summarization models for radiology reports. In this work, we fine-tune a pre-trained BART architecture (Lewis et al., 2019) for the radiology report summarization task.

3 Task Description & Dataset

The objective of this task is to generate summary of a given radiology report. The training data for the MEDIQA 2021 Radiology report summarization shared task is extracted from a subset from the MIMIC-CXR Dataset (Johnson et al., 2019). The training set contains around 91,544 examples of radiology reports and the corresponding summaries. Each example contains three fields; Findings field contains the original human-written radiology findings text, impression contains the human-written radiology impression text and background contains background information of the study in text format. One can use both the findings and the background fields to generate the summary. There are two development sets that come from two different institutes. The first development set from MIMIC-CXR contains around 2000 examples. There is another development set that also contains 2000 examples from the Indiana University radiology report dataset (Johnson et al., 2019). In all our experiments, we first trained our model on the training set and tested on the validation set. For the actual task submissions, we trained our models by combining training set and both the development sets.

4 Method & Results

Our proposed method leverages pretrained summarization models. We finetuned three types of pretrained models for the radiology report summarization; BART (Lewis et al., 2019), T5 (Raffel et al., 2020) and Pegasus (Zhang et al., 2020a). We used Huggingface Transformers (Wolf et al., 2020) library for finetuning.

BART: Developed by Facebook, BART is a denoising autoencoder. Since it uses the standard transformer-based neural machine translation architecture, it is a generalization of both BERT and GPT3 (Brown et al., 2020). For pretraining, it was trained by shuffling the order of sentence (an extension of next sentence prediction) and text infilling (an extension of the language masking). During text infilling, random spans of text are replaced by masked tokens. The job of the model during training is to recreate this span. Due to its flexible transformer architecture, the inputs to the encoder do not need to be aligned with the outputs of the decoder. This enables the BART model to be trained on a variety of tasks such as token masking, token deletion, sentence permutation, document rotation, etc. Since BART has an autoregressive decoder, it is better suited for sequence generation tasks such as summarization.

T5: T5 stands for Text-To-Text Transfer Transformer. It is a sequence-to-sequence model that takes in text and outputs text. This text-to-text framework enables one to use the same model, loss

function, and hyperparameters on any NLP task, which can range from document summarization to classification. As a result, the way that data is fed into the model is quite different from models like BERT. The task description is used as a prefix to the input. For example, to translate a sentence from English to French, the input would be prefixed with "translate English to French:" Similarly, to summarize a passage, you would add the prefix "summarize:" followed by the text to be summarized. This text-to-text framework uses the same model across a range of tasks. T5 model made improvements on a wide range of categories such as model architecture, and pretraining objectives.
T5 uses the standard transformer architecture (Vaswani et al., 2017). For pretraining, T5 was trained on denoising, where spans of text are replaced with the drop token. The model objective is to reproduce the span of text given the drop token.

Pegasus: PEGASUS (Pre-training with Extracted Gap-sentences for Abstractive SUmmarization Sequence-to- sequence) Pegasus starts with the concept that if the pretraining task and fine-tuning task are closely related, then the model will perform better. As a result, they designed a pretraining task specifically for abstractive summarization. This pretraining task, gap-sentence generation, removes entire sentences from documents. The model's learning objective is to recover these sentences in the concatenated model output. Instead of randomly removing sentences, only the important sentences are removed, so that the model can reproduce these sentences that summarize the text. As a result of this pretraining task, Pegasus can achieve results like T5 with 5% of the parameters.

Table 1 shows the model performance of each participant in the leaderboard for the top 10 teams. Only Rouge-2 F1 is shown because that was the metric used to rank the teams in this task. Our method ranked third on the leaderboard.

System	Rouge-2 F1
Baidu	0.436
IBM	0.408
Optum	0.392
QIAI	0.378
Low-rank-AI	0.331
CMU	0.327
ChicHealth	0.324
healthAI	0.308
DAMO-ALI	0.276
Fudan University	0.274

Table 1: Top 10 teams on the leaderboard

Run	Rouge 1	Rouge 2	Rouge L
1	60.51	48.14	57.65
2	52.35	40.98	50.41
3	35.72	22.69	31.53
4	63.47	51.35	60.54
5	56.14	44.65	53.98
6	37.8	24.73	33.80
7	58.59	46.5	56.01
8	62.85	51.22	60.25

Table 2: Evaluation of Radiology Report Summarization on the development set

4.1 Experiments

We propose eight different runs for this task. Table 2 shows the evaluation of different models we experimented with on the development set. We experimented with different versions of BART, T5 and Pegasus on Huggingface Transformers. We ended up using BART-base, T5-small, T5-base and Pegasus-Pubmed due to memory limitations of our GPUs. The following set of hyperparameters are applied for the following runs. Learning rate=5e-05 , number of epochs=15, gradient accumulation steps=5. The evaluations results of various runs for the radiology report summarization task are summarized in Table 2.

1. Our first proposed method is based on BART-base. We finetuned BART-base on the training set and tested on the development set. We used a batch size of 20 for both training and validation sets.

2. In this run, we used T5-small and finetuned on the training set. We used a batch size of 20 for both training and validation sets.

3. In our third run, we finetuned on pegasus-pubmed. We were able to use only a smaller batch size of 2.

4. The fourth run is similar to the first approach, but we also used the background section in

addition to findings. In this case, we were able to use a batch size of 10.

5. This run is same as the fourth one, but we used T5-small as our base model. A batch size of 10 was used.

6. In this run, but we used Pegasus-pubmed as our base model. A batch size of 1 was used.

7. This run is same as the first run, but we used T5-base as the base model. A batch size of 10 was used.

8. In this run also, we used T5-base as the base model except that we also used background section. A batch size of 2 was used.

Overall, the best results on the test set are achieved using the BART-base as the pre-trained model. The model is trained using just the findings section on the test set. But on the development set, using the background section in addition to the findings helped.

5 Conclusion

In this paper, we present all our experiments of fine-tuning pre-trained models for radiology report summarization. Our experiments demonstrate how an encoder-decoder architecture like BART, which achieved state-of-the-art results in text generation tasks outperforms other architectures in this particular task. Our methods proved effective on the summarization task and were ranked third on the leaderboard.

References

Dzmitry Bahdanau, Kyunghyun Cho, and Yoshua Bengio. 2016. Neural machine translation by jointly learning to align and translate.

Tom B. Brown, Benjamin Mann, Nick Ryder, Melanie Subbiah, Jared Kaplan, Prafulla Dhariwal, Arvind Neelakantan, Pranav Shyam, Girish Sastry, Amanda Askell, Sandhini Agarwal, Ariel Herbert-Voss, Gretchen Krueger, Tom Henighan, Rewon Child, Aditya Ramesh, Daniel M. Ziegler, Jeffrey Wu, Clemens Winter, Christopher Hesse, Mark Chen, Eric Sigler, Mateusz Litwin, Scott Gray, Benjamin Chess, Jack Clark, Christopher Berner, Sam McCandlish, Alec Radford, Ilya Sutskever, and Dario Amodei. 2020. Language models are few-shot learners.

Savelie Cornegruta, Robert Bakewell, Samuel Withey, and Giovanni Montana. 2016. Modelling radiological language with bidirectional long short-term memory networks. In *Proceedings of the Seventh International Workshop on Health Text Mining and Information Analysis*, pages 17–27, Auxtin, TX. Association for Computational Linguistics.

Jacob Devlin, Ming-Wei Chang, Kenton Lee, and Kristina Toutanova. 2019. Bert: Pre-training of deep bidirectional transformers for language understanding.

Angela Fan, David Grangier, and Michael Auli. 2018. Controllable abstractive summarization.

C. Friedman, G. Hripcsak, W. DuMouchel, S. B. Johnson, and P. D. Clayton. 1995. Natural language processing in an operational clinical information system. *Natural Language Engineering*, 1(1):83–108.

Esteban F. Gershanik, Ronilda Lacson, and Ramin Khorasani. 2011. Critical Finding Capture in the Impression Section of Radiology Reports. *AMIA Annu. Symp. Proc.*, 2011:465.

Ferris M. Hall. 2012. Language of the Radiology Report. *Am. J. Roentgenol.*

Saeed Hassanpour and Curtis P Langlotz. 2016. Information extraction from multi-institutional radiology reports. *Artificial intelligence in medicine*, 66:29—39.

Alistair E. W. Johnson, Tom Pollard, Roger Mark, Seth Berkowitz, and Steven Horng. 2019. The MIMIC-CXR Database. Type: dataset.

Julian Kupiec, Jan Pedersen, and Francine Chen. 1995. A trainable document summarizer. In *Proceedings of the 18th Annual International ACM SIGIR Conference on Research and Development in Information Retrieval*, SIGIR '95, page 68–73, New York, NY, USA. Association for Computing Machinery.

Mike Lewis, Yinhan Liu, Naman Goyal, Marjan Ghazvininejad, Abdelrahman Mohamed, Omer Levy, Ves Stoyanov, and Luke Zettlemoyer. 2019. Bart: Denoising sequence-to-sequence pre-training for natural language generation, translation, and comprehension.

Chin-Yew Lin. 2004. ROUGE: A package for automatic evaluation of summaries. In *Text Summarization Branches Out*, pages 74–81, Barcelona, Spain. Association for Computational Linguistics.

Yang Liu. 2019. Fine-tune bert for extractive summarization.

H. P. Luhn. 1958. The automatic creation of literature abstracts. *IBM Journal of Research and Development*, 2(2):159–165.

Ramesh Nallapati, Feifei Zhai, and Bowen Zhou. 2016. Summarunner: A recurrent neural network based sequence model for extractive summarization of documents.

Romain Paulus, Caiming Xiong, and Richard Socher. 2017. A deep reinforced model for abstractive summarization.

Colin Raffel, Noam Shazeer, Adam Roberts, Katherine Lee, Sharan Narang, Michael Matena, Yanqi Zhou, Wei Li, and Peter J. Liu. 2020. Exploring the limits of transfer learning with a unified text-to-text transformer.

Alexander M. Rush, Sumit Chopra, and Jason Weston. 2015. A neural attention model for abstractive sentence summarization.

Sajad Sotudeh, Nazli Goharian, and Ross W. Filice. 2020. Attend to medical ontologies: Content selection for clinical abstractive summarization.

Ashish Vaswani, Noam Shazeer, Niki Parmar, Jakob Uszkoreit, Llion Jones, Aidan N. Gomez, Lukasz Kaiser, and Illia Polosukhin. 2017. Attention is all you need.

Thomas Wolf, Lysandre Debut, Victor Sanh, Julien Chaumond, Clement Delangue, Anthony Moi, Pierric Cistac, Tim Rault, Rémi Louf, Morgan Funtowicz, Joe Davison, Sam Shleifer, Patrick von Platen, Clara Ma, Yacine Jernite, Julien Plu, Canwen Xu, Teven Le Scao, Sylvain Gugger, Mariama Drame, Quentin Lhoest, and Alexander M. Rush. 2020. Huggingface's transformers: State-of-the-art natural language processing.

Haoyu Zhang, Jianjun Xu, and Ji Wang. 2019. Pretraining-based natural language generation for text summarization.

Jingqing Zhang, Yao Zhao, Mohammad Saleh, and Peter J. Liu. 2020a. Pegasus: Pre-training with extracted gap-sentences for abstractive summarization.

Yuhao Zhang, Daisy Yi Ding, Tianpei Qian, Christopher D. Manning, and Curtis P. Langlotz. 2018. Learning to summarize radiology findings.

Yuhao Zhang, Derek Merck, Emily Bao Tsai, Christopher D. Manning, and Curtis P. Langlotz. 2020b. Optimizing the factual correctness of a summary: A study of summarizing radiology reports.

QIAI at MEDIQA 2021: Multimodal Radiology Report Summarization

Jean-Benoit Delbrouck and **Han Zhang** and **Daniel L. Rubin**
Laboratory of Quantitative Imaging and Artificial Intelligence
Stanford University
{jbdel, hzhang20, dlrubin}@stanford.edu

Abstract

This paper describes the solution of the QIAI lab sent to the Radiology Report Summarization (RRS) challenge at MEDIQA 2021. This paper aims to investigate whether using multimodality during training improves the summarizing performances of the model at test-time. Our preliminary results shows that taking advantage of the visual features from the x-rays associated to the radiology reports leads to higher evaluation metrics compared to a text-only baseline system. These improvements are reported according to the automatic evaluation metrics METEOR, BLEU and ROUGE scores. Our experiments can be fully replicated at the following address : `https://github.com/jbdel/vilmedic`.

1 Introduction

Radiology report summarization is a growing area of research. Given the Findings and Background sections of a radiology report, the goal is to generate a summary (called an impression section in radiology reports) that highlights the key observations and conclusion of the radiology study. Automating this summarization task is critical because the impression section is the most important part of a radiology report, and manual summarization can be time-consuming and error-prone.

This paper describes the solution of the QIAI lab sent to the Radiology Report Summarization (RRS) challenge at MEDIQA 2021 (Ben Abacha et al., 2021). This challenge aims to promote the development of clinical summarization models that generate radiology impression statements by summarizing textual findings written by radiologists. Since for most reports, the associated x-rays are available, we aim to evaluate if incorporating visual features from x-rays helps our systems for the report summarization task. This task could be defined as Multimodal Radiology Report Summarization (MRRS) as depicted in Figure 1.

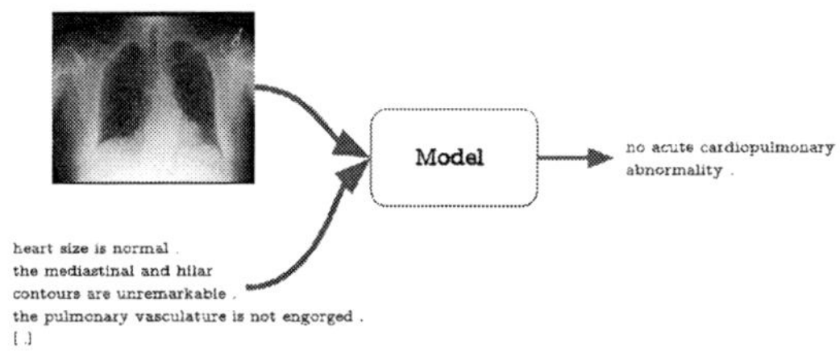

Figure 1: An example of Multimodal Radiology Report Summarization.

2 Data Collection

The training set consists of 91,544 examples taken from the MIMIC-CXR v2.0 dataset (Johnson et al., 2019). Each training example is a free-text chest radiology report that contains the Background, Findings and Impression sections. Two validation sets, each with 2,000 reports were used. One validation set was collected from MIMIC, and the other was collected from the Indiana University Chest X-Rays Report dataset (Indiana-University). The test set contains 300 reports from the Indiana University dataset, and 300 reports from Stanford University School of Medicine. All report sections were tokenized using the Stanford CoreNLP tokenizer (Manning et al., 2014).

split	#report	#report w/o image
mimic-train	91,544	0
mimic-dev	2,000	0
indiana-dev	2,000	53
stanford-test	300	300
indiana-test	300	4

Table 1: Splits statistics from the MEDIQA 2021 challenge.

Proceedings of the BioNLP 2021 workshop, pages 285–290
June 11, 2021.

3 Model

This section describes the two architectures that will be bench-marked in the result section. We start by describing the text-based monomodal architecture at section 3.1. This model only takes as input the findings section and outputs the impression section (the summary). In section 3.2, we incorporate visual information into the monomodal architecture to make it multimodal.

3.1 Monomodal architecture

Given the report's Findings section of M words $\boldsymbol{X} = (\boldsymbol{x}_1, \boldsymbol{x}_2, \ldots, \boldsymbol{x}_M)$, an attention-based encoder-decoder model (Bahdanau et al., 2014) outputs its summary $\boldsymbol{Y} = (\boldsymbol{y}_1, \boldsymbol{y}_2, \ldots, \boldsymbol{y}_N)$. If we denote $\boldsymbol{\theta}$ as the model parameters, then $\boldsymbol{\theta}$ is learned by maximizing the likelihood of the observed sequence $\boldsymbol{Y}$ or in other words by minimizing the cross entropy loss. The objective function is given by:

$$\mathcal{L}(\boldsymbol{\theta}) = -\sum_{t=1}^{n} \log p_{\boldsymbol{\theta}}(\boldsymbol{y}_t | \boldsymbol{y}_{<t}, X) \quad (1)$$

The encoder-decoder model consists of three components : an encoder, a decoder and an attention mechanism.

Encoder At every time-step t, an encoder creates an annotation $\boldsymbol{h}_t$ according to the current embedded word $\boldsymbol{x}'_t$ and internal state $\boldsymbol{h}_{t-1}$:

$$\boldsymbol{h}_t = f_{\text{enc}}(\boldsymbol{x}'_t, \boldsymbol{h}_{t-1}) \quad (2)$$

Every word $\boldsymbol{x}_t$ of the input sequence $\boldsymbol{X}$ is an index in the embedding matrix $\boldsymbol{E}^x$ so that the following formula maps the word to the f_{enc} size S:

$$\boldsymbol{x}'_t = \boldsymbol{W}^x \boldsymbol{E}^x x_t \quad (3)$$

The total size of the embeddings matrix $\boldsymbol{E}^x$ depends on the source vocabulary size $|\mathcal{Y}_s|$ and the embedding dimension d such that $\boldsymbol{E}^x \in \mathbb{R}^{|\mathcal{Y}_s| \times d}$. The mapping matrix $\boldsymbol{W}^x$ also depends on the embedding dimension because $\boldsymbol{W}^x \in \mathbb{R}^{d \times S}$.

The encoder function f_{enc} is a bi-directional GRU (Cho et al., 2014). The following equations define a single GRU block (called f_{gru} for future references) :

$$\begin{aligned}
\boldsymbol{z}_t &= \sigma\left(\boldsymbol{x}'_t + \boldsymbol{W}^z \boldsymbol{h}_{t-1}\right) \\
\boldsymbol{r}_t &= \sigma\left(\boldsymbol{x}'_t + \boldsymbol{W}^r \boldsymbol{h}_{t-1}\right) \\
\underline{\boldsymbol{h}}_t &= \tanh\left(\boldsymbol{x}'_t + \boldsymbol{r}_t \odot (\boldsymbol{W}^h \boldsymbol{h}_{t-1})\right) \\
\boldsymbol{h}_t &= (1 - \boldsymbol{z}_t) \odot \underline{\boldsymbol{h}}_t + \boldsymbol{z}_t \odot \boldsymbol{h}_{t-1}
\end{aligned} \quad (4)$$

where $\boldsymbol{h}_t \in \mathbb{R}^S$. Our encoder consists of two GRUs, one is reading the input sentence from 1 to M and the second from M to 1. Therefore the encoder annotation $\overline{\boldsymbol{h}_t}$ for timestep t is the concatenation of both GRUs annotations $\boldsymbol{h}_t$. The encoder set of annotations $\boldsymbol{H}$ contains the annotations $\overline{\boldsymbol{h}}$ of each timestep and is of size $M \times 2S$.

Decoder At every time-step t, a decoder outputs probabilities $\boldsymbol{p}_t$ over the target vocabulary $\mathcal{Y}_d$ according to previously generated word $\boldsymbol{y}_{t-1}$, internal state $\boldsymbol{s}_{t-1}$ and encoder annotations $\boldsymbol{H}$:

$$y_t \sim \boldsymbol{p}_t = f_{\text{dec}}(\boldsymbol{y}'_{t-1}, \boldsymbol{s}_{t-1}, \boldsymbol{H}) \quad (5)$$

Every word $\boldsymbol{y}_t$ of the summarized report $\boldsymbol{Y}$ is an index in the embedding matrix $\boldsymbol{E}^y$ so that the following formula maps the word in the f_{dec} size D:

$$\boldsymbol{y}'_t = \boldsymbol{W}^y \boldsymbol{E}^y y_{t-1} \quad (6)$$

The decoder function f_{dec} consists of two parts: a conditional GRU (f_{cgru}) and a bottleneck function (f_{bot}).
The following equations describe the cGRU function f_{cgru}:

$$\begin{aligned}
\boldsymbol{s}'_t &= f_{\text{gru}_1}(\boldsymbol{y}'_t, \boldsymbol{s}_{t-1}) \\
\boldsymbol{c}_t &= f_{\text{att}}(\boldsymbol{s}'_t, \boldsymbol{H}) \\
\boldsymbol{s}_t &= f_{\text{gru}_2}(\boldsymbol{s}'_t, \boldsymbol{c}_t)
\end{aligned} \quad (7)$$

where f_{att} is the soft linguistic attention module over the set of source annotation $\boldsymbol{H}$:

$$\begin{aligned}
\boldsymbol{a}'_t &= \boldsymbol{W}^a \tanh(\boldsymbol{W}^s \boldsymbol{s}'_t + \boldsymbol{W}^{\text{H}} \boldsymbol{H}) \\
\boldsymbol{a}_t &= \text{softmax}(\boldsymbol{a}'_t) \\
\boldsymbol{c}'_t &= \sum_{i=0}^{M-1} \boldsymbol{a}_{t_i} \boldsymbol{h}_i \\
\boldsymbol{c}_t &= \boldsymbol{W}^c \boldsymbol{c}'_t
\end{aligned} \quad (8)$$

The bottleneck function f_{bot} projects the cGRU output s_t into probabilities over the target vocabulary. It is defined as such:

$$b_t = \tanh(W^{\text{bot}}[s_t, c_t]$$
$$y_t \sim p_t = \text{softmax}(W^{\text{proj}} b_t) \quad (9)$$

where $[\cdot, \cdot]$ denotes the concatenation operation.

3.2 Multimodal architecture

In the MIMIC dataset, a report can be associated with multiple x-rays images. We pick only one image according to the following priority: PA, AP, LATERAL, AP AXIAL, LL. Using this setting, we can select one image to each report. The Indiana dataset has at most one image associated with each report. In case no image is provided, we input a representation of "zeros" to the pipeline.

For each image, we extract the "pool0" representation of a DenseNet121 (Huang et al., 2017) architecture pretrained on x-rays images made available by the TorchXRayVision library (Cohen et al., 2020). The representation for each image is a vector of 1024 features that we call v in the following equations.

We consider three approaches to integrate the vector v to the monomodal architecture presented in Section 3.1. First, the **encdecinit** policy that consists of initializing both the encoder and decoder state h_0 and s_0 with the visual features as such:

$$h_0 = \tanh(W^{vh0} v)$$
$$s_0 = \tanh(W^{vs0} v) \quad (10)$$

The second one is **ctxmul** that performs the element-wise product of each encoder annotations $\overline{h_i}$ with v:

$$\overline{h_i} = \overline{h_i} \odot W^{vhi} v \text{ for i} = 1 \text{ to } M \quad (11)$$

Finally, the **trgmul** policy consists of the element-wise product of each target embedding of equation 6 with v:

$$y'_t = y'_t \odot W^{vy} v \quad (12)$$

Matrices W^{vh0}, W^{vs0}, W^{vhi}, W^{vy} are trainable weights that transform and map v to right dimension.

Finally, we define a fourth approach, **allv**, using all the aforementioned interactions.

4 Settings

Both monomodal and multimodal architectures use a 2-layered bi-directional GRU for the encoder, and 1-layered GRU for the decoder. Each GRU has a hidden size of 320 units and our embeddings are of size 200. We apply dropout of 0.4 on the source embeddings x'_t, 0.5 on the source annotations H and 0.5 on the bottleneck b_t.

We chose Adam (Kingma and Ba, 2014) as the optimizer with a learning rate of 0.0004 and batch size 64. Model parameters are initialized using the He initialization method (He et al., 2015). We evaluate the model performance using the ROUGE-2 F1 metrics (Lin, 2004), which is commonly used for evaluating machine summarization task. We stop training when the ROUGE score does not improve for 10 evaluations on the validation set. In the experiment section, we also report the METEOR (Banerjee and Lavie, 2005) and BLEU metrics (Papineni et al., 2002).

In the scope of this paper, we only use the findings section as input to our models and discard the background section.

5 Experiments

The experiments are carried out as follows:

1. We define two settings for the dataset splits. The first one is dictated by the challenge as defined in Table 1. We call it **regular-split**. The second setting consists of injecting 1500 out of the 2000 indiana-dev samples into the training set. We keep the remaining 500 for development. This setting allows more training homogeneity compared to regular-split, we refer to it as the **mix-split**;

2. We use our monomodal architecture to predict summarization for both the stanford and indiana test sets. We use the multimodal architecture to predict summarization only on the indiana test set (the stanford test set having no x-rays available). Note that both architectures are trained with the same number of samples.

Figure 2 and 3 depict the results of the best scoring configurations for the monomodal and multimodal models on the development sets. Each results is obtained by using beam-search with width varying from 8 to 12. Finally, E5 means results are from an ensemble of 6 trained models (i.e. model ensembling).

Model	BLEU	METEOR	R2-F1
	indiana-dev		
Mono	13.94	16.47	31.33
Multi allv	13.27	15.19	26.84
Mono E5	**15.88**	**17.67**	**31.37**
Multi E5 allv	15.27	17.12	30.42
	mimic-dev		
Mono	28.67	25.74	47.96
Multi allv	28.90	26.01	48.19
Mono E5	28.66	25.74	48.41
Multi E5 allv	**29.31**	**26.24**	**48.86**

Table 2: Results of our best multimodal and monomodal architectures on the development sets (regular-split).

Model	BLEU	METEOR	R2-F1
	indiana-dev		
Mono	26.93	24.50	52.18
Multi allv	27.21	24.60	51.79
Mono E5	26.61	24.35	52.02
Multi E5 allv	**28.32**	**25.30**	**54.38**
	mimic-dev		
Mono	29.00	25.90	48.10
Multi allv	28.30	25.48	48.47
Mono E5	**28.97**	25.95	48.38
Multi E5 allv	**28.97**	**26.10**	**48.98**

Table 3: Results of our best multimodal and monomodal architectures on the development sets (mix-split).

A few observations can be made. First, three of four best scoring models (highlighted in bold) is the multimodal variant. Each time, the multimodal model is using the *allv* interaction. It means that injecting the visual features from the x-rays in both the encoder and the decoder improves summarization.

Secondly, the only instance where the monomodal variant is better is on the indiana-dev set using the regular-split. One could hypothesize that the multimodal model is sensitive to distribution shift; indeed no indiana samples (and therefore indiana x-rays) are in the training set for this configuration. Though using model ensembling seems to mitigates the performance drop, it is still lower that the monomodal baseline.

Finally, we underline the ROUGE scores from systems that are significantly different (p-value $\leq$ 0.05) than the baseline *mono* models using the approximate randomization test of multeval (Clark et al., 2011). The underlined scores are all from multimodal systems.

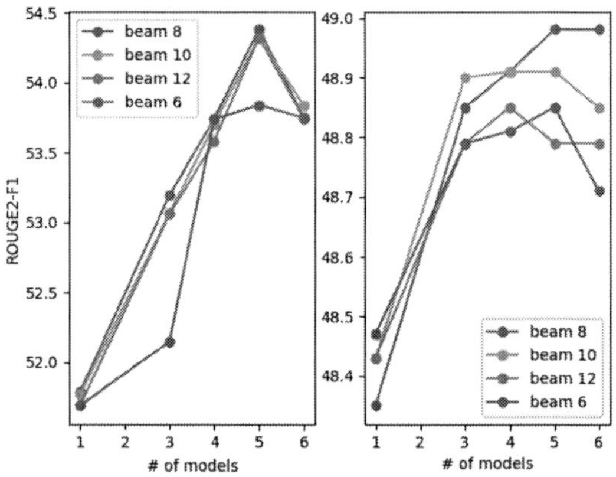

Figure 2: Effect of ensembling and beam-size width for model *Multi E5 allv* (mix-split setting). Left concerns split indiana-dev and right mimic-dev.

6 Related Work

Though relatively new, a few previous work can be denoted in the field of radiology report summarization. Zhang et al. (2018) first studied the problem of automatic generation of radiology impressions by summarizing textual radiology findings, and showed that an augmented pointer-generator model achieves high overlap with human references. This model has been extended with an ontologyaware pointer-generator and showed improved summarization quality (MacAvaney et al., 2019). RL-based approaches have been investigated by Li et al. (2018) and (Liu et al., 2019).

More recently, (Zhang et al., 2020) developed a general framework where the evaluation of the factual correctness of a generated summary is done by factchecking it automatically against its reference using an information extraction module.

To our knowledge, this work is the first attempt to use multimodality for radiology report summarization.

References

Dzmitry Bahdanau, Kyunghyun Cho, and Yoshua Bengio. 2014. Neural machine translation by jointly learning to align and translate. *CoRR*, abs/1409.0473.

Satanjeev Banerjee and Alon Lavie. 2005. METEOR: An automatic metric for MT evaluation with improved correlation with human judgments. In *Proceedings of the ACL Workshop on Intrinsic and Extrinsic Evaluation Measures for Machine Translation and/or Summarization*, pages 65–72, Ann Arbor, Michigan. Association for Computational Linguistics.

Asma Ben Abacha, Yassine Mrabet, Yuhao Zhang, Chaitanya Shivade, Curtis Langlotz, and Dina Demner-Fushman. 2021. Overview of the mediqa 2021 shared task on summarization in the medical domain. In *Proceedings of the 20th SIGBioMed Workshop on Biomedical Language Processing, NAACL-BioNLP 2021*. Association for Computational Linguistics.

Kyunghyun Cho, Bart van Merriënboer, Çağlar Gülçehre, Dzmitry Bahdanau, Fethi Bougares, Holger Schwenk, and Yoshua Bengio. 2014. Learning phrase representations using rnn encoder–decoder for statistical machine translation. In *Proceedings of the 2014 Conference on Empirical Methods in Natural Language Processing (EMNLP)*, pages 1724–1734, Doha, Qatar. Association for Computational Linguistics.

Jonathan H. Clark, Chris Dyer, Alon Lavie, and Noah A. Smith. 2011. Better hypothesis testing for statistical machine translation: Controlling for optimizer instability. In *Proceedings of the 49th Annual Meeting of the Association for Computational Linguistics: Human Language Technologies*, pages 176–181, Portland, Oregon, USA. Association for Computational Linguistics.

Joseph Paul Cohen, Joseph Viviano, Paul Morrison, Rupert Brooks, Mohammad Hashir, and Hadrien Bertrand. 2020. TorchXRayVision: A library of chest X-ray datasets and models. *https://github.com/mlmed/torchxrayvision*.

Kaiming He, Xiangyu Zhang, Shaoqing Ren, and Jian Sun. 2015. Delving deep into rectifiers: Surpassing human-level performance on imagenet classification. In *Proceedings of the IEEE international conference on computer vision*, pages 1026–1034.

Gao Huang, Zhuang Liu, Laurens van der Maaten, and Kilian Q Weinberger. 2017. Densely connected convolutional networks. In *Proceedings of the IEEE Conference on Computer Vision and Pattern Recognition*.

Indiana-University. Indiana university - chest x-rays.

Alistair E. W. Johnson, Tom J. Pollard, Nathaniel R. Greenbaum, Matthew P. Lungren, Chih-ying Deng, Yifan Peng, Zhiyong Lu, Roger G. Mark, Seth J. Berkowitz, and Steven Horng. 2019. MIMIC-CXR-JPG, a large publicly available database of labeled chest radiographs. *arXiv e-prints*, page arXiv:1901.07042.

Diederik P Kingma and Jimmy Ba. 2014. Adam: A method for stochastic optimization. *arXiv preprint arXiv:1412.6980*.

Yuan Li, Xiaodan Liang, Zhiting Hu, and Eric P Xing. 2018. Hybrid retrieval-generation reinforced agent for medical image report generation. In *Advances in Neural Information Processing Systems*, volume 31. Curran Associates, Inc.

Chin-Yew Lin. 2004. ROUGE: A package for automatic evaluation of summaries. In *Text Summarization Branches Out*, pages 74–81, Barcelona, Spain. Association for Computational Linguistics.

Guanxiong Liu, Tzu-Ming Harry Hsu, Matthew McDermott, Willie Boag, Wei-Hung Weng, Peter Szolovits, and Marzyeh Ghassemi. 2019. Clinically accurate chest x-ray report generation. In *Proceedings of the 4th Machine Learning for Healthcare Conference*, volume 106 of *Proceedings of Machine Learning Research*, pages 249–269, Ann Arbor, Michigan. PMLR.

Sean MacAvaney, Sajad Sotudeh, Arman Cohan, Nazli Goharian, Ish Talati, and Ross W. Filice. 2019. Ontology-aware clinical abstractive summarization. In *Proceedings of the 42nd International ACM SIGIR Conference on Research and Development in Information Retrieval*, SIGIR'19, page 1013–1016, New York, NY, USA. Association for Computing Machinery.

Christopher D. Manning, Mihai Surdeanu, John Bauer, Jenny Finkel, Prismatic Inc, Steven J. Bethard, and David Mcclosky. 2014. The stanford corenlp natural language processing toolkit. In *In ACL, System Demonstrations*.

Kishore Papineni, Salim Roukos, Todd Ward, and Wei-Jing Zhu. 2002. Bleu: a method for automatic evaluation of machine translation. In *Proceedings of the 40th Annual Meeting of the Association for Computational Linguistics*, pages 311–318, Philadelphia, Pennsylvania, USA. Association for Computational Linguistics.

Yuhao Zhang, Daisy Yi Ding, Tianpei Qian, Christopher D. Manning, and Curtis P. Langlotz. 2018. Learning to summarize radiology findings. In *EMNLP 2018 Workshop on Health Text Mining and Information Analysis*.

Yuhao Zhang, Derek Merck, Emily Tsai, Christopher D. Manning, and Curtis Langlotz. 2020. Optimizing the factual correctness of a summary: A study of summarizing radiology reports. In *Proceedings of the 58th Annual Meeting of the Association for Computational Linguistics*, pages 5108–5120, Online. Association for Computational Linguistics.

A Multimodal results

Model	BLEU	METEOR	R2-F1
		indiana-dev	
Multi E5 allv	15.27	17.12	**30.42**
Multi E5 encdecinit	**15.37**	**17.32**	30.05
Multi E5 ctxmul	14.57	16.75	29.85
Multi E5 trgmul	15.26	16.75	29.75
		mimic-dev	
Multi E5 allv	**29.31**	**26.24**	**48.86**
Multi E5 trgmul	29.0	26.10	42.56
Multi E5 ctxmul	28.90	26.02	48.47
Multi E5 encdecinit	28.38	25.58	48.42

Table 4: Results of our multimodal architectures on the development sets (regular-split).

Model	BLEU	METEOR	R2-F1
		indiana-dev	
Multi E5 allv	**28.32**	**25.30**	**54.38**
Multi E5 trgmul	27.50	24.72	53.90
Multi E5 ctxmul	26.86	24.53	53.20
Multi E5 encdecinit	26.65	24.52	52.10
		mimic-dev	
Multi E5 allv	**28.97**	**26.10**	**48.98**
Multi E5 trgmul	28.83	25.90	**48.98**
Multi E5 ctxmul	28.83	25.87	48.81
Multi E5 encdecinit	28.31	25.65	48.39

Table 5: Results of our multimodal architectures on the development sets (mix-split).

NLM at MEDIQA 2021: Transfer Learning-based Approaches for Consumer Question and Multi-Answer Summarization

Shweta Yadav*, Mourad Sarrouti*, and Deepak Gupta*
LHNCBC, U.S. National Library of Medicine, MD, USA
{shweta.shweta, mourad.sarrouti, deepak.gupta}@nih.gov

Abstract

The quest for seeking health information has swamped the web with consumers' health-related questions, which makes the need for efficient and reliable question answering systems more pressing. The consumers' questions, however, are very descriptive and contain several peripheral information (like patient's medical history, demographic information, etc.), that are often not required for answering the question. Furthermore, it contributes to the challenges of understanding natural language questions for automatic answer retrieval. Also, it is crucial to provide the consumers with the exact and relevant answers, rather than the entire pool of answer documents to their question. One of the cardinal tasks in achieving robust consumer health question answering systems is the question summarization and multi-document answer summarization. This paper describes the participation of the U.S. National Library of Medicine (NLM) in Consumer Question and Multi-Answer Summarization tasks of the MEDIQA 2021 challenge at NAACL-BioNLP workshop. In this work, we exploited the capabilities of pre-trained transformer models and introduced a transfer learning approach for the abstractive Question Summarization and extractive Multi-Answer Summarization tasks by first pre-training our model on a task-specific summarization dataset followed by fine-tuning it for both the tasks via incorporating medical entities. We achieved the second, sixth and the fourth position for the Question Summarization task in terms ROUGE-1, ROUGE-2 and ROUGE-L scores respectively.

1 Introduction

Healthcare consumers often query over the web to find a quick and reliable answer to their healthcare information needs. On average, 6 million people only in the United States seek health-related information on the Internet every day (Fox and Rainie). One way to facilitate such information-seeking activities is to build a natural language question answering (QA) system that can extract precise answers from the myriad of health-related information sources (Sarrouti and Alaoui, 2020). Though existing search engines respond to the general health-related queries to some extent, users often reach out to specialized medical websites or online health communities for seeking personalized high-quality, and trustworthy answers for their complex health questions. Moreover, consumers while expressing their medical concern on these sources except the involvement of healthcare professionals (HPs) for a quality suggestion and virtual observation (Kummervold et al., 2002). However, the participation of HPs in large-scale discussion forums or medical websites is time-consuming and expensive.

Furthermore, the consumers' questions are very descriptive and contain several peripheral information (like patient's medical history), which contributes to the challenges of understanding natural language questions for automatic answer retrieval (Demner-Fushman et al., 2020). These elaborated details are often not required for providing the relevant answers. Hence, novel strategies should be devised for automatic question simplifications and answer retrieval.

Towards this, we study the tasks of Question Summarization (QS) and Multi-Answer Summarization (MAS) as a part of MEDIQA 2021 (Asma Ben Abacha, 2021) shared task challenge. For the task of Question Summarization (QS), we proposed the transfer learning approach by utilizing multiple pre-trained Transformer (Vaswani et al., 2017) models. In our best run, we fine-tuned the pre-trained models on a variety of question summarization datasets and proposed a medical entities coverage technique to select the best question summary from the pool of question summaries obtained

*All the authors contributed equally to this work.

Proceedings of the BioNLP 2021 workshop, pages 291–301
June 11, 2021.

from the various transformer models.

We also explored the transfer learning approach for the Multi-Answer Summarization task. Specifically, the proposed method uses the Text-to-Text Transfer Transformer (T5) relevance-based re-ranking model (Raffel et al., 2020). In our best system, we first fine-tuned T5 on MSMARCO passage and then MEDIQA-QA 2019 datasets. It first ranks the sentences of the answers and then rejoins the top-k sentences as a summary.

2 Related Work

Existing works on the summarization can be broadly categorized into (i) extractive and (ii) abstractive approach which are discussed as follows:

Extractive Summarization: The recent development in the neural network and transformer based models has led to the significant progress in extractive document summarization. Majority of the models focus on the encoder-decoder model (Cheng and Lapata, 2016; Jadhav and Rajan, 2018; Nallapati et al., 2017), recurrent neural network (Nallapati et al., 2017; Zhou et al., 2018), and state-of-the-art Transformers encoders (Zhong et al., 2019b; Liu and Lapata, 2019). For instance, Cheng and Lapata (2016) and Nallapati et al. (2016b) proposed an encoder-decoder model as a binary classifier to decide whether the input sentence will be part of the summary or not. Chen and Bansal (2018) utilize a pointer generator network (Vinyals et al., 2015) to sequentially select sentences from the document for generating the extractive summary. Other decoding techniques, such as ranking (Narayan et al., 2018) has also been utilized for content selection. Recently several studies have explored pre-trained language models in summarization for contextual word representations (Zhong et al., 2019a; Liu and Lapata, 2019).

Abstractive Summarization (AS): With the development of large-scale datasets on abstractive summarization, there has been a significant advancement in AS techniques in the open domain, from traditional sequence to sequence (seq2seq) models, pointer generator network to Transformer based models. Few earlier studies utilize the seq2seq learning approach, trained on the large corpus of news articles for AS (Takase et al., 2016; Rush et al., 2015; Chopra et al., 2016). Later, Li et al. (2018) exploited the seq2seq models on multi-sentence document summarization. However, it was observed that the seq2seq model often generates out-of-vocabulary (OOV) words, factually incorrect details, and repetitions. To mitigate the issues of the seq2seq model, the pointer generator network was introduced that has the capability of handling OOV words with the copy mechanism (Gu et al., 2016; Nallapati et al., 2016a). Further, to address the repetition problem, Chen et al. (2016) proposed Distraction-based attention model. The additional coverage mechanism (See et al., 2017) ensures the generation of non-hallucinated summaries. Although these methods are good at generating readable summaries to a certain extent, the problem of factual inconsistencies persists with them. To alleviate this issue, several new methods (Lebanoff et al., 2020; Huang et al., 2020) has been proposed to generate more factually correct summaries. Few other recent works (Falke et al., 2019; Kryściński et al., 2019; Wang et al., 2020a) have exploited question answering and natural language inference (NLI) models to identify factual coherence in the generated summary. Recently several new models (Gehrmann et al., 2019) have been proposed that investigates the use of the transfer learning approach. Most recently the pseudo-self attention method (Ziegler et al., 2019) has been developed, which enables transfer learning to be applied in abstractive summarization.

Recently, with the availability of benchmark clinical data sets (MIMIC-CXR, and OpenI), there have been some prominent advancements in abstractive summarization of radiology reports. Zhang et al. (2018) utilized the pointer-generator network to generate the summary of radiology impressions and observed very high overlap with the human summaries. MacAvaney et al. (2019) further advanced the performance of the pointer generator model by augmenting medical-ontologies. Ben Abacha and Demner-Fushman (2019) has focused on the consumer health question summarization task. They created the corpus of $1,000$ question summaries and exploited seq2seq and pointer generator model to generate the consumer-health question summaries.

This work advances the pre-trained models for the summarization of consumers' questions and introduces new approaches to preserve the intent and the salient medical entities of the original questions.

3 Methods

3.1 Question Summarization

We tackle the first task of MEDIQA 2021, consumer health questions (CHQ) summarization with the goal of generating summarized questions that contain the key focus and semantics of the original question. Formally, given a consumer health question Q having m words $q_1, q_2, \ldots, q_m$, the task is to generate the summary sentence $\hat{S}$ having a sequence of n words $\hat{S} = \{s_1, s_2, \ldots, s_n\}$ expressing the key focus and semantics of the original question Q. Mathematically,

$$\begin{aligned} \hat{S} &= \arg\max_{S} prob(S|Q;\phi) \\ &= \arg\max_{S} prob(S|q_1, q_2, \ldots q_m; \phi) \end{aligned} \tag{1}$$

where ϕ are network parameters.

Pre-trained Transformer Models: We utilized the following pre-trained models and uses the transfer learning-based approach to fine-tune them on the task of question summarization.

- **ProphetNet** (Qi et al., 2020): It is a sequence-to-sequence model which is pre-trained using the self-supervised objective called future n-gram prediction. The ProphetNet is pre-trained by predicting the next n tokens simultaneously based on previous context tokens at each time step thus optimizing n-step ahead predictions of the model. The n-step ahead predictions encourage the model to plan for the future tokens and prevent overfitting on strong local correlations. We chose ProphetNet because it is specifically designed for sequence-to-sequence training and it has shown near state-of-the-art results on natural language generation tasks.

- **PEGASUS** (Zhang et al., 2020a): It is a large Transformer-based encoder-decoder model which is pre-trained on massive text corpora with a novel self-supervised objective called Gap Sentences Generation. This object is specially designed to pre-trained the transformer model for abstractive summarization. The important sentences from the document are masked and are generated together as one output sequence from the remaining sentences of the document.

- **T5** (Raffel et al., 2020): This is another pre-trained model developed by exploring the transfer learning techniques for natural language processing (NLP) by introducing a unified framework that converts all text-based language problems into a text-to-text format. The T5 model is an Encoder-Decoder Transformer with some architectural changes as discussed in detail in Raffel et al. (2020).

Pre-processing: To summarize the test questions, we followed certain pre-processing steps to transform the input consumer health question into a well-formed question. We applied the following pre-processing steps to the input test questions.

1. **Spelling Correction:** As consumer health questions are often ill-formed and contain multiple misspelled words particularly the medical terms (entities), therefore, we performed spelling correction on the original consumer health questions. Specifically, we utilized the *CSpell*[1], that aims to correct spellings from consumer health text.

2. **Abbreviation Expansion:** In order to generate the factually complete summaries, we first detect the medical entities and later expand the abbreviated entities using the '*Another database of abbreviations in MEDLINE*' (ADAM[2]) (Zhou et al., 2006).

Post-processing: Our analysis on the generated summary from the validation dataset using the pre-trained model reveals the following: **(1)** The T5 model generates a long summary and ended up with better coverage of the key entities present in the original question; **(2)** For the longer and complex questions, the T5 model generates the extractive-type summary; **(3)** Unlike T5, PEGASUS generates the short and succinct summaries which are often abstractive in nature; **(4)** The ProphetNet model often generates the moderate length summaries but approximately cover the key information from the original questions.

The correct summary of the consumer health questions must contain the key medical entities and question semantics of the original question. Motivated by the aforementioned observations, we obtained the generated summary from the pre-trained

[1] `https://lsg3.nlm.nih.gov/LexSysGroup/Projects/cSpell/current/web/index.html`

[2] `http://arrowsmith.psych.uic.edu/arrowsmith_uic/index.html`

Transformer models and performed the following steps to ensure the maximum coverage of medical entities so that it captures the key question-focus, and select the best question summary from the pool of generated summaries.

1. **Medical Entities Extraction:** We extracted the medical entities using the *Metamap*[3] (Aronson and Lang, 2010) and *Scispacy*[4] medical entity recognizer (en_ner_bionlp13cg_md). We removed some false entities ('*False Interventions*', '*False Anatomy*', '*False Problems*') using the Unified Medical Language System (UMLS) (Bodenreider, 2004) based filters[5]. Given a question Q, we obtained the list of medical entities as follows:

$$ent(Q) = MetaMap(Q) \cup Scispacy(Q)$$
$$entities(Q) = ent(Q) - False(ent(Q)) \tag{2}$$

   where, $MetaMap(;)$ and $Scispacy(;)$ are the medical entities extracted using MetaMap and Scispacy respectively, $False(;)$ is a method which provided the list of *False* entities. The final entities of the question is obtained using the $entities(;)$ method, which filters the false entities from the union of the list of both the entities.

2. **Medical Entities Coverage:** Given the original question Q and candidate question summary C, we extracted the medical entities E_Q and E_C using the approach discussed in Eq 2. We computed the medical entities coverage as follows:

$$coverage(Q, C) = \frac{|E_Q \cap E_C|}{|E_Q|} \tag{3}$$

   where $|x|$ is the cardinality of the set $x \in \{E_Q, E_Q \cap E_C\}$. We computed the coverage score for each candidate question summary generated using the different pre-trained Transformer models. We sort the candidate question summary based on the coverage score and passed the list to check the sanity of generated questions.

3. **Checking well-formed Question:** We check the list of generated questions against the well-formedness of the questions. Formally, we check:

   (a) Whether the generated questions starts with Wh words[6] or not.

   (b) Whether the generated question ends with the question word ('?').

   If the generated question having maximum coverage score is a well-formed question then we select the generated question as the final summary of the original question. Otherwise, we skip the non-well-formed candidate question and check against the next candidate question. In the case of the same coverage score among all three models, we selected the summary generated from PEGASUS, as it is more abstractive in nature.

3.2 Multi-Answer Summarization

To address the Multi-Answer Summarization (MAS) task at the MEDIQA 2021 challenge, we introduce an extractive method based on the T5 relevance-based re-ranking model (Raffel et al., 2020). The proposed method consists of extracting important and most relevant sentences from the answers and rejoining them to form a summary. To evaluate the importance of a sentence, we used T5 relevance-based ranking model. To do so, we first split the multiple answers of a given question into sentences using NLTK[7], and then ranked these sentences based on the relevance score that determines how relevant a candidate sentence is to a question. The sentences are ranked by a pointwise re-ranker (Nogueira et al., 2020) which uses T5, a sequence-to-sequence model that uses traditional transformer architecture, and BERT's masked language modeling (Devlin et al., 2019). We adopt the approach to sentence ranking by using the following input sequence:

$$Question: q\ Sentence:\ s\ Relevant: \tag{4}$$

The model is first fine-tuned to generate the tokens "true" when the sentence is relevant to the question and "false" when the sentence is not relevant to the question. It then applies softmax on

[3] https://metamap.nlm.nih.gov/
[4] https://allenai.github.io/scispacy/
[5] https://gist.github.com/h4ste/14b10d412d0d3c043c1d123c75c6ad29
[6] https://en.wikipedia.org/wiki/Interrogative_word
[7] https://www.nltk.org/

the logits of the "true" and "false" words and ranks the sentences using the probabilities of the "true" token. More details about this approach appear in (Nogueira et al., 2020).

The model is fine-tuned on (1) MS MARCO passage (Bajaj et al., 2018), (2) MS MARCO MED (MacAvaney et al., 2020), and (3) MEDIQA-QA 2019 dataset (Ben Abacha et al., 2019). We used the question-answer pairs in MEDIQA-QA with scores 1 and 2 (i.e., incorrect and related answers) as negative instances and the question-answer pairs with scores 3 and 4 (i.e., incomplete and excellent answers) as positive instances.

We form the summary by rejoining the selected top-k sentences. We also used Metamap[8] (Aronson and Lang, 2010) to replace the abbreviations by their definitions.

4 Experimental Results and Discussion

4.1 Evaluation Metrics

The performance of the question summarization and multi-answer summarization are evaluated against the ROUGE (Lin, 2004) score. We reported the results in terms of ROUGE-1 (R-1), ROUGE-2 (R-2), ROUGE-L (R-L). The organizer also release scores using the BERTScore (Zhang et al., 2020b) and HOLMS (Mrabet and Demner-Fushman, 2020).

4.2 Datasets

Question Summarization: For the task of question summarization, we use the following dataset to fine-tuned the pre-trained Transformer models.

1. **MeQSum** (Ben Abacha and Demner-Fushman, 2019): We use the $1,000$ consumer health question summarization dataset created by the medical experts. The questions are selected from a collection distributed by the U.S. National Library of Medicine (Kilicoglu et al., 2018).

2. **Clinical Questions** (Ely et al., 2000): We also utilized the $4,655$ clinical questions dataset, which contains the clinical questions and their short summaries.

3. **MEDIQA-RQE** (Ben Abacha et al., 2019): This dataset is released in the BioNLP 2019 shared task. The dataset is derived from consumer health questions (CHQs) and frequently asked questions (FAQs) from the U.S. National Library of Medicine and National Institute of Health respectively. We use the MEDIQA-RQE training dataset and choose only the entailed question pairs to form the silver-standard training dataset. We choose the longer question as the source question and the other as the target question. With this process, we formulated the $4,655$ additional training question pairs to train the question summarization model.

4. **MedNLI** (Romanov and Shivade, 2018): We also used the MedNLI - a dataset annotated by doctors, performing a natural language inference task, grounded in the medical history of patients. We augment training, validation, and test datasets and choose only the entailed question pairs to form the silver-standard training dataset. Similar to MEDIQA-RQE, we choose the longer question as the source question and the other as the target question. We obtained the $4,683$ question pairs from this dataset to include in the question summarization training dataset.

5. **LiveQA17** (Ben Abacha et al., 2017): We also utilized the 104 questions and their summary from the LiveQA17 test dataset as it contains the gold summaries of the source questions.

Multi-Answer Summarization: We used the following datasets to fine-tuned the T5 model for the multi-answer summarization task:

1. **MS MARCO Passage** (Bajaj et al., 2018): It is a large dataset for passage ranking. It contains 8.8M passages retrieved by Bing search engine for around 1M natural language questions.

2. **MSMARCO MED** (MacAvaney et al., 2020): This dataset contains the medical subset of MS MARCO. It includes only medical-related queries.

3. **MEDIQA-QA 2019** (Ben Abacha et al., 2019): It is a dataset for medical question answering obtained by submitting medical questions to the consumer health QA system CHiQA. The answers for the questions were manually ranked by medical experts.

[8]https://metamap.nlm.nih.gov/

4.3 Implementation Details

For question summarization task, we used the T5-large[9], ProphetNet-large-uncased[10] and pegasus-large[11] pre-trained models. The models are fine-tuned with maximum source question length of 120 and target summary length of 20. We train the model for 10 epochs and choose the best model based on the model performance (in terms of ROUGE-2) on the MEDIQA 2021 validation dataset. In our MAS experiments, we used the T5-base implementations provided in HuggingFace's Transformers package version 2.10 (Wolf et al., 2020). All models were trained with a batch size of 8 and a maximum sequence length of 512 tokens for 20 epochs using single P100 GPUs (16 GB VRAM) on a shared cluster. We use the beam search method to generate the summarized questions. For both the task Adam optimizer (Kingma and Ba, 2015) with a learning rate of 1e-5 was used for the parameters updates.

4.4 Results and Discussion

We devise multiple runs to assess **(1)** the ability of pre-trained Transformer model to summarize consumer health questions, **(2)** the role of additional datasets to improve the performance of CHQs summarization systems, and **(3)** the effect of the medical entities coverage to effectively select the best summarized questions from the pool of multiple summarized questions generated by pre-trained Transformer models. For the Question Summarization task, we submitted multiple runs which are described below:

1. **Run-1:** In this run, we fine-tuned the MiniLM (Wang et al., 2020b) model on the MeQSum (only 500 question-summary pairs) and Clinical Questions datasets. The summaries are generated using a beam of size 4.
2. **Run-2:** This run is similar to the **Run-1**, except we generated the summaries with the beam of size 6.
3. **Run-3:** For this run, we fine-tuned the ProphetNet model on the MeQSum and Clinical Questions datasets. The summaries are generated using a beam of size 4.
4. **Run-4:** We fine-tuned the T5 model on the MeQSum and Clinical Questions datasets. The summaries are generated using a beam of size 4.
5. **Run-5:** The PEGASUS model is fine-tuned on the MeQSum and Clinical Questions datasets. The summaries are generated using a beam of size 4.
6. **Run-6:** The T5 model is fine-tuned on the MeQSum, Clinical Questions, and MEDIQA-RQE datasets. The summaries are generated using a beam of size 4.
7. **Run-7:** We fine-tuned the T5, PEGASUS, ProphetNet models on the MeQSum, Clinical Questions, MEDIQA-RQE, LiveQA17, and MedNLI datasets. We also performed the pre-processing and post-processing steps (without well-formed questions) discussed in Section 3.1. The summaries are generated using a beam of size 4.
8. **Run-8:** The PEGASUS model is fine-tuned on the MeQSum, Clinical Questions, MEDIQA-RQE, LiveQA17, and MedNLI datasets. We also performed the pre-processing step discussed in Section 3.1. The summaries are generated using a beam of size 4, Top-K Sampling (Fan et al., 2018) with $K = 50$ and Top-p (nucleus) Sampling (Holtzman et al., 2019) with $p = 0.97$.
9. **Run-9:** The run is similar to **Run-7** however, we performed both the pre-processing and post-processing steps as described in Section 3.1 and the beam of size 5 is used to generate the summaries.
10. **Run-10:** This is final run similar to **Run-9**, however, we also included a subset ($10,324$) of questions from Quora duplicate question detection dataset[12] to fine-tuned the pre-trained models. We choose only those questions from the Quora dataset which are duplicates. We consider the question having more than 2 sentences and longer than the associated duplicate question as the source question and other duplicate question as target summary question.

For all our runs, we kept the maximum length of generated summary is 20. We have shown the detailed performance evaluation based on different metrics in Table 1. Our best submission (Run-9) achieved the maximum of ROUGE-1 (35.58), ROUGE-2 (15.14), HOLMS (56.59) and

[9]https://huggingface.co/t5-large

[10]https://huggingface.co/microsoft/prophetnet-large-uncased

[11]https://huggingface.co/google/pegasus-large

[12]http://qim.fs.quoracdn.net/quora_duplicate_questions.tsv

Run#	ROUGE-1	ROUGE-2	ROUGE-L	HOLMS	BERTScore
1	26.24	9.06	23.68	53.74	63.07
2	25.88	8.76	23.23	53.27	63.25
3	30.38	11.25	26.58	54.54	65.62
4	33.01	12.91	27.61	52.26	65.58
5	33.24	13.87	28.77	55.69	67.35
6	34.10	13.71	29.65	55.17	68.54
7	35.58	15.12	31.16	56.51	68.90
8	33.73	14.38	29.79	56.21	68.22
9	35.56	15.14	31.10	56.49	68.92
10	35.28	15.08	30.79	56.59	68.94
Our Best Run	**35.58**	**15.14**	**31.16**	**56.59**	**68.94**
Best Participants	35.80	16.08	31.49	57.87	70.27
Average Participants	29.55	11.59	26.60	53.25	64.93

Table 1: Official results of MEDIQA 2021: NLM runs for the Question Summarization task.

Run#	ROUGE-1	ROUGE-2	ROUGE-L	HOLMS	BERTScore
1	0.524	0.410	0.322	0.674	0.758
2	0.504	0.414	0.302	0.640	0.772
3	0.507	0.417	0.303	0.643	0.773
4	0.547	0.468	0.328	0.657	0.764
5	0.524	0.446	0.309	0.633	0.786
Our Best Run	**0.547**	**0.468**	**0.328**	**0.657**	**0.764**
Best Participants	0.585	0.508	0.435	0.704	0.803
Average Participants	0.524	0.422	0.353	0.668	0.751

Table 2: Official results of MEDIQA 2021: NLM runs for the Multi-Answer Summarization task.

BERTScore (68.94). Run-7 achieves the maximum ROUGE-L score of 31.16. Our best run achieved the ROUGE-2 score of 15.14, which is slightly (0.94) lower than the best run submitted for the Question Summarization task in MEDIQA 2021. Similarly, our best run obtained the improvement of 3.55 ROUGE-2 points over the average ROUGE-2 score obtained by all the participant's runs. We achieved the second-best result (35.58) in terms of the ROUGE-1 score over all the submitted runs for the Question Summarization task in MEDIQA 2021. We also show the best and average results among all the participants against various evaluation metrics in Table 1.

Qualitative Analysis: We carried out an in-depth analysis of the generated summaries of the models (Run 3,4,5,7,9) as shown in Table-3 for the question summarization task. We randomly selected 20 summaries from the test set and manually evaluated the summaries generated by the models. Table-3 shows that for question #1 and #2, our best run (#9) generates the readable summaries with the correct question focus and type. However, for the question #3, our best run is only able to capture partial question type and therefore generated the partially correct summary. We also observed that though T5 and PEGASUS generate factually correct summaries, sometimes it fails to generate a fully correct summary. Overall, the pre-trained models generate readable and succinct summaries which can be further enhanced by integrating the information about question focus and types.

Discussion: Our results confirms the role of additional datasets in fine-tuning the pre-trained Transformer models to improve the performance of CHQs summarization task. Run-1 and Run-2 which are based on fine-tuning the pre-trained Transformer models with the MeQSum dataset achieve low R-2 scores (8.76 and 9.06). The additional Clinical Questions dataset helps to improve the performance of the pre-trained model in Run-3, 4, and 5. The additional datasets (MEDIQA-RQE,

Question #1	Hi to whom this may concern: I have a non alcoholic fatty liver and when I seen my consultant he say I don't have to see Him again for 5 years and advised me to drink 3 cups of coffee a day, But I can't drink coffee at all, So I email a consultant asking if he new of a coffee tablet, So he email me back saying it properly glutathione S-transferases (GSTs) in the coffee so am looking for a cheap tablet which contains.these enzymes to help repair my liver I also only have one kidney which am on long term antibiotics for I do hope you can help. I look forward to your reply
Run-3	how can i find physician(s) who specialize in non alcoholic fatty liver disease?
Run-4	Where can I buy a tablet containing glutathione-s-transferase enzymes to
Run-5	Is there a coffee tablet which contains glutathione-s-transferase enzymes?
Run-7	Where can I buy cheap tablets containing glutathione-s-transferase enzymes?
Run-9	Where can I buy cheap tablets containing glutathione-s-transferase enzymes?
Question #2	Can Irritable bowel syndrome (IBS) be cured with the patent Medicines say Mebverine or others as prescribed the Doctor. Does Probiotics are helpful in controlling the Irritable bowel syndrome (IBS). Some says Peeper Mint can also be suitable. These are few general questions as regards Irritable bowel syndrome (IBS) and not personal.
Run-3	can ibs be cured with the patent medicines mebverine or others as prescribed?
Run-4	Can IBS be cured with the patent Medicines Mebverine or others as prescribed the Doctor. Does Pro
Run-5	Can IBS be cured with the patent Medicines say Mebverine or others?
Run-7	Can Irritable bowel syndrome (IBS) be cured with the patent Medicines say Mebverine or others?
Run-9	Can Irritable Bowel Syndrome (IBS) be cured with the Mebverine ?
Question #3	SUBJECT: I NEED PARENT TRAINING. MESSAGE: My 9 year old son has oppositional defiant disorder (ODD) and attention deficit disorder (ADHD). I'm in desperate need of an extensive parent training class. I don't know what to do anymore and really need help. My son says I do nothing for him after spending the entire day taking him places. He yells and screams at me and tells me I'm a terrible mother. He refuses to listen and I have to ask 10 times before he reacts. He will not do homework.. I have tried positive reinforcement...... Please I need someone that know their stuff..
Run-3	what is the treatment for a child with odd and adhd?
Run-4	Is there a parent training program for ODD and ADHD that I can take with my 9-year-old?
Run-5	What are the treatments for obsessive-compulsive disorder and attention-deficit-hyperactivity disorder?
Run-7	what are the treatments for oppositional defiant disorder (odd) and attention deficit disorder (adhd)?
Run-9	what are the treatments for oppositional defiant disorder (odd) and attention deficit disorder (adhd)?

Table 3: Qualitative Analysis: Examples showing models generated summaries.

LiveQA17, and MedNLI) with the pre-processing and post-processing steps further boost the performance of the question summarization as shown in Run-7 and Run-9. We also fine-tuned the Transformers model with the Quora duplicate question detection dataset in Run-10, in order to generate more diverse summaries. However, it could not improve the question summarization performance compare to the Run-9. It is because Quora dataset is a open domain dataset, which may not be well suited for the medical summarization task.

Multi-answer Summarization Task: We submitted the following runs for the multi-answer summarization task at MEDIQA 2021:

- **Run-1**: We fine-tuned the T5 model on the MSMARCO passage. We ranked the sentences of the answers based on the T5 relevance score and rejoined the top-10 sentences as a summary. We also identified the long-form of abbreviations in the test set.
- **Run-2**: We fine-tuned the T5 model on the MSMARCO passage. We ranked the sentences of the answers based on the T5 relevance score and then concatenated the top-10 sentences to form the summary.
- **Run-3**: We fine-tuned the T5 model on the MSMARCO passage. We ranked the sentences of the answers based on the T5 relevance score and rejoined the top-20 sentences as a summary.
- **Run 4**: We fine-tuned the T5 model on MSMARCO passage and then MEDIQA-QA 2019 dataset. The top-20 sentences are concatenated to form the summary.
- **Run-5**: We fine-tuned the T5 model on MEDMSMARCO and then MEDIQA-QA 2019 dataset. The top-20 sentences are concatenated to form the summary.

Table 2 presents the official results of our systems in the multi-answer summarization task of the MEDIQA 2021 challenge. Out of the five runs, our best result was obtained by the run #4, achieving 0.547, 0.468, and 0.328 in terms of ROUGE-1, ROUGE-2, and ROUGE-L respectively. In terms of BERTScore, our run #5 achieved the best results among our runs. On the other hand, run #1 achieved the highest HOLMS. The obtained results also showed that our T5-based system is more competitive in terms of various evaluation metrics over the other participant's systems.

5 Conclusion and Future Work

In this paper, we describe our submissions for the tasks of Question Summarization and Multi Answer Summarization at MEDIQA 2021 shared task. For the Question Summarization task, our best run achieved the second-best ROUGE-1 score among all the submitted runs in the shared task. We also obtained the competitive scores in terms of various evaluation metrics over the other participant's runs. For the Multi-Answer Summarization task, our T5-based approach achieved good performances compared to participants' systems. In the future, we will explore the techniques to integrate the medical entities and semantics in the pre-trained transformer models for the task of question summarization. Further, we will also explore the abstractive approaches for multi-answer summarization.

References

Alan R Aronson and François-Michel Lang. 2010. An overview of metamap: historical perspective and recent advances. *Journal of the American Medical Informatics Association*, 17(3):229–236.

Yuhao Zhang Chaitanya Shivade Curtis Langlotz Dina Demner-Fushman Asma Ben Abacha, Yassine Mrabet. 2021. Overview of the mediqa 2021 shared task on summarization in the medical domain. In *Proceedings of the 20th SIGBioMed Workshop on Biomedical Language Processing, NAACL-BioNLP 2021*. Association for Computational Linguistics.

Payal Bajaj, Daniel Campos, Nick Craswell, Li Deng, Jianfeng Gao, Xiaodong Liu, Rangan Majumder, Andrew McNamara, Bhaskar Mitra, Tri Nguyen, Mir Rosenberg, Xia Song, Alina Stoica, Saurabh Tiwary, and Tong Wang. 2018. Ms marco: A human generated machine reading comprehension dataset.

Asma Ben Abacha, Eugene Agichtein, Yuval Pinter, and Dina Demner-Fushman. 2017. Overview of the medical question answering task at trec 2017 liveqa. In *TREC 2017*.

Asma Ben Abacha and Dina Demner-Fushman. 2019. On the role of question summarization and information source restriction in consumer health question answering. *AMIA Summits on Translational Science Proceedings*, 2019:117.

Asma Ben Abacha and Dina Demner-Fushman. 2019. On the summarization of consumer health questions. In *Proceedings of the 57th Conference of the Association for Computational Linguistics, ACL 2019, Florence, Italy, July 28- August 2, 2019, Volume 1: Long Papers*, pages 2228–2234. Association for Computational Linguistics.

Asma Ben Abacha, Chaitanya Shivade, and Dina Demner-Fushman. 2019. Overview of the MEDIQA 2019 shared task on textual inference, question entailment and question answering. In *Proceedings of the 18th BioNLP Workshop and Shared Task*, pages 370–379, Florence, Italy. Association for Computational Linguistics.

Olivier Bodenreider. 2004. The unified medical language system (umls): integrating biomedical terminology. *Nucleic acids research*, 32(suppl_1):D267–D270.

Qian Chen, Xiaodan Zhu, Zhenhua Ling, Si Wei, and Hui Jiang. 2016. Distraction-based neural networks for document summarization. *arXiv preprint arXiv:1610.08462*.

Yen-Chun Chen and Mohit Bansal. 2018. Fast abstractive summarization with reinforce-selected sentence rewriting. *arXiv preprint arXiv:1805.11080*.

Jianpeng Cheng and Mirella Lapata. 2016. Neural summarization by extracting sentences and words. In *Proceedings of the 54th Annual Meeting of the Association for Computational Linguistics (Volume 1: Long Papers)*, pages 484–494.

Sumit Chopra, Michael Auli, and Alexander M Rush. 2016. Abstractive sentence summarization with attentive recurrent neural networks. In *Proceedings of the 2016 Conference of the North American Chapter of the Association for Computational Linguistics: Human Language Technologies*, pages 93–98.

Dina Demner-Fushman, Yassine Mrabet, and Asma Ben Abacha. 2020. Consumer health information and question answering: helping consumers find answers to their health-related information needs. *Journal of the American Medical Informatics Association*, 27(2):194–201.

Jacob Devlin, Ming-Wei Chang, Kenton Lee, and Kristina Toutanova. 2019. BERT: Pre-training of deep bidirectional transformers for language understanding. In *Proceedings of the 2019 Conference of the North American Chapter of the Association for Computational Linguistics: Human Language Technologies, Volume 1 (Long and Short Papers)*, pages 4171–4186, Minneapolis, Minnesota. Association for Computational Linguistics.

John W Ely, Jerome A Osheroff, Paul N Gorman, Mark H Ebell, M Lee Chambliss, Eric A Pifer, and P Zoe Stavri. 2000. A taxonomy of generic clinical questions: classification study. *Bmj*, 321(7258):429–432.

Tobias Falke, Leonardo FR Ribeiro, Prasetya Ajie Utama, Ido Dagan, and Iryna Gurevych. 2019. Ranking generated summaries by correctness: An interesting but challenging application for natural language inference. In *Proceedings of the 57th Annual Meeting of the Association for Computational Linguistics*, pages 2214–2220.

Angela Fan, Mike Lewis, and Yann Dauphin. 2018. Hierarchical neural story generation. *arXiv preprint arXiv:1805.04833*.

Susannah Fox and Lee Rainie. Main report: The search for online medical help.

Sebastian Gehrmann, Zachary Ziegler, and Alexander M Rush. 2019. Generating abstractive summaries with finetuned language models. In *Proceedings of the 12th International Conference on Natural Language Generation*, pages 516–522.

Jiatao Gu, Zhengdong Lu, Hang Li, and Victor OK Li. 2016. Incorporating copying mechanism in sequence-to-sequence learning. *arXiv preprint arXiv:1603.06393*.

Ari Holtzman, Jan Buys, Li Du, Maxwell Forbes, and Yejin Choi. 2019. The curious case of neural text degeneration. *arXiv preprint arXiv:1904.09751*.

Luyang Huang, Lingfei Wu, and Lu Wang. 2020. Knowledge graph-augmented abstractive summarization with semantic-driven cloze reward. *arXiv preprint arXiv:2005.01159*.

Aishwarya Jadhav and Vaibhav Rajan. 2018. Extractive summarization with swap-net: Sentences and words from alternating pointer networks. In *Proceedings of the 56th annual meeting of the association for computational linguistics (volume 1: Long papers)*, pages 142–151.

Halil Kilicoglu, Asma Ben Abacha, Yassine Mrabet, Sonya E Shooshan, Laritza Rodriguez, Kate Masterton, and Dina Demner-Fushman. 2018. Semantic annotation of consumer health questions. *BMC bioinformatics*, 19(1):34.

Diederik P. Kingma and Jimmy Ba. 2015. Adam: A method for stochastic optimization. In *3rd International Conference on Learning Representations, ICLR 2015, San Diego, CA, USA, May 7-9, 2015, Conference Track Proceedings*.

Wojciech Kryściński, Bryan McCann, Caiming Xiong, and Richard Socher. 2019. Evaluating the factual consistency of abstractive text summarization. *arXiv preprint arXiv:1910.12840*.

Per E Kummervold, Deede Gammon, Svein Bergvik, Jan-Are K Johnsen, Toralf Hasvold, and Jan H Rosenvinge. 2002. Social support in a wired world: use of online mental health forums in norway. *Nordic journal of psychiatry*, 56(1):59–65.

Logan Lebanoff, John Muchovej, Franck Dernoncourt, Doo Soon Kim, Lidan Wang, Walter Chang, and Fei Liu. 2020. Understanding points of correspondence between sentences for abstractive summarization. *arXiv preprint arXiv:2006.05621*.

Wei Li, Xinyan Xiao, Yajuan Lyu, and Yuanzhuo Wang. 2018. Improving neural abstractive document summarization with explicit information selection modeling. In *Proceedings of the 2018 conference on empirical methods in natural language processing*, pages 1787–1796.

Chin-Yew Lin. 2004. ROUGE: A Package for Automatic Evaluation of Summaries. In *Text Summarization Branches Out: Proceedings of the ACL-04 workshop*, volume 8. Barcelona, Spain.

Yang Liu and Mirella Lapata. 2019. Text summarization with pretrained encoders. *arXiv preprint arXiv:1908.08345*.

Sean MacAvaney, Arman Cohan, and Nazli Goharian. 2020. Sledge: a simple yet effective baseline for covid-19 scientific knowledge search. *arXiv e-prints*, pages arXiv–2005.

Sean MacAvaney, Sajad Sotudeh, Arman Cohan, Nazli Goharian, Ish Talati, and Ross W Filice. 2019. Ontology-aware clinical abstractive summarization. In *Proceedings of the 42nd International ACM SIGIR Conference on Research and Development in Information Retrieval*, pages 1013–1016.

Yassine Mrabet and Dina Demner-Fushman. 2020. HOLMS: Alternative summary evaluation with large language models. In *Proceedings of the 28th International Conference on Computational Linguistics*, pages 5679–5688, Barcelona, Spain (Online). International Committee on Computational Linguistics.

Ramesh Nallapati, Feifei Zhai, and Bowen Zhou. 2017. Summarunner: A recurrent neural network based sequence model for extractive summarization of documents. In *Proceedings of the AAAI Conference on Artificial Intelligence*, volume 31.

Ramesh Nallapati, Bowen Zhou, Caglar Gulcehre, Bing Xiang, et al. 2016a. Abstractive text summarization using sequence-to-sequence rnns and beyond. *arXiv preprint arXiv:1602.06023*.

Ramesh Nallapati, Bowen Zhou, and Mingbo Ma. 2016b. Classify or select: Neural architectures for extractive document summarization. *arXiv preprint arXiv:1611.04244*.

Shashi Narayan, Shay B Cohen, and Mirella Lapata. 2018. Ranking sentences for extractive summarization with reinforcement learning. *arXiv preprint arXiv:1802.08636*.

Rodrigo Nogueira, Zhiying Jiang, Ronak Pradeep, and Jimmy Lin. 2020. Document ranking with a pretrained sequence-to-sequence model. In *Findings of the Association for Computational Linguistics: EMNLP 2020*, pages 708–718, Online. Association for Computational Linguistics.

Weizhen Qi, Yu Yan, Yeyun Gong, Dayiheng Liu, Nan Duan, Jiusheng Chen, Ruofei Zhang, and Ming Zhou. 2020. ProphetNet: Predicting future n-gram for sequence-to-SequencePre-training. In *Findings of the Association for Computational Linguistics: EMNLP 2020*, pages 2401–2410, Online. Association for Computational Linguistics.

Colin Raffel, Noam Shazeer, Adam Roberts, Katherine Lee, Sharan Narang, Michael Matena, Yanqi Zhou, Wei Li, and Peter J. Liu. 2020. Exploring the limits of transfer learning with a unified text-to-text transformer. *Journal of Machine Learning Research*, 21(140):1–67.

Alexey Romanov and Chaitanya Shivade. 2018. Lessons from natural language inference in the clinical domain. In *Proceedings of the 2018 Conference on Empirical Methods in Natural Language Processing*, pages 1586–1596, Brussels, Belgium. Association for Computational Linguistics.

Alexander M Rush, Sumit Chopra, and Jason Weston. 2015. A neural attention model for abstractive sentence summarization. *arXiv preprint arXiv:1509.00685*.

Mourad Sarrouti and Said Ouatik El Alaoui. 2020. SemBioNLQA: A semantic biomedical question answering system for retrieving exact and ideal answers to natural language questions. *Artificial Intelligence in Medicine*, 102:101767.

Abigail See, Peter J Liu, and Christopher D Manning. 2017. Get to the point: Summarization with pointer-generator networks. *arXiv preprint arXiv:1704.04368*.

Sho Takase, Jun Suzuki, Naoaki Okazaki, Tsutomu Hirao, and Masaaki Nagata. 2016. Neural headline generation on abstract meaning representation. In *Proceedings of the 2016 conference on empirical methods in natural language processing*, pages 1054–1059.

Ashish Vaswani, Noam Shazeer, Niki Parmar, Jakob Uszkoreit, Llion Jones, Aidan N Gomez, Łukasz Kaiser, and Illia Polosukhin. 2017. Attention is all you need. In *Advances in neural information processing systems*, pages 5998–6008.

Oriol Vinyals, Meire Fortunato, and Navdeep Jaitly. 2015. Pointer networks. *arXiv preprint arXiv:1506.03134*.

Alex Wang, Kyunghyun Cho, and Mike Lewis. 2020a. Asking and answering questions to evaluate the factual consistency of summaries. *arXiv preprint arXiv:2004.04228*.

Wenhui Wang, Furu Wei, Li Dong, Hangbo Bao, Nan Yang, and Ming Zhou. 2020b. Minilm: Deep self-attention distillation for task-agnostic compression of pre-trained transformers. *arXiv preprint arXiv:2002.10957*.

Thomas Wolf, Lysandre Debut, Victor Sanh, Julien Chaumond, Clement Delangue, Anthony Moi, Pierric Cistac, Tim Rault, Rémi Louf, Morgan Funtowicz, Joe Davison, Sam Shleifer, Patrick von Platen, Clara Ma, Yacine Jernite, Julien Plu, Canwen Xu, Teven Le Scao, Sylvain Gugger, Mariama Drame, Quentin Lhoest, and Alexander M. Rush. 2020. Huggingface's transformers: State-of-the-art natural language processing.

Jingqing Zhang, Yao Zhao, Mohammad Saleh, and Peter Liu. 2020a. Pegasus: Pre-training with extracted gap-sentences for abstractive summarization. In *International Conference on Machine Learning*, pages 11328–11339. PMLR.

Tianyi Zhang, Varsha Kishore, Felix Wu, Kilian Q. Weinberger, and Yoav Artzi. 2020b. Bertscore: Evaluating text generation with bert. In *International Conference on Learning Representations*.

Yuhao Zhang, Daisy Yi Ding, Tianpei Qian, Christopher D Manning, and Curtis P Langlotz. 2018. Learning to summarize radiology findings. *arXiv preprint arXiv:1809.04698*.

Ming Zhong, Pengfei Liu, Danqing Wang, Xipeng Qiu, and Xuanjing Huang. 2019a. Searching for effective neural extractive summarization: What works and what's next. *arXiv preprint arXiv:1907.03491*.

Ming Zhong, Danqing Wang, Pengfei Liu, Xipeng Qiu, and Xuanjing Huang. 2019b. A closer look at data bias in neural extractive summarization models. *arXiv preprint arXiv:1909.13705*.

Qingyu Zhou, Nan Yang, Furu Wei, Shaohan Huang, Ming Zhou, and Tiejun Zhao. 2018. Neural document summarization by jointly learning to score and select sentences. *arXiv preprint arXiv:1807.02305*.

Wei Zhou, Vetle I Torvik, and Neil R Smalheiser. 2006. Adam: another database of abbreviations in medline. *Bioinformatics*, 22(22):2813–2818.

Zachary M Ziegler, Luke Melas-Kyriazi, Sebastian Gehrmann, and Alexander M Rush. 2019. Encoder-agnostic adaptation for conditional language generation. *arXiv preprint arXiv:1908.06938*.

IBMResearch at MEDIQA 2021: Toward Improving Factual Correctness of Radiology Report Abstractive Summarization

Diwakar Mahajan
IBM Research
dmahaja@us.ibm.com

Ching-Huei Tsou
IBM Research
ctsou@us.ibm.com

Jennifer J Liang
IBM Research
jjliang@us.ibm.com

Abstract

Although recent advances in abstractive summarization systems have achieved high scores on standard natural language metrics like ROUGE, their lack of factual consistency remains an open challenge for their use in sensitive real-world settings such as clinical practice. In this work, we propose a novel approach to improve factual correctness of a summarization system by re-ranking the candidate summaries based on a factual vector of the summary. We applied this process during our participation in MEDIQA 2021 Task 3: Radiology Report Summarization, where the task is to generate an impression summary of a radiology report, given findings and background as inputs. In our system, we first used a transformer-based encoder-decoder model to generate top N candidate impression summaries for a report, then trained another transformer-based model to predict a 14-observations-vector of the impression based on the findings and background of the report, and finally, utilized this vector to re-rank the candidate summaries. We also employed a source-specific ensembling technique to accommodate for distinct writing styles from different radiology report sources. Our approach yielded 2nd place in the challenge.

1 Introduction

The radiology report is a crucial instrument in patient care and an essential part of every radiological procedure, serving as the official interpretation of a radiological study and the primary means of communication between the radiologist and referring physician. According to the American College of Radiology, a radiology report should contain certain components, such as relevant clinical information, imaging findings, limitations of the study, and an impression or conclusion (American College of Radiology, 2020). Of these, the impression is the most important component of the radiology report, containing conclusions based on the pertinent findings and suggestions for additional diagnostic studies if warranted (Wallis and McCoubrie, 2011). Previous studies have shown that oftentimes it is the only part of the report that is read; one previous study found that 43% of referring physicians only read the impression if the report was longer than one page (Clinger et al., 1988), while another study found that 23.1% of clinicians agreed with the statement "I usually only read the conclusion of a radiology report" (Bosmans et al., 2011).

In an effort to support radiologists in writing impressions in radiology reports, Zhang et al. (2018) introduced the task of automatic generation of radiology impression statements by summarizing textual findings written by radiologists. MEDIQA 2021 (Asma Ben Abacha, 2021), as part of NAACL-BioNLP 2021 workshop, aims to further research efforts in summarization in the medical domain. Task 3 of the challenge, Radiology Report Summarization (RRS), focuses specifically on radiology impression generation. The basic task setup is as follows: given the findings and background sections of a radiology report, predict the impression or summary.

In this paper, we detail our participation in MEDIQA 2021 RRS challenge. We developed an approach that utilizes a structured label vector of the impression as our proxy for facts for the impression (predicted using findings and background of the report), to re-rank the generated abstractive summaries from a trained encoder-decoder model. We further employed a source-specific ensembling technique utilizing models fine-tuned to each radiology report source to accommodate for distinct language patterns in each source. Our system performed well in the challenge, placing us 2nd on the leaderboard.

2 Related Work

Abstractive Summarization Systems. Abstractive text summarization has been intensively stud-

Proceedings of the BioNLP 2021 workshop, pages 302–310
June 11, 2021. ©2021 Association for Computational Linguistics

ied in recent literature. Rush et al. (2015) introduces an attention-based sequence-to-sequence (seq2seq) model for abstractive sentence summarization. Recent models (e.g. Lewis et al. (2019); Zhang et al. (2020)) employ techniques like denoising or Gap Sentence Generation task for pretraining, to help generation tasks including summarization. However, there are a few domain-specific versions of these state-of-the-art models. Other works like Liu and Lapata (2019); Rothe et al. (2020) have demonstrated the effectiveness of initializing encoder-decoder models from pre-trained encoder-only models, such as BERT (Devlin et al., 2018) and RoBERTa (Liu et al., 2019), for seq2seq tasks providing competitive results in summarization tasks. Our works builds on these findings and utilizes a pre-existing domain-specific pretrained transformer model in an encoder-decoder setting for our summarization task.

Summarization and Factual Correctness in Radiology Reports. Zhang et al. (2018) first studied the problem of automatic generation of radiology impressions by summarizing textual radiology findings, and showed that an augmented pointer-generator model achieves high overlap with human references. They also found that about 30% of the radiology summaries generated from neural models contain factual errors. Research scholars also integrated Radlex ontology into seq2seq models (MacAvaney et al., 2019) to enhance the clinical validity of automated impression prediction systems within the radiology workflows. In their next work, Zhang et al. (2019b) improved upon the problem of factual correctness in radiology reports by optimizing fact scores defined in radiology reports with reinforcement learning methods. They also introduced a new metric Factual F_1 comparing the predicted summaries using a descriptor vector of the gold summary. In our work, we extend the ideas put forward by Zhang et al. (2019b) by utilizing a descriptor vector (generated using off-the-shelf systems like CheXpert (Irvin et al., 2019) or CheXbert (Smit et al., 2020)) to re-rank the automatically generated summaries.

3 Task Description and Dataset

The MEDIQA-2021 RRS task is defined as follows: given a passage of findings represented as a sequence of tokens $x = \{x_1, x_2, \ldots, x_N\}$, with N being the length of the findings, and a passage of background represented as a sequence of tokens $y = \{y_1, y_2, \ldots, y_M\}$ with M being the length of the background, find a sequence of tokens $z = \{z_1, z_2, \ldots, z_L\}$ that best summarizes the salient and clinically significant findings in x, with L being an arbitrary length of the impression or summary[1].

Type	Source-specific Size		Total Size
	MIMIC-CXR	Indiana	
Training	91,544	0	91,544
Validation	2,000	2,000	4,000
Test	?	?	600

Table 1: Dataset statistics.

Datasets for training and validation of summarization models provided by the MEDIQA organizers consisted of radiology reports with findings, background, and impression sections. The training set consists of 91,544 radiology reports from the MIMIC-CXR database (Johnson et al., 2019), while the validation set consists of an additional 4,000 radiology reports - 2,000 from MIMIC-CXR and 2,000 from the Indiana Network for Patient Care (Indiana) (Demner-Fushman et al., 2016). As part of the shared task rules, the rest of the publicly available MIMIC-CXR and Indiana radiology reports were not allowed for use in training or validation. However, the organizers allowed the use of validation data for training. At the conclusion of the shared task, to evaluate participant systems, a test set of 600 radiology reports containing only findings and background sections were released with their sources unknown at the time of the challenge. Dataset statistics are presented in the Table 1.

4 System Description

Our system is a three-step process in which we (1) utilize pre-trained transformer-based language models in an encoder-decoder setting to get our base summarization models, (2) improve the factual correctness of our base models' predictions by incorporating a re-ranking methodology, and (3) utilize a source-specific ensembling technique which identifies the source of a radiology report, and chooses the prediction of the best performing source-specific model accordingly. We detail the above three steps in the following sections.

[1]Throughout this paper we use terms "impression" and "summary" interchangeably.

4.1 Base Models

Previous work by Liu and Lapata (2019); Rothe et al. (2020) have demonstrated the effectiveness of initializing encoder-decoder models from pre-trained encoder-only models, such as BERT and RoBERTa, for seq2seq tasks. Inspired by this work, we experimented with pre-trained transformer models used as both encoder and decoder with parameters shared between encoder and decoder. Using this setup, we experimented with RoBERTa-large, which showed promising results in Rothe et al. (2020), and BioMed-RoBERTa-base, a domain-specific version of RoBERTa that is publicly available[2] from AllenNLP (Gururangan et al., 2020), and fine-tuned both models using the training set of 91,544 MIMIC-CXR reports. Of the two models, BioMed-RoBERTa-base achieved better results and was therefore used as our initial model for subsequent experiments.

Next, we conducted experiments to evaluate the performance of this initial model on different radiology report sources. As the provided training and validation data contains two sources, MIMIC-CXR and Indiana, each with their distinct language (more details in Section 4.3) and official test data could be any source, we further developed two more base models. Using the initial BioMed-RoBERTa-base model fine-tuned on MIMIC-CXR training set, we further fine-tuned the initial model in two settings: (1) with a subset of reports in the Indiana validation dataset, and (2) with a subset of reports in the Indiana and MIMIC-CXR validation dataset.

Our end result is three base models tuned for 3 source categories:

- $\text{BioRoBERTa}_{(\text{M})}$: BioMed-RoBERTa-base fine-tuned on MIMIC-CXR training data. This is the base model for MIMIC-CXR source.
- $\text{BioRoBERTa}_{(\text{M+I})}$: $\text{BioRoBERTa}_{(\text{M})}$ further fine-tuned on Indiana validation data. This is our base model for Indiana source.
- $\text{BioRoBERTa}_{(\text{M+M+I})}$: $\text{BioRoBERTa}_{(\text{M})}$ further fine-tuned on both MIMIC-CXR and Indiana validation data. This is our base model for unknown sources.

[2]https://huggingface.co/allenai/biomed_roberta_base

4.2 Fact-Aware Re-ranking (FAR)

Previous works in extracting structured labels from free-text radiology reports have identified 14 observations based on clinical relevance and the prevalence in the reports, and have developed automated systems to predict a 14-observations-vector for an impression summary of a radiology report (Irvin et al., 2019; Smit et al., 2020). The 14 observations are: "Atelectasis", "Cardiomegaly", "Consolidation", "Edema", "Enlarged Cardiomediastinum", "Fracture", "Lung Opacity", "Lung Lesion", "No Finding", "Pneumonia", "Pneumothorax", "Pleural Effusion", "Pleural Other", and "Support Devices". "Pneumonia", despite being a clinical diagnosis, was included as a label in order to represent the images that suggested primary infection as the diagnosis. The 13 observations (excluding "No Finding") take on one of the following classes: blank, positive, negative, and uncertain. The 14th observation, "No Finding", is intended to capture the absence of all pathologies, and takes on only one of the two following classes: blank or positive.

Utilizing this 14-observations-vector we developed an approach to improve the factual correctness of our base models by incorporated a factual re-ranking component that re-ranks our N highest scoring summaries predicted from a base model. We achieve this in the following steps, we (1) first fine-tune a transformer-based language model to predict the 14-observation-vector of the impression given the finding and background of a radiology report, (2) obtain top N highest scoring candidate summaries predicted from our base encoder-decoder model (3) use CheXbert to obtain the 14-observation-vector for each of the N candidate summaries, and (4) use a similarity function between predicted 14-observation-vector for impression (obtained in step 1) and each vector for N candidate summaries obtained in step 3 to re-rank these summaries. Finally, we use the highest similarity scoring candidate summary as our impression summary. We detail our impression 14-observation-vector prediction and our similarity function in the following sections.

We apply our FAR methodology on the three base models introduced in section 4.1 to get our three source-specific models, and denote the new models as $\text{BioRoBERTa}_{(\text{M}),\text{FAR}}$, $\text{BioRoBERTa}_{(\text{M+I}),\text{FAR}}$, and $\text{BioRoBERTa}_{(\text{M+M+I}),\text{FAR}}$, respectively.

Source	Finding	Background	Impression
MIMIC-CXR	There is hyperexpansion of both lungs with severe underlying emphysema. Minimal blunting of the right costophrenic angle may reflect underlying atelectasis. No pleural effusion or pneumothorax identified. The size the cardiomediastinal silhouette is within normal limits.	INDICATION: ___ year old woman with COPD exacerbation // evaluate lung sizes, look for PNA TECHNIQUE: AP portable chest radiograph COMPARISON: No prior radiographs available. Comparison is made to the CT torso from ___	No radiographic evidence of acute cardiopulmonary disease. Hyperexpanded lungs with severe underlying emphysema.
Indiana	Heart size and mediastinal contours appear within normal limits. Hyperinflated lungs with flattening of diaphragms, compatible with emphysema. No focal consolidation, pleural effusion or pneumothorax. No acute bony abnormality.	Indication: Short of breath. Comparison: None.	1. Emphysema. 2. No acute cardiopulmonary abnormality.

Table 2: Example depicting the difference in language between MIMIC-CXR and Indiana reports for findings, background and impression sections.

4.2.1 Impression 14-Observations-Vector Prediction

We utilize the 14-observation-vector representation of the impression section of a radiology report predicted by CheXbert as our ground truth label in a prediction task given the finding and background section of the report as inputs. In this process, for each given radiology report that has findings, background and impression section, we (1) first utilize CheXbert to obtain 14-observations-vector representation of the impression section, (2) convert the multiple values of each of the 14 observations to be binary (i.e. presence or absence of the observation)[3], (3) train a transformer-based language model using finding and background (concatenated) as input to predict 14-observations-vector of the impression section.

4.2.2 Similarity Function

Among the 14 observations categories predicted in CheXbert, "No Finding" is intended to capture absence of all pathologies, i.e. if "No Finding" is positive then all other observations must be negative. Therefore, we constructed our similarity function in cases where (1) "No Finding" is not matched, we assign a similarity score of 0, (2) "No Finding" is a match, the similarity score is the cosine similarity between the rest of the vector representing the 13 other observations.

[3]CheXbert outputs for 13 observations one of the following classes: blank, positive, negative, and uncertain. For the 14th observation corresponding to No Finding, the labeler only outputs one of the two following classes: blank or positive. We convert uncertain to positive and blank to negative to get binary positive and negative output for all 14 observations.

4.3 Source-specific Ensemble

We observed in the provided training and validation data that MIMIC-CXR and Indiana reports use distinctly different language when expressing findings, background, and impression, even when the conveyed content is very similar. As shown in Table 2, although both the MIMIC-CXR report and Indiana report convey the same two key findings in their impression, "emphysema" and "no acute cardiopulmonary disease", the MIMIC-CXR report describes these findings with more detail in prose form, while the Indiana report lists the findings more concisely using a numbered list form. This variation in language between different healthcare organizations is common in the clinical NLP domain, resulting in a need to adapt algorithms depending on the applicable dataset (Carrell et al., 2017).

To address this, we trained a BERT-based source-specific classifier which predicts the source given the findings and background as input. We trained this model using a subset MIMIC-CXR and Indiana reports. However, during prediction or evaluation phase, we chose a higher threshold of 0.7 for predicting a source i.e. if an input is predicted to be Indiana or MIMIC-CXR with a probability of 0.7 or higher, we predict it to be Indiana or MIMIC-CXR respectively, otherwise it is marked to be of an unknown source. Based on the predicted source of a test sample (MIMIC-CXR, Indiana or unknown), the source-specific models' output is chosen as the prediction for that sample.

4.4 Evaluation Metrics

We use two sets of metrics to evaluate model performance at the corpus level, ROUGE and Factual

Model	MIMIC$_{200}$				Indiana$_{200}$				Combined$_{400}$			
	R-1	R-2	R-L	F-F_1	R-1	R-2	R-L	F-F_1	R-1	R-2	R-L	F-F_1
RoBERTa-large$_{(M)}$	0.634	0.509	0.602	0.768	0.425	0.259	0.415	0.634	0.533	0.390	0.516	0.725
BioRoBERTa$_{(M)}$	0.642	0.513	0.617	0.770	0.449	0.273	0.437	0.638	0.541	0.391	0.520	0.729
BioRoBERTa$_{(M),FAR}$	**0.647**	**0.524**	**0.623**	**0.781**	0.455	0.276	0.442	0.665	0.546	0.394	0.523	0.734
BioRoBERTa$_{(M+I)}$	0.499	0.356	0.472	0.694	0.691	0.605	0.677	0.678	0.594	0.480	0.574	0.709
BioRoBERTa$_{(M+I),FAR}$	0.507	0.362	0.481	0.717	**0.701**	**0.626**	**0.685**	**0.685**	0.596	0.480	0.577	0.716
BioRoBERTa$_{(M+M+I)}$	0.585	0.463	0.563	0.712	0.685	0.597	0.671	0.660	0.642	0.539	0.623	0.719
BioRoBERTa$_{(M+M+I),FAR}$	0.592	0.469	0.570	0.719	0.687	0.601	0.676	0.667	0.647	0.544	0.629	0.726
Ensemble	0.632	0.519	0.611	0.768	0.692	0.604	0.672	0.679	**0.670**	**0.568**	**0.650**	**0.741**

Table 3: Results of our base model, factually correct re-ranking and source-specific ensembling experiments on our internal test data of 200 MIMIC-CXR and 200 Indiana radiology reports. Combined presents results for each model when both the sources (400 reports) are considered together. R-1, R-2, R-L and F-F_1 represent ROUGE-1, ROUGE-2, ROUGE-L and Factual F_1 scores respectively.

F_1. The organizers used ROUGE and CheXbert F_1 metrics for evaluation. ROUGE-2 F_1 metric was used for the task leaderboard.

ROUGE We use the standard ROUGE scores (Lin, 2004), and report the F_1 scores for ROUGE-1, ROUGE-2 and ROUGE-L, which compare the word-level unigram, bigram and longest common sequence overlap with the reference summary, respectively.

Factual F_1 For factual correctness evaluation, we use a Factual F_1 score as proposed by Zhang et al. (2019b). The Factual F_1 scores are calculated by 1) running the CheXbert labeler on both the reference and generated summaries to obtain the binary presence values of a collection of disease variables 2) calculating the F_1 score for each of the variables over the entire test set, using reference values as oracle; and 3) obtaining the macro-averaged F_1score over all variables. Following the process in Zhang et al. (2019b), we exclude some variables due to their small sample sizes (with less than 5% positive ratio in the entire dataset). We included only Cardiomegaly, Lung Opacity, Lung Lesion, Pneumonia, Atelectasis, Pleural Effusion and No Finding in our calculation of Factual F_1 scores.

CheXbert F_1 The organizers used CheXbert F_1 score to calculate the factual correctness, which follows the same process as Factual F_1. However, in their calculation they considered a different set of observations which were found prominent in the official test data: Cardiomegaly, Lung Opacity, Edema, Pneumonia, Atelectasis, Pleural Effusion and No Finding.

5 Experiments & Results

5.1 Data

As noted in section 3, training and validation datasets provided in MEDIQA 2021 can be combined and re-split. We set aside 200 radiology reports each, randomly chosen from MIMIC-CXR validation dataset and Indiana validation dataset, to form our combined internal test dataset. The remaining 1,800 reports each from MIMIC-CXR validation data and Indiana validation data, along with 91,544 of MIMIC-CXR training data are utilized for training.

For the clarity of reading, from here onward, we will refer to the original MIMIC-CXR dataset with 91,544 reports as MIMIC$_{train}$. The 200 reports randomly selected each from the original MIMIC-CXR and Indiana validation sets will be denoted as MIMIC$_{200}$ and Indiana$_{200}$, respectively. Together, these 2 new sets formed our internal test set Combined$_{400}$. The remaining reports from the original MIMIC-CXR and Indiana validation sets will be denoted as MIMIC$_{1800}$ and Indiana$_{1800}$, respectively. We present results on this internal test data under 3 settings (1) results on MIMIC$_{200}$, (2) results on Indiana$_{200}$, and (3) results on the combined internal test dataset, Combined$_{400}$. Most of the following results (Tables 3, 4 & 5) are presented on the internal test dataset. The official results presented in Table 6 are on the official external test data of 600 radiology reports.

5.2 Base Models

We conducted four experiments to get our three base models specific to MIMIC-CXR, Indiana and unknown sources. We utilized MIMIC$_{train}$ to train our first two models, RoBERTa-large$_{(M)}$

and BioRoBERTa$_{(M)}$. We used Indiana$_{1800}$ for the model BioRoBERTa$_{(M+I)}$, and used Indiana$_{1800}$ and MIMIC$_{1800}$ for the model BioRoBERTa$_{(M+M+I)}$. In each setting we split the available dataset into 90/10 for training and validation splits. We evaluated all our models on the internal test set of 400 radiology reports. Each of our models uses a seq2seq architecture with encoder and decoder both composed of Transformer layers. For both encoder and decoder, we inherited the RoBERTa Transformer layer implementations. We also added an encoder-decoder attention mechanism. All models were fine-tuned on the target task using Adam optimizer with a learning rate of 0.05. We used Huggingface's transformers library[4] (Wolf et al., 2019) for executing our experiments. In our encoder-decoder setup, our input was capped at 128, output summary at 40, beam size was 10, our length penalty was set as 0.8. Finally, in our summary generation, trigram and higher length phrases were not repeated.

Table 3 presents results of the 4 experiments. Between the 2 models that were trained using only MIMIC$_{train}$, BioRoBERTa$_{(M)}$ consistently outperform RoBERTa-large$_{(M)}$ in this task, likely due to BioRoBERTa$_{(M)}$ utilizing a domain adapted version of RoBERTa. Among the 3 BioMed-RoBERTa-base based models, BioRoBERTa$_{(M)}$ performs better for MIMIC$_{200}$, and BioRoBERTa$_{(M+I)}$ provides better performance for Indiana$_{200}$. BioRoBERTa$_{(M+M+I)}$ fine-tuned on both MIMIC-CXR and Indiana provides better performance on the Combined$_{400}$ but performs poorly when we consider each source separately.

5.3 Fact-aware Re-ranking (FAR)

For the prediction of the 14-observations-vector we combined MIMIC$_{train}$, MIMIC$_{1800}$, and Indiana$_{1800}$ to form our training and validation splits. Table 4 presents our F_1 scores for our impression 14-observations-vector prediction model evaluated on the internal test dataset Combined$_{400}$. We utilized Smit et al. (2020)'s publicly available implementation[5] to train the domain-specific RoBERTa model (BioMed-RoBERTa-base) for predicting impression 14-observations-vector. In this setup, the transformer architecture was modified with 14 linear heads, corresponding to 14 observations. We concatenate Findings and background of a radiology report to be our input, which is then tokenized and the input is capped at 128. The hidden state of the CLS token is fed as input to each of the linear heads. The model is trained using cross-entropy loss and Adam optimization with a learning rate of 2×10^{-5}. The cross-entropy losses for each of 14 observations are added to produce the final loss. During training, the model was periodically evaluated and the best performing model averaged over 14 observations was saved.

Category	Macro F_1	Micro F_1
Atelectasis	0.839	0.915
Cardiomegaly	0.803	0.943
Consolidation	0.809	0.973
Edema	0.930	0.963
Enlarged Cardiom.	0.634	0.990
Fracture	0.783	0.988
Lung Opacity	0.848	0.911
Lung Lesion	0.829	0.982
No Finding	0.881	0.881
Pneumonia	0.898	0.950
Pneumothorax	0.939	0.996
Pleural Effusion	0.899	0.950
Pleural Other	0.640	0.990
Support Devices	0.918	0.969
Average	0.832	0.957

Table 4: Impression observations-vector prediction results.

Label	P	R	F_1
MIMIC$_{200}$	0.987	0.993	0.989
Indiana$_{200}$	0.993	0.987	0.990

Table 5: Source-specific classifier results

For fact-aware re-ranking we utilize the model trained above to re-rank the top 10 (N=10 was empirically determined) generated summaries from our three base models presented in section 5.2. Table 3 presents results for our following three factually correct re-ranking experiments, BioRoBERTa$_{(M),FAR}$, BioRoBERTa$_{(M+I),FAR}$, and BioRoBERTa$_{(M+M+I),FAR}$. As BioRoBERTa$_{(M),FAR}$ shows best performance for MIMIC-CXR radiology reports (MIMIC$_{200}$), BioRoBERTa$_{(M+I),FAR}$ exhibits best performance for Indiana radiology reports (Indiana$_{200}$) and the combined BioRoBERTa$_{(M+M+I),FAR}$ shows best performance for the combined test data (Combined$_{400}$), these models are chosen to be our source-specific models for MIMIC-CXR, Indiana and unknown sources respectively.

[4] https://github.com/huggingface/transformers

[5] https://github.com/stanfordmlgroup/CheXbert

Model	R-1	R-2	R-L	CheXbert F_1
Ensemble	0.5252	0.4002	0.5060	0.6823
+post-processing	0.5328	0.4082	0.5134	0.6774

Table 6: Official submission and results.

5.4 Source-specific Ensemble

For training our source-specific classifier we used a downsampled subset of $\text{MIMIC}_{\text{train}}$ of 10,000 radiology reports and Indiana_{1800} and formed 90/10 training and validation splits. We evaluated the model on MIMIC_{200} and Indiana_{200} and present our results in Table 5. We again utilized Huggingface transformers library to conduct our experiments. In this setup, we used the BERT-base architecture with a single linear head for our classification of the source. We concatenate Findings and background for a radiology report to be our input, which is then tokenized and input is capped at 512. The model is trained using cross-entropy loss and Adam optimization with a learning rate of 2×10^{-5}. Our model was trained for 3 epochs.

Utilizing the above model we identify the source of a radiology report and apply the source-specific models. Ensemble results in Table 3 presents our results after we apply source-specific ensembling technique to our internal test dataset. Our ensembled results show a slight drop in performance for individual source MIMIC_{200} and Indiana_{200} (due to classification errors), but show best performance on the combined dataset (Combined_{400}).

5.5 Official Submissions & Results

Table 6 presents our top 2 official submission results. Ensemble presents our best performing source-specific ensemble technique applied to the official test data. In our another submission (Ensemble + post-processing) we remove certain tokens (like "1.", "2.", "__") to clean up our source-specific ensemble technique output which slightly improved the rouge scores.

6 Discussion

In this section, we present two major findings of our approach. First, we find that radiology reports from different sources have distinct language, and fine-tuning a model trained on source A with a small amount of data from source B provides significant gains in performance on source B, allowing the model to be transferable. As it can be seen in Table 3, zero-shot application of our model $\text{BioRoBERTa}_{(\text{M})}$, which is fine-tuned only on MIMIC-CXR ($\text{MIMIC}_{\text{train}}$), shows lower performance on the Indiana dataset. However, on further fine-tuning $\text{BioRoBERTa}_{(\text{M})}$ on a small dataset of 1,800 Indiana reports (Indiana_{1800}) leads to huge gains in performance on Indiana dataset (model $\text{BioRoBERTa}_{(\text{M+I})}$ on Indiana_{200}).

Second, fact-aware re-ranking methodology improves performance of the models on natural language metrics (ROUGE) as well as factual correctness of our predictions, but metrics beyond lexical overlap are needed. As shown in Table 3, models using FAR outperform the base models when measured in ROUGE even through FAR's objective is not to optimize ROUGE. Table 7 shows examples of the most probable predictions from base model compared with the predictions after FAR, and the human-generated ground-truth impressions. ROUGE scores for both predictions compared to the ground-truth are shown at the end of each example. In the first example, FAR chooses a better ROUGE scoring prediction over the most probable prediction by the base model. However, in the second example, FAR doesn't choose the higher ROUGE scoring prediction but rather the more factually correct one. With the current evaluation metric ROUGE, this would lead to a drop in performance. Developing and adopting new metric that consider both lexical as well as factual correctness jointly (Mrabet and Demner-Fushman, 2020) is crucial to steer the research community to develop systems that ensure factual correctness as well as readability.

Limitations and Future Work. We acknowledge several limitations to our work. First, we recognize our dependence on an external structured label generator. As we use CheXbert labels as our proxy for ground truth for training our 14-observations-vector predictor, as well as in our similarity function, any errors in CheXbert have a direct impact on our system's performance. Second, though FAR methodology has shown significant gains in performance in Factual F_1 and ROUGE scores, the system is limited by the generated candidate summaries. We aim to build on this approach by incorporating this methodology during training as a modified version of beam search. Third, all of our presented results are evaluated using a relatively small set of internal test data, due to the limitations on data during the challenge. Though

Base Model's Prediction	Prediction after FAR	Human-generated Impression
No acute cardiothoracic process. R-1: 0	No acute cardiopulmonary process. Tiny right pleural effusion. R-1: 0.6	Tiny right pleural effusion.
No acute cardiopulmonary process. R-1: 0.6	Normal chest radiograph. Mild cardiomegaly. R-1: 0.3	Mild cardiomegaly, new since ___. No acute cardiopulmonary process.

Table 7: Examples depicting the most probable prediction from base model, re-ranked prediction using our FAR methodology compared to the ground truth (human-generated impression).

our approach has translated into similar good performance on the official test data, we aim to further evaluate our approach on an increased test data. Finally, as ROUGE has been shown to be an imperfect metric for radiology report summarization evaluation (Zhang et al., 2019b), we aim to further evaluate our system (1) using other automated metrics such as BERTScore (Zhang et al., 2019a), BLEURT (Sellam et al., 2020), and HOLMS (Mrabet and Demner-Fushman, 2020), (2) by conducting qualitative evaluation of our system's predictions by involving human annotators such as radiologists or subject matter experts.

7 Conclusion

We have presented our system developed during our participation in MEDIQA 2021 RRS challenge. We found that radiology reports from different sources have distinct language and fine-tuning a trained model with a small amount of data from another source leads to gains in performance and allows the models to be transferable. Further, techniques like fact-aware re-ranking, which utilizes a factual vector of the summary to re-rank candidate summaries, not only improves factual correctness of the summary but also improves the performance of the model on the traditional natural language metrics like ROUGE. We have also identified limitations of our work, and discussed promising areas of future research.

References

American College of Radiology. 2020. Acr practice parameter for communication of diagnostic imaging findings. Available at www.acr.org Accessed March 2020.

Yuhao Zhang Chaitanya Shivade Curtis Langlotz Dina Demner-Fushman Asma Ben Abacha, Yassine Mrabet. 2021. Overview of the mediqa 2021 shared task on summarization in the medical domain. In *Proceedings of the 20th SIGBioMed Workshop on Biomedical Language Processing, NAACL-BioNLP 2021*. Association for Computational Linguistics.

Jan ML Bosmans, Joost J Weyler, Arthur M De Schepper, and Paul M Parizel. 2011. The radiology report as seen by radiologists and referring clinicians: results of the cover and rover surveys. *Radiology*, 259(1):184–95.

David S Carrell, Robert E Schoen, Daniel A Leffler, Michele Morris, Sherri Rose, Andrew Baer, Seth D Crockett, Rebecca A Gourevitch, Katie M Dean, and Ateev Mehrotra. 2017. Challenges in adapting existing clinical natural language processing systems to multiple, diverse health care settings. *Journal of the American Medical Informatics Association*, 24(5):986–991.

Neal J. Clinger, Tim B. Hunter, and Bruce J. Hillman. 1988. Radiology reporting: attitudes of referring physicians. *Radiology*, 169(3):825–826.

Dina Demner-Fushman, Marc D Kohli, Marc B Rosenman, Sonya E Shooshan, Laritza Rodriguez, Sameer Antani, George R Thoma, and Clement J McDonald. 2016. Preparing a collection of radiology examinations for distribution and retrieval. *Journal of the American Medical Informatics Association*, 23(2):304–310.

Jacob Devlin, Ming-Wei Chang, Kenton Lee, and Kristina Toutanova. 2018. Bert: Pre-training of deep bidirectional transformers for language understanding. *arXiv preprint arXiv:1810.04805*.

Suchin Gururangan, Ana Marasović, Swabha Swayamdipta, Kyle Lo, Iz Beltagy, Doug Downey, and Noah A. Smith. 2020. Don't stop pretraining: Adapt language models to domains and tasks. In *Proceedings of ACL*.

Jeremy Irvin, Pranav Rajpurkar, Michael Ko, Yifan Yu, Silviana Ciurea-Ilcus, Chris Chute, Henrik Marklund, Behzad Haghgoo, Robyn Ball, Katie Shpanskaya, et al. 2019. Chexpert: A large chest radiograph dataset with uncertainty labels and expert comparison. In *Proceedings of the AAAI Conference on Artificial Intelligence*, volume 33, pages 590–597.

Alistair EW Johnson, Tom J Pollard, Seth J Berkowitz, Nathaniel R Greenbaum, Matthew P Lungren, Chih-ying Deng, Roger G Mark, and Steven Horng. 2019. Mimic-cxr, a de-identified publicly available database of chest radiographs with free-text reports. *Scientific data*, 6(1):1–8.

Mike Lewis, Yinhan Liu, Naman Goyal, Marjan Ghazvininejad, Abdelrahman Mohamed, Omer Levy, Ves Stoyanov, and Luke Zettlemoyer. 2019. Bart: Denoising sequence-to-sequence pre-training for natural language generation, translation, and comprehension. *arXiv preprint arXiv:1910.13461*.

Chin-Yew Lin. 2004. Rouge: A package for automatic evaluation of summaries. In *Text summarization branches out*, pages 74–81.

Yang Liu and Mirella Lapata. 2019. Text summarization with pretrained encoders. *arXiv preprint arXiv:1908.08345*.

Yinhan Liu, Myle Ott, Naman Goyal, Jingfei Du, Mandar Joshi, Danqi Chen, Omer Levy, Mike Lewis, Luke Zettlemoyer, and Veselin Stoyanov. 2019. Roberta: A robustly optimized bert pretraining approach. *arXiv preprint arXiv:1907.11692*.

Sean MacAvaney, Sajad Sotudeh, Arman Cohan, Nazli Goharian, Ish Talati, and Ross W Filice. 2019. Ontology-aware clinical abstractive summarization. In *Proceedings of the 42nd International ACM SIGIR Conference on Research and Development in Information Retrieval*, pages 1013–1016.

Yassine Mrabet and Dina Demner-Fushman. 2020. Holms: Alternative summary evaluation with large language models. In *Proceedings of the 28th International Conference on Computational Linguistics*, pages 5679–5688.

Sascha Rothe, Shashi Narayan, and Aliaksei Severyn. 2020. Leveraging pre-trained checkpoints for sequence generation tasks. *Transactions of the Association for Computational Linguistics*, 8:264–280.

Alexander M Rush, Sumit Chopra, and Jason Weston. 2015. A neural attention model for abstractive sentence summarization. *arXiv preprint arXiv:1509.00685*.

Thibault Sellam, Dipanjan Das, and Ankur P Parikh. 2020. Bleurt: Learning robust metrics for text generation. *arXiv preprint arXiv:2004.04696*.

Akshay Smit, Saahil Jain, Pranav Rajpurkar, Anuj Pareek, Andrew Y Ng, and Matthew P Lungren. 2020. Chexbert: combining automatic labelers and expert annotations for accurate radiology report labeling using bert. *arXiv preprint arXiv:2004.09167*.

A Wallis and P McCoubrie. 2011. The radiology report—are we getting the message across? *Clinical radiology*, 66(11):1015–1022.

Thomas Wolf, Lysandre Debut, Victor Sanh, Julien Chaumond, Clement Delangue, Anthony Moi, Pierric Cistac, Tim Rault, Rémi Louf, Morgan Funtowicz, et al. 2019. Huggingface's transformers: State-of-the-art natural language processing. *arXiv preprint arXiv:1910.03771*.

Jingqing Zhang, Yao Zhao, Mohammad Saleh, and Peter Liu. 2020. Pegasus: Pre-training with extracted gap-sentences for abstractive summarization. In *International Conference on Machine Learning*, pages 11328–11339. PMLR.

Tianyi Zhang, Varsha Kishore, Felix Wu, Kilian Q Weinberger, and Yoav Artzi. 2019a. Bertscore: Evaluating text generation with bert. *arXiv preprint arXiv:1904.09675*.

Yuhao Zhang, Daisy Yi Ding, Tianpei Qian, Christopher D Manning, and Curtis P Langlotz. 2018. Learning to summarize radiology findings. *arXiv preprint arXiv:1809.04698*.

Yuhao Zhang, Derek Merck, Emily Bao Tsai, Christopher D Manning, and Curtis P Langlotz. 2019b. Optimizing the factual correctness of a summary: A study of summarizing radiology reports. *arXiv preprint arXiv:1911.02541*.

UETrice at MEDIQA 2021: A Prosper-thy-neighbour Extractive Multi-document Summarization Model

Duy-Cat Can[1], Quoc-An Nguyen[1]*, Quoc-Hung Duong[1], Minh-Quang Nguyen[1]
Huy-Son Nguyen[1], Linh Nguyen Tran Ngoc[2], Quang-Thuy Ha[1] and Mai-Vu Tran[1]†
[1]VNU University of Engineering and Technology, Hanoi, Vietnam.
{catcd, 18020106, 18020021, 19020405}@vnu.edu.vn
{18021102, thuyhq, vutm}@vnu.edu.vn
[2]Viettel Big Data Analytics Center, Viettel Telecommunication Company, Viettel Group.
linhntn3@viettel.com.vn

Abstract

This paper describes a system developed to summarize multiple answers challenge in the MEDIQA 2021 shared task collocated with the BioNLP 2021 Workshop. We propose an extractive summarization architecture based on several scores and state-of-the-art techniques. We also present our novel prosper-thy-neighbour (PtN) strategies to improve performance. Our model has been proven to be effective with the best ROUGE-1/ROUGE-L scores, being the shared task runner-up by ROUGE-2 $F1$ score (over 13 participated teams).

1 Introduction

Biomedical documents are available with the tremendous amount on the Internet, together with several search engines (e.g., Pubmed®[1]) and question-answering systems (e.g., CHiQA[2]) developed. However, the returned results of these systems still contain a lot of noise and duplication, making them difficult for users without medical knowledge to quickly grasp the main content and get the necessary information. Hence, generating a shorter condensed form with important information would benefit many users as it saves time and can retrieve massive useful information. This motivation leads to the growing interest among the research community in developing automatic text summarization methods. The BioNLP-MEDIQA 2021 shared task[3] (Ben Abacha et al., 2021) aims to attract further research efforts in text summarization and their applications in medical Question-Answering (QA). This shared task is motivated by a need to develop relevant methods, techniques, and gold standards for text summarization in the medical domain and their application to improve the domain-specific QA system. Task 2 - Summarization of Multiple Answers focuses on developing multi-document summarization approaches that could synthesize and compress information from answers to a medical question.

According to Radev et al. (2002) a summary is defined as *'a text that is produced from one or more texts, that conveys important information in the original text(s), and that is no longer than half of the original text(s) and usually, significantly less than that'*. Automatic text summarization is the task of condensing the document(s) and generating a compressed summary, which is shorter but preserves key information content and overall meaning. A summary can be generated through extractive or abstractive approaches (or hybrid). Typically, to produce an abstractive summarization, we need to use advanced linguistic techniques to 'understand' the text as well as re-generate the summary in natural language from useful information. Up to now, the research community is focusing more on extractive summarization. This approach tries to achieve coherent and meaningful summaries in a more simple and faster way than the abstractive approach. Extractive summarization chooses important sentences (or phrases) from the original documents (without any modification) and merges them to generate a summary.

Our proposed model for the multi-answer summarization task follows extractive summarization approaches. We try to select sentences containing the most important information in the original answers. Our novel contributions are: (i) Proposing the question-driven scores to ensure that the summary is the answer to the question, (ii) Proposing Prosper-thy-neighbour (PtN) strategies, which increase the constraint of neighbouring sentences, to take advantage of paragraph information in the answer. (iii) Combining several scores that successfully applied for summarization problem, includ-

*Contributed equally & Names are in alphabetical order
†Corresponding author

[1]https://pubmed.ncbi.nlm.nih.gov/
[2]https://chiqa.nlm.nih.gov/
[3]https://sites.google.com/view/mediqa2021

Proceedings of the BioNLP 2021 workshop, pages 311–319
June 11, 2021.

ing TF-IDF, Lexrank, and Textrank with optimized weights, (iv) Improving the maximal marginal relevance technique (MMR) for multi-document summarization with BERT-based embedding to improve the performance.

The remaining of this paper is organized as follows: Section 2 gives a brief introduction to some state-of-the-art related works. Section 3 describes task data and our proposed model. Section 4 is the experimental results and our discussion. And finally, the conclusion.

2 Related works

From the early 1950s, various methods have been proposed for extractive summarization (Allahyari et al., 2017). Some of them are based on the idea of using scores to choose the most important phrases in the documents. Term Frequency-Inverse Document Frequency (TF-IDF) (Hovy et al., 1999; Christian et al., 2016) is a frequency-based score to detect important sentences by calculating the scores of its words. Lexrank (Erkan and Radev, 2004) and Textrank (Mihalcea and Tarau, 2004) are two graph-based methods that rank sentences/words using their degree centrality. Maximal Marginal Relevance (MMR) (Carbonell and Goldstein, 1998; Bennani-Smires et al., 2018) is one of the most well-known approaches for multi-document summarization. It is a diversity-based re-ranking method based on the document similarities and can be used to remove redundancy in the summaries. Although encouraging results have been reported, most of these scores are applied individually. Since each score type has its unique contribution, combining them may help to improve the performance. Hence, we propose an architecture to take advantage of several scores with weights and calculate a final combined score.

With the advent of machine learning techniques in NLP, many research projects tried to apply machine learning methods to extractive summarization tasks, from the Naive Bayes, Decision tree, Support vector machine (Gambhir and Gupta, 2017) to deep learning models. Most recently, Savery et al. (2020) improved the Bidirectional auto regressive transformer (BART) with a question-driven approach, but it is more well-known for abstractive summarization, which is not discussed in-depth in this paper.

3 Materials and Methods

3.1 Shared task data

The MEDIQA-AnS Dataset (Savery et al., 2020) is used as the training data set. The validation and the test sets are the summaries that were created by the experts from the original answers generated by the question-answering system namely CHiQA[4]. Table 1 gives our statistics on the given datasets (see (Ben Abacha et al., 2021) for detailed description of shared task data).

An important observation is that answers often tend to have related sentences in a passage that makes an important 'point'. Some adjacent sentences are structured in a deductive manner (e.g., several explanatory sentences follow after a stated sentence) or inductive (e.g., the last sentence is the conclusion of previous sentences). Extracting these whole pieces of text ensures a complete summary while enhancing fluency and natural language resemblance. Our prosper-thy-neighbour strategies are proposed to take advantage of this characteristic.

Table 1: Statistics of the datasets.

Statistic aspects	Training Article	Training Section	Validation	Test
Questions	156	156	50	80
Average				
A per Q	3.54	3.54	3.85	3.80
Sent per A	84.93	29.07	14.50	13.03
Sent per SSum	6.31	6.31	-	-
Sent per MSum	10.30	10.30	11.06	-
Compression ratio				
SSum	0.12	0.49	-	-
MSum	0.06	0.18	0.33	-

A: Answer, Q: Question, Sent: Sentence, SSum: Single-answer Summary, MSum: Multi-answer Summary

3.2 Proposed model

The overall architecture of our Prosper-thy-Neighbour (PtN) summarization model is shown in Figure 1. It comprises four main phases: preprocessing, single document summarization, multi-document summarization and post-processing phases.

3.2.1 Pre-processing

The pre-processing phase receives question Q and a set of corresponding answers (documents) $D = \{d_i\}_{i=1}^{n}$ as the input. ScispaCy (Neumann et al., 2019), which is based on SpaCy (Honnibal et al.,

[4]https://chiqa.nlm.nih.gov

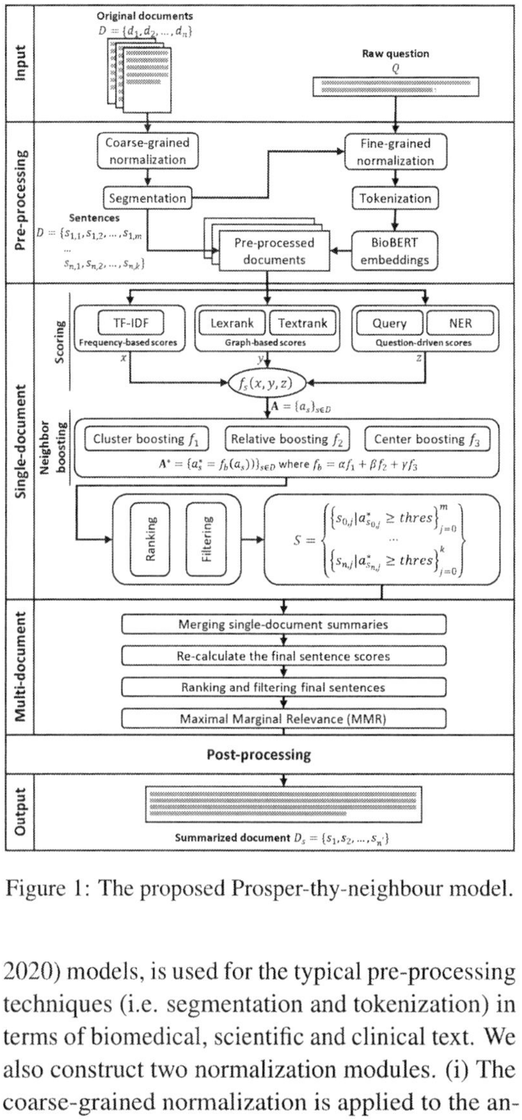

Figure 1: The proposed Prosper-thy-neighbour model.

2020) models, is used for the typical pre-processing techniques (i.e. segmentation and tokenization) in terms of biomedical, scientific and clinical text. We also construct two normalization modules. (i) The coarse-grained normalization is applied to the answer only. It removes noise from the raw text (non-ASCII characters, HTML tags, duplicate spacing, etc.) (ii) The fine-grained normalization includes stop-words removing, lower-casing, stemming, and full form generation (Schwartz and Hearst, 2002) for biomedical abbreviations. Finally, BioBERT (Lee et al., 2020), which is designed for multiple biomedical text mining tasks, is used for part-of-speech tagging, named entities/keywords recognizing and embedding generating. BioBERT-based embeddings are $768-$ dimensional vectors used for calculating the similarity of words and sentences.

3.2.2 Single-answer extractive summarization

Using information from the pre-processing phase, the single-document extractive summarization phase generates the summary for every single answer. Our extractive summarization model tries to determine which sentences are important to the document by sentence scoring.

Sentences scoring: Since it is difficult to identify the importance of sentences from a single point of view, hence, we use three different types of scores: Frequency-based scores, graph-based scores and question-driven scores.

Frequency-based score: *Term Frequency - Inverse Document Frequency (TF-IDF)* (Salton and McGill, 1986) is the probabilistic method that reflects the importance of words in a set of documents by a float number. The TF-IDF score of a word w contained in document d of document set D is defined as $tfidf(w, d, D)$. We apply two rules to improve TF-IDF: (i) Boosting the TF-IDF score of keywords, and (ii) Assigning TF-IDF score to 0 if it is lower than a pre-selected threshold. The TF-IDF score of a sentence is the cumulative TF-IDF scores of its component words.

Graph-based scores are used to determine which sentences and words seem to be the core of a document. Lexrank and Textrank are two of the most well-known methods of this approach.

Lexrank (Erkan and Radev, 2004) computes sentence importance based on the concept of eigenvector centrality in a graph representation of sentences. A document is considered as a graph, each node represents a sentence. Two nodes have a weighted edge depending on the similarity of their corresponding sentences. Cosine similarity is used to calculate the similarity between two sentences x and y (see Formula 1). In which, x and y are represented by TF-IDF vectors of n dimensions, i.e., X and Y respectively (n is the number of distinguished tokens in two sentences).

$$sim(x, y) = \frac{X \cdot Y}{\|X\| \times \|Y\|} \tag{1}$$

To calculate the centrality of a node, we analyze the weight of its connected edges and the centrality of adjacent nodes (Formula 2). If a sentence is similar to many other sentences, it has higher centrality and conceived having a certain ability to represent other sentences.

$$p(u) = \frac{d}{n} + (1-d) \sum_{v \in adj_u} \frac{sim(u, v)}{\sum_{z \in adj_v} sim(z, v)} p(v) \tag{2}$$

where adj_u is the set of nodes that adjacent to u, n is the number of nodes and d is the damping factor.

Textrank (Mihalcea and Tarau, 2004) is mostly similar to Lexrank. It calculates the centrality of terms instead of the centrality of sentences as in Formula 3. In the PtN model, if the Textrank score is lower than a predefined threshold, we assign it to 0. The Textrank score of a sentence is the sum of Textrank scores of its participated terms.

$$sim(X,Y) = \frac{|w|w \in X \text{ and } w \in Y|}{log(|X|) + log(|Y|)} \quad (3)$$

in which w is the token and X and Y are two terms.

Question-driven scores are used to give higher priorities to sentences that are related to the questions. These scores are proposed to focus on the answer summarization task, ensuring that the summary is a suitable answer to the question.

Question-similarity score uses the BioBERT and Cosine distance (Formula 1) to calculate the similarities between the question and sentences in all of its answers. Formally, qb(sentence), the question-similarity score of a sentence is defined as:

$$qb(\text{sentence}) = sim(\text{sentence}, \text{question}) \quad (4)$$

Keyword-based score is determined by the percentage of question keywords that appear in a sentence. Let K is the set of question keywords, kw(sentence) is the keyword-based score of a sentence, it is defined by the following formula:

$$kw(\text{sentence}) = \frac{|\{k : k \in K\}|}{|K|} \quad (5)$$

Scores combination: All scores are normalized in the range $[0 - 1]$ by using `Min-Max` normalization. We then combine them into a final sentence score by using optimized weights (see Formula 6.

$$\begin{aligned} score = & w_1 \times tfidf \\ & + w_2 \times lexrank + w_3 \times textrank \\ & + w_4 \times querybase + w_5 \times keywords \end{aligned} \quad (6)$$

in which, w_i is the weight of each score. They are fine-tuned on the validation set.

Prosper-thy-neighbour strategies:
As described in Section 3.1, an important sentence may need some adjacent sentences to clarify or support it. Hence, answers often tend to have continuous segments of sentences that make important 'points'. Since the aforementioned scores do not consider the neighbours of a sentence, our prosper-thy-neighbour strategies are proposed to take advantage of this characteristic. There are three different prosper-thy-neighbour strategies: cluster-boosting, relative-boosting and centre-boosting.

Cluster-boosting: We calculate the averaged scores of n continuous sentences ($n = 3, 4, 5$) as cluster scores. We then select top-k clusters with the highest average scores. The sentence score is set equal to its highest cluster score. Sentences that are not selected in any clusters are assigned the score of 0.

Relative-boosting is performed by three steps:

- Step 1: Find top-n highest-score sentences with their original orders.
- Step 2: For consecutive selected sentences, let L is the position of the preceding sentence, R is the position of the following sentence. If $R - L + 1 \leq k$ (k is predefined), step 3 is executed.
- Step 3: Let $score_i$ be the score of the i-th sentence. The final scores $final_i$ of all sentences having the position between L and R are updated by the following formula:

$$final_i = max_{j=L}^{R}(score_j) \quad (7)$$

Centre-boosting: Let $score_i$ be the score of i-th sentences. The final score $final_i$ of sentence i-th is updated by the following formula:

$$final_i = max_{j=max(i-L+1,1)}^{min(i+R-1,n)} score_j \quad (8)$$

in which, n is the number of sentences, L and R is the number of sentences that impact the current sentence i in two directions: left and right. With centre-boosting, the important sentence strongly affects its adjacent sentences.

However, with these prosper-thy-neighbour strategies, the selected neighbour sentences can bring redundant information, i.e., we may keep too many sentences to the left/right of an important sentence. Those redundancies can be cut off in the post-processing phase (Section 3.2.4).

Ranking and and Filtering Sentences We use the final score boosted by the prosper-thy-neighbour strategy to rank the sentences. There are several ways to choose sentences for the single-document extractive summary: getting top-n or top-p% of sentences, using the threshold to filter unimportant sentences. In the proportion- and threshold-based approach, the number of sentences depends on the document length and sentence scoring. It might probably cause an unexpected bias in the next multi-document summarization phase. Based on the experimental results on the validation set, we fix the number of selected sentences in each document.

3.2.3 Multi-answer extractive summarization

Multiple extractive single-answer summaries from the previous phase are merged into a single document. Since the previous phase chooses an equal number of sentences for all answers, there might be some redundant sentences. Since the current sentence scores are based on separate documents, we re-calculate them as in the merged document by using the proposed score described in Section 3.2.2. The filtering step then removes some lowest-score sentences.

Maximal Marginal Relevance (MMR): (Carbonell and Goldstein, 1998) is also used to reduce redundancy while maintaining query relevance. MMR works in the selected appropriate sentence in merged documents. It is the combination of the relevance and diversity concepts, in a controllable way. Let S_i is the i-th sentence, its MMR score is calculated based on the similarities between S_i, the answer D and the question Q (Formula 9). The similarity to the question and the duplication with other sentences affects the MMR score through the ratio λ. In which, BioBERT is used to represent sentences and question and Cosine distance is used to calculate the similarities. We use the MMR score to discard duplicated and question-irrelevant sentences, i.e., remove m sentences having the lowest MMR score.

$$\begin{aligned} \text{MMR}_i = \arg\max_{S_i \in D}[\lambda(sim(S_i, Q) \\ - (1-\lambda)max_{j \neq i} sim(S_i, S_j))] \end{aligned} \quad (9)$$

3.2.4 Post-processing

For each segment of continuously selected sentences, we find the position of the most important sentence which has the highest combined score. Then, for other sentences in the segment, if the distance from their position to the important sentence exceeds a predefined k parameter, those should be eliminated in the final multi-document extractive summary.

4 Experimental results

4.1 Evaluation metrics

We adopt the official task evaluations with ROUGE scores (Lin and Och, 2004) including ROUGE-1, ROUGE-2 and ROUGE-L. ROUGE-n Recall (R), Precision (P) and $F1$ between predicted summary and referenced summary are calculated as in Formulas 10, 11 and 14, respectively. Choosing correct sentences help to increase ROUGE-n R and P.

$$\text{ROUGE-}n\ P = \frac{|\text{Matched N-grams}|}{|\text{Predict summary N-grams}|} \quad (10)$$

$$\text{ROUGE-}n\ R = \frac{|\text{Matched N-grams}|}{|\text{Reference summary N-grams}|} \quad (11)$$

$$\text{ROUGE-L}\ P = \frac{\text{Length of the LCS}}{|\text{Predict summary tokens}|} \quad (12)$$

$$\text{ROUGE-L}\ R = \frac{\text{Length of the LCS}}{|\text{Reference summary tokens}|} \quad (13)$$

ROUGE-L recall (R), precision (P) and $F1$ are calculated as in Formula 12, 13 and 14, respectively. ROUGE-L uses the Longest Common Subsequence (LCS) between predicted summary and referenced summary and they are normalized by the tokens in the summary.

$$F1 = 2 \times \frac{R \times R}{P + R} \quad (14)$$

4.2 Comparative models

We use the official results of the MEDIQA shared task as a comparison to other participated teams on the multi-answer summarization task.

For a detailed evaluation of the effectiveness of the single-answer summarization phase, we also make some comparisons with related works:

- Lead-3: First three sentences of an article were taken as a summary.
- k-random sentences: k random sentences were selected as a summary.
- k-best ROUGE: k sentences with the highest ROUGE-L score relative to the question were selected.

Table 2: Official results of the MEDIQA 2021: Task 2 - Multi-Answer Summarization.

Team	ROUGE-1			ROUGE-2			ROUGE-L	HOLMS	BERTscore
	P	R	F1	P	R	F1	F1		F1
paht_nlp	0.471	**0.878**	0.585	0.407	**0.767**	**0.508**	0.435	0.706	**0.804**
UETrice	**0.528**	0.814	**0.611**	**0.432**	0.680	0.504	**0.441**	0.738	0.796
XIaoHouZi	0.464	0.864	0.577	0.395	0.748	0.495	0.431	0.699	0.797
ChicHealth	0.474	0.842	0.578	0.398	0.718	0.489	0.426	0.703	0.792
I_have_no_flash	0.472	0.843	0.573	0.397	0.719	0.488	0.425	**0.745**	0.791

Only show results of top-5 participated teams.
The highest results in each column are highlighted in bold.

- Bidirectional long short-term memory (BiLSTM) network (Hochreiter and Schmidhuber, 1997): The most relevant sentences in an article were selected by a BiLSTM.
- Pointer-generator network (See et al., 2017): A hybrid sequence-to-sequence attention model which creates summaries with two approaches: copying text and create new text from the source documents.
- Bidirectional auto-regressive transformer (BART) (Savery et al., 2020): A transformer-based encoder-decoder model improved with a question-driven approach.

The results of these comparative models are taken from experimental results reported in Savery et al. (2020).

Table 3: The comparative results of single-document summarization models.

Model	ROUGE-1 F1	ROUGE-2 F1	ROUGE-L F1
Lead-3	0.23	0.11	0.08
3-random sentences	0.20	0.08	0.06
3-best ROUGE	0.16	0.08	0.06
BiLSTM	0.22	0.10	0.08
Pointer-generator	0.21	0.09	0.07
BART	0.24	0.10	0.07
BART + Query-based	0.29	0.15	0.12
PtN model w/o post-processing	0.26	0.22	0.24
PtN model	**0.30**	**0.22**	**0.25**

All results are reported on the training data set.
The highest results in each column are highlighted in bold.

4.3 Task final results and comparison

Based on the validation set experiments, the number of selected sentences in single-answer summarization is 7 per answer. In the multi-answer summarization phase, the score-based filter selects top-20 sentences in the merged document, then MMR removes 5 lowest-score sentences. Therefore, our multi-answer document summaries have 15 sentences (or less, based on the length of the original answers). Post-processing with distance value $k = 3$ often removes 2-4 sentences. The final outputs often have $\sim$13 sentences. Since both cluster-boosting and relative-boosting show their drawbacks with the lower F1-score performance on the validation set, we use the centre-boosting strategy in our optimal model.

4.3.1 Official results of the multi-answer extractive summarization

Table 2 shows the shared task official results of top-5 competitors. ROUGE-2 $F1$ is used as the main metric to rank the participating teams. We also show several other evaluation metrics for detailed results: ROUGE-1 $F1$, ROUGE-L $F1$, HOMLS $F1$ and BERT-based $F1$. We are the runner-up in the leader board, with ROUGE-2 $F1$ at 0.504 (0.004 less than the rank No.1 team). However, our ROUGE-1 $F1$ and ROUGE-L $F1$ are the highest of all participating teams.

4.3.2 Result of the single-answer extractive summarization

Table 3 shows the performances of our model and comparative models at the single-answer level. Because the results of the comparative models are reported in the training dataset, all results are reported on the training dataset. To ensure the comparisons are fair, we report both model results with and without the post-processing phase. The results show that our model outperforms all comparative models. To ensure the comparisons are fair, we report both model results with and without the post-processing phase. The results show that our model outperforms all comparative models.

4.4 Contribution of model components

We study the contribution of each model component to the system performance by ablating each of them in turn from the model and afterward evaluating the model on the validation set. Validation data are used for evaluation because we use validation data to optimize the model's hyperparameters. We compare these experimental results with the full system results and then illustrate the changes of ROUGE-2 $F1$ in Figure 2. The changes of ROUGE-2 $F1$ show that all model components help the system to boost its performance (in terms of the increments in ROUGE-2 $F1$).The contribution, however, varies among components, TF-IDF and MMR have the biggest contribution while Lexrank/Textrank brings the smallest contribution. The prosper-thy-neighbour strategy also demonstrates its effectiveness to improve the ROUGE-2 $F1$. Centre-boosting seems to be the most suitable strategy for this task since the results increase dramatically if we replace it with cluster-boosting or relative-boosting.

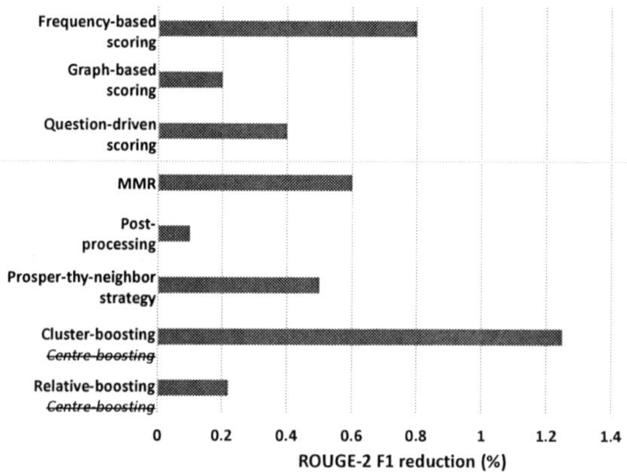

Figure 2: Ablation test results on validation data set for various components and Prosper-thy-neighbour strategies. Cluster-boosting and relative-boosting: Replace centre-boosting by another strategy.

We also investigate the change of results at different compression ratios. Figure 3 shows the change of ROUGE-2 P, R and $F1$ on the validation set when taking 2-20 sentences to the summary (excluding the post-processing step). We observed that P and F have trade-off results while increasing the number of sentences. $F1$ got the best results at 15 sentences, due to the balance between P and F. Therefore, we choose this configuration for our official runs on the test set.

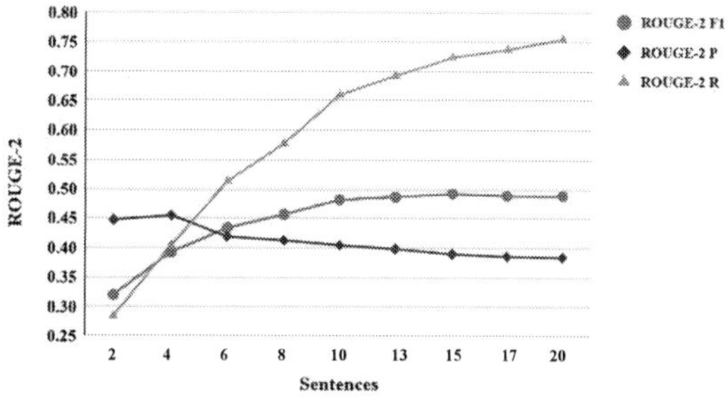

Figure 3: System performance with different compressed ratios.

4.5 Errors analysis

To further evaluate the performance of the proposed system, we have analyzed the results of the best model on the validation set. Table 4 provides some examples of the model problems and their effects.

Firstly, because of using a fixed statistical-based maximum number of output sentences, we ran into problems with too long or too short documents. Question #56 is an example of the redundancy in the output summary that there are only 5 important sentences but our model keeps fixed 13 sentences. On the contrary, in Question #91, the answer to '*How can I stop being allergic to caffeine?*' are summarized in 23 sentences. However, many relevant sentences have been filtered out to ensure a fixed size of the output.

Although we have combined many different ranking methods for tokens and sentences, some final scores did not meet our expectation. The frequency-based scores (TF-IDF) are failed in Question 82, in which the token *'Hirschsprung'* is over-weighted due to repeated occurrence. In addition, the popular keywords like *'treatment'*, *'medicine'* have too low weight. As a result, in Question #19 about *'the cure for pulsatile tinnitus'*, all of the sentences related to treatment and medicine were filtered out.

Some other issues related to the driven question are illustrated in Question #22 and Question #36. In the first example, the question analyzer failed to extract the keyword *'safe'*. For this reason, the summary phase went in the wrong direction – the content is only related to *'defibrillator'*. In the second one, the proposed model did not focus on the driven question so that the summary does not contain the desired information.

Besides the problems related to the model components, we also noticed some problems related to

Table 4: Examples of some errors in validation set.

#	Question	Problems	Effect
56	How can we improve fertility in Klinefelter syndrome karyotype 47 XXY?	Fixed number of output sentences	Redundant output sentences **(low precision)**
91	How can I stop being allergic to caffeine?	Fixed number of output sentences	Missing output sentences **(low recall)**
82	Where can i find information for adults with Hirschsprung's disease?	Imperfect ranking scores	Ranking of irrelevant sentences are too high **(low precision)**
19	Is there a cure for pulsatile tinnitus?	Imperfect ranking scores	Ranking of important sentences are low **(low recall)**
22	Is it safe to have ultrasound with a defibrillator?	Missing keywords and NER	Summary is on the wrong direction **(poor precision and recall)**
53	Is there a way to improve kidneys in a person on twice-weekly dialysis?	Not focus on driven-question	Summary is not contain the desired information **(poor precision and recall)**
36	Are there herbal medicines for rheumatoid arthritis?	Problem in chiQA answers	Not enough information to summarize
78	Can spinal surgery cause hydrocephalus and blindness in adults?	Problem in neighbour boosting	Adding some irrelevant sentence **(decreasing precision)**
28	Can you help me find a clinic that specializes in treatment for atopic eczema?	Problem in post-processing	Removal of important sentence **(decreasing recall)**

the input data for which Question #36 is an example. The question is about *'herbal medicines for rheumatoid arthritis'* while the chiQA answers do not mention this topic. Therefore, our model as well as other machine learning models do not have enough linguistic information to summarize these documents.

Some other errors seem attributable to our model's limitations (Example #28 and #78). We listed here some highlight problems to prioritize future researches: (i) The neighbour boosting method needs to be improved to only increase the weight of related sentences instead of all neighbouring sentences; (ii) Post-processing rules need to be stricter to avoid eliminating important sentences.

5 Conclusions

This paper presents a systematic study of our extractive approach to the MEDIQA 2021 - Task 2: Multi-answer summarization. We combined and optimized several scoring criteria such as TF-IDF, Lexrank, Textrank, query-based, keywords-based and MMR scores. We also developed a strategy called Prosper-thy-neighbour to take advantage of adjacent sentences in the answers. The proposed model has a potential performance, being the runner-up of the shared task. Our best performance achieved a ROUGE-2 $F1$ is 0.504, comparable to that of the highest-ranked system with 0.507.

Experiments were also carried out to verify the rationality and impact of model components and the compressed ratio. The results demonstrated the contribution and robustness of all techniques and hyper-parameters. The error analysis was made to analyze the sources of the errors. The evidence pointed to some imperfection of the sentence selecting strategy, the ranking score combination and the question analyzer. Our proposed system is extensible in several ways: applying machine learning model, deeply question-analyzing, sentences clustering, etc. We will release our source code on the public repository to support the re-producibility of our work and facilitate other related studies.

Acknowledgement

We would like to thank the organizing committee of MEDIQA NAACL-BioNLP 2021 shared task. We also thank the anonymous reviewers for thorough and helpful comments.

References

Mehdi Allahyari, Seyedamin Pouriyeh, Mehdi Assefi, Saeid Safaei, Elizabeth D Trippe, Juan B Gutierrez, and Krys Kochut. 2017. Text summarization techniques: A brief survey. *International Journal of Ad-*

vanced Computer Science and Applications (ijacsa), 8(10).

Asma Ben Abacha, Yassine Mrabet, Yuhao Zhang, Chaitanya Shivade, Curtis Langlotz, and Dina Demner-Fushman. 2021. Overview of the mediqa 2021 shared task on summarization in the medical domain. In *Proceedings of the 20th SIG-BioMed Workshop on Biomedical Language Processing, NAACL-BioNLP 2021*. Association for Computational Linguistics.

Kamil Bennani-Smires, Claudiu Musat, Andreea Hossmann, Michael Baeriswyl, and Martin Jaggi. 2018. Simple unsupervised keyphrase extraction using sentence embeddings. In *Proceedings of the 22nd Conference on Computational Natural Language Learning*, pages 221–229.

Jaime Carbonell and Jade Goldstein. 1998. The use of mmr, diversity-based reranking for reordering documents and producing summaries. In *Proceedings of the 21st annual international ACM SIGIR conference on Research and development in information retrieval*, pages 335–336.

Hans Christian, Mikhael Pramodana Agus, and Derwin Suhartono. 2016. Single document automatic text summarization using term frequency-inverse document frequency (tf-idf). *ComTech: Computer, Mathematics and Engineering Applications*, 7(4):285–294.

G¨unes Erkan and Dragomir R Radev. 2004. Lexrank: Graph-based lexical centrality as salience in text summarization. *Journal of artificial intelligence research*, 22:457–479.

Mahak Gambhir and Vishal Gupta. 2017. Recent automatic text summarization techniques: a survey. *Artificial Intelligence Review*, 47(1):1–66.

Sepp Hochreiter and J¨urgen Schmidhuber. 1997. Long short-term memory. *Neural computation*, 9(8):1735–1780.

Matthew Honnibal, Ines Montani, Sofie Van Landeghem, and Adriane Boyd. 2020. spaCy: Industrial-strength Natural Language Processing in Python.

Eduard Hovy, Chin-Yew Lin, et al. 1999. Automated text summarization in summarist. *Advances in automatic text summarization*, 14:81–94.

Jinhyuk Lee, Wonjin Yoon, Sungdong Kim, Donghyeon Kim, Sunkyu Kim, Chan Ho So, and Jaewoo Kang. 2020. Biobert: a pre-trained biomedical language representation model for biomedical text mining. *Bioinformatics*, 36(4):1234–1240.

Chin-Yew Lin and FJ Och. 2004. Looking for a few good metrics: Rouge and its evaluation. In *Ntcir Workshop*.

Rada Mihalcea and Paul Tarau. 2004. Textrank: Bringing order into text. In *Proceedings of the 2004 conference on empirical methods in natural language processing*, pages 404–411.

Mark Neumann, Daniel King, Iz Beltagy, and Waleed Ammar. 2019. Scispacy: Fast and robust models for biomedical natural language processing. *arXiv preprint arXiv:1902.07669*.

Dragomir R Radev, Eduard Hovy, and Kathleen McKeown. 2002. Introduction to the special issue on summarization. *Computational linguistics*, 28(4):399–408.

Gerard Salton and Michael J McGill. 1986. Introduction to modern information retrieval.

Max Savery, Asma Ben Abacha, Soumya Gayen, and Dina Demner-Fushman. 2020. Question-driven summarization of answers to consumer health questions. *Scientific Data*, 7(1):1–9.

Ariel S Schwartz and Marti A Hearst. 2002. A simple algorithm for identifying abbreviation definitions in biomedical text. In *Biocomputing 2003*, pages 451–462. World Scientific.

Abigail See, Peter J Liu, and Christopher D Manning. 2017. Get to the point: Summarization with pointer-generator networks. In *Proceedings of the 55th Annual Meeting of the Association for Computational Linguistics (Volume 1: Long Papers)*, pages 1073–1083.

MNLP at MEDIQA 2021: Fine-Tuning PEGASUS for Consumer Health Question Summarization

[1]Huong Ngoc Dang* [1]Jooyeon Lee* [2]Samuel Henry [1]Özlem Uzuner
[1] Department of Information Science and Technology
George Mason University, Virginia, United States
[2] Department of Physics, Computer Science and Engineering
Christopher Newport University, Virginia, United States
[1]{hdang20,jlee252,ouzuner}@gmu.edu, [2]samuel.henry@cnu.edu

Abstract

This paper details a Consumer Health Question (CHQ) summarization model submitted to MEDIQA 2021 for shared task 1: Question Summarization. Many CHQs are composed of multiple sentences with typos or unnecessary information, which can interfere with automated question answering systems. Question summarization mitigates this issue by removing this unnecessary information, aiding automated systems in generating a more accurate summary. Our summarization approach focuses on applying multiple pre-processing techniques, including question focus identification on the input and the development of an ensemble method to combine question focus with an abstractive summarization method. We use the state-of-art abstractive summarization model, PEGASUS (Pre-training with Extracted Gap-sentences for Abstractive Summarization), to generate abstractive summaries. Our experiments show that using our ensemble method, which combines abstractive summarization with question focus identification, improves performance over using summarization alone. Our model shows a ROUGE-2 F-measure of 11.14% against the official test dataset.

1 Introduction

The MEDIQA 2021 shared task consists of several independent tasks: task 1 is Question Summarization, task 2 is multi-answer summarization, and task 3 is Radiology Report Summarization. We participated in task 1, Question Summarization. We approached the task by developing an ensemble learning method that combines information from automatic question focus identification with information from a state-of-the-art summarization model. We also studied the effects of different preprocessing techniques for this challenge. The descriptions of the dataset are shown in the task guidelines (Ben Abacha et al., 2021). The training datasets are from Ben Abacha and Demner-Fushman (2019b) along with the focus of each question. The test dataset contains consumer health questions only.

2 Related Works

The goal of Consumer Health Question Answering (CHQA) is to construct an automated question answering system aimed toward answering questions from individuals who are unlikely to possess professional medical knowledge. Typical consumer health questions include requests for information regarding symptoms of particular diseases, queries regarding possible diseases from individuals experiencing symptoms, and whether an individual would be safe to mix specific medications and so forth. In this field, there are circumstances in which individuals submit straightforward questions, but there are many cases where people list extra background and other unnecessary information which are not required to answer their question. In fact, this additional information can essentially serve as a source of noise which can reduce the effectiveness of the QA system as a whole.

Recent CHQA systems employ pipeline architectures that utilize Question Understanding, Information Retrieval and Answer Generation components sequentially (Demner-Fushman et al., 2019). This architecture facilitates modular optimization. Furthermore, it allows individual components to be swapped, either for need or to provide special features. This allows the entire QA system to adapt to the specific nature of the problem at hand. As previously mentioned, many CHQs possess extraneous information in addition to the primary question. Therefore, the Question Understanding component of such an architecture is especially important, and improvements to it can be particularly beneficial to the overall CHQA system. Facilitating Ques-

*These authors contributed equally to this work

Proceedings of the BioNLP 2021 workshop, pages 320–327
June 11, 2021. ©2021 Association for Computational Linguistics

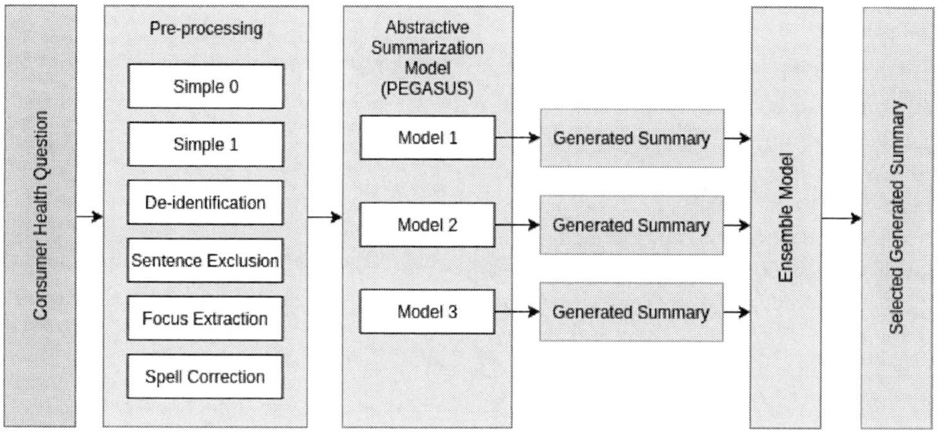

Figure 1: System Architecture.

tion Understanding through summarizing the consumer health questions has demonstrated significant improvement as shown by Ben Abacha and Demner-Fushman (2019b) and Ben Abacha and Demner-Fushman (2019a). Thus, given the benefits to the overall QA system, improving upon existing summarization methods was selected as task 1 in MEDIAQA 2021. In this component, the extraneous information can be removed via preprocessing prior to inputting into further stages of the QA system. Different preprocessing methods have also been explored to perform this task, and a performance improvement on deep learning models has been shown (Camacho-Collados and Pilehvar 2017; Husain et al. 2020).

2.1 Consumer Health Question Understanding

Robust CHQA systems could serve as a component in a broader solution to inform the public of the latest medical updates and breakthroughs, leading to more optimal outcomes for both individuals and the public as a whole.

In Question Understanding, recent breakthroughs relevant to CHQA have included: Ben Abacha and Demner-Fushman (2019), which demonstrated that retrieving entailment answers for CHQA systems many not gather any answers; Ben Abacha and Demner-Fushman (2019b), which studied the role of summarization on CHQA; and Roberts et al. (2014) proposes decomposition methods and techniques for consumer health datasets.They suggest decomposing the questions into focus of the question, exemplification, question sentence(s), background sentence(s) and “ignore” sentence(s).

2.2 Abstractive Summarization

Abstractive Summarization aims to re-write the given input in a shorter form. This is opposed to Extractive Summarization, which aims to select essential sentences from the given input only. There are different approaches to Abstractive Summarization, such as structured-based, semantic-based, deep learning-based, discourse, and rhetoric-based (Gupta and Gupta, 2019).

In this paper, we selected a deep learning approach. Deep learning methods include Pointer Generator Networks See et al. (2017), Pre-training with Extracted Gap-sentences for Abstractive Summarization (PEGASUS) (Zhang et al., 2019), Multi-Document Summarization by Niu et al. (2017) and others (Kouris et al., 2019; Khatri et al., 2018). We selected Pointer Generator Networks as our baseline method, because it showed high performance in summarizing consumer health questions Ben Abacha and Demner-Fushman (2019b) and compare the results with PEGASUS.

3 Methodology

Our model follows a traditional language generation pipeline: pre-processing, abstractive summarization, and post-processing. We experimented with several different combinations of preprocessing to generate multiple summaries from a single given question. We selected the best summary in the post-processing stage from these numerous generated summaries, which use ensemble learning.

3.1 Dataset

The MEDIQA 2021 event organizers provided three different datasets: a training set, a validation set, and a testing set. The training dataset, called MeQSum, is from Ben Abacha and Demner-Fushman (2019b) and consists of 1000 pairs of consumer health questions and corresponding summaries. Of the questions in the training dataset, 658 questions had both SUBJECT and MESSAGE entered by users, while 342 Questions lacked a SUBJECT. Information such as [SUBJECT], [CONTACT], [NAME] and [LOCATION] were de-identified. The validation dataset consists of 50 raw consumer health questions with corresponding summaries, focus, and type for each question. The testing dataset consists of 100 raw consumer health questions.

3.2 Pre-processing

The goal of pre-processing is 1) to make the abstractive summarization model focus on the important information by removing the redundant strings from both the training and validation/test set, and 2) minimize the difference between the training dataset versus both the validation and test datasets as described in 3.1. We try multiple pre-processing techniques for the training dataset and both the validation and test datasets. The outputs generated by the different combinations of pre-processing techniques served as inputs into our ensemble post-processing stage.

3.2.1 Simple Pre-processing

We employed two different simple pre-processing steps:

1. “Simple0” which removes the text "SUBJECT: " and "MESSAGE: ", replaces "\n" by " ", and removes already tagged named entities: [LOCATION], [NAME], [CONTACT], [DATE], [PROFESSION], [AGE], [ID] from the training set.

2. “Simple1” which removes the text "SUBJECT: " and "MESSAGE: " and replace "\n" by " " from the training set.

3.2.2 De-identification

[SUBJECT], [CONTACT], [NAME] and [LOCATION] terms are de-identified in training set, but not in the validation/test set. For consistency and to reduce variation between these terms, we apply de-identification on the dataset with Spark NLP (Kocaman and Talby, 2021). The Spark De-identification model was trained on n2c2 2014: De-identification and Heart Disease Risk Factors Challenge (Stubbs and Uzuner, 2015). This model allows us to mask information such as [LOCATION], [NAME], [CONTACT], [DATE], [PROFESSION], [AGE] and [ID], which were de-identified. To prevent inadvertently masking essential medical terms, we used stanza Bio NER models (Zhang et al., 2020) to identify these medical terms and omit them from masking.

3.2.3 Sentence Exclusion

Sentences such as "Hi", "Thank you in advance, regards", "kindly advise me" and others do not improve summarization performance, yet also exhaust the computational time and resources by increasing the input sequence size. Thus, before input into the summarization model, we remove these sentences. For this effort, we used 10 different Stanza Bio NER models. The differentiating factors between these models are the datasets they were trained on. The datasets consist of one of 8 biomedical datasets or 2 clinical datasets, specifically: i2b2-2010, Radiology, NCBI-Disease, BC5CDR, BioNLP13CG, JNLPBA, AnatEM, BC4CHEMD, Linnaeus, and S800. If none of these 10 models found any medical terms in a sentence, we excluded that sentence from the dataset. The models are ordered by priority, high to low, and once an entity was found using one model, we kept the sentence and began processing the next.

3.2.4 Focus Extraction

Roberts et al. (2014) defines *Focus* as a Noun Phrase indicating the theme of the consumer health question. We believe that by incorporating the focus into our summarization model, we can increase the overall performance. We test focus impact on both pre-processing and post-processing. We added the focus in front of the question during pre-processing and used the combined strings as an input of the abstractive summarization model. During post-processing, we used focus to rank the output accuracy as described in more detail in Section 3.4.

To extract a focus, we explored two different methods: Focus Detection and Focus Generation. For Focus Detection, we employed Named Entity Recognition (NER) with hybrid of two neural networks suggested by Chiu and Nichols (2015). The paper shows high performance with bidirectional

Long Short-term Memory (Bi-LSTM) and Convolutional Neural Network(CNN) architecture with the CoNLL-2003 dataset (Tjong Kim Sang and De Meulder, 2003). The architecture automatically detects word and character level features by having the CNN extract features from words and by using a Bi-LSTM to tag Named Entities. We test Bi-LSTM with CNN and Recurrent Neural Network (RNN) with CNN, LSTM with CNN, and Gated Recurrent Unit (GRU) with CNN.

The risk of using NER methods to detect focus is that there is a possibility of not extracting any focus from a given question. However, focus generation uses language generation techniques, which ensures there is a focus for each given question, though the accuracy of focus is often lower compared to NER techniques. We chose Pointer Generator Networks (PG) (Ben Abacha and Demner-Fushman, 2019b) for focus generation. The model is hybrid of a sequence-to-sequence model (Sutskever et al., 2014) and a pointer network (Vinyals et al., 2017). This hybrid model allows copying words from the source text via pointing that handles out-of-vocabulary words efficiently while retaining the ability of generating new words. The Question Decomposition dataset provided by Roberts et al. (2014) was used to train and evaluate the focus extraction. The dataset includes manually annotated 1496 questions.

3.2.5 Spell Correction

Consumer health questions tend to include misspelled words. This can lead to many problems in downstream question processing. A problem unique to summarization models is that summarization models generate summaries based on words it has seen in the dataset before. Therefore the model may generate summaries with misspellings. To reduce incorrect word generation, we use Microsoft Bing Spell Check API (Microsoft, 2016) to correct misspelled words. This API recognizes misspelled words in the input sentence and provides suggestions with confidence scores. We replace these words with the suggested words with the highest confidence score.

3.3 Abstractive Summarization

We compare two different abstractive summarization models, Pointer Generator (PG) networks and PEGASUS.

3.3.1 PEGASUS

PEGASUS (Zhang et al., 2019) is a Sequence-to-Sequence model based on Transformer. It is pre-trained on massive text corpora with a self-supervised objective called Gap Sentences Generation (GSG). This objective is tailored for abstractive text summarization because the authors of PEGASUS model hypothesize that a pre-training objective that more closely resembles the downstream task leads to better and faster fine-tuning performance. In fact, PEGASUS model using this GSG objective pre-trained on newswire C4 and HugeNews corpora push forward state-of-the-art models on 12 summarization tasks.

In real-world practice, to generate summaries on a specific domain such as news, science, emails, and patents, PEGASUS should be fine-tuned using some supervised samples in that specific domain. Particularly in our shared task, the biomedical questions which need summarizing are related to the biomedical domain. To generate summarized answers on the validation dataset, we use a pretrained model that is fine-tuned on the PubMed dataset by continuing training the model with the MedQSum dataset to obtain a biomedical question summarizer. Hyperparameters we used to fine-tune PEGASUS are described in Table 1.

3.3.2 Pointer Generator Networks

We compare PEGASUS with the Pointer Generator Network described in Section 3.2.4. We train this model with the pre-processed dataset. Pointer Generator Networks (Ben Abacha and Demner-Fushman, 2019b) generate summaries using 128 dimensions of word embedding trained with the summary dataset, hidden state vectors of 256 dimensions, a learning rate of 0.15, and with beam search of size 4. For our experiment, we use the hidden vector size of 256 dimensions, learning rate of 0.01 and 210 size of word vectors. We use pre-trained word vectors with the size of 200. The vectors are from BioWordVec (Yijia et al., 2019), which are trained on PUBMED and MIMIC-III. 10 vectors are zeros and ones of Named Entities (NE). If a word is a medical-related entity, it is set as ones. Otherwise, it is set as zeros. The NEs are decided using spaCy pretrained NER models. Detailed hyperparameters are shown in Table 1.

3.4 Post-processing

For the post-processing, we employ an ensemble learning technique. Ensemble learning aims to re-

Dataset	PG Networks QD	PG Networks MeQSum	PEGASUS MeQSum
LR	0.01	0.01	1e-4
Batch #	25	25	1
Training steps	-	-	20 K
Beam size	8	8	8
Beam α	N/A	N/A	0.8
Max input	155	155	512
Max target	6	35	64
Min target	1	6	N/A

Table 1: Hyperparameters used in the PG Network (Baseline) and PEGASUS model. In PG Networks, QD indicates Question Decomposition Dataset used to train the question focus model, MeQSum is used to train the abstractive summarization model. LR is Learning Rate. We stop training PG Networks when the loss score converges less than 0.1, where the number of epochs varies from 5K to 150K depends on the architecture of Neural Network or different input pre-processing. Thus the epoch number is omitted for PG Networks.

duce the variance of a single model by training multiple models with different parameters or dataset and then selecting the optimal result. This method is widely used in predictive models (Huang et al. 2020; Dang et al. 2020).

Our ensemble method generates multiple outputs by training our model numerous times on the same data. We vary the number of training steps and create one model trained for 80,000 steps, and another model trained for 150,000 steps. Due to the limitation of our resources, we do not further increase the number of steps. We consider the outputs of these two systems and select the optimal output based on Equation 1. We hypothesized that this would balance drawbacks caused by potentially over-fitting and under-fitting the training data.

$$Score = \alpha * Similarity(Focus, Y) + \beta * Similarity(X, Y) \quad (1)$$

As mentioned, Equation 1 is used to determine which generated output is optimal. This equation calculates the similarity between the generated output (question summary) and the given question and the generated output and the focus of the question. We do this because, in our error analysis, we found frequent problems where the generated text was syntactically and often factually correct but was the focus of the summary was incorrect. We set α and β set as 0.5 to equally balance the importance of similarity between focus and question.

In Equation 1, the function $Similarity()$ measures the similarity between two strings. X is given question, Y is generated summary of given question X and $Focus$ is Focus phrase extracted from the given question X. α and β were used to impose the weight of each score. The Sum of α and β is 1. We use the same method used in section 3.2.4. We use spaCy (Honnibal et al., 2020), a library for advanced natural language processing, which includes state-of-the-art neural network models for similarity measures and NER. To measure the similarities for our output, which determines the similarity by comparing word vectors, we used the "en_core_web_lg" model for the word vectors. This model has 684,830 unique vectors with 300 dimensions.

4 Results

All experiments are done on Google Colab Pro with Tesla V100 GPU, RAM 25.51 GB0, CPU of Intel(R) Xeon(R) (2.20GHz). Pointer Generator Networks training took up to 1 hour for both Focus Extraction and Abstractive Summarization. To fine-tune PEGASUS took 1.5 hours for 20K training steps, 12 hours for 80K steps, and 23 hours for 150K steps.

4.1 Performance of the Summarization

4.1.1 Pre-processing Combination Testing

We train our models on the training dataset and report results on the provided validation dataset (Table 2). We withheld the test set from all model development and hyper-parameter tuning and report results in Table 3.

Accuracy is measured using Recall-Oriented Understudy for Gisting Evaluation (ROUGE) (Lin, 2004) measuring overlapping words between reference and summaries. We use ROUGE-1 (R-1), ROUGE-2 (R-2) and ROUGE-L (R-L). R-L refers to the longest common subsequence-based ROUGE score, R-1 is 1-gram based, and R-2 is bi-gram based ROUGE score.

Among all pre-processing combinations, we found Simple1, Spell Check, De-identification, and Sentence Exclusion applied on both validation and training datasets produced the highest score across all ROUGE metrics. Pre-processing with PEGASUS output provides higher accuracy generally compared to pre-processing with PG Networks. We choose the PEGASUS model for abstractive summarization to generate output with the official test dataset. During the manual evaluation, we found not only ROUGE scores are higher with PEGASUS, but also outputs of PG Networks tend to generate repetitive words within an output, while PEGASUS outputs mostly have grammatically correct form. Scores of all 11 experiments can be found in a Table 2.

4.1.2 Official Results

Our highest-scoring submission produced an R2-F score of 11.14%. This submitted system consists of 3 steps: (1) pre-processing, which corrects misspelling words, removes sentences without biomedical/clinical related terms, (2) abstractive summarizing by PEGASUS with 150k training iteration, (3) post-processing where we ensembled the outputs of two systems together. System 1 was trained for 80K training steps, while system 2 was trained for 150K training steps. Our other submitted system performed only the first two steps. No ensemble method was used. As shown in Table 3, we see there is an increase in performance by ensembling the two outputs rather than relying on the output of a single model. We used both the training and validation dataset to train the models to generate the summaries for the test dataset.

4.2 Performance of Focus Extraction

4.3 Performance of Focus Extraction

We measure Focus Detection with precision, recall, and f-measure, and Focus Generation with ROUGE-1, ROUGE-2, and ROUGE-L. The exact scores are shown in Table 4. Experiments No 1, 2, and 3 are Focus Detection results, and No 4 and 5 are Focus Generation results. Model No.5 is Focus Generation using PG Network with duplicated term removal resulted in an accuracy of 85%. We choose the Focus Generation method over Focus Detection, even though Focus Detection accuracy is considerably high to avoid the possibility of not detecting any focus, which may occur for some questions if the NER technique were to be used. As mentioned previously, Focus Generation will always generate focus for every question.

5 Discussion and Future Work

We found many incorrect summaries with the wrong focus during the experiment of different combinations of pre-processing. For example, given input question "I have **chronic renal disease** and worry that **Magnesium silicofluoride** treatment of moth infestation of a large living room rug will be harmful to my health. If the rug is treated in house how long before any toxic fumes or skin contact would be a hazard .", PEGASUS generated output (a) and (b):

(a) What are the side effects of **silicofluoride** treatment?

(b) What is the treatment for **moth infestation of a rug**?

In the given question, we see that the person is concerned with the effect of Magnesium silicofluoride on individuals with chronic renal disease. Thus, the focus of the given question would be **chronic renal disease** and **Magnesium silicofluoride**. In contrast, both generated output (a) and (b) summaries are built on incorrect focus. We believe extracting the correct focus and studying how to incorporate the focus would improve accuracy. In this paper, we applied focus in the post-processing step. We ranked the output using the Equation 1, and then select the output with the highest score.

The limitation of Equation 1 is that the extracted focus may not be accurate. If the extracted focus is not correct, the ensemble model may choose a non-relevant output. For example, input "**hydroxychloroquine for rheumatoid arthritis**. Can you tell me if this medication that my doctor put me on could make me sweat profusely at the slightest little strenuous activity I'm also **methotrexate** 6 2.5 mg once a week ." gives the following answers:

(a) Can **hydroxychloroquine** and **methotrexate** be taken together?

(b) What are the dosage side effects and drug interactions for **rheumatoid arthritis**?

The question asks if the **hydroxychloroquine for rheumatoid arthritis** and **methotrexate** be taken together. The generated summary (a) shows a reasonably accurate answer. In contrast, the focus extraction model assumed **rheumatoid arthritis** to be a focus, which leads the model to choose summary (b) over (a).

Despite this limitation, our experiments showed performance improvements after applying focus detection and the ensemble method in post-processing. The post-processing effect is limited to the performance of the summarization model, accuracy of focus for each question, and the number of outputs from the summarization models. Due to time limitations, we use two outputs in the ensemble process, while typical ensemble learning

No.	Preprocessing on Training Data	Preprocessing on Validation Data	R-1	R-2	R-L
1	Simple0	-	29.69	13.23	28.91
2	-	Simple1 + Deid	29.28	13.48	28.72
3	Simple1 + Deid + SE	Simple1 + Deid	31.03	14.46	29.26
4	Simple1 + Deid + SE + Merge(Subject, Message)	Simple1 + Deid + Merge(Focus(FD), Question)	28.35	11.69	26.96
5	Simple1 + Deid + SE + Merge(Focus, Subject, Message)	Simple1 + Deid + Merge(Focus(FD), Question)	27.47	11.25	26.31
6	**Simple1 + Spell Check + Deid + SE**	*	**31.60**	**13.93**	**30.55**
7	Simple1 + Deid + SE + Merge(Focus(FG), Question)	*	28.41	11.90	26.27
8	Simple1 + Deid + SE + Merge(Focus(Gold), Question)	*	26.26	10.65	25.51
9	Simple1 + Spell Check + Deid + SE + Message Only	*	28.93	12.47	27.83
10	Simple1 + Spell Check + Deid + SE	*	18.98	10.50	17.32
11	Simple1 + Spell Check + Deid + SE + Message Only	*	19.22	10.31	18.23

Table 2: Performance testing with validation dataset. '-' indicates no pre-processing technique applied, '*' indicates that the method applied for the training dataset was applied to validation data. SE is an abbreviation of Sentence Exclusion. Deid is De-identification. Merge() is concatenating strings. FD is Focus Detection, and FG is focus generation. Experiments 1-9 are done with PEGASUS, and 10-11 are done with PG Networks.

No.	Method	R1-P	R1-R	R1-F1	R2-P	R2-R	R2-F1	RL-R	RL-F1
1	Pre-processing + PEGASUS (150K)	**0.321**	0.285	0.283	0.120	0.105	0.106	0.257	0.257
2	**Pre-processing + PEGASUS (150K & 80K) + Ensemble**	0.315	**0.291**	**0.284**	**0.123**	**0.112**	**0.111**	**0.265**	**0.259**

Table 3: Performance testing with official test dataset. Experiment 1 output is with 150K training epochs. Experiment 2 were ensemble of outputs of the model trained for 80K epochs and the model trained for 150K epochs.

No.	FD Method	P	R	F
1	GRU + CNN	75.56	81.13	78.24
2	RNN + CNN	67.89	64.21	66
3	**LSTM + CNN**	**78.79**	**82.21**	**80.47**
No.	**FG Method**	**R-1**	**R-2**	**R-L**
4	PG	0.64	0.37	0.63
5	**PG-duplicates**	**0.85**	**0.58**	**0.84**

Table 4: Performance of Focus Extraction. Experiment No 1, 2, 3 are the results of Focus Detection and Experiment No 4 and 5 are results of Focus Generation. PG is short for PG Network and PG-duplicates is PG Network with removal of duplicated terms in generated output.

models use considerably larger numbers than 2. Thus, we believe there is a significant potential improvement by investigating: 1) Methods to generate more trained models with different parameters and datasets 2) Method to generate multiple models with less training time 3) Method to increase the performance of the focus extraction 4) Develop better methods for incorporating question focus information into the summary generation system. The current ensemble method is applied at a fairly late stage in the process. Study the effect of incorporating ensembling as early as the training step is an area of exploration.

6 Conclusion

In this paper, we present our Question Summarization system for Consumer Health Questions(CHQ). We explored effect of multiple pre-processing methods (De-Identification, Sentence Exclusion and Focus Extraction) and on state-of-art Abstractive Summarization. Our results show the best F-measure score of 11.14% through applying Ensemble Learning to different combinations of the pre-processing outputs. In our analysis, we identified future directions, including investigating the use of Extracted Focus, Ensemble Learning for the generative model.

References

Asma Ben Abacha and Dina Demner-Fushman. 2019a. On the role of question summarization and information source restriction in consumer health question answering. *AMIA Joint Summits on Translational Science proceedings. AMIA Joint Summits on Translational Science*, 2019:117–126.

Asma Ben Abacha and Dina Demner-Fushman. 2019b. On the summarization of consumer health questions. In *Proceedings of the 57th Annual Meeting of the*

Association for Computational Linguistics, pages 2228–2234, Florence, Italy. Association for Computational Linguistics.

Asma Ben Abacha and Dina Demner-Fushman. 2019. A question-entailment approach to question answering. *CoRR*, abs/1901.08079.

Asma Ben Abacha, Yassine Mrabet, Yuhao Zhang, Chaitanya Shivade, Curtis Langlotz, and Dina Demner-Fushman. 2021. Overview of the mediqa 2021 shared task on summarization in the medical domain. In *Proceedings of the 20th SIG-BioMed Workshop on Biomedical Language Processing, NAACL-BioNLP 2021*. Association for Computational Linguistics.

José Camacho-Collados and Mohammad Taher Pilehvar. 2017. On the role of text preprocessing in neural network architectures: An evaluation study on text categorization and sentiment analysis. *CoRR*, abs/1707.01780.

Jason P. C. Chiu and Eric Nichols. 2015. Named entity recognition with bidirectional lstm-cnns. *CoRR*, abs/1511.08308.

Huong Dang, Kahyun Lee, Sam Henry, and Özlem Uzuner. 2020. Ensemble BERT for classifying medication-mentioning tweets. In *Proceedings of the Fifth Social Media Mining for Health Applications Workshop & Shared Task*, pages 37–41, Barcelona, Spain (Online). Association for Computational Linguistics.

Dina Demner-Fushman, Yassine Mrabet, and Asma Ben Abacha. 2019. Consumer health information and question answering: helping consumers find answers to their health-related information needs. *Journal of the American Medical Informatics Association : JAMIA*, 27.

Som Gupta and S. K Gupta. 2019. Abstractive summarization: An overview of the state of the art. *Expert Systems with Applications*, 121:49–65.

Matthew Honnibal, Ines Montani, Sofie Van Landeghem, and Adriane Boyd. 2020. spaCy: Industrial-strength Natural Language Processing in Python.

Tongwen Huang, Qingyun She, and Junlin Zhang. 2020. Boostingbert:integrating multi-class boosting into bert for nlp tasks.

Fatemah Husain, Jooyeon Lee, Sam Henry, and Ozlem Uzuner. 2020. SalamNET at SemEval-2020 task 12: Deep learning approach for Arabic offensive language detection. In *Proceedings of the Fourteenth Workshop on Semantic Evaluation*, pages 2133–2139, Barcelona (online). International Committee for Computational Linguistics.

Chandra Khatri, Gyanit Singh, and Nish Parikh. 2018. Abstractive and extractive text summarization using document context vector and recurrent neural networks. *CoRR*, abs/1807.08000.

Veysel Kocaman and David Talby. 2021. Spark nlp: Natural language understanding at scale.

Panagiotis Kouris, Georgios Alexandridis, and Andreas Stafylopatis. 2019. Abstractive text summarization based on deep learning and semantic content generalization. In *Proceedings of the 57th Annual Meeting of the Association for Computational Linguistics*, pages 5082–5092, Florence, Italy. Association for Computational Linguistics.

Chin-Yew Lin. 2004. ROUGE: A package for automatic evaluation of summaries. In *Text Summarization Branches Out*, pages 74–81, Barcelona, Spain. Association for Computational Linguistics.

Microsoft. 2016. Bing spell check.

J. Niu, H. Chen, Q. Zhao, L. Su, and M. Atiquzzaman. 2017. Multi-document abstractive summarization using chunk-graph and recurrent neural network. In *2017 IEEE International Conference on Communications (ICC)*, pages 1–6.

Kirk Roberts, Halil Kilicoglu, Marcelo Fiszman, and Dina Demner-Fushman. 2014. Decomposing consumer health questions. In *Proceedings of BioNLP 2014*, pages 29–37, Baltimore, Maryland. Association for Computational Linguistics.

Abigail See, Peter J. Liu, and Christopher D. Manning. 2017. Get to the point: Summarization with pointer-generator networks. *CoRR*, abs/1704.04368.

Amber Stubbs and Özlem Uzuner. 2015. Annotating risk factors for heart disease in clinical narratives for diabetic patients. *Journal of biomedical informatics*, 58 Suppl:S78—91.

Ilya Sutskever, Oriol Vinyals, and Quoc V. Le. 2014. Sequence to sequence learning with neural networks. *CoRR*, abs/1409.3215.

Erik F. Tjong Kim Sang and Fien De Meulder. 2003. Introduction to the CoNLL-2003 shared task: Language-independent named entity recognition. In *Proceedings of the Seventh Conference on Natural Language Learning at HLT-NAACL 2003*, pages 142–147.

Oriol Vinyals, Meire Fortunato, and Navdeep Jaitly. 2017. Pointer networks.

Zhang Yijia, Qingyu Chen, Zhihao Yang, Hongfei Lin, and Zhiyong lu. 2019. Biowordvec, improving biomedical word embeddings with subword information and mesh. *Scientific Data*, 6.

Jingqing Zhang, Yao Zhao, Mohammad Saleh, and Peter J. Liu. 2019. PEGASUS: pre-training with extracted gap-sentences for abstractive summarization. *CoRR*, abs/1912.08777.

Yuhao Zhang, Yuhui Zhang, Peng Qi, Christopher D. Manning, and Curtis P. Langlotz. 2020. Biomedical and clinical english model packages in the stanza python nlp library. *arXiv preprint arXiv:2007.14640*.

UETfishes at MEDIQA 2021: Standing-on-the-Shoulders-of-Giants Model for Abstractive Multi-answer Summarization

Hoang-Quynh Le[1], Quoc-An Nguyen[1], Quoc-Hung Duong[1], Minh-Quang Nguyen[1]
Huy-Son Nguyen[1], Tam Doan Thanh[2], Hai-Yen Thi Vuong[1] and Trang M. Nguyen[1]
[1]VNU University of Engineering and Technology, Hanoi, Vietnam.
{lhquynh, 18020106, 18020021, 19020405}@vnu.edu.vn
{18021102, yenvth, trangntm}@vnu.edu.vn
[2]doanthanhtam283@gmail.com

Abstract

This paper describes a system developed to summarize multiple answers challenge in the MEDIQA 2021 shared task collocated with the BioNLP 2021 Workshop. We present an abstractive summarization model based on BART, a denoising auto-encoder for pre-training sequence-to-sequence models. As focusing on the summarization of answers to consumer health questions, we propose a query-driven filtering phase to choose useful information from the input document automatically. Our approach achieves potential results, rank no.2 (evaluated on extractive references) and no.3 (evaluated on abstractive references) in the final evaluation.

1 Introduction

In the past several decades, biomedicine and human health care have become one of the major service industries. They have been receiving increasing attention from the research community and the whole society. The rapid growth of volume and variety of biomedical scientific data make it an exemplary case of big data (Soto et al., 2019). It is an unprecedented opportunity to explore biomedical science and an enormous challenge when facing a massive amount of unstructured and semi-structured data. The development of search engines and question answering systems has assisted us in retrieving information. However, most biomedical retrieved knowledge comes from unstructured text form. Without considerable medical knowledge, the consumer is not always able to judge the correctness and relevance of the content (Savery et al., 2020). It also takes too much time and labour to process the whole content of these documents rather than extracting the useful compressed content. *Automatic summarization* is a challenging application of biomedical natural language processing. It generates a concise description that captures the salient details (called summary) from a more complex source of information (Mishra et al., 2014). Summarization can be particularly beneficial for helping people easily access electronic health information from search engine and question answering systems.

MEDIQA 2021[1] (Ben Abacha et al., 2021) tackles three summarization tasks in the medical domain. Task 2- Summarization of Multiple Answers challenge aims to promote the development of multi-answer summarization approaches that could simultaneously solve the aggregation and summarization problems posed by multiple relevant answers to a medical question.

There are two approaches to summarization: extractive and abstractive. Extractive summarization, i.e., choose important sentences from the original text, is extensively researched but have several limitations: (i) it is unable to keep the coherence of the answer, (ii) the information compressed may be incomplete because information may take many sentences to expose, and (iii) it must include non-relevant part of a relevant sentence. Recently, the research has shifted towards more promising approaches, i.e. abstractive summarization, which can overcome these problems give higher precision than extractive summaries (Gupta and Gupta, 2019). Abstractive text summarization is the task of generating a short and concise summary that captures the salient ideas of the source text. The generated summaries potentially contain new phrases and sentences that may not appear in the source text. Abstractive summarization helps resolve the dangling anaphora problem and thus helps generate readable, concise and cohesive summaries. In abstractive summary, we can merge several relate sentences or make them shorter, i.e., removing the redundancy part.

Our proposed model for the multi-answer summarization task follows abstractive summarization

[1]https://sites.google.com/view/mediqa2021

Proceedings of the BioNLP 2021 workshop, pages 328–335
June 11, 2021.

approaches. We try to process original answers as a shorter representation while preserving information content and overall meaning. We take advantage of BART, a pre-trained model combining bidirectional and auto-regressive transformers (Lewis et al., 2020). We construct an architecture with two filtering phases to choose the more concise input for BART. Since the summary should be question-oriented, the coarse-grained filtering phase removes question-irrelevant sentences. The fine-grained filtering phase is then used to cut-off noise phases.

The remaining of this paper is organized as follows: Section 2 gives brief introduction of some state-of-the-art related work. Section 3 describes task data and our proposed model. Section 4 is the experimental results and our discussion. And finally, the Conclusion.

2 Related work

Because of the complexity of natural language, abstractive summarization is a challenging task and has only been of interest in recent years. Gerani et al. (2014) proposed an abstractive summarization system for product reviews by taking advantage of their discourse tree structure. A important subgraph in the discourse tree were then selected by using PageRank algorithm. A natural language summary was then generated by applying a template-based NLG framework.

According to current research trends, witnessing the success of deep learning in other NLP tasks, researchers have started considering this framework as an promising solution for abstractive summarization. Nallapati et al. (2016) used an attentional encoder-decoder recurrent neural networks and several models such as key-words modeling, sentence-to-word hierarchy structure, and emitting rare words, etc. Song et al. (2019) proposed an LSTM-CNN based ATS model to construct new sentences by exploring fine-grained phrases from source sentences (of CNN and DailyMail) and combining them. Gehrmann et al. (2018) used a bottom-up attention technique to improve the deep learning model by over-determining phrases in a source document that should be part of the summary. Inspired by the successful application of deep learning methods for machine translation, abstractive text summarization is specifically framed as a sequence-to-sequence learning task. BART is a transformer-based pretrained denoising encoder-decoder model that is applicable to a very wide range of end tasks, includes summarization. It combines a bidirectional encoder and an auto-regressive decoder (Lewis et al., 2020). There are several BART-based model, example includes DistilBart[2] and Question-driven BART (Savery et al., 2020). Question-driven BART re-trained BART on objectives designed to improve its general ability to understand the content of text (including document rotation, sentence permutation, text-infilling, token masking and token deletion) and fine-tuned the model for biomedical data. Another recently published abstractive summarization framework is PEGASUS (Zhang et al., 2020), it masks important sentences and generates those gap-sentences from the rest of the document as an additional pre-training objective.

3 Materials and Methods

3.1 Shared task data

The shared task suggested to use the MEDIQA-AnS Dataset (Savery et al., 2020) as the training Data. The validation and test sets includes the original answers are generated by the medical question answering system system CHiQA[3] . In these data sets, extractive and abstractive summaries are manually created by medical experts. Table 1 gives our statistics on the given datasets (see (Ben Abacha et al., 2021) for detailed description of shared task data).

Table 1: Statistics of the datasets.

Statics aspects	Training		Valid-ation	Test
	Article	Section		
Question	156	156	50	80
Average				
A per Q	3.54	3.54	3.85	3.80
T per A	152.35	532.83	219.44	240.22
T per SSum	70.51	70.51	-	-
T per MSum	119.04	119.04	81.18	-
Compression radio				
SSum	0.07	0.32	-	-
MSum	0.04	0.13	0.15	-

A: Answer, Q: Question, T: Token
SSum: Single-answer summary,
MSum: Multi-answer summary.

3.2 Proposed model

As a team participating in MEDIQA - Task 2, we proposed an abstractive summarization sys-

[2]https://huggingface.co/sshleifer/DistilBart-cnn-12-6

[3]https://chiqa.nlm.nih.gov

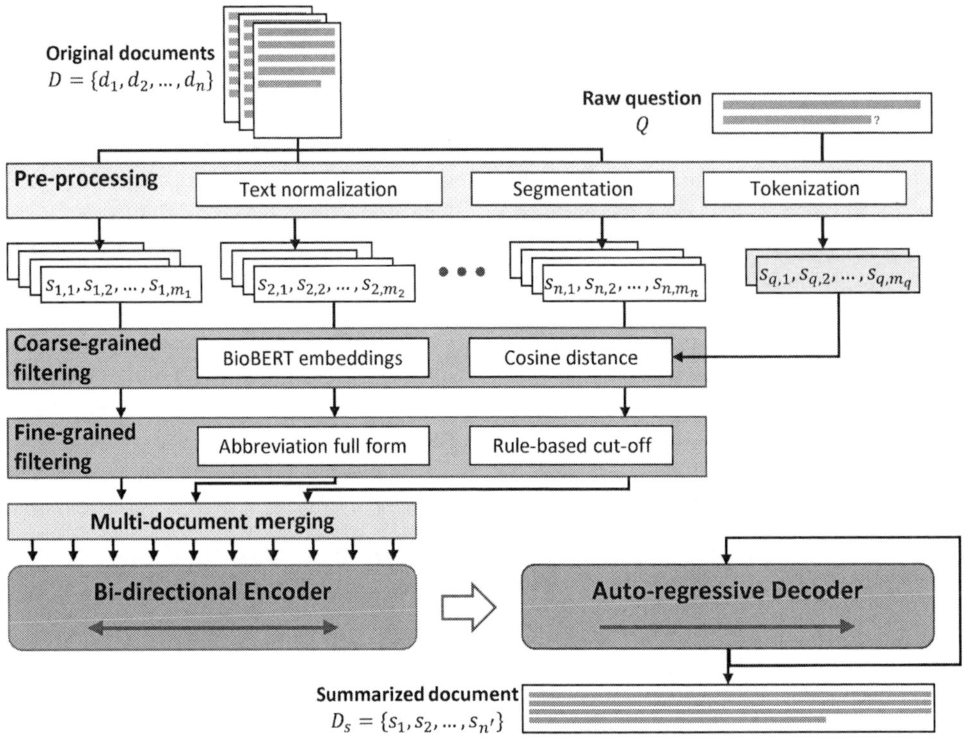

Figure 1: The proposed 'Standing-on-the-Shoulders-of-Giants' model.

tem based on BART - the denoising sequence-to-sequence model. We designate this as a 'Standing-on-the-Shoulders-of-Giants' (SSG) model because BART is the recently state-of-the-art model for abstractive summarization task. To improve the performance, we propose to apply two filtering phases to make the condensed question-driven input for BART. In addition, the BART-based model only receives a limited length document *(with 1024 tokens)*, and our original input is too large to fit. Our model requires a cut-off strategy to reduce length. The overall architecture of the system is described in Figure 1 which includes five main phrases: pre-processing, coarse-grained filtering, fine-grained filtering phase and BART-based summary generation.

3.2.1 Pre-processing

The pre-processing phase receives question Q and a set of corresponding answers (documents) $D = \{d_i\}_{i=1}^{n}$ as the input. The pre-processing phase removes html tags, non-utf-8 characters and redundant signs/spaces. scispaCy (Neumann et al., 2019), a powerful tool for biomedical natural language processing, is also used for the typical pre-processing steps (i.e. segmentation and tokenization).

3.2.2 Coarse-grained filtering

The original BART summarizes a text by generating a shorter text with the same semantic. It processes all information with the same priority and does not take the question into account. Therefore, its output may lose the function of answering the question. We orient BART to question-driven by filtering out less valuable sentences, increasing the rate of question-related sentences in the BART input. There are two strategy to choose sentences that are highly related to the questions:

(i) **Top-n query-driven sentences:** The main idea of this method is to choose sentences that most likely can answer the questions. We calculate the cosine similarity between two bioBERT embedding vectors (Lee et al., 2020) of the question and each sentence. We assume that the sentence with higher cosine similarity might be a good answer for the question. The top-n sentences of each answer with the highest scores are kept with their original orders.

(ii) **Top-n query-driven passages:** Some passages are structured in an deductive manner (e.g., several explanatory sentences follow after a stated sentence) or inductive (e.g., the last sentence is the conclusion of previous sentences). Extracting these whole text pieces may help an important sentence have some adjacent sentences to clarify or support it, making it more coherence and informative. There are three factors to determine an important passage:

- *Central sentence:* A passage is chosen if and only if it has at least one sentence likely answering the question. Cosine similarity with BioBERT embedding vector is used to find these sentences.
- *Passage length:* A passage must not exceed k sentences.
- *Break point:* If the similarity between two adjacent sentences is lower than a pre-defined threshold, a breakpoint is addressed.
- *Passage score:* is calculated by the sum of its sentences similarity scores.

Top-n best passages are then combined with their original order.

In addition to two aforementioned strategies, we also use two other simple strategies as the baseline:

(iii) n **first sentences:** Taking n first sentences from each answers.

(iv) n **random sentences:** Taking n random sentences from each answers.

In which, the number of passages/sentences is not limited which satisfies that the whole length of final document is fit of smaller than the allowed input size of BART model. It should take as much information as possible.

3.2.3 Fine-grained filtering

The nature of BART is to convert one piece of text into another with the same semantics. If the input contains too much noise and is difficult to understand, it may negatively affect the output quality. Therefore, we try to filter out the noise phrases to get the most concise input to BART, thereby getting better results. Through the data surveying, there are two approaches to reduce noises and ambiguous information:

(i) Biomedical text uses many abbreviations, of which many do not follow a standard convention and are only used locally within the scope of authors' articles. Unfortunately, these local abbreviations might be the keywords and lead to the ambiguous to the system. We identify and generate the full form of all local abbreviation use the Ab3P tool (Sohn et al., 2008).

(ii) we apply some rules to cut redundant elements of sentences. Examples include:

- Cut-off listed text that follows *'such as'*.
- Cut-off text that follows *'for example'*.
- Cut-off text that appears in the brackets *()*.
- Cut-off text that follows a colon and is not in enumerated form.

3.2.4 BART-based summary generation

All sentences are selected and cut-off from aforementioned filtering phases are then combined into a single document. This is the input to the BART-based summary generation phase.

BART is implemented as a standard sequence-to-sequence Transformer-based model. It is a denoising autoencoder that maps a corrupted document to the original document it was derived from (Lee et al., 2020). Special power of this model is that it can map the input string and output string with different lengths. BART consists of two components: Encoder and Decoder that combines the advantages of BERT and GPT.

Encoder: BART uses a bidirectional encoder over corrupted text taken from BERT (Devlin et al., 2019). As the strength of BERT lies in capturing two-dimensional contexts, BART can encode the input string in both directions and get more context information. In the abstractive text summarization problem, the input sequence is the collection of all token in the answers. Each word is represented by x_t, where i is its ordinal. The h_t hidden states are calculated with the formula:

$$h_t = f(W^{hh} \cdot h_{t-1} + W^{hx} \cdot x_t) \qquad (1)$$

in which, the hidden states are computed by the corresponding input x_t and the previous hidden state h_{t-1}. Encoder vector is the hidden state at the end of the string, calculated by the encoder. It then acts as the first hidden state of the decoder.

Decoder: BART uses a left-to-right auto-regressive decoder. Its decoder is similar to GPT (Radford et al.) with the capability of self-regression, can be used to reconstruct the input noise. A stack of subnets is the element of the RNN that predicts the output y_t at time t. Each of these words takes input as the previously hidden state and produces its own output and hidden state.

For the abstractive text summarization problem, the output sequence is the set of words of the summarized answer. Each word is represented by y_t where i is the word order. The hidden state is calculated by the preceding state. So, the h_i hidden states are calculated by the formula:

$$h_t = f(W^{hh} \cdot h_{t-1}) \quad (2)$$

We compute the output using the corresponding latency at the present time and multiply it by the corresponding weight W^S. Softmax is used to create a probability vector that helps us to determine the final output. The output y_t are calculated by the formula:

$$y_t = softmax(W^S \cdot h_t) \quad (3)$$

BART uses Beam Search algorithm for decoding.

4 Experimental results

4.1 Evaluation metrics

We adopt the official task evaluations with ROUGE scores (Lin and Och, 2004) including ROUGE-1, ROUGE-2 and ROUGE-L. ROUGE-n Recall (R), Precision (P) and $F1$ between predicted summary and referenced summary are calculated as in Formular 4, 5 and 8, respectively. Choosing correct sentences help to increase ROUGE-n R and P.

$$\text{ROUGE-}n\ P = \frac{|\text{Matched N-grams}|}{|\text{Predict summary N-grams}|} \quad (4)$$

$$\text{ROUGE-}n\ R = \frac{|\text{Matched N-grams}|}{|\text{Reference summary N-grams}|} \quad (5)$$

$$\text{ROUGE-L}\ P = \frac{\text{Length of the LCS}}{|\text{Predict summary tokens}|} \quad (6)$$

$$\text{ROUGE-L}\ R = \frac{\text{Length of the LCS}}{|\text{Reference summary tokens}|} \quad (7)$$

ROUGE-L recall (R), precision (P) and $F1$ are calculated as in Formular 6, 7 and 8, respectively. ROUGE-L uses the Longest Common Subsequence (LCS) between predicted summary and referenced summary and normalized by the tokens in summary.

$$F1 = 2 \times \frac{\text{R} \times \text{R}}{\text{P} + \text{R}} \quad (8)$$

4.2 Comparative models

We use the official results of the MEDIQA shared task as a comparison to other participated teams on the multi-answer summarization task. For a further comparison, we also make the comparisons with three state-of-the-art abstractive summarization models:

- The orginal BART (Lewis et al., 2020).
- DistilBart[4]: A very effective model for text generation task release by HuggingFace.
- PEGASUS (Zhang et al., 2020) is state-of-the-art abstractive summarization model provided by Google AI.

4.3 Task final results and comparison

Based on the experimental results on the validation set, we choose top-n query-driven passages as a coarse-grained filter to run our official output. In our model, Beam Search uses $beamwidth = 5$ and uses sampling instead of greedy decoding. Beam Search is stopped when at least 5 sentences finished per batch. After two filtering phases, the input often have 10-15 sentences and less than 1024 tokens. On average, the total token in a summary is equal to ∼75% of the number of tokens in the BART input.

4.3.1 Official results of the multi-answer abstractive summarization

Table 2 show the shared task official results of accepted competitors. ROUGE-2 $F1$ is used as the main metric to rank the participating teams. We also show several other evaluation metrics for further comparison: ROUGE-1 $F1$, ROUGE-L $F1$, HOMLS $F1$ and BERT-based $F1$. The organizers offer two rankings, one on the extractive references, the other on the abstractive references. Evaluated on extractive references, our team is the runner-up. On the evaluation using abstractive references, we ranked third.

4.3.2 Comparison with other state-of-the-art models

Table 3 shows the comparison between our proposed model and two other state-of-the-art text generation models, i.e., DistilBart and Pegasus. Our SSG model yields much better results than DistilBart and PEGASUS in this data. Since both models

[4] `https://huggingface.co/sshleifer/distilbart-cnn-12-6`

Table 2: Official results of the MEDIQA 2021: Task 2 - Multi-Answer Summarization

Team	ROUGE-1 F1	ROUGE-2 F1	ROUGE-L F1	HOLMS	BERTscore F1
Evaluated on extractive references					
paht_nlp	0.585	**0.508**	**0.436**	**0.554**	**0.653**
UETfishes	0.572	0.470	0.400	0.520	0.646
UCSD-Adobe	**0.592**	0.460	0.417	0.493	0.632
yamr	0.516	0.445	0.384	0.536	0.636
I_have_no_flash	0.523	0.422	0.360	0.542	0.615
Evaluated on abstractive references					
paht_nlp	**0.386**	**0.162**	**0.232**	**0.554**	**0.653**
UCSD-Adobe	0.384	0.160	0.212	0.494	0.632
UETfishes	0.381	0.147	0.202	0.520	0.647
I_have_no_flash	0.384	0.133	0.222	0.478	0.615
yamr	0.271	0.131	0.160	0.388	0.636

Only show results of top-5 participated teams for each type of evaluation. The highest results in each column are highlighted in bold.

Table 3: Comparison with other state-of-the-art models.

Model	ROUGE-2 P	ROUGE-2 R	ROUGE-2 F1
DistilBART	0.0825	0.1031	0.0874
Pegasus	0.0401	0.0597	0.0450
Our SSG	0.0977	0.1274	0.1062

All results are reported on the validation data set.

are very strong competitors, our higher outcome may because they are not suitable with the characteristics of the data (biomedical domain, question-driven answers).

4.4 Contribution of model components

We study the contribution of each model component to the system performance by ablating each of them in turn from the model and afterwards evaluating the model on the validation set. We compare these experimental results with the full system results and then illustrate the changes of ROUGE-2 $F1$ in Figure 2. The changes of ROUGE-2 $F1$ show that all model components help the system to boost its performance (in terms of the increments in ROUGE-2 $F1$). The contribution, however, varies among components. The coarse-grained filtering phase has the biggest contribution, while abbreviation processing and cut-off rules of the fine-grained phase bring very small effectiveness. We also investigate the effectiveness of components/configures in the BART-based summary generation. Components that have a pronounced effect on the result are shown in Figure 2 : Preventing 3-gram repeater, sampling, early stopping and beam search. Preventing 3-gram repeater and using sampling also improves results.

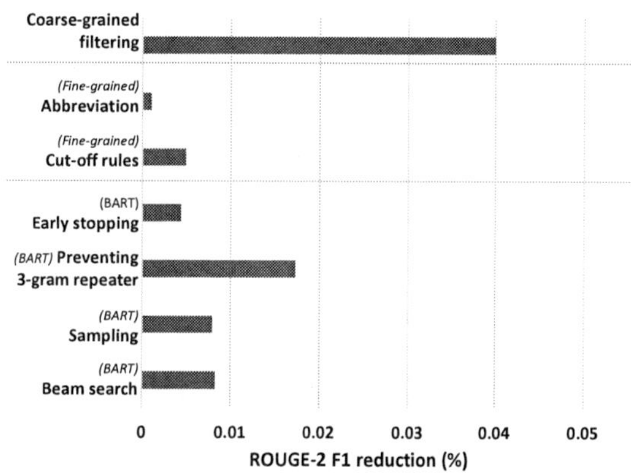

Figure 2: Ablation test results for model components.

Considering the results of three different approaches in the *coarse-grained filtering phase* (Figure 3), top-n question-driven passage seems the most promised way. Other approaches do not take advantages of the semantic relation between adjacent sentences, which leads to losing important information.

4.5 Error analysis

In order to improve the proposed model, we have analyzed the output on the validation set to find out problems that need to be taken into account. All the evidence points to five biggest problems, including content generalization, synonyms and antonyms, paraphrasing, cosine similarity problem, and aggressive cut-out strategy.

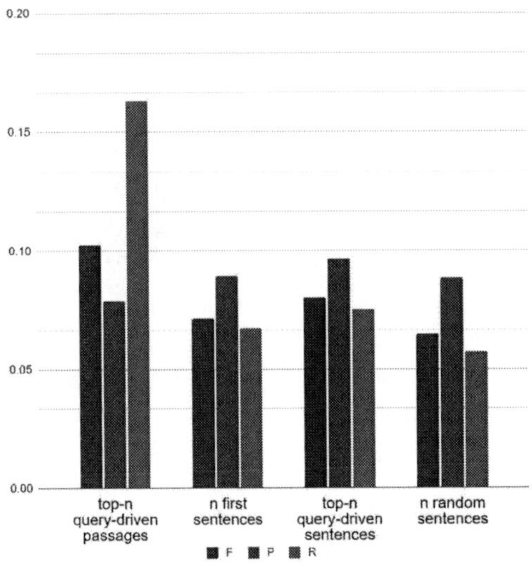

Figure 3: Comparison of different coarse-grained filtering strategies based on ROUGE-2 scores.

The biggest problem with our proposal model and other text summary models is the generalization of the input content. In particular, for the answer summary system, this issue is emphasized more and more. The responses may contain a variety of content related to the directional question. However, the summary should draw conclusions to answer that question. For example, in Question #22, to answer the question *'Is it safe to have ultrasound with a defibrillator?'*, our model performed well that carried out the summary *'Most of the time, ultrasound procedures do not cause discomfort. The conducting gel may feel a little cold and wet. Current ultrasound techniques appear to be safe.'* However, the expected outcome was *'There are no known risks or contraindications for ultrasound tests.'* For which, our model gets a 0.0 ROUGE-2 F1 score for this example.

Another problem is that golden data depends on the style and language usage of the abstractor. The writer may use different expressions, synonyms, antonyms to paraphrase and summarise, leading to the inconsistency of ground truth data. Take Question #8 for example, the sentence *'This treatment leads to remission in 80% to 90% of patients'* is paraphrased into *'Remission is possible in up to 90% of the patients.'*

The analysis process also raises some imperfections of the proposed model in sentence selection and sentence cutting strategies. Cosine similarity metric does not really perform well with documents containing many sentences. In particular, many sentences contain important content but do not have high similarity to the question. Besides, fine-grained filtering strategies also filter some important information in the sentence. We leave these problems to be addressed in future work.

5 Conclusion

This paper presents a systematic study of our abstractive approach to question-driven summarization problem, specifically depending on MEDIQA 2021 - Task 2: Multi-answer summarization. We present a model improved and optimized based on BART - a state-of-the-art method for abstractive summarization called SSG (Standing on the shoulders of giants). The proposed model has a potential performance, being the runner-up of the shared task. Our best performance achieved a ROUGE-2 $F1$ is 0.470 evaluated on extractive summarization references and 0.147 evaluated on abstractive summarization references .

Experiments were also carried out to verify the rationality and impact of model components and the compressed ratio. The results demonstrated the contribution and robustness of all techniques and hyper-parameters. Besides, the error analysis was made to analyze the sources of the errors. The evidence pointed out some imperfection of the sentence selecting strategy, the ranking score combination, and the question analyzer. In further works, there could be several ways: applying machine learning model, deeply question-analyzing, sentence clustering, etc. applied to extend the ability of the model.

Our source code will be released publicly to support the reproducibility of our work and facilitate other related studies.

Acknowledgements

We would like to thank the organizing committee of MEDIQA NAACL-BioNLP 2021 shared task. We also thank the anonymous reviewers for thorough and helpful comments.

References

Asma Ben Abacha, Yassine Mrabet, Yuhao Zhang, Chaitanya Shivade, Curtis Langlotz, and Dina Demner-Fushman. 2021. Overview of the mediqa 2021 shared task on summarization in the medical domain. In *Proceedings of the 20th SIG-BioMed Workshop on Biomedical Language Processing, NAACL-BioNLP 2021*. Association for Computational Linguistics.

Jacob Devlin, Ming-Wei Chang, Kenton Lee, and Kristina Toutanova. 2019. Bert: Pre-training of deep bidirectional transformers for language understanding. In *Proceedings of the 2019 Conference of the North American Chapter of the Association for Computational Linguistics: Human Language Technologies, Volume 1 (Long and Short Papers)*, pages 4171–4186.

Sebastian Gehrmann, Yuntian Deng, and Alexander M Rush. 2018. Bottom-up abstractive summarization. In *Proceedings of the 2018 Conference on Empirical Methods in Natural Language Processing*, pages 4098–4109.

Shima Gerani, Yashar Mehdad, Giuseppe Carenini, Raymond Ng, and Bita Nejat. 2014. Abstractive summarization of product reviews using discourse structure. In *Proceedings of the 2014 conference on empirical methods in natural language processing (EMNLP)*, pages 1602–1613.

Som Gupta and SK Gupta. 2019. Abstractive summarization: An overview of the state of the art. *Expert Systems with Applications*, 121:49–65.

Jinhyuk Lee, Wonjin Yoon, Sungdong Kim, Donghyeon Kim, Sunkyu Kim, Chan Ho So, and Jaewoo Kang. 2020. Biobert: a pre-trained biomedical language representation model for biomedical text mining. *Bioinformatics*, 36(4):1234–1240.

Mike Lewis, Yinhan Liu, Naman Goyal, Marjan Ghazvininejad, Abdelrahman Mohamed, Omer Levy, Veselin Stoyanov, and Luke Zettlemoyer. 2020. Bart: Denoising sequence-to-sequence pre-training for natural language generation, translation, and comprehension. In *Proceedings of the 58th Annual Meeting of the Association for Computational Linguistics*, pages 7871–7880.

Chin-Yew Lin and FJ Och. 2004. Looking for a few good metrics: Rouge and its evaluation. In *Ntcir Workshop*.

Rashmi Mishra, Jiantao Bian, Marcelo Fiszman, Charlene R Weir, Siddhartha Jonnalagadda, Javed Mostafa, and Guilherme Del Fiol. 2014. Text summarization in the biomedical domain: a systematic review of recent research. *Journal of biomedical informatics*, 52:457–467.

Ramesh Nallapati, Bowen Zhou, Cicero dos Santos, Çağlar GuÌ‡lçehre, and Bing Xiang. 2016. Abstractive text summarization using sequence-to-sequence RNNs and beyond. In *Proceedings of The 20th SIGNLL Conference on Computational Natural Language Learning*, pages 280–290, Berlin, Germany. Association for Computational Linguistics.

Mark Neumann, Daniel King, Iz Beltagy, and Waleed Ammar. 2019. Scispacy: Fast and robust models for biomedical natural language processing. In *Proceedings of the 18th BioNLP Workshop and Shared Task*, pages 319–327.

Alec Radford, Karthik Narasimhan, Tim Salimans, and Ilya Sutskever. Improving language understanding by generative pre-training.

Max Savery, Asma Ben Abacha, Soumya Gayen, and Dina Demner-Fushman. 2020. Question-driven summarization of answers to consumer health questions. *Scientific Data*, 7(1):1–9.

Sunghwan Sohn, Donald C Comeau, Won Kim, and W John Wilbur. 2008. Abbreviation definition identification based on automatic precision estimates. *BMC bioinformatics*, 9(1):1–10.

Shengli Song, Haitao Huang, and Tongxiao Ruan. 2019. Abstractive text summarization using lstm-cnn based deep learning. *Multimedia Tools and Applications*, 78(1):857–875.

Axel J Soto, Piotr Przybyła, and Sophia Ananiadou. 2019. Thalia: semantic search engine for biomedical abstracts. *Bioinformatics*, 35(10):1799–1801.

Jingqing Zhang, Yao Zhao, Mohammad Saleh, and Peter Liu. 2020. Pegasus: Pre-training with extracted gap-sentences for abstractive summarization. In *International Conference on Machine Learning*, pages 11328–11339. PMLR.